AUTOMATION
AND ROBOTICS
IN CONSTRUCTION X

AUTOMATION AND ROBOTICS IN CONSTRUCTION X

Proceedings of the 10th International Symposium on
Automation and Robotics in Construction (ISARC)
Houston, Texas, U.S.A., 24-26 May, 1993

Organized by The Construction Industry Institute

edited by

GEORGE H. WATSON
Amoco Corporation
Naperville, IL U.S.A.

RICHARD L. TUCKER
JEWELL K. WALTERS
Construction Industry Institute
Austin, TX, U.S.A.

1993
ELSEVIER
AMSTERDAM • LONDON • NEW YORK • TOKYO

ELSEVIER SCIENCE PUBLISHERS B.V.
Sara Burgerhartstraat 25
P.O. Box 211, 1000 AE Amsterdam, The Netherlands

ISBN: 0 444 81523 6

This book is printed on acid-free paper.

Printed in The Netherlands.

FOREWORD

We are pleased to welcome you to the Tenth International Symposium on Automation and Robotics in Construction (ISARC). It is appropriate that Houston should be the location for this symposium as it is the home for several of our nation's major construction organizations.

The 1993 ISARC covers both recent accomplishments and visions for the future in practical applications of construction robots and automated systems. The symposium structure this year varies slightly from previous years. Some features for the symposium include the following:

1. The plenary sessions will cover "The Need for Automation."

2. The "Current Status of Automation in Construction" will be covered during the technical sessions.

3. The "Process in Applications" discussions will be covered both by video sessions and during the final plenary sessions.

We are pleased to announce that the technical proceedings have been published in a hardback, library volume. This will allow for easier storage and retrieval. We have made arrangements with the publisher to allow future purchases of technical session proceedings. The plenary session papers are bound as a second volume. This has allowed speakers with a later submission deadline to incorporate the latest developments in their summaries.

Future developments in automation and robotics will occur as information is shared and cooperative bridges are built, thus avoiding needless duplication. To that end, we hope you will:

1. Learn from the progress in other regions of the world.

2. Make contact with others working in your areas of interest.

3. Put into practice where possible that which you learn.

Jack F. Browder
President
Brown & Root Power
Co-Chair
1993 ISARC Organization Committee

George H. Watson
General Manager, Technology Support
Amoco Corporation
Co-Chair
1993 ISARC Organizing Committee

CONTENTS

2C – Working Examples of Finishing Operations – B

3A – Machines that Climb Building Exteriors

3B – New Models for Construction Automation

3C – Working Examples in Roadway Construction

4A – Designs and Tests of Large-Scale Manipulators

4B – New Approaches to Automated Construction Planning – A

5C – Working Examples in Heavy Construction – B

6A – Positioning Devices for Site Activities

6B – New Simulations and Walk-Throughs

6C – Working Examples in Tunneling and Underground Construction

Automation and Robotics in Construction X
G.H. Watson, R.L. Tucker and J.K. Walters (Editors)
1993 Elsevier Science Publishers B.V.

Study on Simulation for ROOF PUSHUP CONSTRUCTION METHOD

Masahiro Morita, Assistant Chief Researcher
Eiji Muro, Senior Chief Researcher
Tetsuo Kanaiwa, Chief Researcher
Hiroyuki Nishimura, Researcher

Takenaka Technical Research & Development Institute, 2-5-14 Minamisuna, Koto-ku, Tokyo

Abstract

In the construction industry, both construction and management technology have been developed, as well as industrization, automation. However, many problems were found in the process of applying such technologies for each construction projects in which it is required to move to the different work place and forced to construct at outside.

Based on this, aiming to improve the environment of the work place, we have developed ROOF PUSHUP METHOD CONSTRUCTION which requires to gather main constructive tasks beneath the top floor, while the construction itself is being managed by the computer, and this method was applied to construction of an office building. This paper describes the outline of ROOF PUSHUP CONSTRUCTION METHOD, and the actual result of applying it. This paper discusses the necessity of the construction simulation system for ROOF PUSHUP CONSTRUCTION METHOD, and the current status and the examples of such simulation systems. Furthermore the paper discusses the outline and the result of applying the simulation system developed based on the actual requirements.

1. Introduction

To resolve the problem of labor shortage (present and future) in the construction industry, The new construction method (the Roof Push-up Construction Method) has been developed. This method can automatize, robotize, and industrialize various aspects and phases of construction work in all weather conditions. Takenaka Corporation has applied the method to a real construction project and obtained positive results in shorter construction time and manpower savings. To evaluate the benefits of the Roof Push-up Method to determine its applicability, however, a model system for analyzing, simulating and evaluating the method must be devised. Only with such a system will it become possible to determine, in advance, the optimum installation order of construction materials and the most effective material flow.

The purpose of this paper is: (1) to provide general information on the Roof Push-up Method and its application in the Yanagibashi Mitsui Building in Japan and (2) to present a study on simulation in determining the benefits of the Roof Push-up Method.

2. An Outline of Roof Push-up Construction Method

2.1 The Positioning and the Benefit of Roof Push-up Construction Method

It is obvious that labor shortages will become more and more serious in the future. To overcome this problem, an automated construction method has been devised. In our estimation, a fully automated and industrialized construction method can reduce construction time and the number of processes involved 40% in comparison with the same construction methods and at the same construction cost. With this goal in mind, Takenaka Corporation has developed the Roof Push-up Construction Method (Fig. 2.1), which exemplifies the industrialization of building construction. It unifies various aspects of construction work via robotic, computer science, and automation technologies. With the Roof Push-up Method, a safe and comfortable working environment can be provided under all weather conditions.

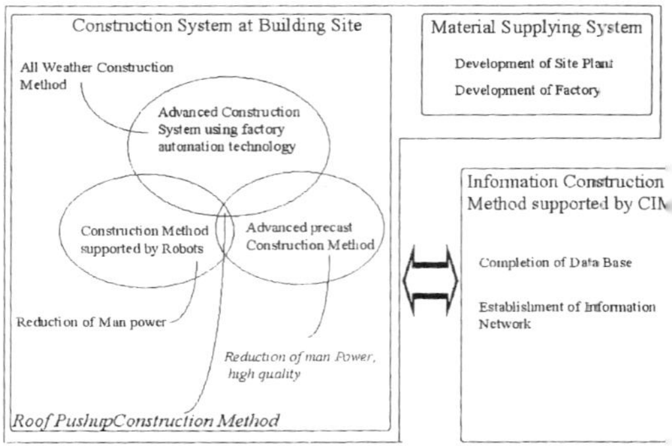

<Fig. 2.1> Conceptual Illustration of the Roof Push-up Method

The Roof Push-up Method differs from conventional construction methods in methods in which steel frames are assembled up to a specified height, and then the ceiling and floor of each level are assembled. The Roof Push-up Method was applied in construction of the Yanagibashi Mitsui Building in Nagoya in 1990 for the first time.

<Fig. 2.2> Automated Building Construction

In this method the roof floor (push-up floor) is constructed first. This is done in the sequence shown in Fig. 2.3. The benefits of benefits of this method are summarized in Fig. 2.4

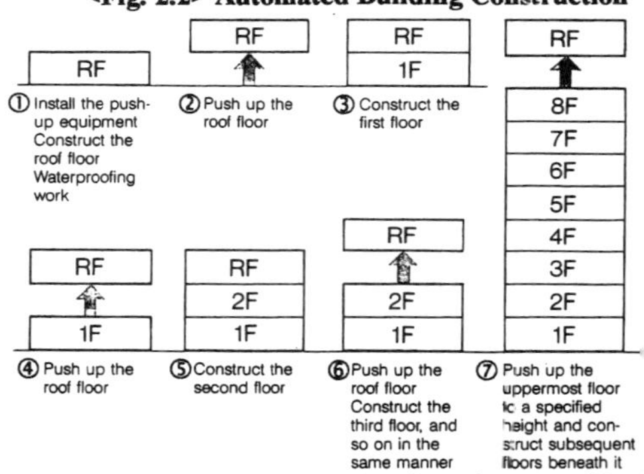

<Fig. 2.3> Sequence and Mechanism of the Roof Push-up Method

2.2 Mechanism of the Roof Push-up Method

Fig. 2.5 shows an actual application of the Roof Push-up Method. The method utilizes a post jacking system, an overhead circular gantry crane system, and a central control system.

(1) Jack-up Mechanism

Our post jacking system has two types of jacks (floor jacks and column jacks).

- The construction work cannot be effected by the weather
- Labor is elevated places is reduced. This improves safety
- Construction can be automatized due to the integration of various steps of construction.
- The building can be constructed in a narrow or restricted area.

<Fig. 2.4> Benefit of Roof Push-up Method

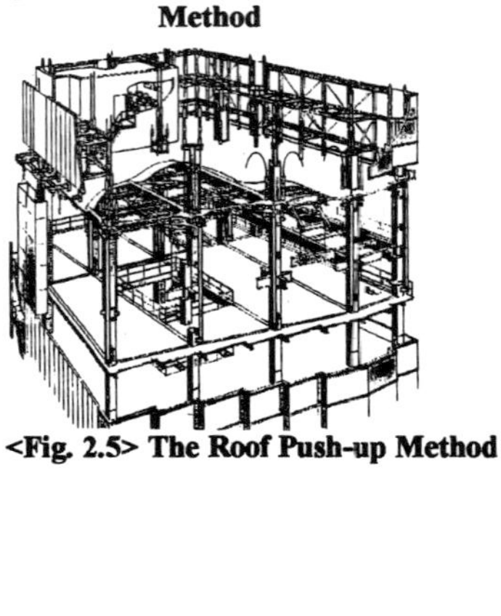

<Fig. 2.5> The Roof Push-up Method

These jacks are used interchangeably to push up the movable floor (roof floor) as floors below are constructed. As a fixed number of movable columns are jacked up, temporary support columns will be installed. This is to make the building under construction earthquake and wind resistant.

(2) Overhead Circular Gantry Crane System

An overhead circular gantry crane system has been developed to lift and accurately place building frame components for structural support. In addition, to move lifted materials horizontally, Material transporters have also been developed. These transporters are air-cushioned and used to deliver construction materials to their destinations.

(3) Central Control System

Operations in the Roof Push-up Method are controlled by a computer system located in the central operation booth on the site. The system consists of three subsystems. The operation monitor subsystem monitors work in progress in the whole construction site. The data from this system is used to inform workers. The communication subsystem is used to give direct instructions to workers working on the movable floor or floors below. The precision monitor subsystem can provide a real-time monitor on floor and column installation.

2.3 Sequence of the Roof Push-up Method

First of all, the Roof Push-up Method uses its overhead circular gantry crane system to lift up movable posts, beams, and other materials for the topmost floor. After the materials for the roof portion are assembled, the push-up equipment is installed. Then, facing materials for weatherproofing are installed. The basic installed. The basic construction process after this initial process, are summarized in Fig 2.3.

<Table 2.1>

Construction Site	Nakamura-ku Nagoya Japan
Building Area	682.19 m^2
Total Floor Area	7938.87 m^2
Structure	RC, SRC, S/B2, F12, 91
Term of Works	1898,11,1 ~ 1991,3,22

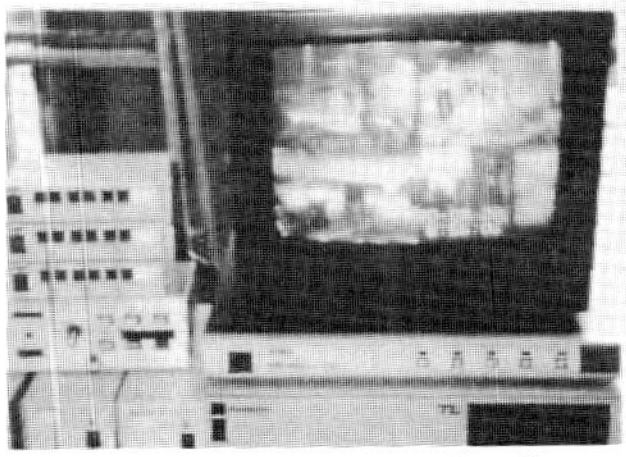

<Photo. 2.1> Monitoring System

2.4 Application Example in Construction of the Yanagibashi Mitsui Building

(1) Building Outline

As shown in Table 2.1, the site is near Nagoya station and surrounded by houses on three sides. Moreover, there was an overhead microwave line above the tower of the building to be constructed. This means

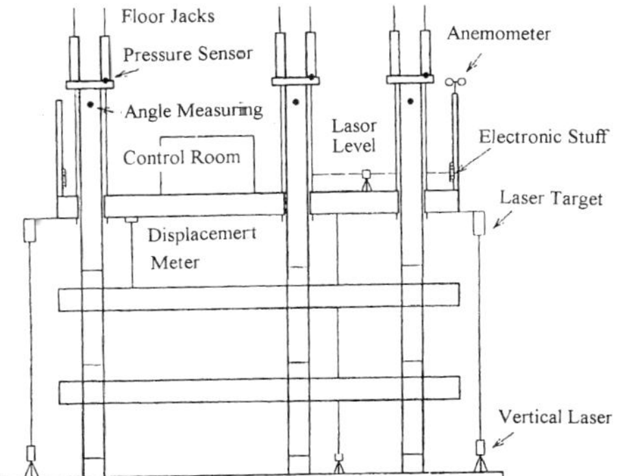

<Fig. 2.6> Sensors

that a tower cranecould not be used because of the danger of interfering with the line. As a result, the Roof Push-up Method was adapted to floors 10-12.

By using floor jacks (up-down post jacks), the roof floor (250 tons) was pushed up. Then, wire jacks were used to push up movable posts. After the roof floor was erected, 4 circular gantry cranes were used to install movable columns.

As shown in Photo. 2.1, operations were monitored through remote controlled TV cameras. The precision monitoring was done through sensors installed at various sites in the building (shown in Fig. 2.6). The data was collected in the central operation unit and observed.

(2) Results

The horizontal precision of the movable floor push-up, the vertical precision of the movable column installation, and the floor jack load were all maintained within the set ranges.

The work unit ratio based on the number of construction processes was 0.25 (man/square meters). This is very close to the average figure of the steel frame structure construction method This result can be represented in

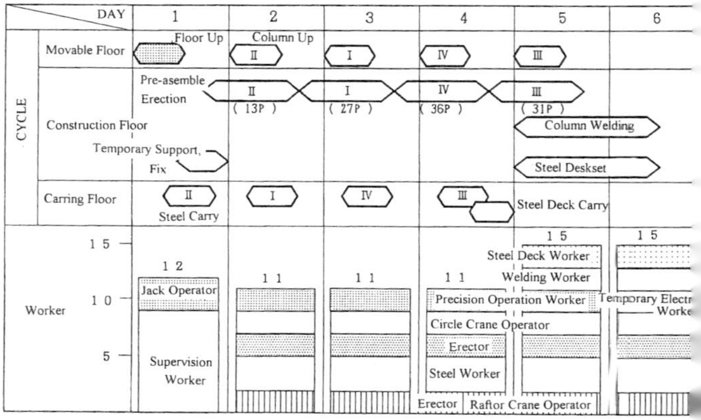

<Fig. 2.7> Overall Work Process Chart

result can be represented in terms of a work process cyclic chart as shown in Fig. 2.7. It takes 6 days to complete a floor with the work unit ratio of 0.17, in our estimation. Therefore, it was concluded that the overall productivity can be improved.

3. Simulation System for Roof Push-up Construction

3.1 Need for Simulation Software

As shown in Fig. 3.1, for the Roof Push-up Method to yield the maximum result, it is necessary to figure out the number of construction processes, the number of workers, the number of temporary storages and their locations, the optimal material flow, and the types and number

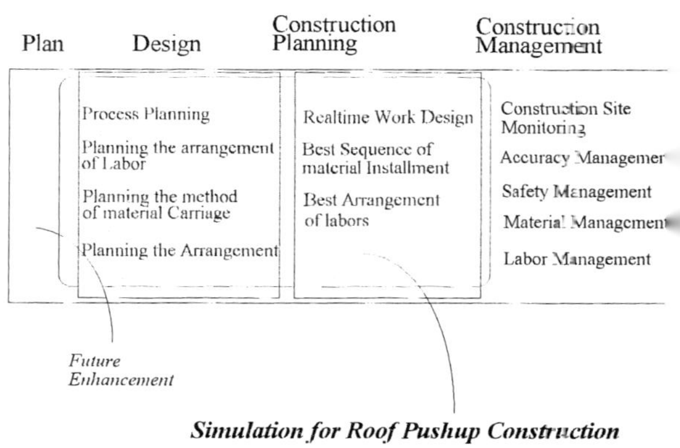

<Fig. 3.1> Areal Map for Simulation

of machines needed for a given construction to obtain the maximum productivity. Since these variables have to be determined prior to the erection of a building, it is necessary to devise a reliable computer software system, by which an application of the Roof Push-up Method under various conditions can be simulated. Such a simulation system, equipped with adequate database, can provide vital information in the future.

3.2 Special Purpose Simulation System

The preliminary study using a general purpose simulation package has shown that in order to generalize the Roof Push-up Method, it is necessary to devise a simulation system which figures out, in the most effective way, (1) the sequence and interval for transporting construction materials, (2) the number of worker teams, (3) the number of cranes and air-cushioned transporters, and (4) routes for material lift-ups.

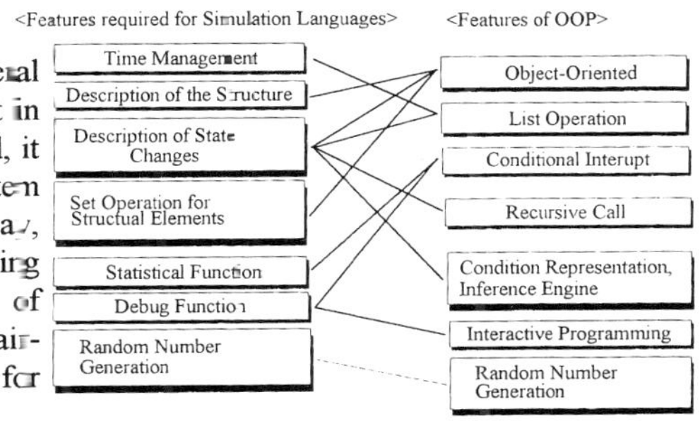

<Fig. 3.2> Functional Correspondence between Simulation Language and Object-Oriented Language

3.2.1 Language Selection

Object-Oriented programming (OOP) manipulates objects which can be either base classes or derived classes. Classes are building blocks consisting of variables and functions that associate with variables. Thus, the class mechanism of Object-Oriented programming allows encapsulation of different data types and operations. Derived classes can be created from base classes and other derived classes through inheritance. The foremost benefit of the inheritance mechanism is that it allows for the reuse of existing code to create a new type. In other words, one class is used as a building block for another, which in turn enables a hierarchical representation of knowledge in simulating different models for a given problem. Also the inheritance mechanism can eliminate redundancies in programming. As shown in Fig.3.2, most of the functions needed in our simulation are offered by OOP.

Therefore Common Lisp Object System (CLOS) has been selected for the simulation system. The CLOS was made by reference to Daniel G. Bobrow's system[2], as none of CLOS package was available.

3.2.2 Object-Oriented Simulation System

3.2.2.1 Classes for the Simulation

Considering the ease of maintenance and enhancement, All the simulation entities were defined in terms of classes. For example, every construction material has its unique ID number and its delivery destination (floor number and the attachment position). We need to know whether a material has been delivered to the site or not. The status of delivered material can also vary. Because there can be more than one way for a material to be lifted, it is necessary to keep track of cranes involved in lifting. Because a material has to be joined or connected with other materials, we need to encode how it has to be installed in relation to other materials. All these details can be grasped and manipulated easily if simulation entities can be expressed in terms of classes as shown in Fig. 3.3.

In Fig. 3.3, the Carrier class is shown as the top node. This class contains such members as `Name of carrier, Movable or not, Maximum speed, Maximum Load, Number of operation workers`. Because all these bits of information are encapsulated in a Carrier Class Object, it is possible to grasp the state of a carrier in every stage of the simulation.

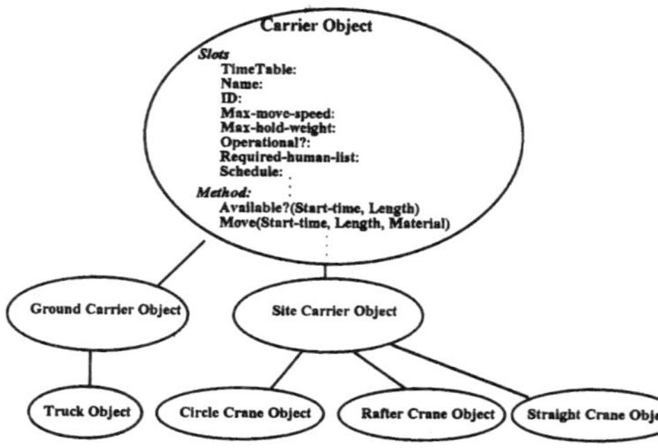

<Fig. 3.3> Example of Object Declarations

3.2.2.2 Inferences used in Our Simulation

Our simulation program performs logical inferences as shown in Fig. 3.4. In other words, upon taking input data specifying the following information, it performs simulation through various inferences to produce an output, which is a work schedule showing the optimum rate of operation for the cranes, the optimum work unit ratio and the shortest construction period on the given input. For example, the schedule chart in Fig. 3.5 has been obtained by setting the lift-up intervals as 3830 and 1915 seconds and by specifying the number of worker teams as 1 - 3 teams.

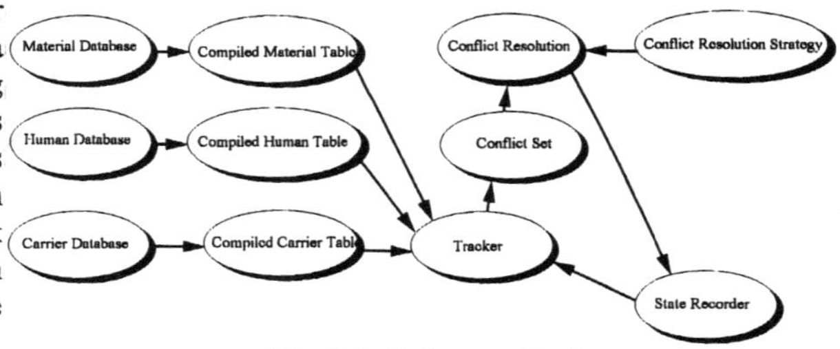

<Fig. 3.4> Inference Engine

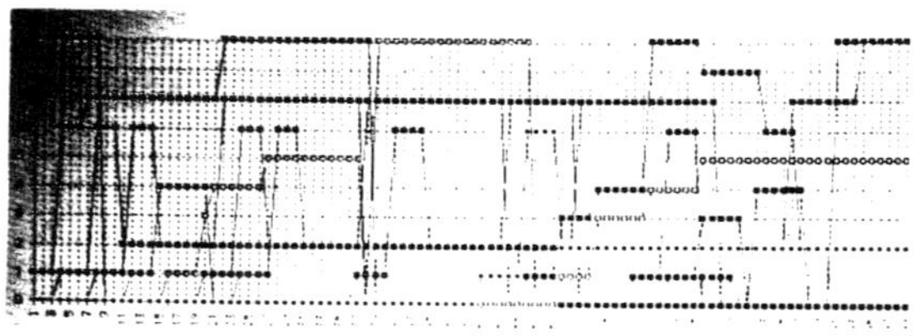

<Fig. 3.5> Simulated Schedule Chart

Information on construction materials, workers, and machines are compiled into a database which is stored within the simulation program as various types of tables. Our simulation program keeps track of every carrier object. When there is at least one unused crane and there are available workers, liftable materials are grouped for each unused crane to make a combination set of carrier objects. This combination set can be processed in accordance with our inference strategy, i.e., make the selection in such a way that it will shorten the construction period, select the carrier object that is placed furthest from the transporter. Then the specified

operation of a selected carrier object will be performed and the work progress status record will be updated.
This process will be repeated until all members of the combination set are used. Let us take a more concrete example. Consider the first selection of the day when twenty-nine materials (ID numbers (1) - (29)) are waiting to be lifted up. These twenty-nine items will be selected from the material table to form a combination set. In selecting the first carrier object to be processed, the inference strategy requires consideration of the number of cranes to use, the installation position, the distance from the transporter, the number of workers available, and the installation manner, etc. Suppose ID 1 is selected. Then the portion of the work progress status record for ID 1 will be updated. Now the combination set contains ID numbers (2) - (29). Thus, the whole process will be repeated on that set.

3.2.3. Simulation Results

Based on the model used for a commercial simulation package, Simulations under various conditions were performed. It was important to find out (1) the most efficient worker team setup, (2) the number of storable materials which leads to minimization of the daily manpower hours, and (3) the most efficient crane setup to obtain the optimum rate of operation for the cranes. The time needed for the simulation was 18 - 24 minutes on our Macintosh IIfx. Only half of that time was needed on a SPARC2 workstation.

Assuming that it takes 3830 seconds to lift up an installation post, the simulation was performed by altering the number of worker teams, the capacity of the storage place in terms of the number of storable items, and the number and types of cranes available. The simulation results are given in Tables 3.1, 3.2, and 3.3.

3.2.3.1 Worker Team Setup

As shown in Table 3.1, the amount of completion time with the 10 worker team setup was 5.1 hours. With the 45 worker team setup it was 3.6 hours. (We postulated 45 is the maximum number of workers we could put in one work team). As far as the manpower per hour is concerned, we concluded that the 10
concerned, we concluded that the 10 worker setup was the most efficient.

<Table 3.1> No. of workers and their time required

No of workers	Work time (Hour)	Man-time
1 Team (10)	5.15	51.5
2 Teams (12)	4.85	58.2
3 Teams (15)	3.99	59.9
Unlimited (45)	3.36	128.7

3.2.3.2 Storage Capacities

As shown in Table 3.2, with three storable items the amount of time required for completion was in the 10-13 hour-range. When the number of storable items increased to 5 or more, the amount of time required was in the 4-8 hour-range.
range. Thus, we concluded that the temporary storage place needs to be wide enough to keep at least 5 items at one time.

<Table 3.2> Time required depending on No. of workers and Max. No. of items in stock-yields (Hours)

No of workers	Maximum 3 in stock-yields	Max 5 in stock-yields	Unlimited in stock-yields
1 Team (10)	13.26	8.22	7.50
2 Teams (12)	12.42	6.06	5.76
3 Teams (15)	11.34	5.52	5.16
Unlimited (45)	10.44	3.60	3.30

3.2.3.3 Crane Setups

<Table 3.3> Time required when different cranes are used by 1 team

Case1: Use 2 slide cranes instead of 4 circle cranes	Case2: Case1 + one Rafter crane	Case3: 4 circle crane + one Rafter crane
11.16	4.50	3.96

As shown in Table 3.3, with the two slide-crane setup, the amount of the crane operation ended up with 11.16 hours. With the two-slide crane and one rafter crane setup, it was 4.5 hours. With the four circular gantry crane and one rafter crane setup, it was 3.96 hours.

3.2.3.4 Evaluation of Simulation Results

The simulation experiment has shown that the productivity of the Roof Push-Up Method depends on the worker team setup and the crane setup. Thus, it was concluded that it is possible to make the Roof Push-Up Method cost-productive by carefully choosing the most effective combination of the worker team setup and the crane setup.

4. Summary and conclusions

In this paper Takenaka Corporation's Roof Push-Up Method has been introduced, which offers automation and all-weather operation in building construction. This method was employed partially in the Yanagibashi Mitsui Building construction in Nagoya, Japan. From that trial, it was confirmed that the Roof Push-Up Method can result in shorter construction time and manpower savings. The Yanagibashi Mitsui project also showed that the method can be augmented with a simulation system capable of figuring out, in advance, the optimum material installation sequence and the optimum material delivery routes. Therefore, a simulation system was developed in CLOS to take care of the problems which became clear in the preliminary simulation study with a commercial simulation package. Our CLOS simulation experiment confirmed that the Roof Push-Up Method can be made effective and productive by choosing parameters that affect the crane setup and the worker team setup.

The developed simulation program offers a reasonable foundation for further design improvement. Currently, the following areas of enhancement are being prepared in order to make a program capable of a real-time simulation (limiting the maximum time required at 3 minutes).

- Development of tables for data indicating installation start time;
- Development of a parallel processing system for inferences;

Also planned is a graphic user interface which can make the simulation system more user-friendly and practical.

<References>

1) ISTEL: "WITNESS User Manual Version 6.0 Issue 2", 1989
2) Daniel G. Bobrow: "Common Lisp Object System Specification", ANSI at the X3J13, 1988
3) APLLE Computer, INC.: "Allegro Common Lisp$_{TM}$ 1.4", 1990

Automation and Robotics in Construction X
G.H. Watson, R.L. Tucker and J.K. Walters (Editors)

AUTOMATED CONSTRUCTION IN THE ATLSS INTEGRATED BUILDING SYSTEMS

B. Vincent Viscomi, Research Professor
William D. Michalerya, Manager of Industry Liaison and Technology Transfer
Le-Wu Lu, Professor of Civil Engineering

NSF Center for Advanced Technology for Large Structural Systems (ATLSS)
Lehigh University, Bethlehem, PA 18015 USA

ABSTRACT

The AIBS (ATLSS Integrated Building Systems) program was developed to coordinate ongoing research projects in automated construction and connections systems. The objective of this program is to design, fabricate, erect, and evaluate cost-effective building systems with a focus on providing a computer integrated approach to these activities. A family of structural systems, called ATLSS connections, are being developed with enhanced fabrication and erection characteristics. These ATLSS connections possess the capability of being erected by automated construction techniques. This feature minimizes human assistance during construction and will result in quicker, less expensive erection procedures in which workers are less susceptible to injury or fatalities. The technology for automated construction is heavily dependent on the use of Stewart platform cranes which are controlled by a system of six cables to allow precise movement in six directions. A scale-model Stewart platform crane has been constructed in the ATLSS laboratory to test the feasibility and limitations of automated construction with these connections.

INTRODUCTION

A comprehensive research program in automated construction is underway at the National Science Foundation-sponsored Center for Advanced Technology for Large Structural Systems (ATLSS) at Lehigh University. The broad objective of the "ATLSS Integrated Building Systems (AIBS)" project is to develop connections and structural systems that are economical in terms of overall construction costs, easier to erect, are effective in resisting gravity and lateral loads together with the development of automated systems (see Figure 1) [1].

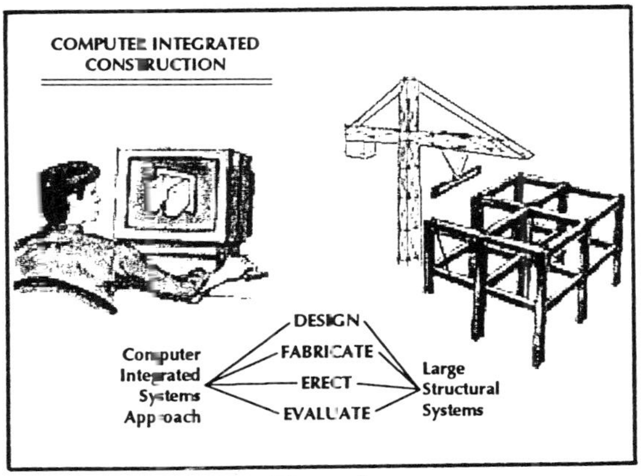

Figure 1

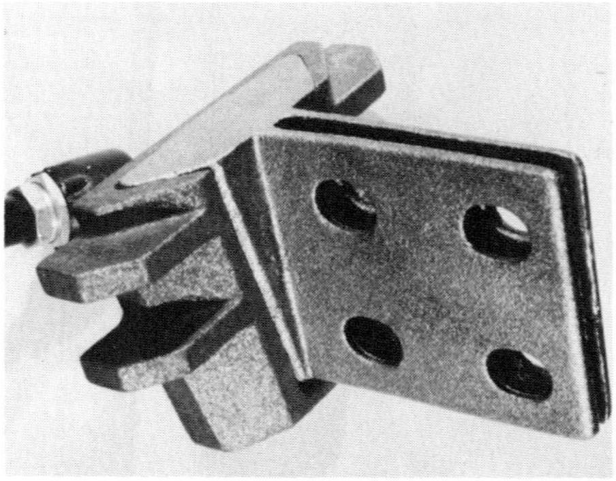

Figure 2

Connections are an expensive part of structural systems and are typically the location of potential structural distress and failure. A series of new beam-to-column connections, known as ATLSS connections, are currently under development. The emphasis of these new designs is on a self-guiding feature for use in automated construction. This feature will minimize human assistance during construction and should result in quicker, safer and less expensive erection procedures. The concept is based on using a tapered solid "tenon" piece on the beam which slips into a three-dimensional "mortice" guide mounted on the column (see Figure 2). The analytical analysis of this connection is being performed using finite element procedures. Experimentation, conducted concurrently with the theoretical work, has been performed by isolating the various loadings the connection would experience. This connection concept can cover a large range of structural needs, including shear, partial- and full-moment connections.

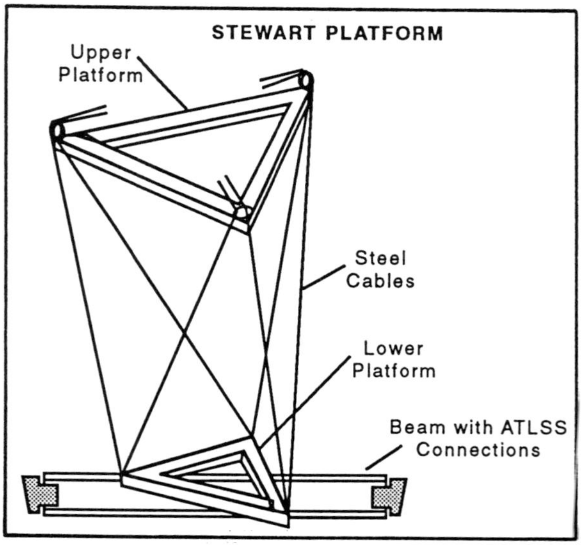

Figure 3

The technology for automated construction at the ATLSS Center is heavily dependent on the use of Stewart platform cranes. The Stewart Platform is actually two platforms connected by a series of six individually controlled linkages. At the ATLSS Center it has been designed and assembled with six cables used as the linkages to move the lower platform and payload relative to the upper platform position (see Figure 3). Orientation of the cables is such that the system has properties of a space frame, and this configuration not only provides excellent translational and rotational stiffness when compared to a boom crane but can also adjust the position and orientation of the lower platform with precision. The lower platform can move with six degrees-of-motion control to make insertion of the ATLSS connections possible.

BUILDING CONSTRUCTION

Construction, a $300 billion dollar industry, comprises approximately 8% of the United States Gross National Product and employs approximately four million people annually, nearly 6% of the American work force. In recent years, the total volume of construction awards has been declining. Moreover, the U.S. share of the market has shown a significant decrease. The primary mission of the ATLSS Center is to serve as a focal point for research and education that will lead to technological developments which increase the competitiveness of the U.S. construction industry. Another goal of the AIBS team, by concentrating on the design of alternative connection methods, is to reduce construction costs to a minimum while maintaining or improving the quality of the constructed project.

The frames of steel structures are constructed in two basic steps: fabrication of structural steel members within the fabricating shop and erection of these members at the construction site. The fabrication of the structural members involves such operations as surface preparation, cutting, drilling, machining, assembly and painting. These operations are usually performed in an industrial shop environment by heavy equipment. The layout of the equipment is such as to facilitate the flow of material and information on the shop floor and will often incorporate up-to-date computerized operations. On the other hand, site erection requires the use of light equipment and labor which are constantly moved and repositioned in various parts of the structure, stipulating a product oriented layout.

In reality, the procedure for erecting building structures has changed very little over the past seventy five years. Field work requires workers to perform complicated and strenuous tasks in a highly dangerous environment. Workers are often required to position themselves at great heights, in difficult positions, and then must guide and secure structural members that typically weigh many times more than they do (see Figure 4). In most cases, a construction crane is involved and it is not unusual for the above described activities to take place out of the crane operator's sight.

Figure 4

Thus, it is not surprising that, in the United States alone, between one and two thousand fatalities occur each year. During a five year period,the largest causes were due to falling from an elevation (33%) and from being struck by falling objects (22%). When the statistical pool is limited to building erection, the percentage of fatalities from falling climbs to 43%. It should be clear to the reader that a primary focus of the AIBS concept is to address this issue.

ATLSS CONNECTION

The ATLSS Connection concept is based on using an ATLSS Connector, comprised of a tapered tenon piece on the beam which slips into a mortise guide mounted on the column. The final ATLSS Connector configuration developed into a three-dimensional, conical solid tenon piece, and a three-dimensional mortise piece containing a conical cavity complementary to the tenon piece. To be useful for automated construction, both pieces must be attached to their respective members in the shop or on the ground at the site. The ultimate goal is to have a limited assortment of mass-produced connectors with a standard shop fitting operation and quick, automatic erection capabilities [2].

The development of the ATLSS Connection involved the evaluation of a number of self-aligning, self-securing connection ideas. These ideas were conceived and evaluated by a team of ATLSS engineers and an industry panel, who exchanged ideas on structural and erection performance. The key concerns of the industry panel were the handling of fabrication and erection tolerances; the connection's ability to be adjusted during plumbing; and the elimination of as much field fastening as possible.

Manufacturing options were considered next. Some trial finite-element analyses were performed on a typical mortise to examine the possibility of fabricating the pieces out of steel plate. However, cast steel was determined as the cost-effective solution for this stage of development. It was realized that cast steel does not have universal industry acceptance and there would have to be continued attempts to improve the manufacture of the connection. Stamping and cold-forming have been acknowledged as possibilities.

The experimental program for the ATLSS Connection has been extensive. The ATLSS Connection's behavior is complex due to nonlinearities in geometry and material. The complex nature of the connection dictates that the analytical models be verified experimentally. The experimental approach was one of response isolation. Each step of the experimental program was an attempt to assess the behavior of the connection due to a single component of load. The first step was to test the connection by subjecting it to a total shear loading. This was followed by loading the connection in tension. Finally, the connection was loaded in a flexural mode. From the results of these tests, the prototype ATLSS connectors were modified to included a seat, seating bolt, and flutes (see Figure 5). The seat and bolt were added to provide a physical connection between the tenon and mortise whereas the flutes were added for tension/flexural resistance.

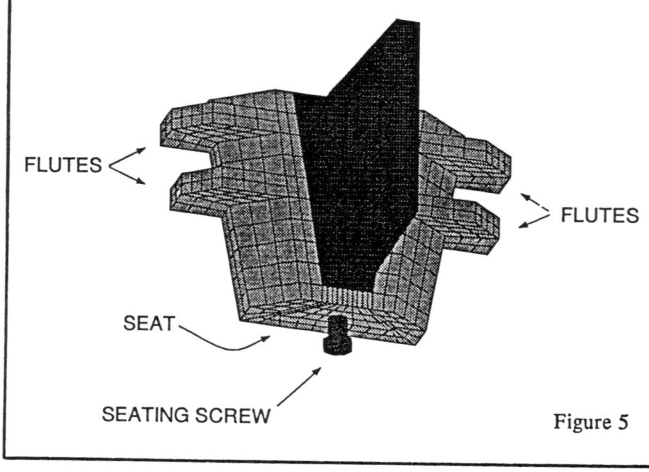

Figure 5

The approach taken in the analytical work has been consistent with the response isolation philosophy. The highly complex three-dimensional, compounded non-linear state of stress has been approximated as a 2-D planar problem by examining one loading situation at a time. Though the change is uniform and the shape remains the same, the cross-sections through the connection do vary in dimension. Also, the connection is not very deep with respect to its width, i.e., its aspect ratio is in the order of one. A representative analytical solution is shown in Figure 6.

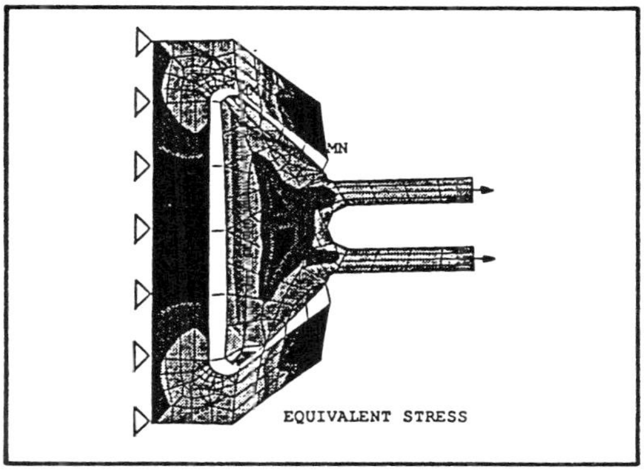

Figure 6

Given the objective of AIBS to provide cost-effective building systems, the ATLSS connector should be a modular connector, capable of multiple structural capabilities. Thus, the research program has been designed to examine the connector in a number of building situations: as a shear connection, as a composite partial-moment connection, as a full-moment connection and as a component in a tubular system. Research is ongoing in these areas and a future goal of AIBS is to design and test a three-dimensional structure using these connections in the ATLSS Center's multidirectional testing facility.

Recently, the ATLSS Connection was implemented in an actual building. The application was a low-elevation roof bay of a industrial plant. The dimensions of the bay were approximately 20' by 30'. The bay was designed to carry gravity load only. There were plans to possibly extend the bay upward, so removable connections would be advantageous. Since there were a large number of individual members in each bay, it was decided that the bay be pre-assembled on the ground using traditional connections, and then erected using an ATLSS Connector on each corner.

The entire bay was hoisted and installed with a boom crane, two iron workers with guy wires, and a foreman. The assembly on the ground took 20 minutes, the attachment of the tenons to the corners of the bay took 5 minutes, and the lift took less than 4 minutes. Figure 7 shows a photograph of the large bay section being installed. After the large piece was placed, a three-point end piece was installed. This piece took approximately 1 minute to erect. Following successful erection, it took the iron workers 7 minutes to install the securing bolts.

Figure 7

The total time of erection for the bay was 30 minutes. When compared with the more than one hour time to erect a similar bay in the adjacent span, a significant erection time savings was realized.

STEWART PLATFORM

A general Stewart platform consists of two platforms and six linkages. Each of the six linkages are connected one end to each platform. Each platform has the six linkages connected to it at a total of three connection points to which the six linkages are connected - two linkages being connected to each connection point. These connection points are non-colinear, and are almost exclusively arranged to form an equilateral triangle. The linkages are connected to the platforms to form a kind of 'daisy chain' format. (See Figure 3.) This arrangement has many useful properties when compared to a single cable construction site crane or a hook suspended from several parallel cables, such as an industrial hoist. These are:

a) lateral stability - allows the Stewart platform the ability to apply lateral force where required.
b) rotational stability - allows the Stewart platform to control the angle in the plane perpendicular to the direction of gravity of the lower platform - thus affording precise positioning and orientation abilities.
c) positioning capabilities - with control of the linkage lengths the Stewart platform is provided with limited positioning and orientationing capabilities of the lower platform with respect to the top platform. These limited capabilities are referred to as fine motion control and are useful as fine adjustments of the position and orientation of the lower platform.

A scale-model Stewart platform crane, shown in Figure 8, has been constructed at the ATLSS research facility [4]. Six servoelectric motors are used to control the cable lengths and to lift a design payload of up to 2 KN. The upper platform is attached to a motorized trolley yielding a work volume of a 4 m longitudinal run, a 4 m lift and a 1 -2 m of transverse movement. The ACES crane can be controlled directly at the computer keyboard or by use of a specially designed joystick. This joystick is actually a small scale Stewart platform which outputs the position and orientation as the command signal for the control loop. Force sensors, attached to each platform cable, are being used within cable slack and platform stability control algorithms.

Figure 8

Planned future activities include research in the sensor fusion area and the development of a vision system as illustrated in Figure 9. Automatic operation then becomes possible by integrating the vision system into the manual operation control loop. The installation of a vision system will not also permit the development of an automated crane erection system but also will assist with on-site material storage and inventory management.

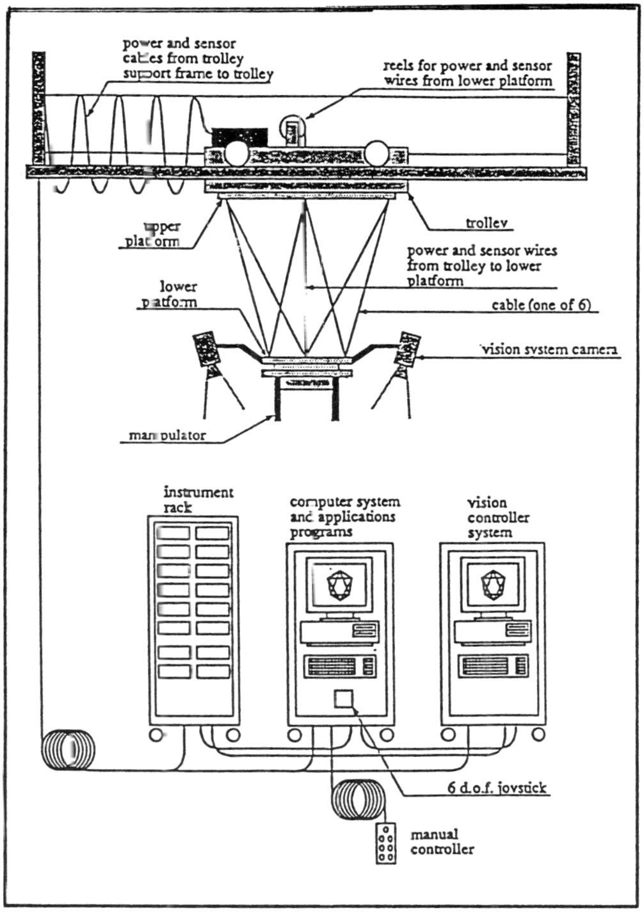

Figure 9

An automated framing system equipped with a vision system would be an important component in a program for the automated monitoring of construction quality control. Potential applications include quality control monitoring of connection placement, structural member alignment and actual dimensions. Most current quality control processes produce after-the-fact rejection with consequential delays, interruptions and expensive rework. Immediate feedback would enable remedial action to be taken while that construction activity is underway and thus minimize the consequences of defects.

ATLSS is engaged in a collaborative research effort with the National Institute of Standards and Technology (NIST) combining the ATLSS Connection and Automated Crane Erection System (ACES) activities with the automated crane developments at the NIST Robot Systems Division. Both the ACES system and the NIST ROBOCRANE are based on the Stewart platform.

ANTICIPATED OUTCOMES

The end users of the AIBS technology will be design and construction firms in charge of building the structure and the manufacturer of the automated construction system used for the erection of the structure. ACES is being-designed to assist erection firms in locating, securing, moving and placing the structural elements. It is also being designed as the primary material handling system capable of locating, securing, moving and placing non-structural elements (such as windows, facade, and the like) throughout the construction site. The key technological barriers to automated construction being overcome are: connection and structural design for constructability, implementation of mechanical/control hardware for construction crane application, development of active sensing such as active vision and force reflective joysticks, and the development of the design and as-built construction database.

TECHNOLOGY TRANSFER

A consortium of government/industry/university partners is envisioned as the most effective mechanism to move AIBS from the laboratory to the field. Currently ATLSS is forming this consortium of national and international partners and a commercialization strategy has been developed by Competitive Technologies, Inc., a new technology transfer subsidiary of Lehigh University. A patent application has been filed for the ATLSS Connection, and is nearing issue.

ACKNOWLEDGEMENT

The studies reported here were conducted in the Engineering Research Center for Advanced Technology for Large Structural Systems (ATLSS) at Lehigh University. The ATLSS Center was created by a grant from the National Science Foundation, and Dr. John W. Fisher is the ATLSS Director.

REFERENCES

1. ATLSS, SEVENTH ANNUAL REPORT TO NATIONAL SCIENCE FOUNDATION, Volume I: Programs and Plans, Center for Advanced Technology for Large Structural Systems, Bethlehem, PA, February 15, 1993.

2. Kaczinski, M.R., Viscomi, B.V., Lu, L-W., and Fleischman, R.B., An Approach to Integrated Building Systems. Proceedings of the Tenth Structures Congress, ASCE, San Antonio, TX, April 1992.

3. Fleischman, R.B., Viscomi, B.V., and Lu, L.W., Development, Analysis, Experimentation and Implementation of ATLSS Connection for Automated Construction, Proceedings of the National Steel Construction Conference, AISC, Orlando, FL, March, 1993.

4. Viscomi, B.V. and Perreira, N.D., An Automated Construction Erection System, Proceedings of the 9th International Symposium on Automation and Robotics in Construction, Tokyo, Japan, June 1992.

Automation and Robotics in Construction X
G.H. Watson, R.L. Tucker and J.K. Walters (Editors)

A LIFT UP INTELLIGENT CONTROL SYSTEM

Takehito TEZUKA and Dr.Hiroo TAKADA

Institute of Technology, SHIMIZU Corporation
4-17, Etchujima 3-chome, Koto-ku, Tokyo 135, Japan

ABSTRACT

The important subject that the construction industry should be accomplished today and in the future is in the rationalization of the construction processes and how to improve the labor-intensiveness of construction sites.

The authors have carried out research on the rationalization of the construction of a structure having a long-span of over 10 m. In the course of this research, we came to the conclusion that by using waffle slabs it is best to adopt large-size precasting in floor size and to use a production process in which such precast members are lifted up.

As hardware technology used in such a construction system we fabricated a prototype device of a Lift Up Intelligent Control System using a new logic, and in lift-up operation experiments we confirmed that it is possible to control the relative position at each point during lift-up and to do synchronized lifting of large-size precast members.

INTRODUCTION

It is pointed out that one reason for the low productivity in the construction industry is the waste of excess labor in construction work. What makes production in the construction industry different from production with factory-line systems is that 1)the location where construction is carried out changes with each work, 2)dimensions vary with each structure, 3)the production is a field job, and 4)the working environment is easily affected by weather and other conditions. Because of this, it is characteristic of construction-industry production that 5)production equipment is temporary, and 6)relative to labor costs, it is difficult to rationalize operations by investing in equipment and mechanizing production processes.

Construction methods in which the steps in production are systematized have been proposed as a means for achieving production rationalization in construction work. In this technique, the production steps in construction work are analyzed and, while taking into account the constraints on each operation, each step of the operation is studied to determine whether it is really necessary (study of whether the step can be eliminated), whether other operations can be incorporated into it (integration study), and whether it can be transposed with another step (study of change of sequence). Finally, the step is simplified with the expectation of making it easier to become skilled at. In this, automation is adopted for the construction step if the work would be difficult without it, and the adoption of automation is rational.

The demand for building space that can flexibly accommodate changes in layout has led us to focus our attention on, and do research in, buildings in which long-span structures have often come to be adopted.

The Lift Up Intelligent Control System described in this report differs f)m the conventional lift-up construction method in the following respects.

(1)It is not necessary, as it is in convèntional lift-up systems, to install jacks on column capitals, and the system is capable of self-climbing with the built-in columns.

(2)In floor-size and other large-size structures, the distributed load varies depending on the thickness, and the balance of the structure is disrupted if the same lifting power is applied to each lifting point. Use of a control device for the lift-up made it possible to do synchronized lifting in accordance with the load balance. This makes it possible to lift up, with near horizontal precision, even for a large-size structure whose load distribution is unbalanced.

THE CONSTRUCTION OF A LONG-SPAN STRUCTURE

A Construction System

A waffle slab combines plate and ribs having high flexural and torsion resistance and distributes the load in two directions. Also, except for the concrete portion that bears the tensile stress in the lower part of the member, waffle slabs can be light in weight. This makes the slab advantageous for long-span structures. Moreover, the longer the span in a waffle slab, the lower the construction costs.

In the construction procedure, first, as the half-precast member, the rib portion of the waffle slab is made by multi-layer casting on the ground. Next, with a method that produces no time-dependent deflection and little flexural cracking, by using the unbonded tendons, the ribs of the waffle slab are introduced to a prestress of 10-20 kgf/cm^2 in average axial compression using the post tension method. Then the building structure is built by lifting up the ribs by the Lift Up Intelligent Control System with respect to the built-in columns and by placing the topping concrete.

This building construction system not only rationalizes the construction by large-size precasting in which floor-size slabs are fabricated on the ground, but it can also achieve rationalization of the construction work, including saving labor and the reduction of the construction period. This system is precise the building of long-span structures and promotes the safe of construction work.

Figure 1 shows how construction is done with a Lift Up Intelligent Control System.

Prototype of The Lift Up Intelligent Control Sytem

In a prefabricated construction system, the larger the structure, the greater the effect of construction rationalization. In this case, the larger the size, the greater the massive force, but there is a tradeoff with size and weight from the standpoint of construction work.

In lifting large-size precast members, the following problems occur.

(1)In order to minimize construction stress during the lift-up of a large-size structure, the lifting must be done with the member kept in horizontal balance. This requres synchronized control so that all the lift-up points are kept in balance.

(2)With a large-size structure it is difficult to predetermine the structure's load center of gravity. Because the load on each lift-up point is different, the horizontal balance of the structure is lost if the same lifting power is applied at each point.

A Lift Up Intelligent Control System that lifts up a large-size precast member with massive force has the following characteristics.

(1)The lift-up system is in self-climbing the columns of the building structures as rods.

(2)With a central control unit, synchronized lifting is done in such a way that the lift-up of each point is kept in balance.

(3)Multi-point cylinder units are controlled by a single operator system.

The lift-up system consists of hydraulic jacks for lifting and hydraulic jacks for chucks that hold them in at the top and bottom. The lift-up and chuck jacks consist of multiple hydraulic cylinders which can respond to changes in lift-up load and chuck force, depending on the friction conditions of the column surface.

In lifting the structure, the top chuck system is seated and the lift-up jacks are retracted with the bottom chuck system in an open state. When a lift-up jack reaches the end of its stroke, the bottom chuck sytem is seated, and the top chuck sytem goes into the open state. In this state, the lift-up jacks are extend. Thereafter these steps are repeated.

The central control unit consists of a main box and sub-boxes. The main box ascertains the lift-up level from the standard points by the voltage output singnals of a linear encorder then issues instructions using a microprocessor so as to synchronously control the relative position of each lift-up unit within prescribed values. The sub-box receives instructions from the main box and controls the electromagnetic valve of the pump of each lift-up unit, balancing relative position.

Experimental Mock-Up Test

The Lift Up Intelligent Control System used in the mock-up test has a lift capacity of 6 tons, with a total of four lift-up cylinders. The cylinder stroke is 100 mm, the raising speed is 380 mm/min, and the lowering speed is 190 mm/min. The top and bottom chuck system are each clamped from two directions, and the clamp capacity totals 60 tons in the two directions.

The Lift Up Intelligent Control System is a two-part assembly system attached to the columns corresponding to the rods. It becomes operational when the assembly pins are inserted.

Figure 2 shows the experimental results when the Lift Up Intelligent Control System raises and lowers, with a synchronous width of 30 mm and 50 mm, an approximately 5-ton precast floor slab a lift-up distance of 2.5 m, corresponding to a steel-pipe column filled with concrete.

From the experimental results, the level error from a exact distance of 2.5 m is 7~8 mm, and it was possible to control the lift-up while controlling the clamping force for variations in friction force associated with changes in the shape of the surface of the column.

Figure 3 shows the results when one linear encoder was made to emit simulated voltage output signals of the lift-up level and the Lift Up Intelligent Control System changed to synchronized lifting.

From the experimental results, synchronized control of each lift-up point was possible so as to keep the member in horizontal balance.

The Lift Up Intelligent Control System used in the mock-up test is shown in Photo1.

ANTICIPATED OUTCOMES

The new Lift Up Intelligent Control System can be applied to the synchronized lifting of large-size precast members which have an unbalanced load. This device makes possible the high-precision lifting of large-size precast members, minimizing the construction stress on such members. The application of the Lift Up Intelligent Control System to construction work helps systematize construction processes, minimizes temporary work such as the reinforcing of members during construction, and contributes to the rationalization of the construction processes of a building having a long-span structure.

Application of this system accomplishes the following improvements compared with conventional construction systems.

(1)Figure 4 compares the amount of labor expended on construction work under conventional methods of in-place concrete casting and precast construction. Construction methods that use half-precast concrete can save approximately 45% of the labor needed with conventional construction methods. The greater the level of processing, from raw material to finished product, by using the precasting production processes on the ground work, the less amount of labor expended on the construction work. It is clear that an amount of labor is saved with the precast construction method.

The proposed construction system requires labor for the precasting of long-span slabs on the ground. When the amount of labor for the proposed construction system is considered in the data of Figure 4, taking into account the amount of labor for on-site prefabrication, this construction system can save construction labor by 30% as compared with conventional construction methods.

(2)Figure 5 shows the number of days by which the work schedule is reduced with the precast construction method as compared with conventional construction methods.

In the precast construction method, a 40-50% reduction of the construction period is achieved. Although the proposed construction system involves the work of building-in precast columns and fabricating slabs by on-site prefabrication, compared with conventional construction methods it can reduce the construction period for structural parts of this construction system by 40%, based on the data given in Figure 5.

The proposed construction system promises to be effective for the rationalization of the construction of long-span structures through the introduction of large-size precasting and automated equipment, and it is expected to contribute to the improvement of the labor-intensiveness of production processes in the construction industry.

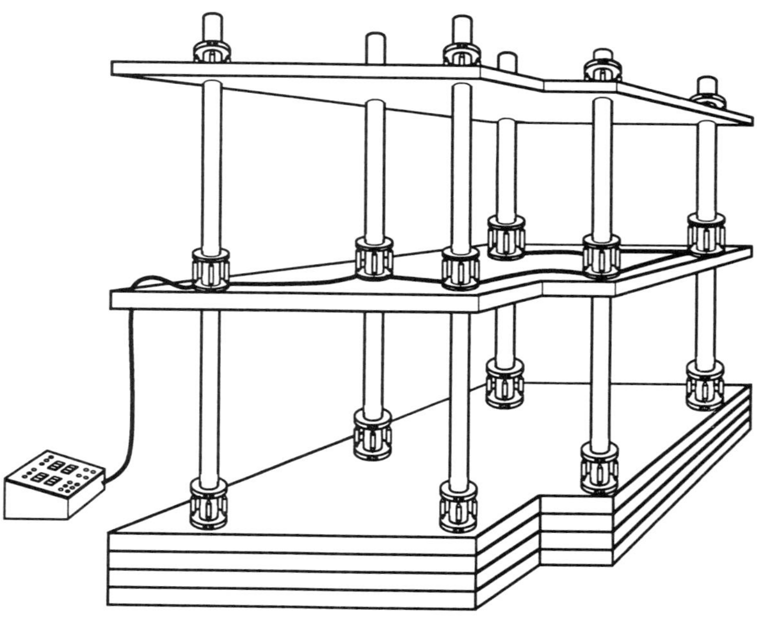

Figure 1 Construction using a Lift Up Intelligent Control System

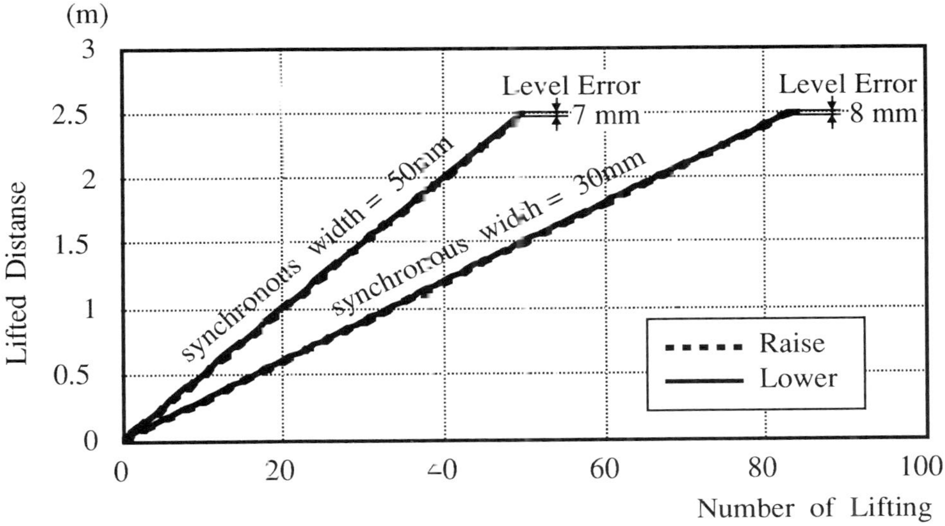

Figure 2 Operation results of a Lift Up Intelligent Control System

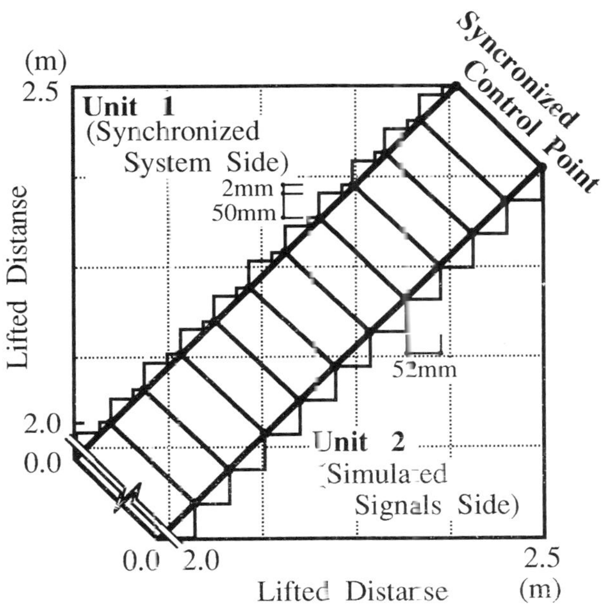

Figure 3 Synchronized lifting (synchronous width = 50mm)

Photo 1 The Lift Up Intelligent Control System used experiment

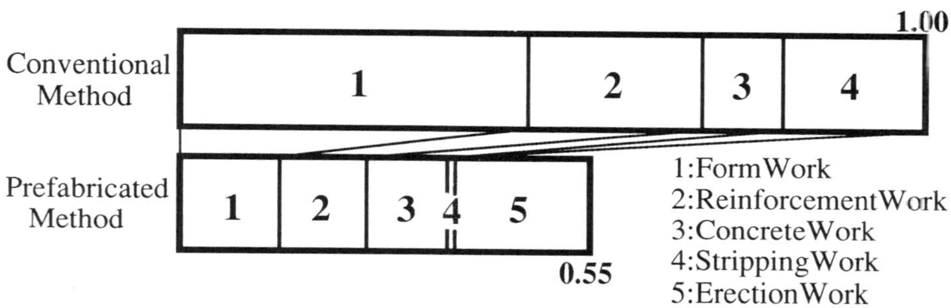

Figure 4 Comparison of the amount of labor

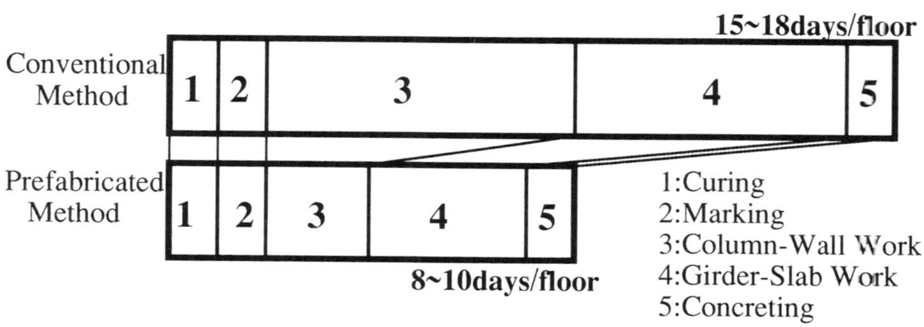

Figure 5 Comparison of the construction period

Automation and Robotics in Construction X
G.H. Watson, R.L. Tucker and J.K. Walters (Editors)

Functional Profile of a Wall Assembly Robot System and Different Solution Approaches

Thomas Bock[a], Friedrich Gebhart[a] and Werner Steck[b]

[a]Division for Automation in Construction, Institute for Mechanical Engineering in Construction Management, University of Karlsruhe (TH), P.O. Box 6980, 7500 Karlsruhe, Germany

[b]Institute for Machine Tools and Production Science, University of Karlsruhe (TH), P.O. Box 6980, 7500 Karlsruhe, Germany

Abstract

The presented robotic system is based on the idea of Computer Integrated Construction (CIC). This means the integration of all trades and processes from architectural planning to execution of construction work. It contains computer aided design and computer aided generation of robot programs, custom made prefabrication of small building components including logistics planning and roboterized wall assembly on the site. Within the framework of the presented project a handling system adapted to the construction-site environment is to be developed and realized. It enables with an autonomous vehicle and suitable grippers and sensors flexible masonry assembly and improves working conditions and safety on the construction-site. Parallel to the machine development, the information integration as well as the planning of the required logistics and construction-site organization is to be studied, developed and implemented.

1. INTRODUCTION

The structural problems of the construction industry can be solved long-term only through flexible automation of construction to releave the people on the construction site from the high physical demands and to increase the productivity of the building industry.

As a test case of a technology carrier for the building construction industry the processes of masonry work was chosen because of the clear definable and separatable sub-tasks and the demand and willingness of this industry to introduce advanced technologies to explore possible solutions for specific problems:

First there is the necessity to increase the production due to a strong demand in housing particular in Germany caused by the structural change from an emigrant country to an immigrant country.

A first approach to solve the problem was to enlarge the size of the blocks. However, this caused the second problem, which was adressed through mechanization: too heavy weights for manual processing necessitated the use of mechanical lifting aids, a pre-step towards robot-automation.

The third problem is the customization of each building. In combination with small assembly units like blocks a flexible automation of the assembly process seems possible to meet the individual demands.

The whole project is embedded into an European research program with a strong industrial participation and a continous feed back of potential end-users to enable fast and goal oriented R&D work.

2. MASONRY APPLICATION ANALYSIS

2.1 Decision Process

In a first step, the decision process should be illustrated, which occures, when a potential customer wants to build a house (see fig. 1).

Through the demanded flexibility with regard to the building materials resulting from the position of our potential user at the lower end of the decision process, first an analysis of the most frequently used building materials was carried out (see fig. 2).

Based on the results, the system should handle at least sub-groups of building systems of each of the four big material groups. In each of the groups are at least some systems existing, which meet the below mentioned requirements for a simplified automated assembly.

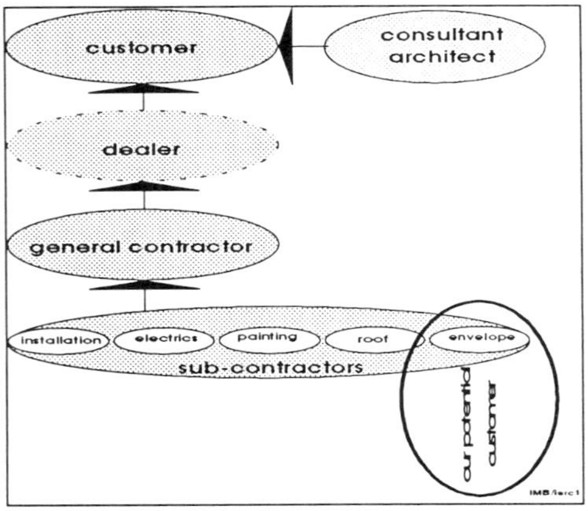

Fig 1: decision chain to build a house

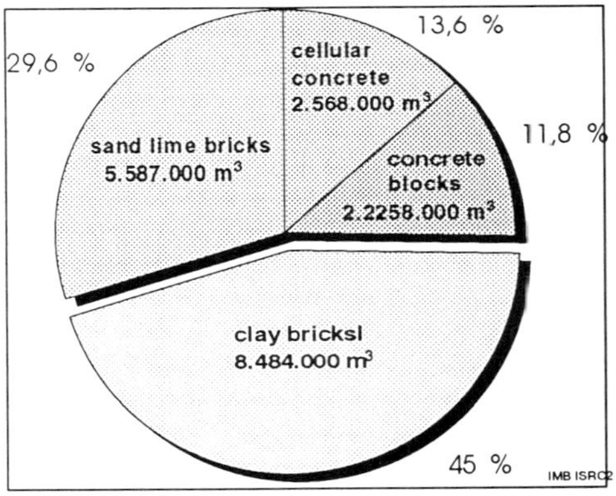

Fig 2: market shares of different building materials in Germany

2.2 Working Process

The second step was the analysis of the working process to integrate an automated tool in a complex environment with different simultanous activities on the construction site. In close cooperation with participating construction companies, the analysis was carried out.

It turned out, that different sub-tasks should still carried out manually, particulary the preparation of the construction site and the preparation of the respective walls to be built with an automated masonry assembly system. This includes the set-up of a first levelling layer to compensate the inaccuracies of the concrete floor and the cleaning of the floor and removing of obstacle to enable an easier movement of the mobile robot.

A second part of work, which should be carried out manually are after-works like closing of joints with mortar or the assembly of bigger concrete parts like large lintels.

The third activity, to be carried out manually in a first step, are the construction site logistics to provide the blocks for the robot. The use of the hereby necessary crane should be simultanously minimized according to the construction companies.

To have as less as possible lost time for the automated assembly, the approach is a segment oriented cycle process with parallel and synchronized manual and automated working parts for each wall segment.

3. FUNCTIONAL PROFILE OF A MASONRY ROBOTIC SYSTEM

3.1 Operational Task of the Robot System

The goal is the automated erection of vertical walls with the required openings for doors and windows. The material to build the vertical walls are bricks made of different raw materials (see fig. 2) and in different sizes.

As a restriction in the first approach only walls and openings are regarded, which are describable as three-dimensional volumes with rectangular, trapeze or triangle surfaces. No round walls or openings in any dimension are regarded. The walls may meet each other at any given angle and any kind of gable should be able to be erected.

An additional restriction for a first approach is the demand of gripping the blocks only from above. Therefore the last layer underneath a ceiling cannot be assembled. But according to building companies, this is not a significant restriction for the benefit of the system, because most of the walls are erected before concreting the next floor. In addition, this implies that the blocks are being delivered vertivally put on pallets, which actually happens in most of the cases.

The goal of the automated masonry process is first to grip the block directly from one of the pallets, which are attached to the system to calibrate them, then to make a quality control, to apply the thin bed mortar to the underside of the block, to move the block to the wall and finally to assemble the block on the wall with a final accuracy of 2 mm in each horizontal direction.

3.2 Boundary Conditions on the Construction Site

A robot, operating at a construction site is naturally exposed to all weather conditions. The operating requirements take this into account. These conditions are very diifferent to the conditions in a factory building:

- Operating temperature: -5°C to 60°C
- Storing temperature: -20°C to 60°C
- Protection class: IP 65 or better
- Employment in the vicinity of the construction area: wind, rain, dust, dirt

In addition, the operation at a construction site requires a mobile system, which is able to move on the concrete surface of a floor. The minimum requirement is the moveability on a surface with following features:

- Exactness of surface: ± 3 cm height difference refering to the length of support (≈ 3 m)
- Shoulders and steps: up to a height of 3 cm
- Floor loading capacity: 0,5 kg/cm² (with loads, that have their points of application in 2 m intervals, one can assume 16 m² as the carrying surface)

The maximum weight of the robot, therefore, is essentially limited by the loading capacity of the crane. The following weight is possible without any difficulties using a normal building site crane:

- Maximum crane payload: 2000 kg

The next limitation is caused by the necessity of transporting the robot system on a conventional truck. The size limits are:

- length: 3 m
- width: 2 m
- height: 2 m

The current security regulations for construction site equipment have to be considered. For the use of automatic systems at construction sites, however, there aren´t any clear rules or instructions yet. Therefore, the requirements, along with the development, have to be discussed with the responsible administrations.

3.3 Useable Block Types

With the intended robotic system, one should be able to use block types, which are normally used today for the above mentioned reasons of flexibility. On the other hand, it is important for an automatic system to have minimum tolerances of each assembly part.

According to different building material suppliers, they guarantee a sufficient parallelism of the horizontal brick surfaces. Particularly the sand-lime brick and cellular concrete block industries have an acceptable accuracy due to the material and production process. In the clay brick and concrete block industries only few suppliers have already installed post-processing equipment, which gives the blocks the necessary accuracy. But another problem occures now, due to the post-processing, a high percentage of blocks has small breaks or quarry faces. Therefore in the first approach, one should use sand-lime bricks or cellular concrete blocks for automated masonry.

This accuracy enables the use of thin bed mortar for joining the bricks. The advantage is an easier application of mortar on the bricks and different ways to level out the assembled bricks (see fig. 3). Additionally, thin bed mortar has better heat and noise insulation features than normal mortar.

In order to work with thin bed mortar, the inaccuracies of the floor have first to be levelled out by a first layer. This layer may consist of normal mortar and eventually the first layer of bricks. It is important to have an accurate first layer: The tilting of 1° leads to a deviation of 6 cm in 3,5 m height, if the blocks were assembled without correction. Therefore the first layer should be accurate to ±0,15° to have a maximum deviation of 1 cm in 3,5 m height.

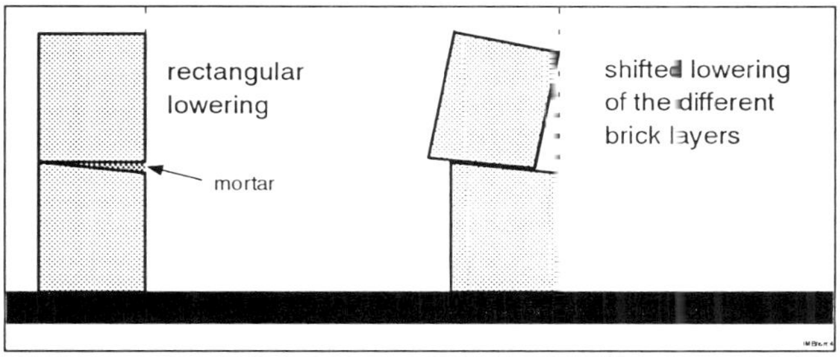

Fig 3: two ways of vertical levelling

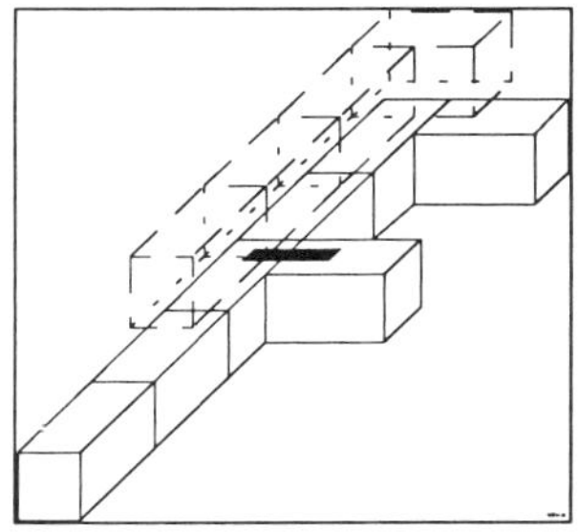

Fig 4: butt joint technique and composite construction

Due to the different requirements to put the first layer (absolute measurement and accuracy, working with normal mortar) in contrary to the further masonry process, this research section does not intend to develop the necessary equipment, which would be significantly different.

With all block types, there are fundamentally different wall thicknesses and other fitting lengths and heights. It can be assumed, that 30% of the blocks are to be regarded as non-standard blocks. The concept is based on a maximum block size of 100 x 50 x 36,5 cm and a minimum block size of 12,5 x 12,5 x 12,5 cm.

As the second important parameter, the weight of the blocks should be regarded: Normally used blocks for manual masonry have a maximum weight of 30 kg, blocks to be masoned with mechanical aids have a maximum weight of up to 300 kg.

As a first approach, we used a two step development: first a high speed and wider working space for the normal blocks and second a reduced speed and/or working space for the heavy elements.

3.4 Wall Connections

With wall connections, one generally distinguishes the butt-joint technique and a composite construction (see fig. 4).

The regulations and traditions in the masonry work state that it must be possible to connect two rectangular walls with a composite construction. To build a T-connection of two walls, it is possible to juse the butt-joint technique, which is easier to handle and has advantages in noise and heat insulation against a T-composite construction.

For an automated system, possibly a different device than the wall clamps for the butt joint technique has to be developed to ease the automated assembly process.

3.5 Working Space

Two aspects need to be taken into account: first, a required maximum assembly range and second a required minimum gripping range.

According to an examination of building sites, with a maximum assembly altitude of 4,5 m, 86% of all floors can be assembled completly. Therefore, a higher working height does not seem reasonable.

As mentioned earlier, it has got to be possible to brick up the exterior walls of buildings with a composite construction technique. Bigger walls must be assembled as segments in the forms of stairs at each working point. To use the system efficiently it must be possible to border this staircase on a wall part of 1,5 to 2 m width up to an height of 3,5 m. This requirement in combination with two different block sizes is shown in figure 5 and 6.

On the previous pictures the required range of the operating device is illustrated. With this range it is definitely possible to brick up salient corners. Furthermore it is possible to brick up the wall crossings to a certain overlap degree.

At the required maximum working altitude of 4,5 m, it is sufficient, if two standard sized blocks can be assembled from one working position due to the rare occurence of these heights.

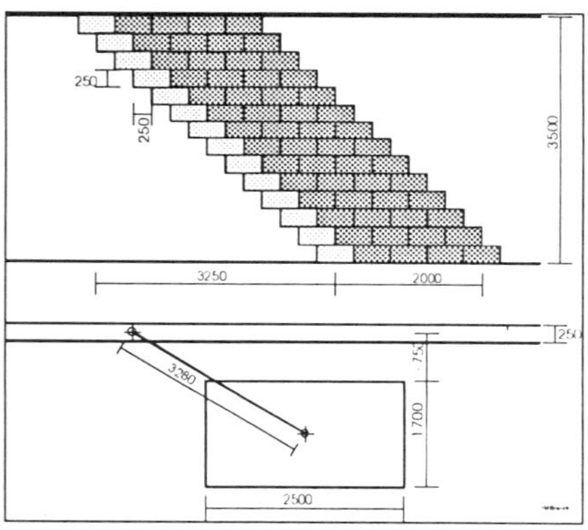

Fig. 5: wall segment with standard sized blocks (50 x 25 x 25 cm)

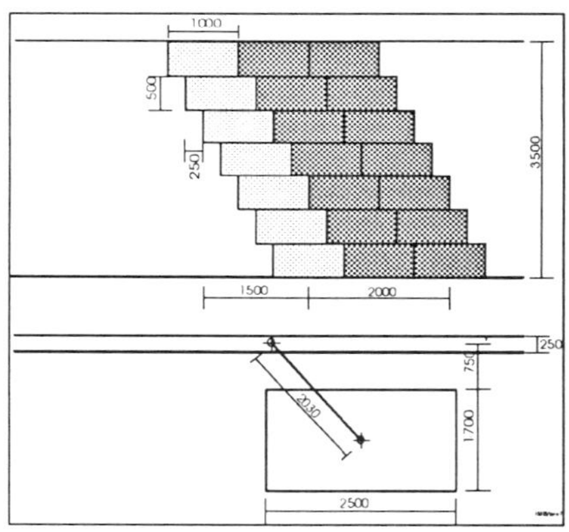

Fig. 6: wall segment with bigger blocks (100 x 50 x 25 cm)

4. CONCEPT FOR THE MASONRY ROBOT

4.1. Operation Sequence

The total operation can be devided into different tasks carried out by the components of the system:

1. Calibration of the absolute measurement system.
2. Calibration of the robot system on the respective working points.

3. Movement from one working point to the next one.
4. Support of the robot system with the outriggers.
5. Docking and calibration of the block pallets.
6. Gripping of the next block.
7. Application of mortar and quality control.
8. Collision free movement of the block from the pallet to the wall.
9. Assembly of the block with the desired accuracy.

The task no. 1 must be carried out one time after each movement of the absolute measurement system. The tasks no. 2 to no. 5 have to be done one time for each working point. The tasks no. 6 to no. 9 are the actual assembly procedure of each block.

4.2. Different solution approaches

To perform the tasks mentioned above we developed both a set of individual components to carry out particular tasks and a set of different approaches to attack specific types of problems.

The first task can be carried out by the absolute measurement system itself by measuring fixed points on the floor surface.

The second task is very similar to the first one with a replacement of the fixed floor points by fixed points on the robot system. For a first approach, a manual tachymeter system is chosen for the absolute measurement. It is possible to upgrade it step by step to an auto-tracking system based on the first test results. The measurement procedure is carried out after the support of the robot system.

The third task of moving from one working point to the next one will be carried out by a 4-wheel steered vehicle. In the first approach, it is intended to move only with the help of relative sensors like encoders. The vehicle will be able to move in each direction without the necessity of shunting due to the choosen kinematic. The positioning errors detected by the absolute measurement system, are not compensated by the vehicle but by the robot kinematic. Therefore it is necessary to keep the positioning errors as small as possible, because they are influencing directly the working space of the robot system.

The fourth task of supporting the robot system will be carried out by 4 outriggers. It is planned to measure the tilting errors but not to compensate them with the outriggers.

The construction site logistics as the fifth task are done manually with the help of the crane and a small manual fork lift. It is necessary to provide docking points at the vehicle to bring the pallets into a defined position relative to the robot system.

There are several restrictions to grip the blocks from the pallets as the sixth task. First one has to compensate the positioning errors of the blocks on the pallet due to pre-production and transportation influences. Second only two of the 6 surfaces of the block are reachable simultanously and in every case. Therefore passive complient elements are necessary for the first restriction. For the second one a new gripping concept has to be developed with extensive building material analysis and tests.

With the intended use of thin bed mortar, it is possible to realize the first part of task seven by immersing the block into the mortar during the movement from the pallet to the wall. The second part of the task, the quality control, is more difficult to perform. In the first approach, mechanical sensors integrated in the gripper shall recognize material damages.

The collision free motion (task no. 8) from the pallet through the mortar to the wall including the gripping and assembly process will be guaranteed by a computer integration concept that covers the information flow from the architectural plan, to the working sequence planning, to the palletising planning, to the automated robot program generation (for the vehicle and the robot) and highly sophisticated simulation and on-line control interface. The necessary computing power is provided by the separate control components off-line station, on-line coordinator, robot control and vehicle control.

Mechanically, it is intended to employ a 4-DOF kinematic with rotatory joints and hydraulic drives due to the heavy load and large workspace. After an analysis of different kinematic approaches, it turned out, that four degrees of freedom are enough for the intended tasks. The blocks are vertically stacked on the pallet and have to be vertically assembled on the wall. Therefore, only the three main axis and one orientation axis are necessary to reach each point in the working space and to assemble wall corners with different orientations. To keep the blocks vertical, a parallelogram design of the kinematic is sufficient. The only errors, which the kinematic is not able to compensate, are the rotational components of the vehicle´s tilting errors. A vehicle tilting of 1° causes a tilting of the block of 1°. It is intended to compensate these errors with the passive complient systems mentioned above.

To assemble the blocks with a stated accuracy of 1 mm as the last and most important task, a two step concept will be realized in this approach. First by moving the robot kinematic to a position near the end position of the block with an intended accuracy of 5 cm. This way, the inaccuracies are always in one direction. Secondly by performing the fine positioning with active compensation elements integrated in the gripper. In first tests a concept without additional sensors will be developed, were in further steps according to the results necessary sensors can be integrated step by step.

5. CONCLUSION

The current state of the research shows the whole amount and spectrum of difficulties, occuring while trying to automate as simple tasks as putting a block onto another. Yet different solutions approaches for the different sub-tasks are presented, which encourage us to go the chosen way further. Ia addition the parallel carried out economic calculations and market analysis are positive enough to let us hope for a competive system in the future, which helps us to humanize and automate hard and dangerous construction working tasks.

6. REFERENCES

1. Bock, T., Blaser C., Gebhart, F.: *Automatisierungsgerechtes Planen und Konstruieren für computerintegriertes Bauen.* in: Bautechnik 69 (1992), No. 3, Ernst & Sohn Berlin.
2. Bock, T., Gebhart, F., Steck, W.: *Automatisierung von Hochbauprozessen.* in: Baumaschinen- technik, April 1992 ff., Bauverlag Wiesbaden.
3. Laukemper J.: *Automation im Mauerwerksbau: Voraussetzungen, Verfahren, Wirtschaftlichkeit.* 1992 Expert Verlag, Ehningen b. Böblingen.

Automation and Robotics in Construction X
G.H. Watson, R.L. Tucker and J.K. Walters (Editors)
1993 Elsevier Science Publishers B.V.

Analysis of the Development and Application of Automation Technology in Building Production

Yuji Hirabayashi

Architectural Institute of Japan, Research Committee on Building Materials and Construction Procedures; The Committee on Technology for Robotized Building Production.

Shimizu Corporation, Building Technology Development Department, Technology Division, No. 2–3, Shibaura 1-chome, Minato-ku, Tokyo Japan 105-07

Abstract

The purpose of this research is to examine current and future trends in the use of automation technology at advanced construction sites. At the start of this study, we established basic concepts to enable a working group to share a common understanding of construction automation. Based on these concepts, research was then carried out to determine what kinds of automation technology are presently being developed by general contractors in Japan to increase the efficiency of each production process starting from the production design stage. We then carried out various analyses on the introduction of automation technology to actual construction sites This research was conducted as part of the long-term study on robotized building production being carried out by the Subcommittee on Technology for Robotized Building Production, which belongs to the Architectural Institute of Japan's Research Committee on Building Materials and Construction Procedures.

1. INTRODUCTION

The Subcommittee on Technology for Robotized Building Production was formed in April 1987 as part of the Architectural Institute of Japan's Research Committee on Building Materials and Construction Procedures. This subcommittee was established to analyze the present state of robot development in the building construction industry, and to clarify guidelines needed for the development of robotized execution systems and construction automation (hereafter referred to as "CA"). The subcommittee also established a working group (hereafter referred to as "WG") in April 1990 to focus on the automation technology that will be needed for CA, as well as a WG to concentrate on the building technology involved in CA. This paper reports on the results of the above research.

2. SUMMARY OF ACTIVITIES

The WG has been carrying out its research and study activities according to the following procedures:

1) Confirmation of methods for CA research

The procedures and methods for doing CA research were discussed, then the WG confirmed the research agenda and prepared a schedule for the entire project.

2) Study of the common perceptions of CA

The WG tentatively established the CA concept and fields related to CA in order to ensure a common understanding of CA among members of the WG.

3) Research on current automated production technology

The WG researched the current status of automation technology in building production by general contractors in areas subject to CA. The research results were then arranged in a table according to individual classification codes.

4) Research on technology related to CA

Based on the above table, the WG conducted its research on technology related to CA in questionnaire format. The goal was to determine the current status of automation technology as it is used in the work sites of general contractors that employ advanced construction technology.

5) Analysis of research results and preparation of report

The accumulated technological research results were entered into a data base, and the work of analyzing each research item began. The frequencies of use of the technology were totalled, and the effect of the site conditions and the interrelationships between the research items were analyzed.

This paper reports on steps 3, 4, and 5 as described above.

3. CURRENT STATUS OF AUTOMATED PRODUCTION TECHNOLOGY

The WG conducted research on the current use of automation technology for building production. Automation technology in the building production process was classified into

Table 1
Automation technology classification code chart

A. Production design →	B. Construction planning →	C. Construction control →	D. Construction work →	E. Maintenance
a. Estimation	a. Project budget	a. Cost control	a. Building frame construction	a. Maintenance
b. Construction planning	b. Scheduling	b. Scheduling control	b. Finishing work	b. Inspection
c. Construction method planning	c. Construction method planning	c. Safety control	c. Electrical and mechanical installation work	c. Repair work
d. Structural planning	d. Temporary work plannning	d. Quality control	d. Others	d. Demolition work
e. Others	e. Shop drawing	e. Others		e. Others
	f. Others			

five stages: production design, construction planning, construction control, construction work, and maintenance. We further divided each stage into smaller categories, and prepared a classification code chart (see Table 1) and table showing the distribution of technologies that are currently in possession of general contractors and those that are actually in use.

The WG collected 190 responses from personnel at eight general contractors. The distribution of the answers for each classification code is shown in Table 2. The response, "construction work," was the most frequent. There were comparatively fewer types of automation technology used in the "upstream" portion of the production system. We found that automation technology tends to be concentrated in the areas most closely related to on-site execution work.

4. RESEARCH ON CA TECHNOLOGY

The objective of this research can be summarized to understand the current use of automation in the work sites of general contractors that use advanced technology.

The WG received responses from 159 architectural technicians and 117 work sites. This research was conducted in September and October 1991. Eighty percent of respondents were members of a work site. The typical work site involved in this research would be a steel frame combined structure office building with 15 above ground stories and two basement floors, a total floor area of approximately 50,000 m^2, and a construction term of about 26 months.

Table 2
Comparison of distribution for technologies in possession and in actual use.

	A. Production design			B. Construction planning			C. Construction control			D. Construction work			E. Maintenance		
a	Estimation	5.8%	1.0%	Project budget	2.1%	4.9%	Cost control	2.1%	6.6%	Building frame construction	18.9%	18.0%	Maintenance	5.8%	0.3%
b	Construction planning	0.5%	0%	Scheduling	2.1%	1.3%	Scheduling control	3.7%	4.3%	Finishing work	7.9%	9.8%	Inspection	3.7%	0.3%
c	Construction method planning	3.2%	2.3%	Construction method planning	2.6%	1.6%	Safety control	7.4%	14.4%	Electrical and mechanical installation work	0%	4.6%	Repair work	2.6%	1.0%
d	Structural planning	1.1%	1.6%	Temporary work planning	2.6%	3.9%	Quality control	8.9%	8.5%	Others	1.6%	0.7%	Demolition work	4.2%	0.7%
e	Others	10.5%	4.9%	Shop drawing	7.9%	10.8%	Others	2.6%	1.6%				Others	0.5%	0%
f				Others	2.1%	1.6%									
Total		10.5%	4.9%		19.5%	24.3%		24.7%	35.4%		28.4%	33.1%		16.8%	2.3%

Notes: Left value: Percentage of technologies in possession.
Right value: Percentage of technologies in actual use.

5. TOTALLING OF RESEARCH RESULTS

Research results for a total of 259 cases of applied technology were gathered from 117 work sites. The work sites were separated into groups according to their total floor areas. The number of applied technologies used per work site for each group was as shown in Table 3. The results show that as the scale of construction increases, more advanced technologies are applied.

Next, the applied technologies listed in the 259 responses were grouped according to the technology classification codes shown in Table 1. Since some technologies cover two or more classification codes, the table shows a total of 305 cases of technologies. We prepared a distribution for the 305 cases, as shown in Table 2.

When the applied technologies were observed according to technology classification codes, the most frequent application of technology on site was "building frame construction" in construction work, followed by "safety control" in construction control, "shop drawing" in construction planning, and "finishing work" in construction work.

6. ANALYSIS OF RESULTS CONCERNING APPLIED TECHNOLOGIES

The WG analyzed construction technology where automation was actually employed at the work site. The following five items were considered for each type of technology.

6.1. Analysis of applied technologies according to their stage of development

Of the applied technologies mentioned in the answers, established technologies (those with long records of actual use) constituted about 63% of all applied technologies listed, so only 37% were new technologies. In the trial stages, new technologies made up only 10% of all applied technologies.

In relation to the technology classification codes, most of these technologies belong to "construction planning," "construction control," and "construction work." A large proportion of technology still in the trial stages, such as the use of robots and automated machinery, belonged to the category of "construction work." Well established technologies were mainly found in the areas of "construction planning" and "construction control" (see Figure 1).

Table 3
Relationships between applied technology and floor area

Total floor area	Number of applied technologies	Number of work sites	Number of applied technologies per work site
5,000 m^2 or less	18	14	1.29
10,000 m^2 or less	14	10	1.40
20,000 m^2 or less	23	19	1.21
50,000 m^2 or less	75	32	2.34
100,000 m^2 or less	70	24	2.92
Over 100,000 m^2	59	18	3.28
Total	259	117	(Average) 2.21

6.2. Analysis of applied technologies according their rates of application

We checked the percentage of work done by using each applicable technology, relative to the total work volume or entire work term. Technologies with application rates over 90% were mentioned in about 40% of all responses (the largest single category), and technologies for which the application rates were low were mentioned in fewer responses. These results suggest that many of the technologies have a high application rate.

Observation of the applied technologies in relation to the technology classification codes showed that the application rates were high for technologies in the area of "construction control," while those for "construction work" technologies were low. (See figure 2)

6.3. Analysis of applied technologies relative to the reasons for their adoption

When we checked the reasons for the adoption of technologies, three were especially notable: the labor shortage, a shortage of on-site supervisors, and technological interest. Together, these reasons constituted over 60% of all answers (see Figure 3).

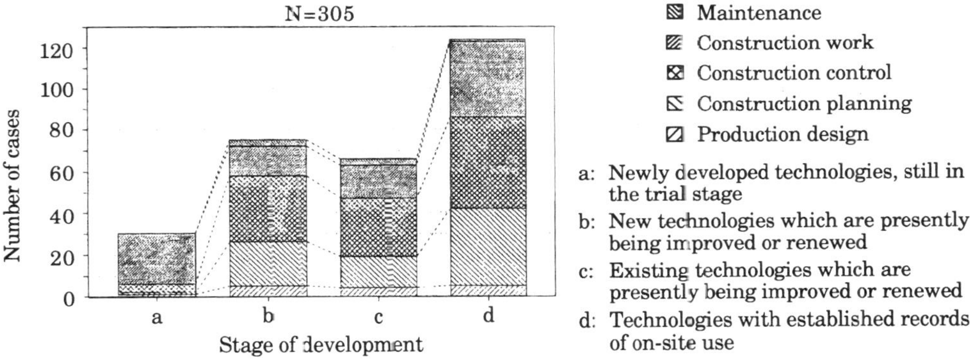

Figure 1. Relationships between development stages and classification codes

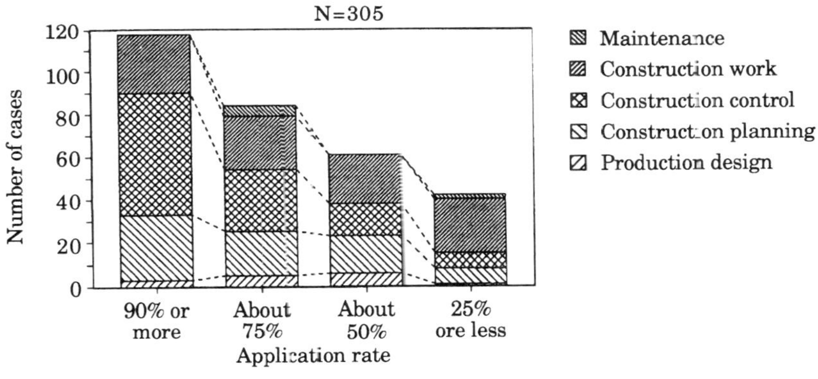

Figure 2. Relationships between application rates and classification codes

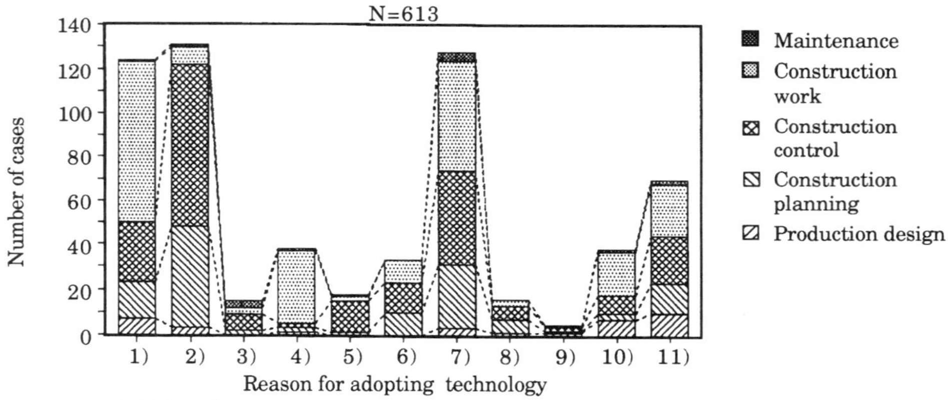

1) Labor shortage
2) Shortage of on-site supervisors
3) Owner's demand
4) Demand by a cooperating company
5) Requested by nearby parties
6) Ordered by a superior
7) Technological interest
8) Demand from designer
9) Company policy
10) Demand for safety assurance
11) Other

Figure 3. Relationships between reasons for adopting a technology and its application rates

6.4. Analysis of reasons for adopting technologies versus their results

We totalled the reasons for adopting technologies as stated in the 259 answers. Of these, 193 (75%) said the goal was to save labor and energy; 134 (52%) cited "shortening of the construction term and shortening of work hours" as the reason, followed by "improvement of the work environment," "elimination of dangerous jobs," and "quality control." Only 38 (15%) mentioned "cost reduction", the least common reason. Next, we collated the results relative to each goal. Most responses indicated that the results were almost as expected. However, those who said that their reason for adopting the technology was to reduce costs gave the lowest evaluations, indicating how difficult it is for automation to contribute to cost reduction.

Consideration of the reasons for adopting technology for each classification code indicates a shortening of the construction term, labor saving, and energy saving in high percentages for construction planning and construction control. The rates for elimination of dangerous jobs and improvement of the work environment are high for construction work (see Figure 4).

Over 90% of the responses gave comprehensive evaluations of the technologies adopted of "satisfactory", with 10% of these giving an evaluation of "very satisfactory." In the comprehensive evaluation of results according to technology classifications codes, the results for applied technologies in the areas of temporary work planning, quality control, scheduling control, and building frame construction were evaluated highly.

6.5. Analysis of conditions for the accelerated use of applicable technologies

As shown in Figure 5, the following six items constituted about 60% of all results:

1. Improving performance of equipment and systems
2. Making systems easier to operate
3. Providing sufficient time for prior study before use

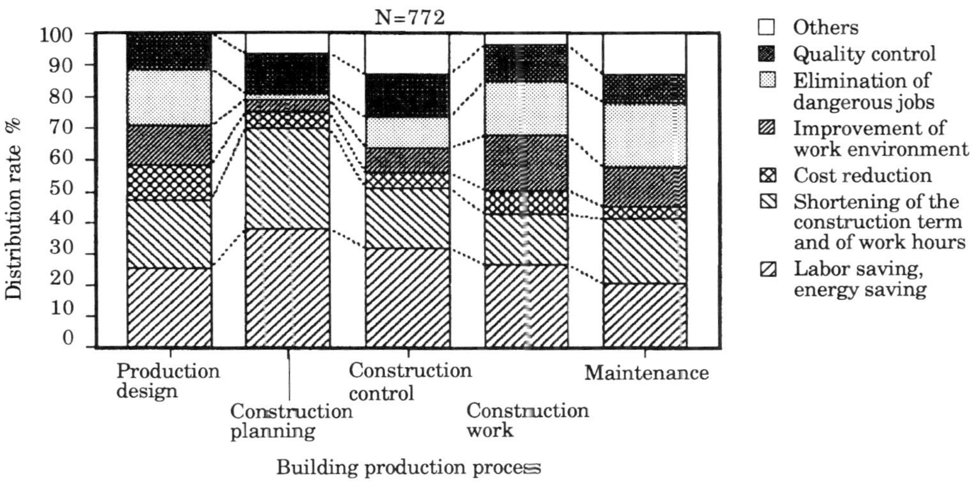

Figure 4. Distribution of the reasons for adopting technology for each classification code

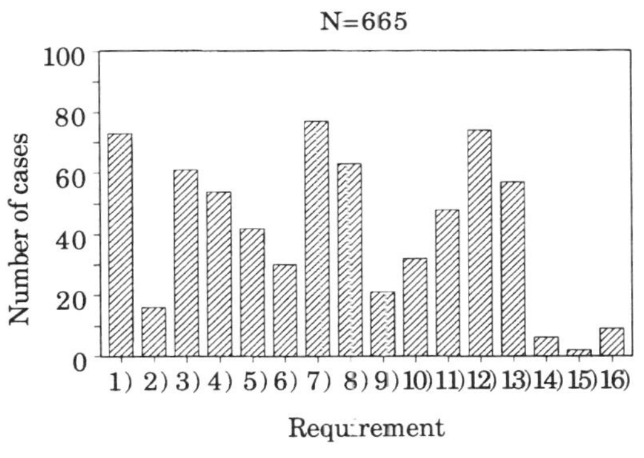

1) Providing sufficient time for prior study before use
2) Improving knowledge among top executives
3) Assembling complete reference materials
4) Achieving thorough communications with specialists on introductory training, planning, operation guidance, and system maintenance.
5) Preventing excessive expenditures on preparatory work or on arrangements and preparation for production and design changes
6) Preventing excessive expenditures of time and money on inspection and repair work
7) Improving performance of equipment and systems
8) Ensuring skillful operation of equipment and systems
9) Taking measures to gain the cooperation of operators
10) Maintaining sufficient numbers of high-quality operators
11) Taking measures to minimize the time and labor needed to input data when operating the system
12) Making systems easier to operate
13) Making systems more widely applicable
14) Designing lightweight and compact system
15) Achieving cost reductions for equipment and systems
16) Other

Figure 5. Requirements for accelerating the active use of CA technology

4. Ensuring skillful operation of equipment and systems
5. Assembling complete reference materials
6. Making systems more widely applicable

We noticed that there was great demand for improved performance of equipment and systems, as well as improved operability.

7. CONCLUSIONS

The following points were clarified as a result of our research:

1) The enthusiastic attitude of persons in charge of automation at work sites where advanced technologies are applied was evident. Shortages of staff members and construction workers and a strong interest in technology were also notable.
2) Quite a large percentage of applied technologies met their expected purpose, and these systems received satisfactory evaluations. In particular, technologies with an established record of use in shortening the construction term and saving energy received excellent evaluations.
3) We have clarified some of the actual problems involved with applied technologies, including technologies with long records of use. We have also clarified the requirements for accelerating the use of these technologies.

In the future, we will carry out further evaluations of applied technologies that are aimed at the achievement of CA and will continue our efforts to develop new systems that incorporate CA technology. At the same time, we will further explore CA through concrete evaluations of overall effectiveness.

In closing, we would like to express our gratitude to the many people who contributed to the present report.

8. REFERENCES

1 Junichiro Maeda, Research Study of the Building Construction Technologies Aimed at Automated Construction, 7th Construction Robot Symposium (1993)

2 Kazuhiko Arai, Images of the Future Development of Automated Construction Techonology, 10th ISARC (1993)

List of Subcommittee Members

Technology for Robotized Building Production, AIJ

[Chairman]
Yasushi Tamura (Waseda University)

[Secretary]
Kazuhiko Arai (Kajima Corporation)
Nobuyasu Miura (Kokushikan University)

[Members]
Norimoto Kashino (Building Research Institute, Ministry of Construction)
Junichiro Maeda (Shimizu Corporation)
Eiji Muro (Takenaka Corporation)
Hiroshi Naito (Japan Association of Representative General Contractors)
Shigeru Sakamoto (Taisei Corporation)
Kunio Shiina (Tokai University)
Toyokazu Shiire (Kanagawa University)
Tooru Shinozaki (Toda Corporation)
Tatsuo Terai (Chiba Inst. of Technology)
Masashi Wada (Kumagai Gumi Co, Ltd.)
Tatsuya Wakisaka (Obayashi Corporation)
Ryoji Yoshitake (Fujita Corporation)
Yasuo Inoue (Obayashi Corporation)
Katumi Kobayashi (Kumagai Gumi Co., Ltd.)
Toshifumi Sato (Kajima Corporation)
Hiroshi Fujita (Nihon Bisoh Co., Ltd.)
Muneyoshi Matuo (Fujita Corporation)
Takashi Miyazaki (Takenaka Corporation)
Takashi Shiokawa (Obayashi Corporation)
Osamu Nagayasu (Fujita Corporation)
Yukio Haga (Taisei Corporation)
Yuji Hirabayashi (Shimizu Corporation)
Masahiro Morita (Takenaka Corporation)

Automation and Robotics in Construction X
G.H. Watson, R.L. Tucker and J.K. Walters (Editors)

AUTOMATION NEEDS DETERMINATION USING AHP APPROACH

Sy-Jye Guo and R. L. Tucker

Department of Civil Engineering

University of Texas at Austin, TX 78712, U.S.A.

Abstract

There is a consensus among researchers that identification of automation needs is necessary for any further hardware developments. To accomplish this goal, a clear division of automation levels, a comprehensive breakdown of construction activities to task levels, and a method for measuring automation concern are essential. This paper presents an automation spectrum with five automation levels: hand tools, manual-controlled devices, tele-controlled devices, pre-programmed devices, and cognitive robots. Forty-two basic construction tasks are defined to assist in identification and development of automation needs. An Automation Concern Index (ACI) for construction task using the Analytical Hierarchy Process (AHP) is presented to assist in identifying automation needs. The Automation Concern Index serves as a base for automation needs evaluation and can be revised to fit in various construction environments by using the AHP approach.

1. Automation level

Automation development and implementation should be defined on an appropriate level to become successful based on the current state of technology. It is essential to determine the appropriate level of automation in the task identification phase. Otherwise, a proposed development will become too difficult to implement and turn out to be unfeasible. Tucker [13] has proposed a man-machine line to show the relationship between human conditions and mechanical contributions. On the left end side, manual labor is the only input for the construction task, while on the right end side, a fully roboticized device with cognitive ability is used to complete the job. Both ends represent an extreme status of automation level. Manual labor has been used since the beginning of human civilization and, in many areas of construction work, it is still the major source for construction activity. With the advanced technology in computers, robotics, and new equipment developed in the last decade, roboticization undoubtedly will have many applications in the construction process. Meanwhile, many construction tasks which are labor intensive are still located near the left end of the spectrum. There is potential that these tasks can be significantly improved through some way of automation or partial roboticization.

The automation spectrum is shown in Fig. 1. Typical examples of concrete floor troweling devices are presented to fit in the appropriate intervals. As is still used in many job sites for concrete floor finishing, the manual labor with hand trowel fits in the hand tools interval. Walk-behind and ride-on trowels have been widely used in the U.S. because of the advancement and improvement of both performance and productivity. Both these two power trowels fit in the manual-controlled devices interval. The prototype concrete floor troweling robots developed by Shimizu Corp.[2] and Kajima Corp.[12] in Japan fit in the tele-controlled and pre-programmed intervals respectively.

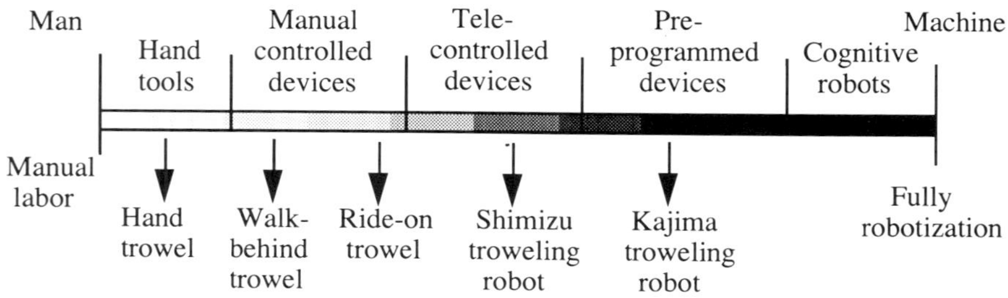

Fig. 1 Automation spectrum

Only a few cases exist in the tele-controlled interval and most of them are prototype devices currently under development. Concrete floor troweling robots developed by Shimizu, Takenaka, and Obayashi Corp. are typical examples in this interval. However, the tele-controlled technology has been widely used in our daily life, such as TV's, VCR's and etc. More and more construction equipment will be upgraded with the tele-controlled technique in the near future. Much more effort and techniques will be needed for developing a pre-programmed device, especially in the control system and sensing ability. Very few prototype devices exist in this category. The Automated Guided Vehicle (AGV) developed by the Caterpillar Corp. is one of the pioneer devices. Kajima's floor troweling robot is another example. The full roboticization with cognitive ability device is the optimal goal of construction automation. However, none of this type of device has yet been developed.

To advance the construction task on the automation spectrum, the appropriate level of automation should first be considered. It may be more feasible to upgrade the level by one interval at a time, e.g., advance hand tools to the manual-controlled level, or upgrade manual-controlled devices to the tele-controlled level. Only in extreme situations, such as in a radiation polluted area, the deep sea or air space, will fully roboticized devices become the better option. Many difficulties and obstacles will occur if a current hand tool is upgraded directly to the tele-controlled or even pre-programmed level. Not only are difficulties concerned about technology problems, but also the labor resistance results from fear of losing jobs, suspiciousness of performance, or lacking of special skills for new technologies, will inhibit automation progress.

2. Division of work

Another significant aspect of construction automation is the division of work to be designed for automation. Because of the complicated nature of construction work, it is important to define the type of task an automated device should perform. For more complex tasks, a device will require more sophistication. The manufacturing industry has utilized many robot applications to replace human workers in production lines and perform single tasks such as painting, welding and bolting. It will be very productive to adapt the experience of the manufacturing industry for developing automation applications in construction tasks.

Some researchers such as Kangari and Halpin [8], Fazio [5], Basford and Askew [1] have identified automation opportunities on a wide range of construction processes. Their subjects range from the task level up to a whole construction division. Tucker and Peterson, et al. [14] have proposed a division definition between areas, activities, and tasks. A typical breakdown example of construction work is shown as below.

Table 1 Area-Activity-Task example

AREA	ACTIVITY	TASK
Concrete	Rebar fabrication	Bend
		Position
		Tie
	Concrete placement	Spread Concrete
		Vibrate
		Finish

The task level is the simplest and most appropriate level for an automation device. The activity level, which consists of several tasks, becomes more complicated for a single device to complete the work. Not only does the problem of the control system arise, but also the different end effectors, sensors, mobility, and the material supply problems must be addressed. The area level, which is too broad and vague for automation definition, is probably unrealistic during the identification phase.

Some other researchers have tried to identify automation opportunities at the task level to make them more feasible based on the current state of technology. Warsawski and Sangrey [16] presented 10 basic construction "activities" in building construction. Their presentations are actually at the task level. Everett [4] has expanded the list to 18 basic construction tasks which purport to include all the construction activities. Demsetz [3] has utilized a conceptual design process to identify construction tasks for automation. Her study investigates the division of labor between man and machine based on the forces and displacements required to carry out construction work

Based on the experience of the manufacturing industry and the discussion above, the task level is an appropriate level for construction automation . A comprehensive breakdown of construction activities to the task level will be significant in assisting identification of automation opportunities. A list of basic construction tasks has been generated by a panel consisting of professors, researchers, and graduate students in construction Engineering. Forty-two generic construction tasks have been identified and defined as listed in Table 2.

Table 2 Basic Tasks Involved in Construction Activities

	Basic Task	**Definition**	**Examples**
1	Arrange	Put a number of objects in a proper order	Arrange rebar
2	Align	Keep objects in a straight line or orientation	Align steel columns
3	Bend	Deform the shape of an object	Bend steel rebar
4	Caulk	Inject sealant between two adjacent objects	Seal concrete joints
5	Clean	Remove unwanted dirt, material or impurities	Sweep floor
6	Coat	Apply a layer of liquid on an object's surface	Paint wall
7	Communicate	Talk or use hand signals to transfer information	Ask or answer
8	Compact	Condense soil or other material	Compact soil refill
9	Connect	Join or fasten two objects to each other	Nail, bolt, tie, weld
10	Cover	Unroll sheet material on an object's surface	Unroll carpet
11	Cut	Divide one object into two or more pieces	Saw wood, cut tile
12	Disconnect	Break the connection between two objects	Strip forms, unbolt
13	Dismantle	Demolish or break down an undesired portion	Dismantle concrete
14	Drill	Make a hole by rotation	Drill hole
15	Excavate	Dig out by remove soil or material	Excavate tunnel

16	Fill	Put soil or material in place	Dump soil
17	Finish	Apply continuous mechanical treatment to surface	Trowel, grind
18	Hit	Strike hardly to push an object	Hit piles
19	Hold	Keep an object in a position temporarily	Hold hose, tag line
20	Identify	Recognize an appropriate member	Identify steel column
21	Inlay	Set small flat pieces next to each other	Set tiles.
22	Insert	Push an object into another one	Insert form ties
23	Inspect	Examine and detect flaws or verify correctness	Visually check
24	Install	Put an object into final position	Install light fixture
25	Level	Keep material on a horizontal plane	Screed concrete
26	Lift	Move an object upward for transporting	Lift concrete panel
27	Lay	Set objects next to or on top of each other	Lay bricks
28	Measure	Determine or layout correct dimensions	Measure rebar layout
29	Operate	Control an equipment for work	Operate crane
30	Position	Move an object to the correct location	Position steel beam
31	Pour	Cast concrete into forms or slabs	Pour concrete slabs
32	Prepare	Make material ready for future use	Get material
33	Pull	Draw electrical wire through conduit	Pull cable
34	Pump	Transport material by air pressure	Pump concrete
35	Roll	Move an object on wheels along a surface	Roll the flooring
36	Shape	Modify the shape of an object to fit into position	Trim wood
37	Spray	Jet liquid or particles without contact with surface	Spray paint
38	Spread	Apply semi-liquid material to various location	Spread mortar
39	Tap	Strike or touch an object gently	Tap tiles or bricks
40	Transport	Move material to designated location	Transport bricks
41	Vibrate	Shake or tremble to consolidate material	Vibrate concrete
42	Write	Make note or mark to indicate a specific purpose	Write notes

3. Automation concern factors

For automation needs concerns, researchers have developed five major factors which motivate automation development. They are safety, productivity, worker utilization, superhuman handling and quality [14]. All these factors are further divided into several sub-factors. The hierarchy structure for automation concern with sub-factors is shown in Fig. 2.

Safety concerns arise from the high accident rate of the construction industry. Thus, any task involving a safety concern should be considered for automation. Hazardous to health, physical dangerous, and elevated work are sub-factors defined in this concern. The productivity concern involves aspects such as repetitive, dirty or unpleasant, boring, tedious, and exhaustive work. The manufacturing industry has utilized many robot applications to improve human productivity. Construction tasks with similar natures undoubtedly will gain essential productivity improvement if automation development can be applied. The worker utilization concern deals with the supply and skill level of human labor. As construction jobs are considered to be unsafe, unpleasant, and low-knowledge work, the younger generation is more unwilling to become construction workers. For construction tasks requiring special skills, this situation becomes worse. Special skill requirements and labor intensive are the sub-factors in this category. The super-human handling concern implies tasks which are not appropriate for human workers to handle. For tasks where the requirements exceed the physical capabilities of human labor, some kind of equipment or automation device will be needed to assist in completing the task. Heavy lift, high lift and meticulous tasks are typical cases in this category. The quality concern considers both the tolerance level and consistency requirement of a construction task. Much rework commonly occurs on a job site because of poor quality control. Automation development is expect to relieve this problem by achieving a better working quality through the use of machines.

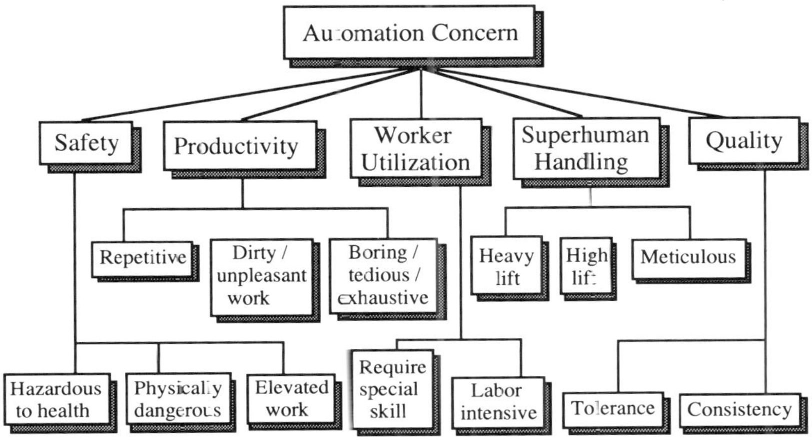

Fig 2 Hierarchy of Automation Concern

4. Analytic hierarchy process

An Analytic Hierarchy Process (AHP) was developed by Saaty [10] in the 1980's and utilized in broad application areas, such as decision making, resource allocation, optimization, and etc. [7]. The basic concept of AHP is to develop an hierarchy structure of the factors which contribute to the final goal. By comparing these factors in a matrix format, a priority vector can be derived through mathematical procedures. The coefficients of the priority vector represent an indication of the weight that should be assigned to each factor. Some researchers have also proposed mathematical mechanisms for checking the consistency of these weights. By utilizing AHP, concerns about non-quantitative factors can be effectively presented in a systematic way. Traditionally, researchers intuitively assign a subjective weight to factors which are difficult to quantify. AHP resolves the argument of subjective weights and reaches an objective distribution of priority.

The basic procedure involved in AHP is described as follows:

1. Develop an hierarchical structure of factors and sub-factors contributing to the final goal
2. Rank these factors in order and put them as headings of both rows and columns in the comparison matrix. The comparison matrix will have 1 at all the diagonal cells.
3. Compare these factors relatively with a scale of 1 to 5 and fill in the upper half of the comparison matrix. The scale 1 to 5 represents a ratio comparison of two factors with respect to,
   1: two factors contribute equally
   2: one factor is slightly favorable than another
   3. one factor is moderately favorable than another
   4. one factor is strongly favorable than another
   5. one factor dominates another
4. Put the reciprocal of each cell to the symmetric cell of the lower half of the matrix.
5. Calculate $\sum$ cell value/column sum to reach a combined weight of each factor.
6. Normalize the combined weight to become a priority vector.

The coefficient of the priority vector implies the weight of each factor. The second part of the AHP is to aggregate the weight of each alternative based on different factors. A similar matrix is used to compare the alternatives pair-wisely based on only one factor. Combining the priority vector of factors with the priority vector of alternatives based on different factors yields the final priority vector for all the alternatives [10]. Formal procedures for checking the consistency of the comparison matrix are also provided by the AHP approach. Details and discussions of the philosophy of AHP can be found in the references [7,10]. The purpose of this research is to elicit the priority weight of each automation concern factor instead of deciding which alternative should be automated. The second part of the AHP is not utilized in this study. However, if automation alternatives have been available, the AHP can be utilized to decide which alternative is the best choice.

5. Assessment of weights

To accomplish an appropriate weight for these five factors, a questionnaire has been distributed to researchers and experts. Discussions and judgment based on the replies of the survey yield the scales to be filled in the comparison matrix.

	Safety	Product.	Super.	Worker.	Quality
Safety	1	2	2	3	5
Productivity		1	2	2	5
Super. handl.			1	2	5
Worker util.				1	4
Quality					1

Once the comparison scales are determined, the reciprocal of each cell is put into the lower half of the matrix, and the column sums are calculated. Then, the row sum is calculated by $\sum$ cell value$_i$ / column sum$_i$. Finally, normalize the row sum vector to become the priority weight vector.

	Safety	Product.	Super.	Worker.	Quality	Row sum	Weight
Safety	1	2	2	3	5	1.84	0.37
Productivity	1/2	1	2	2	5	1.28	0.26
Super. handl	1/2	1/2	1	2	5	0.98	0.20
Worker util.	1/3	1/2	1/2	1	4	0.66	0.13
Quality	1/5	1/5	1/5	1/4	1	0.24	0.05
Column sum	2.53	4.20	5.70	8.25	20.00	5.00	1.00

Following the mathematical procedures listed above, a preliminary weight of each factor has been reached as follows :

Automation Concerns Index (ACI)
=0.37*Safety + 0.26*Productivity + 0.20*Super. handl.
+ 0.13*Worker util. + 0.05*Quality

6. Automation Concern Index

To reach a final index, all the sub-factors for each concern are assigned a value of 0 or 1 and summed to be the input for each specific concern. The value 1 indicates that the task is concerned with this sub-factor, and the value 0 means the task is not associated with this sub-factor. For example, vibrating concrete is concerned with the following sub-factors:

repetitive, dirty and unpleasant, boring and tedious, labor intensive, and consistency required. No safety sub-factors are related but three productivity sub-factors are associated. Thus, the input values of safety and productivity for vibrating concrete are 0 and 3. Similarly, the super-human handling, worker utilization, and quality concern are assigned as 0, 1, and 1 respectively. The Automation Concern Index for vibrating concrete can be calculated as

$$ACI = 0.37*0 + 0.26*3 + 0.20*0 + 0.13*1 + 0.05*1 = 0.96$$

This binary decision is quite rational and easy to follow. However, this approach reveals the relative concern of different tasks for each sub-factor. To implement this approach, an interview with foremen on a viaduct girder construction job site was held to decide if each sub-factor was related to different tasks. Judging from the responses of the foremen, 16 task examples were given automation indices in Table 3 according to the equation above.

The ACI derived from the equation ranges from 0 to a possible maximum value of 2.85 if all sub-factors apply. For tasks listed in the Table 3, the sample mean (μ) is 1.24 and standard deviation (σ) is 0.29. Because few construction task will involve all these five concerns at the same time, most tasks will have an index between 1 to 2. Hence, index greater than 1.43, half of the maximum value, is recommended for automation consideration. Weld rebar, position parapet segment, plaster concrete surface, grind concrete surface, and fireproof steel frame are tasks qualified for this consideration.

The Automation Concern index (ACI) is designed to quantify the intangible motivation for automation development. It does not consider the cost impact in the concern hierarchy, neither are the technology factors considered. The cost impact is a quantifiable prospect and should be compared with the concern index while identifying automation needs. To identify potential candidates for automation development, the technical consideration should then also be addressed and investigated. Definition of an appropriate level in respect to the automation spectrum and clear division of construction tasks will definitely be helpful to further hardware development.

Table 3 Automation concern indices for tasks

Task	Safety	Product.	Super.	Worker.	Quality	ACI
Vibrate concrete	0	3	0	1	1	0.96
Spread concrete	0	3	0	1	0	0.91
Level concrete	0	3	0	2	1	1.09
Weld rebar	2	2	0	2	1	1.57
Arrange rebar	1	2	1	1	1	1.27
Position veneer	1	2	2	1	0	1.42
Tap compacting	0	2	0	1	1	0.70
Chip pile head	1	3	0	1	0	1.28
Position parapet	2	1	2	2	0	1.66
Bolt parapet	2	1	0	1	1	1.18
Clean form	1	2	0	1	0	1.02
Nail form	1	1	0	2	1	0.94
Plaster concrete	1	3	0	2	1	1.46
Grind concrete	2	3	0	1	1	1.70
Bush hammering	1	2	0	2	1	1.20
Fireproof steel	2	2	0	1	1	1.44

7. Conclusions

Identification of automation needs and opportunities are essential for any hardware development. To identify a feasible automation candidate, an appropriate automation level should be defined, the working content should be broken down to the task levels, and the motivation for automation concern should be evaluated. An automation spectrum defined with five levels is presented in this study. Advancement of the automation level of current construction practice by one level is recommended. Forty-two basic construction tasks have been defined to assist in identification of automation needs. Finally, the Automation Concern Index (ACI) using the Analytic Hierarchy Process is presented. The AHP is an excellent approach for evaluating intangible concerns. The Automation Concern Index serve as a base for automation needs evaluation and can be utilized and revised to fit any construction environment by using the AHP approach. Combining the ACI with further cost impact concerns allows potential automation candidates to be identified [6].

References

1. Bashford, S. F., Askew, W.H. et al.,"Initial Task Selection for Detailed Robotic Evaluation", The 9th ISARC; 1992, Japan, pp 141-150.
2. Cranmer, Lee R. and Tucker, R. L.; *Operational Test and Evaluation of Automated Construction Device,* Construction Industry Institute, SD-59, 1990.
3. Demsetz, L.; "Task Identification for Construction Automation"; The 6th ISARC; 1989, San Francisco, pp 95-102.
4. Everett, J.; "Back to Basics in Construction Automation"; The 7th ISARC; 1990, London, pp 583-590.
5. Fazio, P., Moselhi, O., and Hason, S.; "Automation Index for Evaluation of Robotics Implementation in Construction"; The 6th ISARC; 1989, San Francisco, pp 317-324.
6 Guo, Sy-Jye; "*Automation Needs Identification for the Taipei Rapid Transit System Project* ", Dissertation, University of Texas at Austin, May 1993.
7. Harker, P.; "The Art and Science of Decision Making: The Analytical Hierarchy Process"; *The Analytic Hierarchy Process Applications and Studies*, Springer-Verlag Berlin Heidelberg, New York,1989, pp 3-58.
8. Kangari, R. and Halpin, D. W.; "Automation and Robotics in Construction : A Feasibility Study"; The 5th ISARC; 1988, Tokyo, pp 161-167.
9. Kangari, R. and Halpin, D. W.; "Potential Robotics Utilization in Construction"; Journal of Construction Engineering and Management, Vol. 115, No. 1, March, 1989, ASCE, pp 126-143.
10. Saaty, T. L.; *Decision Making for Leaders : The Analytical Hierarchy Process for Decisions in a Complex World*, Lifetime Learning Publications, California, 1982
11. Skibniewski, M. J.; *Robotics in Civil Engineering*, New York, Van Nostrand Reinhold Publisher, 1988.
12. Tanaka, N. et al, Kajima Corporation; "The Development of the 'Mark II' Mobile Robot for Concrete Slab Finishing", CAD and Robotics in Architecture and Construction, Proceedings of the Joint International Conference at Marseilles,France,1986,pp 249-257.
13. Tucker, Richard L.; "High Payoff Areas for Automation Applications"; The 5th ISARC; 1988, Tokyo, pp 9-16.
14. Tucker, Richard L., et al; *A Methodology for Identifying Automation Opportunities in Industrial Construction.* Construction Industry Institute, SD-56, 1990.
15. Warszawski, A.; *Industrialization and Robotics in Building--A Managerial Approach*, New York, Harper & Row Publishers, 1990.
16. Warszawski, A. and Sangrey, D. A.; "Robotics in Building Construction"; Journal of Construction Engineering and Management, Vol. 111, No. 3, March, 1985, ASCE, pp 260-280.

Automation and Robotics in Construction X
G.H. Watson, R.L. Tucker and J.K. Walters (Editors)

Lean production in construction

Lauri Koskela

Technical Research Centre of Finland, Laboratory for Urban Planning and Building Design, P.O. Box 209, SF-02151 Espoo, Finland.

Abstract

In manufacturing, great gains in performance have been realized by a new production philosophy, which leads to "lean production". This new production philosophy is a generalization of such partial approaches as JIT, TQM, time-based competition, and concurrent engineering. In construction, lean production is little known. The concepts, principles and methods of lean production are reviewed, and their applicability in construction is analyzed. The implications of lean production to construction practice and research are considered.

1. INTRODUCTION

The chronic problems of construction are well-known: low productivity, poor safety, inferior working conditions, and insufficient quality. A number of solutions or visions have been offered to relieve these problems in construction. Industrialization (i.e. prefabrication and modularization) has for a long time been viewed as one direction of progress. Currently, computer integrated construction is seen as an important way to reduce fragmentation in construction, which is considered to be a major cause of existing problems. The vision of robotized and automated construction, closely associated with computer integrated construction, is another solution promoted by researchers.

Manufacturing has been a reference point and a source of innovations in construction for many decades. For example, the idea of industrialization comes directly from manufacturing. Computer integration and automation also have their origin in manufacturing, where their implementation is well ahead compared to construction.

Currently, there is another development trend in manufacturing, the impact of which appears to be much greater than that of information and automation technology. This trend, which is based on a new production philosophy, rather than on new technology, stresses the importance of basic theories and principles related to production processes (Shingo 1988, Schonberger 1990, Plossl 1991). However, because it has been developed by practitioners in a process of trial and error, the nature of this approach as a philosophy escaped the attention of both professional and academic circles until the end of 1980's.

In construction, there has been rather little interest in this new production philosophy. The goal of this paper is to assess whether or not the new production philosophy has implications for construction. The paper is based on a more detailed study (Koskela 1992a).

2. LEAN PRODUCTION

2.1 Origins of lean production and the new production philosophy

Since the end of 1970's, a confusingly long array of new approaches to production management has emerged: JIT, TQM, time based competition, value based management, process redesign, lean production, world class manufacturing, concurrent engineering.

After closer analysis, it transpires that the above mentioned management approaches have a common core, but view this from more or less different angles. This common core is made up by a conceptualization of production or operations in general; the angle is determined by the design and control principles emphasized by any particular approach. For instance JIT stresses the elimination of wait times whereas TQM aims at the elimination of errors and related rework but both apply this angle to a flow of work, material or information.

Thus, a new production philosophy is emerging through generalization of these partial approaches, as has been suggested recently by various authors (Schonberger, 1990; and Plossl, 1991). The new production philosophy, regardless of what term is used to name it (world class manufacturing, lean production), is the emerging mainstream approach practised, at least partially, by major manufacturing companies in America, Europe and Japan. The new philosophy has already had a profound impact in such industries as car manufacturing and electronics. The application of the approach has also diffused to fields like customized production, services, administration and product development.

The conception of the new production philosophy evolved through three stages : It was viewed as a tool (like kanban and quality circles), as a manufacturing method (like JIT) and as a general management philosophy (referred to, for example, as world class manufacturing or lean production). The conceptual and theoretical aspects of the new production philosophy are least understood. However, without conceptual and theoretical understanding the application of methods is bound to remain inefficient and haphazard.

In Figure 1, an attempt for a consolidation of the new production philosophy is presented. The various levels are analyzed in the following.

2.2 Conceptual framework

The core of the new production philosophy is in the observation that there are two aspects in all production systems: conversions and flows. While all activities expend cost and consume time, only conversion activities add value to the material or piece of information being transformed into a product. Thus, the improvement of non value adding flow activities (inspection, waiting, moving), through which the conversion activities are bound together, should primarily be focused on reducing or eliminating them, whereas conversion activities should be made more efficient. In design, control and improvement of production systems, both aspects have to be considered. Traditional managerial principles have considered only conversions, or all activities have been treated as though they were value-adding conversions.

Due to these traditional managerial principles, flow processes have not been controlled or improved in an orderly fashion. We have been preoccupied with conversion activities. This has led to complex, uncertain and confused flow processes, expansion of non value-adding activities, and reduction of output value.

Material and information flows are thus the basic unit of analysis in the new production philosophy. Flows are characterized by time, cost and value.

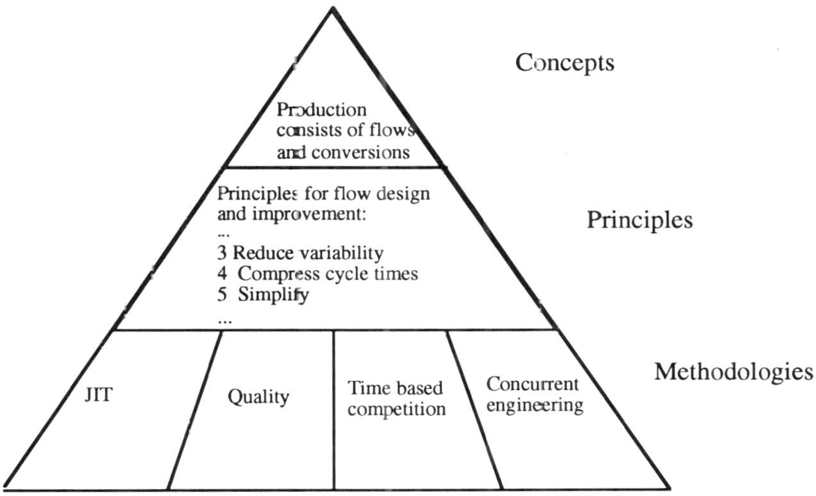

Figure 1. Different levels of the new production philosophy

2.3 Principles

In various subfields of the new production philosophy, a number of heuristic principles for flow process design, control and improvement have evolved. There is ample evidence that through these principles, the efficiency of flow processes in production activities can be considerably and rapidly improved. The principles may be summarised as follows (Koskela 1992a):

1. Reduce the share of non value-adding activities (also called **waste**).
2. Increase output value through systematic consideration of customer requirements.
3. Reduce variability.
4. Reduce cycle times.
5. Simplify by minimizing the number of steps, parts and linkages.
6. Increase output flexibility.
7. Increase process transparency.
8. Focus control on the complete process.
9. Build continuous improvement into the process.
10. Balance flow improvement with conversion improvement.
11. Benchmark.

In general, the principles apply both to the total flow process and to its subprocesses. In addition, the principles implicitly define flow process problems, such as complexity, intransparency or segmented control.

Experience shows that these principles are universal: they apply both to purely physical production and to informational production, like design. Also, they seem to apply both to mass production and one-of-a-kind production.

2.4 Methodologies and tools

Among the methodologies for attaining lean production are the following most important:

- Just In Time (JIT)
- Total Quality Management (TQM)
- Time Based Competition
- Concurrent Engineering
- Process Redesign (or Reengineering)
- Value Based Management
- Visual Management
- Total Productive Maintenance (TPM)
- Employee Involvement.

Most of these methodologies have originated around one central principle. Even if they usually acknowledge other principles, **their approach is inherently partial.** Thus, for example, the quality approach has variability reduction as its core principle. Time based management endeavors to reduce cycle times. Value based management aims at increasing output value.

In the framework of all these methodologies, useful techniques, tools and procedures have been developed. For example, such techniques as quality circles and the 7 quality tools (fishbone diagram, Pareto-diagram etc.) are used in TQM.

2.5 Comparison between conventional production and lean production

What is the conventional production philosophy being now replaced by the new philosophy? It is the paradigm of industrial mass production, which evolved in the beginning of this century. The most important differences between the conventional and the new philosophy are summarized in Table 1.

Table 1. The conventional and the new production philosophy.

	Conventional Production Philosophy	*New Production Philosophy*
Conceptualization of production	Production consists of conversions (activities); all activities are value-adding	Production consists of conversions and flows; there are value-adding and non-value-adding activities
Focus of control	Cost of activities	Cost, time and value of flows
Focus of improvement	Increase of efficiency by implementing new technology	Elimination or suppression of non-value adding activities, increase of efficiency of value adding activities through continuous improvement and new technology

The results of the implementation of the conventional and the new production philosophy are schematically illustrated in Figure 2.

Conventional production is improved by implementing new technology, primarily in value adding activities, to some extent also in non-value adding activities (like automated storages, transfer lines and computerized control systems). However, with time, the cost share of non-value adding activities, which are not explicitly controlled, tends to grow: production becomes more complex and prone to disturbances.

In lean production, non value-adding activities are explicitly attended. Through measurements and the application of the principles for flow control and improvement, it is possible to initially reduce the costs of non value-adding activities considerably. Value adding activities are first improved through internal continuous improvement and finetuning of existing machinery. Only after these improvement potentials are realized, major investments in new technology are considered. The implementation of new technology is easier in lean production, because less investments are needed and the production is better controlled. Thus, after the initial phase, increase of efficiency of value adding activities should also be more rapid in lean production than in conventional production.

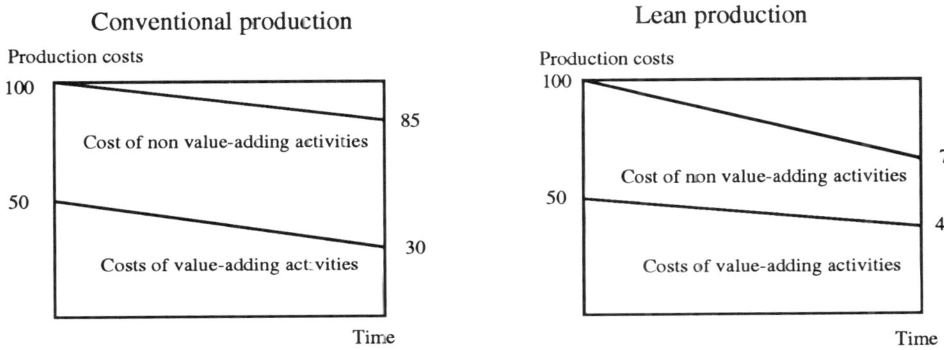

Figure 2. Conventional and lean production focus of development efforts

3. LEAN PRODUCTION IN CONSTRUCTION

3.1 Preliminary implementation

In the construction industry, the overall diffusion of the new philosophy seems to be rather limited and its applications incomplete. Quality assurance and TQC have been adopted by a growing number of organizations in construction, first in construction material and component manufacturing, and later in design and construction. The new approach, in its JIT-oriented form, has been used by component manufacturers, for example in window fabrication and prefabricated housing.

Why has the diffusion of the new production philosophy been so slow in construction? The most important barriers to the implementation of these ideas in construction seem to be the following:

- Cases and concepts commonly presented to teach about and diffuse the new approach have often been specific to certain types of manufacturing, and thus not easy to internalize and generalize from the point of view of construction.
- Relative lack of international competition in construction.
- Lagging response by academic institutions.

However, it seems that these barriers are of a temporary nature. On the other hand, the slow diffusion is not explained by an inadequacy of the new philosophy with respect to construction. This is justified by following analyses of waste and peculiarities in construction.

3.2 Waste in construction

To what degree do the problems associated with the conventional production view, as observed in manufacturing, also exist in construction? If the flow aspects in construction have been historically neglected, it logically follows that current construction would demonstrate a significant amount of waste (non value-adding activities). Thus, it is appropriate to check whether the existing information supports this hypothesis.

There has never been any systematic attempt to observe all wastes in a construction process. However, partial studies from various countries can be used to indicate the order of magnitude of non value-adding activities in construction. The compilation presented in Table 1 indicates that a considerable amount of waste exists in construction. However, because conventional measures do not address it, this waste is invisible in total terms, and is considered to be unactionable.

Table 1. Waste in construction: compilation of existing data (Koskela 1992a).

Waste	*Cost*	*Country*
Quality costs (non-conformance)	12 % of total project costs	USA
External quality cost (during facility use)	4 % of total project costs	Sweden
Lack of constructability	6 - 10 % of total project cost	USA
Poor materials management	10 - 12 % of labor costs	USA
Excess consumption of materials on site	10 % on average	Sweden
Working time used for non-value adding activities on site	Appr. 2/3 of total time	USA
Lack of safety	6 % of total project costs	USA

3.3 Problems of construction are caused by neglect of flows

Analysis (Koskela 1992a) shows that, as in manufacturing, the conceptual basis of construction engineering and management is conversion or activity oriented. The construction process is seen as a set of activities, each of which is controlled and improved as such. Conventional managerial methods, like the sequential method of project realization or the CPM network method, deteriorate flows by violating the principles of flow process design and improvement. They concentrate on conversion activities. The resultant problems in construction tend to compound and self-perpetuate. In project control, fire-fighting current or looming crises consumes management resources and attention so totally that there is little room for planning, let alone improvement activities. As a consequence, there is considerable waste in construction.

3.4 New conceptualization of construction

Following the lead of manufacturing, the next task is to reconceptualize construction as flows. The starting point for improving construction is to change the way of thinking, rather than seeking separate solutions to the various problems at hand.

Thus, it is suggested that the information and material flows as well as work flows of design and construction be identified and measured, first in terms of their internal waste (non value-adding activities), duration and output value. For improving these flows, it is a prerequisite that new managerial methods, conducive to flow improvement, are developed and introduced.

Such methods have already been developed to varying degrees. Not unexpectedly, they try to implement those flow design and improvement principles which are violated by the respective conventional method. However, lacking a sound theory, these efforts have remained insufficient.

Generally, taking flows as the unit of analysis in construction leads to profound changes of concepts and emphasis.

3.5 Construction peculiarities

Construction peculiarities refer especially to following features: one-of-a-kind nature of projects, site production, and temporary multiorganization. Because of its peculiarities, the construction industry is often seen in a class of its own, different from manufacturing. These peculiarities are often presented as reasons when well-established and useful procedures from manufacturing are not implemented in construction.

Indeed, these peculiarities may prevent the attainment of flows as efficient as those in stationary manufacturing. However, the general principles for flow design and improvement apply for construction flows in spite of these peculiarities: **construction flows can be improved**. Consider, for example, the one-of-a-kind nature of construction projects. The same peculiarity is shared by many - if not most - product development projects in manufacturing. However, it has been possible to shorten the development time and to improve the output quality in such projects by implementing principles of the philosophy.

On the other hand, these construction peculiarities can be overcome. Initiatives in several countries, like "Sequential Procedure" in France, "Open Building System" in the Netherlands and "New Construction Mode" in Finland try to avoid or alleviate related problems:

- one-of-a-kind features are reduced through standardization, modular coordination and widened role of contractors and suppliers
- difficulties of site production are alleviated through increased prefabrication, temporal decoupling and through specialized or multi-functional teams
- the number of temporary linkages between organizations is reduced through encouragement of longer term strategic alliances.

However, the **elimination of construction peculiarities is not any solution itself**: it just brings construction to the same level as manufacturing. Unfortunately, a large amount of waste also exists in manufacturing before process improvement efforts begin. Thus, only a starting point for effective process improvement is provided.

Thus it is concluded that the construction peculiarities do not diminish the significance of the new philosophy for this industry.

4. IMPLICATIONS

The implications of the new production philosophy for construction will be far-reaching and broad, as they are in manufacturing. The renewal of manufacturing has been realized in a feverish burst of conceptual and practical development. This might also happen in construction.

4.1 Implications for academic research

Current academic research and teaching in construction engineering and management is founded on an obsolete conceptual and intellectual basis. It is urgent that academic research and education address the challenges posed by the new philosophy. The first task is to explain the new philosophy in the context of construction. Formalization of the scientific foundations of construction management and engineering should be a long term goal for research.

4.2 Implications for major development efforts in construction

Current development efforts like industrialized construction, computer integrated construction and construction automation have focused primarily on the efficiency of value-adding and to some extent also non value-adding activities. They have to be redefined in order to acknowledge the needs for flow improvement. For example, the following guidelines for construction automation can be derived from the principles presented above (Koskela 1992b):

- automation should be primarily focused on value-adding activities
- construction process improvement should precede automation
- continuous improvement should be present in all stages of development and implementation of automation.

4.3 Implications for the industry

Every organization in construction already can initially apply the generic principles, techniques and tools of the new production philosophy: defect rates can be reduced, cycle times compressed, and accident rates decreased. Examples of pioneering companies show that substantial, sometimes dramatic improvements are realizable in a few years after the shift to the new philosophy. Given the presently low degree of penetration, there are ample opportunities for early adopters to gain competitive benefits.

However, for continued progress, new construction specific managerial methods and techniques are needed; presumably they will emerge from pratical work, as occurred in manufacturing.

5. REFERENCES

Koskela, Lauri. 1992a. Application of the New Production Philosophy to Construction. Technical Report # 72. Center for Integrated Facility Engineering. Department of Civil Engineering. Stanford University. 75 p.

Koskela, Lauri. 1992b. Process Improvement and Automation in Construction: Opposing or Complementing Approaches? The 9th International Symposium on Automation and Robotics in Construction, 3 -5 June 1992, Tokyo. Proceedings. Pp. 105-112.

Plossl, George W. 1991. Managing in the New World of Manufacturing. Prentice-Hall, Englewood Cliffs. 189 p.

Schonberger, Richard J. 1990. Building a chain of customers. The Free Press, New York. 349 p.

Shingo, Shigeo. 1988. Non-stock production. Productivity Press, Cambridge, Ma. 454 p.

Automation and Robotics in Construction X
G.H. Watson, R.L. Tucker and J.K. Walters (Editors)

Applications of Large-Scale Manipulators in the Construction Environment

Hsieh, Ting-Ya, Research Assistant
Dr. Haas, Carl T., Assistant Professor

Department of Civil Engineering
University of Texas at Austin
Austin, Texas 78712, USA

ABSTRACT

Large-Scale Manipulators (LSMs) are gaining attention in construction automation research because of their long reach and positioning capabilities. A case study based on the Model Plant data compiled by the Construction Industry Institute indicates that LSMs have very high potential in industrial construction. They can be used for pipe, cable tray and structural steel erection as well as elevated concrete placement, painting and sandblasting, which on average constitute 33% of the total project work-hours. Use of LSMs can reduce the amount of non-value added tasks, increase productivity, shorten construction time, and improve workers' safety. A task selection framework which uses a decision matrix to evaluate the uses of LSMs is developed. The same framework could also be used for evaluating the usefulness of other types of automated equipment or machinery in various construction projects.

1. INTRODUCTION

Historically, large-scale manipulators (LSMs) have been of major interest in construction automation and robotics. The most promising capability of LSMs is in independently and quickly positioning and orienting special tooling, inspection devices, concrete pump hoses, and construction materials in a large 3-D space. Despite significant developments world wide, a comprehensive study of LSMs' potential applications still does not exist. Many existing LSMs are designed for only one application and their limited utility in the construction environment may render them uneconomical particularly due to the high initial costs.

Therefore, maximizing the use of LSMs in construction is a critical step forward achieving a high profitability, and as this paper illustrates, maximizing the use of LSMs is best achieved by examining their potential applications during the design stage. The purpose of this paper is to present a method for identifying suitable applications of LSMs in construction. A task selection procedure to systematically pinpoint the applications of LSMs on a project basis is developed. Based on a case study, this framework is implemented for a petrochemical construction project to demonstrate some of the potential applications of LSMs. While the scope of this study is limited to petrochemical construction projects, more LSM applications exist in other types of projects such as commercial construction and highway construction.

2. CONSTRUCTION MANIPULATORS

Due to structure and complexity differences in the production processes from industry to industry, the definition of manipulators varies. In manufacturing where most operations are repetitive and can be performed with very few variations, and where the working environment is often well structured, manipulators are defined as machinery that are re-programmable and assisted with some sensor guidance in order to move materials, parts, tools or specialized devices through various fixed paths with high accuracy and at high speed. In military and aerospace, their definition changes to machinery which perform non-repetitive tasks in an unknown and often ill-structured environment with good accuracy, therefore, these manipulators often have to be equipped with a highly complex sensory system, a sophisticated mobility system, and learning capabilities.

Construction shows some characteristics of both areas. Often, its environment is ill-structured and complex. Tasks are relatively repetitive but can seldom be performed without changes in terms of process, material and dimension. The source of errors are numerous with an accumulative nature. However, precision requirements are normally not as critical as those in manufacturing. Given these conditions, construction manipulators may be described as machinery which bear the following characteristics:

- Large force capability
- Often hydraulically powered
- With many (sometimes more than six) degrees of freedom
- With human involvement in the control loop (man-machine interface)
- Usually with computer and sensor control assistance

To further define construction manipulators, it is useful to generalize the construction environment into an internal environment (inside the building envelope) and an external environment (outside the building envelope) [1]. In the internal environment where the work space is relatively small and confined, construction manipulators are used to assemble partitions and fixtures, and finish interior walls and ceiling. In the external environment where the work space is open and large, construction manipulators are applied to erect large bore pipes and steel beams, and finish exterior building closure. The two groups of manipulators are dissimilar mainly due to three parameters imposed by the environment as well as the tasks which they perform: (1) the reach, (2) the payload and (3) the mobility.

The focus of this paper is on the latter group of construction manipulators which are used in the external environment. They are called *large-scale manipulators* (LSMs) in this paper as opposed to smaller-scale manipulators used in the internal environment. Due to different environment characteristics and task requirements, LSMs impose rather distinct considerations in component design such as the manipulation/actuating system, the end effector, the motion system, the sensory system, and the control system [2]. Consequently, the core technology required by LSMs differs largely from that in industrial and smaller-scale construction manipulators. Further discussion of the core technology is outside the scope of this paper.

3. TASK SELECTION FRAMEWORK

To identify the potential applications of LSMs, a task selection procedure is proposed. This procedure is also applicable to opportunity identification of other automated construction systems on a project basis, and it involves seven steps, described as follows:

I. Machine Capability Identification

Generally, the potential of an automated system for construction applications is determined by how well it can satisfy the needs or concerns of construction tasks under four categories: safety, productivity, quality, and human factors[1] [3]. Therefore, a machine's capability can be expressed in terms of the task concerns which it can satisfy. The task concerns that can be used to describe a machine's capabilities are identified in Table 1.

II. Project Selection

The essential project data to implement this procedure include a complete construction task list, labor costs, and other construction-related costs such as supportive material costs. The selection of project type normally depends not on the interest of researchers but the needs of potential users of the system. To collect useful data, the selected project should possess the following qualities:

- representative of its type
- comprising a considerable volume of the construction industry
- exhibiting needs for technological advancements
- feasible to obtain project data

Table 1 Task Concern Checklist

A. Safety	**C. Quality**
A1. Eliminate work that is hazardous to health • breathing problems • mental problems • vision problems • hearing problems • long-lasting back bending A2. Eliminate physically dangerous work • falling more than six feet • falling less than six feet • extreme temperatures • high electrical voltage or current • struck by material or tools • dangerous machinery operation • dangerous hand tool operation	C1. Reduce rework due to errors and omissions • low repeatability • low accuracy • difficult to measure outcome C2. Increase precision of installation • high accuracy • high consistency/repeatability
B. Productivity	**D. Human Factors**
B1. Eliminate physical constraints to productivity • congested work space • coordination among many trades • meticulous connections and fixtures • slow operation required by safety B2. Eliminate cognitive constraints to productivity • large amount of computation • high memory requirement • waiting for information • decision making required • trial-and-error process involved B3. Eliminate non-value added process • overhead and planning effort • preparation work • supportive work • time-consuming safety procedures B4. Improve efficiency of repetitive work • faster completion of repetitive processes • combination of processes	D1. Reduce need for specialized skills in shortage • skills difficult to train • general shortage of labor D2. Eliminate unfavorable work • dirty work • unpleasant work D3. Eliminate boring work • low to no skill required • routine work in isolation D4. Eliminate physically exhaustive work • heavy handling • tiring body posture • wearing heavy safety equipment

III. Machine Potential Evaluation

To determine the potential of an automated system in a particular project, the machine's capabilities identified in Step 1 are compared with the project tasks obtained in Step 2. A convenient way to conduct this evaluation is by using a tabular form in which all project task are listed in the far left column and the machine capabilities in the top row. In this fashion, the value to be assigned to each blank depends on whether a machine's capability meets the concern of a particular task. If the machine's capability can satisfy the concern of a task, the respective blank is marked, otherwise it is left intact[2]. The total number of marks that are given for each project task is then added up, showing the preliminary ranking. After completing the evaluation for all project tasks, a ranked list that shows all project tasks with their concern rating can be generated.

IV. Task Volume Calculation

In this step, the project tasks that show low or no potential in gaining benefits from using the automated system are eliminated from the ranked list. Then, the task volumes of the remaining tasks in the ranked list are calculated. A task volume is defined as the construction-related portion of a project assignable to a particular task. Construction costs, consisting of mainly labor costs and equipment costs, are a good indication of the volume of a task. Material costs being relatively independent of the construction portion of a project are not included within the definition of a task volume.

Accordingly, a task volume is large when completing this task requires much labor (e.g. high labor costs) and/or a high level of mechanized power (e.g. high equipment costs). Because equipment costs are normally comparatively insignificant to labor costs, the labor requirement or the percentage of total project work-hours of a task can well represent a task volume.

V. Decision Matrix Development

Up to this point, two sets of information related to project tasks have been derived. The first set is obtained by evaluating each project task with respect to the machine capabilities. In the evaluation, a project task receives a mark if one of the machine capabilities meets one of its concerns. A second set of information is associated with the task volume. A task volume is large when it requires much labor, and the labor requirement is represented by the percentage of total project work-hours.

Based on the two sets of information, a decision matrix (Figure 1) can be developed. The number of task concerns and task volume by percentage of total project work-hours are each represented by an axis. Each project task is marked on the matrix as a point and is positioned according to its values on both axes.

VI. Task Ranking

Once the decision matrix is developed, the next step is to decide the ranking or priority among the project tasks shown as points in the matrix. Since accuracy in this framework is not critical, there is little sense to reckon the priority for every project task. Rather, it is more convenient to divide the decision matrix into several regions and determine the priority of each region. From experience, four to five divisions are appropriate for comparison. In dividing the decision matrix, a simple distance principle can be followed (Figure 1).

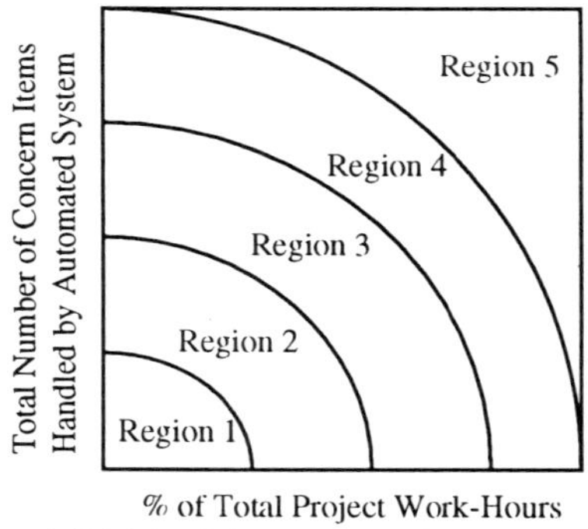

Figure 1 Division Principles of the Decision Matrix

When the two dimensions of the decision matrix are considered to possess the same weight or the relative importance between them is difficult to establish, the distance principle can be used. In this principle, the priority of each region in the matrix is determined by its distance from the origin or the bottom left corner of the matrix. Consequently, Region 2 receives a higher priority than Region 1 but a lower priority than Region 3.

VII. Potential Evaluation

One can easily examine from the decision matrix the potential of an automated system by aggregating the total project work-hours of the tasks that appear in the high-priority regions. A further step is to divide these high-priority tasks under major activity categories before aggregating their work-hours. Since construction activities may have temporal (sequential or parallel) relationships and can be similar in terms of constructing methods, this approach will provide insight for the economic analysis of the actual system implementation.

4. CASE STUDY

In this example, the University of Texas Large-Scale Manipulator (UT LSM) is chosen for evaluation. The UT LSM can be described physically as a computer controlled, electro-hydraulically actuated, duel functional, material handling device attached to the boom nose of a 22-ton, telescopic, rough-terrain crane.

According to the system's characteristics [4], its capabilities are expressed in terms of the task concerns listed in Table 1 and summarized as follows:

A1. to eliminate work which is hazardous to health
- breathing problems
- vision problems
- hearing problems
- long-lasting back bending

A2. to eliminate physically dangerous work
- falling more than 6 feet
- extreme temperatures
- high electrical voltage/current
- dangerous machine operation
- dangerous hand tool operation

B3. to eliminate non-value added construction processes
- supportive work
- time-consuming safety procedures

B4. to improve the efficiency of repetitive construction processes
- faster completion of repetitive construction processes
- combination of processes

D4. to eliminate physically exhaustive work
- heavy handling
- tiring body posture
- wearing heavy safety equipment

The Construction Industry Institute (CII) Model Plant is a baseline against which to measure construction productivity. A typical petrochemical facility was chosen as the physical baseline facility. CII members (contractors and owners), based on their own actual experience, have estimated the amount of construction labor that would be required to build selected parts of the Model Plant utilizing a code of accounts which has been established for the various scopes of work at both a detailed and summary level. The Model Plant data used in this case study are averaged over nine bidder's estimates. In short, this petrochemical project has sixteen categories of activities which contain a total of 121 tasks. Further information regarding the project tasks are not provided in this paper but can be found in CII publications [5,6].

The 121 project tasks under sixteen categories are evaluated against the sixteen machine capabilities listed above under four main categories. In this study, two senior engineers, whose combined expertise in petrochemical construction exceeds seventy years, were asked to perform the evaluation in two separate sessions.

Based on the CII Model Plant data, the work-hours of each project task are calculated. Due to length limitation, only partial results of the machine capability evaluation and the task volume calculation are shown in Table 2. Fifty-seven out of the 121 project tasks have at least one task concern item, representing more than 50% of total project work-hours. The project tasks which show no potential have been eliminated from the original ranked list.

Based on Table 2, the decision matrix containing the task concern dimension and the task volume dimension can be developed. Using the distance division principles, the decision matrix is presented in Figure 2. To maintain neatness of presentation, only the distinguishable points are accompanied with the respective task number.

Table 2 Summary of Ranked List (After Excluding "No Potential" Tasks)

Task No.	Project Tasks	Concerns	% of Total WH	Total WH
01. Site Preparation				
0130	Site Cut and Fill	1	0.82%	4206
0140	Demolition Existing Structures	8	0.01%	65
02. Site Improvements				
0212	Placement of Culverts	4	0.08%	398
0250	Fencing	1	0.07%	380
03. U/G Electrical				
0320	Direct Burial Cable	2	0.47%	2417
0330	Underground Conduit/Duct	3	2.99%	15292
04. U/G Piping				
0411	Pressure Flow < 12"	5	3.86%	19757
0412	Pressure Flow 12" - 30"	4	1.34%	6851
0421	Gravity Flow < 12"	5	0.09%	470
08. Concrete				
0812	Foundations 3 - 50 CY	3	1.67%	8558
0813	Foundations > 50 CY	3	4.34%	22189
0814	Special Slabs on Grade	4	0.17%	875
0840	Elevated Structures & Slabs	7	0.35%	1769
0850	Structures Poured in Place	3	0.03%	174
0860	Walled Structures & Supports	6	5.43%	27770
0870	Drilled Footings	3	0.03%	150
09. Specialized Concrete				
0910	Grout	3	0.57%	2917
0911	Poured Fireproofing	4	0.69%	3511
12. Structural Steel				
1212	Steel Erection (Light)	8	0.12%	602
1214	Steel Erection (Medium)	8	0.10%	531
1216	Steel Erection (Heavy)	8	0.42%	2171
1222	Misc. Support Erection	7	0.14%	741
1242	Erection - Handrails	8	0.05%	272
1244	Erection - Ladders	8	0.03%	169
1247	Erection - Stairs	8	0.03%	164
1248	Grating	8	0.15%	751
1252	Pipe Racks Erection	8	2.41%	12352

Task No.	Project Tasks	Concerns	% of Total WH	Total WH
15. Building Construction				
1511	Exterior Walls & Roof	4	0.01%	38
1521	Foundations	3	0.20%	1021
16. Aboveground Racked Piping				
1631	Spool Erection- <= 2"	7	0.75%	3817
1632	Spool Erection- 2.5" - 12"	7	3.84%	19624
1633	Spool Erection- >= 14"	7	0.42%	2173
1636	Field Run Pipe > 2"	2	2.03%	10382
1637	Hangers and Supports	7	0.40%	2050
1638	Testing, X-Ray, Cleaning, etc.	7	0.09%	468
1696	Piping Material Handling	5	0.22%	1143
1697	Scaffolding	6	0.65%	3304
17. Aboveground Non-Rack Piping				
1735	Field Run Pipe- <= 2" Erection	1	1.48%	7559
1737	Hangers and Supports Erection	7	1.28%	6559
1738	Testing, X-Ray, Cleaning, etc.	7	1.05%	5392
1796	Piping Material Handling	5	0.76%	3899
1797	Scaffolding	6	0.67%	3404
18. Aboveground Electrical				
1821	Aboveground Conduit	7	2.87%	14708
1822	Cable Tray Systems	7	0.47%	2406
1831	Wire/Cable in Tray/Conduit	7	2.89%	14781
1872	Poles/Towers/Guys/Anchors	4	0.27%	1383
1873	Area/Strut Lighting Poles	4	0.02%	89
1877	Feed Rail- Trolley Buss	5	0.08%	425
1897	Scaffolding	6	0.08%	387
19. Instrumentation				
1921	Instrument Conduit System	7	0.23%	1188
1922	Instrument Cable tray System	7	0.05%	246
1931	Instrument Piping	7	0.26%	1305
1941	Install Panels & Racks	6	0.01%	65
21. Painting				
2111	Paint Structural Steel	7	0.02%	100
2112	Paint Pipes	8	0.04%	188
2113	Paint Equipment	7	2.75%	14067
2120	Special Coatings	7	1.19%	6078

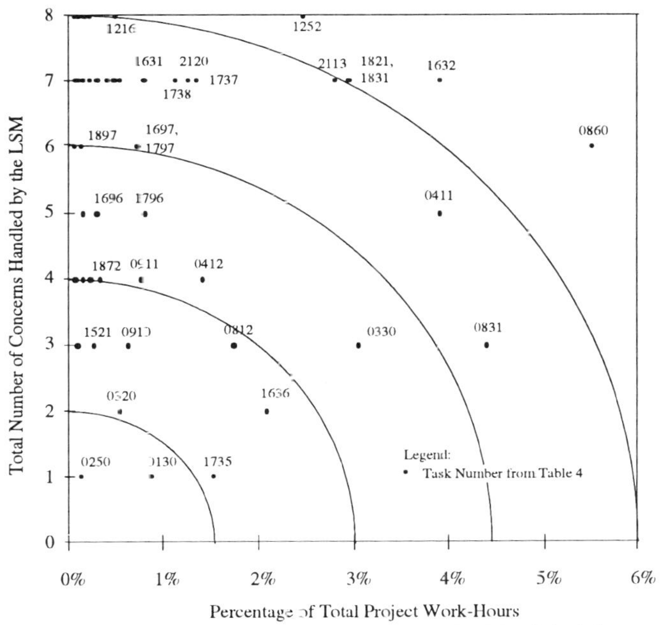

Figure 2 Decision Matrix Using the Absolute Division Principle

From Figure 2, some observations regarding the ranking of potential applications of the UT LSM can be summarized:

(1) Tasks which exhibit very high potential for LSM applications are "Aboveground Racked Piping Spool Erection from 2.5" to 12" in Pipe Diameter", "Concrete Placement for Walled Structures and Supports", "Pipe Rack Erection", "Painting Pipes", "Aboveground Conduit and Cable Tray Erection", and "Underground Pressure Flow Piping, Less Than 12" in Pipe Diameter".

(2) "Underground Conduit and Duct Erection" and "Concrete Placement in Foundations Larger Than 50 Cubic Yards" represent two applications that have high task volumes and low concern ratings.

(3) A large concentration of project tasks in the top-left corner may indicate high potential for LSM application but lack attractive profitability individually. These tasks generally center in the following major areas
- Structural Steel Erection
- Aboveground Piping Erection
- Aboveground Non-Racked Piping Erection
- Instrumentation

(4) By combining the volumes of the project tasks identified in Items (1), (2) and (3), the potential of the UT LSM in petrochemical construction is very high:

• Piping Erection	17.33%
• Elevated Concrete Placement	5.77%
• Painting (Including Sandblasting)	3.99%
• Structural Steel Erection	3.47%
• Scaffolding Erection	1.39%
• Testing, X-Ray, Cleaning, etc	1.15%
• Demolition	0.01%
Total	**33.11%**

5. CONCLUSIONS

The task selection procedure proposed in this paper can serve as a decision tool that will help potential users identify the suitable applications of new systems which they are considering purchasing for a particular type of project. As shown in the case study, the procedure requires fairly moderate effort to implement, in that a majority of concentration is in the collection of project data. The users will then perform their own economic analysis for a final decision.

Based on the CII Model Plant data and the characteristics of the UT LSM, a case study concerning the applications of a LSM in industrial construction was presented. This case study indicates that the combined potential of LSMs in piping erection, elevated concrete placement, painting (including sandblasting), structural steel erection, scaffolding erection and testing of piping and instrumentation is high. Therefore, an ideal LSM that is designed for industrial construction should have the capability to perform most of these tasks. Research at the University of Texas is directed toward developing such multi-functional systems. It is also important for the task selection framework to be applied to other types of construction projects to explore further opportunities. It is likely that they exist.

NOTES

1. This is modified by the five-category system introduced by Construction Industry Institute which includes "Safety", "Quality", "Productivity", "Worker Utilization", and "Superhuman Handling".

2. In stead of the binary scale, a five-point ordinal scale (0,1,2,3, and 4) is also applicable, in that the value "4" means the highest level of satisfaction and "0" the lowest. For higher accuracy, an interval scale can be used in the evaluation, in that the maximum and minimum numbers that can be assigned remain at "1" and "0", respectively. However, depending on an expert's opinion, any real number between 0 and 1 can be given to represent the level of satisfaction.

REFERENCES

1. Warszawski, A., and Sangrey, D.A. (1985). "Robotics in Building Construction.", *Journal of Construction Engineering and Management*, ASCE, 111(3), 260-280.

2. Paul, R. P. (1982). "Robot Manipulators: Mathematics, Programming and Control." *MIT Press,* Cambridge.

3. Tucker, R. L. (1982). "Construction Technology Needs and Priorities." *Report to the Business Roundtable*, University of Texas at Austin, April.

4. Fisher, D. J. (1989). "Piping Erection Constructability Issues in a Semi-Automated Environment." *Dissertation*, University of Texas at Austin, May.

5. Construction Industry Institute (1986). *Model Plant,* Publication 2-1, Austin, Texas, October.

6. Construction Industry Institute (1988)., *Model Plant Update*, Publication 2-2, Austin, Texas, November.

Automation and Robotics in Construction X
G.H. Watson, R.L. Tucker and J.K. Walters (Editors)

A ROBOT FOR LEVELLING AND COMPACTION OF CONCRETE

Yngve Alvarsson and Larissa Molina

Rollit Innovation, Skogshemsvagen 7 B, S-141 41 Huddinge
Stockholm, Sweden FAX +46-8-7794126

Larissa Molina, CBI, S-100 44 Stockholm
Stockholm, Sweden

ABSTRACT

The modern attempt in the construction industry is to reduce and replace heavy labor and messy activities with ergonomic methods. Placing and compaction of fresh concrete on a building site is one of such areas. Today it is difficult to recruit young people for this kind of work. For the future it will be necessary to make this heavy work easier. It has to be more attractive for the comming generation and to compensate for eventual shortage of skilled workers. High-quality concreting involves also securing of a dense, outer skin of a concrete for protecting the entire construction against corrosion and erosion, for promoting higher durability. Repairing of bridgedecks by placing a new upper concrete layer often requires external surface vibration and revibration.

A robot has been developed to travel on the fresh concrete surface, externally vibrate the concrete, densify this outer skin and smooth this surface. This Rollit Robot can move itself through the area of freshely placed concrete on the building site. A prototype has been used for repairing bridgedecks with advantage. The project is supported by the Swedish Work Environment Fund and by the selected contractors.

1. INTRODUCTION

It is well known that vibration and revibration of concrete plays a major role in placing and destributing high-quality concrete /1/. One effect of revibration is to reconsolidate the concrete. Other desirable effects are to improve adjacent of the concrete to the reinforcing bars and to reduce the voids caused by settlement and bleeding /1/.

There are sufficient needs in many countries to repair a large number of bridgedeck surfaces. Those surfaces have been damaged by erosion and corrosion caused by vehicles and by deicing agents. Deterioration of highway bridgedecks continues to be an enormous problem in forms of maintenance and rehabilitation /9/.

When repairing the damaged upper part of the deck, the concrete has to be removed and replaced by a new layer. Good bond requires between the resurfacing layer and the old bridgedeck. The thickness of this layer can vary sufficiently. Especially if the high pressure water-jet is used for the removal of the damaged concrete. In the case with the large bridgedeck the amount varies of removed, damaged concrete. Consequently, the thickness of the new layer will differ. Despite thickness variations with following differences in shrinkage, the good bond, required level and flatness have to be achieved. Beside this problems it is often difficult to get a proper compaction of this new layer.

Sufficient durability of the concrete constructions necessiates a reduction of the contact surface between the concrete and the aggressive compounds (f ex deicing agent). Densifying of the concrete outer skin by external revibration has consequently been a good alternative. Partly it has been achieved due to reduction of shrinkage and strength impovement of the interfacial bonds. Additionally, the bond strength between top-cast bars and concrete has been increased in high-slump concrete or for bars affected by settlement and bleeding /1/.

Poker vibrators and/or surface vibrators guided by trackrails are commonly used for the above mentioned type of repairing work. However it has been difficult to achieve desirable flatness by use of these equipment. Without the spoiling of the surface it is difficult to revibrate with poker vibrators or other , nowadays available equipment.

Therefore the Rollit Robot has been developed, in order to travel on the fresh surface, consolidate and level the surface layer and secure the outer skin of the concrete construction.

2.LEVELLING AND COMPACTION

In the construction industry when producing large slabs or bridge decks of concrete, there is an old and common problem. All professionals are aware of it. If a labor for any reason needs to get back out on a fresh, newly poured area, he will spoil the surface finish when he walks out. While he is walking back in, leaving the wet area, he has to repair not only the small spots he went out for, but also all his own footprints in the concrete. If the cement is starting to set, he can not walk out at all. It will be very difficult to repair a spoiled surface. Instead he has to come back later for grinding and correcting the hardened surface even for the physical shrinkage cracks of the surface layer, see § 3.3.2.

A robot, dealing with placing, levelling, vibration etc, has to be carried by a unit, which can walk on a fresh concrete surface. Such a unit is now possible to design and robots for different applications can be made.

By a new invention the possibility has been opened to design equipment for treating of concrete in many different new ways, compared with the traditional production methods. Equipment based on a completely new concept has been developed. Such equipment can move through the area of wet or early setting concrete mass and thereby treate the concrete in a non-conventional way. In order to improve the concrete quality, simplify production and promote the ergonomic conditions for the labor, it is now possible to design a robot for concrete placement.

A *Rollit Robot* may have any size from a small remote-control operated machine to a large, even drivable vehicle, covering the full width of a road or bridgedeck. It can travel directly on the rough placed concrete without any kind of traditional support from i.e. trackrails or formsides. This robot can level, vibrate and smooth the surface. The design is based on a patented principle using an endless belt. This special belt can either move the equipment forward or backward while it floats very lightly on the surface of the fresh concrete mass. There is a motorvibrator mounted on the robot. The transportation speed and the vibration effect can be regulated during the movement of the machine and these both functions can be adjuasted to the actual consistence and/or degree of setting of the concrete.

A small excavator-blade can be mounted at the front of the machine for equalizing the level of the concrete by moving the concrete from a high spot to a low spot. The production capacity can be very high during the levelling of a large slab.

3. THE INFLUENCE OF VIBRATION

The strength and durability of concrete in freeze-thaw and deicer environments is determined for the most part by the size and distribution of microscopic (entrained) and entrapped air-voids, cracks, connectivity of the pore system, the bond strength between reinforcement and concrete and between the aggregate and cement paste.

Correct vibration and revibration helps to ahieve desirable characteristics as an adequate air content of entrained air, reduced entrapped air content, a certain specific surface area of the entrained air.

3.1. COMPACTION

Efficiency of the vibration generally is given in direct terms of changing porosity of a concrete. Those changes are related to a concrete material itself with the mixture compactability, with and without reinforcement. Compaction will then be measured by f ex degree of compaction. Concrete slump, temperature and bar positions will influent the final effect of vibration on the bond strength, interfacial strength and the quality of the concrete. Low slump, low water-binder ratio-concrete requires sufficient vibration to achieve the desirable degree of consolidation of the concrete /9/. External revibration is to prefer in this case. Otherwise the structure around the poker vibrator will be disturbed (more porous).

Vibration and revibration change the structure of the pore system. The amount of pores between 10,000-50,000 Å reduces, while the amount of micro- and mesopores increases. Totaly, a densifying of a concrete is accompanied by the increased plasticity of the fresh concrete. The fresh concrete receive the better reological characteristics. Small deviation in the compressive strength on the vertical direction occures /1/. If properly afforded, the correct vibration and revibration lead to decreased permeability, increased ultimate and compressive strength /1/ and durability.

3.2. SHRINKAGE

3.2.1. CHEMICAL SHRINKAGE

Mixed water filles the pores in the fresh concrete, beside some entrapped air-filled pores. As time going, some water will be used in the chemical reactions of hydration. Then shrinkage, induced by this chemical water consumption, will be called chemical shrinkage. Many LS-concretes show high capability to chemical shrinkage, specially at lack of mixed water.

3.2.2. PHYSICAL SHRINKAGE

Beside the chemical drying and -shrinkage water will be dried from the surface, primarily in the surrounding air. Wind, high surrounding temperature or concrete temperature increases drying rate and possibility for the physical shrinkage.

In all cases, after concrete placement and initial curing, the drying of the unprotected surface will be of considerable extent. The shrinkage of the outer layer will then be larger than of the deeper layers. If the outer layer can not resist the tension stresses, caused by this difference, the shrinkage cracks of the surface layer release this tension. Generally, LS-concrete resists better this tension without crack formation.

Especially, the shrinkage cracks can often appear on the unprotected concrete surface. There Rollit Robot can be used with advantage in order to close or reduce the wide of those cracks by compacting- and revibrating action.

4. BOND IMPROVEMENT

4.1. INTERFACIAL BOND BETWEEN AGGREGATE AND PASTE

This bond is often depending on the porosity of the interface and its wide. External vibration by Rollit Robot assumes to decrease this porosity and wide by better compaction, hardening activation and more homogeneous distribution of water and air voids through the entire concrete body.

4.2. REINFORCING BOND

Effect of revibration on the bond strength between concrete and horizontal deformed bars assumes be influenced by the following parameters /7/:

- -concrete slump
- -concrete temperature
- -bar position
- -the time interval between initial vibration and revibration.

At increased concrete temperature, f e x from 16-19'C to 26-27'C, more effort requires to screed the surface and reconsolidate the concrete at age of 90 minuts /1/.

Both bond and compressive strengh will increase for concrete with slump below 3-4 in. after applyed revibration /6/.

Increase appears in ultimate bond strength of up to 62% at the delayed external vibration up to 9 hours after placement, compared with nonvibrated concrete /4,5/.

5. IMPROVED QUALITY OF THE SURFACE LAYER

For most bridgedecks, the aim is to prevent deterioration caused by aggressive agents, penetrating into the concrete body, initiating concrete corrosion, reaching steel reinforcement and initiating steel corrosion /9/. Protective system of a dense outer skin at resurfacing is used therefore. The choise of LS-low water to binder ratio overlayer is an additional method. In both cases the lower permeability will reduce penetration of the aggressive ions into the concrete and to the reinforcement /9/.

External revibration and vibration help to achieve the dense surface layer with less and/or smaller shrinkage cracks and to improve strength. Revibration can start very early and be repeated in intervalls. It can be repeated many times with lap up to hours after placing.

In the similar way as a roller compactor is used to make the surface of an asfalt road smooth and finished, the Rollit Robot can correct misstakes or stop marks from a surface vibrator. The surface will after a few passes be flatter and normal manual screeding will be eliminated or reduced..

6. REMOTE-CONTROLLED AND LASER GUIDED PROTOTYPE

The prototype is provided with a long cabel through which it is possible to control the robot at testing. Remote control by radio will be required for the wide bay construction in the future.

A large, drivable vehicle, with a possibility to ride on, has also been tested. It has many advantages. The operator can visually inspect the achieved result from a close position on the top of the vehicle instead of trying to judge the treating result from the far distance at the side of the bay. A ride-on vehicle allows also the operator to make an extra pass over and to correct some surfacing imperfections or irregularities directly and easy.

Rollit Robot has also been tested with an automatic laser-guide system for control of the level or elevation of horizontal slabs.

7.CONCLUSION

It is now possible to design equipment which can improve the traditional production methods in building construction. A robot has been developed to travel on the fresh concrete surface, externally vibrate the concrete, densify and smooth this surface. A proper and adequate compaction and repeated revibration of the surface will improve the durability of the entire constructure unit.

The attack on the reinforcement steel will be delaid, the life time of the structure can be extended by use of the Rollit Robot. Wear resistance may also be improved as well as a thin layer bond to the underlaying structure if LS-concrete is used and a proper and adequate compaction from the surface is accomplished.

Companies manufacturing equipment for handling of concrete are looking for the possibility to develope better equipment or design some kind of a robot to replace the heavy manual work. It is necessary to make this kind of work easier and more attractive for the coming generation.

Poker-immersion vibration or surface vibration will seldom guarantee the optimal compaction of the surface layer. Rollit Robot creates a compacted surface layer, which is dense and has a low permeability. Rollit Robot will help to improve the quality of the concrete and to prolong the life of a structure.

8. FUTURE STUDY

Future study is planned in actual building sites in cooperation with Scandinavian contractors in order to verify results scientifically and propose the guidelines for the future use of this new tool. All kind of tests, including those on the drilled cores from the structural concrete, would help to establish the practical effects of external revibration by Rollit Robot.

The study on a large scale will include the optimal time interval, which is governing by a concrete workability between initial vibration and repeated revibration.

This is a subject of the future study to investigate the propagation of the vibrating waves and their effect on the quality of the hardened concrete for Rollit Robot in comparision with other vibrator types.

9. REFERENCES

1. W. A. K. Altowaiji et al: Bond of Reinforcement to Revibrated Concrete. ACI Journal. November-December 1986.

2. C. A. Vollic: Effects of Revibrating Concrete. ACI Journal. Proceedings. V. 54, No. 9. March. 1958, pp. 721-732.

3. W. J. Larnach: Changes in Bond Strength Caused by Revibration of Concrete and the Vibration of Reinforcement. Magazine of Concrete Research. London. V. 4, No. 10, July 1952, pp. 17-21.

4. R. E. Davis et al: Some Factors Influencing the Bond Between Concrete and Reinforcing Steel. Proceedings. ASTM, V. 38, Part 2, 1938, pp. 394-406.

5. R. C. Danahey et al: Effects of Construction Procedures on Bond in Bridge Decks. SM Report No. 7, University of Kansas Center for Research, Lawrence, January 1983.

6. S. Harsh et al: Traffic Induced Vibrations and Bridge Deck Repairs. Concrete International: Design & Construction. V. 8, No. 5. May 1986, pp. 36-42.

7. W. Altowaiji et al: Preliminary Study of the Effect of Revibration on Concrete-Steel Bond Strength. SL Report No. 84-2. University of Kansas Center for Research, Lawrence, November 1984.

8. Nase-cho, Totsuka-ku, Yokohama. Taisei Technical Research Report No. 18, 0387-2254. March 1986.

9. D. Whiting. Air Contents and Air-Void Characteristics in Low-Slump Dense Concretes. Journal of the ACI. V. 82, No. 5, p.716-723. 1985.

10. GLOSSARY

Delayed vibration is a vibration at some period after the concrete had been placed.
Entrapped air forms the air voids with diameter larger than 0.9-1 mm.
Entrained air is formed when air-entraining admixture are added to the fresh concrete.
HS is high slump, *LS* is low slump
Repeated revibration is an external vibration made with a certain time lap bt the Rollit Robot.
Resurfacing is when the damaged upper part of the concrete has to be removed and replaced by a new concrete layer.

Fig.1 The Rollit Robot is transported to the site...

Fig.2..unloading itself...

Fig.3..walking to the actual concrete placement area...

Fig.4..and starting to work.

Fig.5 The left part of this area is revibrated and levelled with Rollit Robot-the last pass 3 hours after placement.

Automation and Robotics in Construction X
G.H. Watson, R.L. Tucker and J.K. Walters (Editors)

PRACTICAL TILE-SETTING ROBOT FOR EXTERIOR WALLS

Kazuya KIKAWADA[a], Shigeaki ASHIKAGA[b], Seiichirou ISHIKAWA[c] and Shigeomi NISHIGAKI[a]

[a] Technical Research Institute, Hazama Corporation, 515-1 Nishimukai Karima, Tsukuba-shi, Ibaraki 305, Japan

[b] Research Division, Komatsu Ltd., 1200 Manda, Hiratsuka-shi, Kanagawa 254, Japan

[c] Japan Ceramic Tile Association, 39-18 Daikancyoh, Higashi-ku, Nagoya-shi, Aichi 461, Japan

Abstract

Hazama Corporation, the Japan Ceramic Tile Association and Komatsu Ltd. have been and are developing tile-setting robots to automatize the process of tile-setting operations on the exterior wall of building. This paper describes an improved tile-setting robot and reports the observations in the indoor and outdoor experiments of its operations.

1. Introduction

Faced by a need to improve the durability of concrete surface and to add the beauty of its appearance, tiled exterior walls of buildings are very popular in Japan. Tiles, which are set on exterior walls, most likely are very thin such as 108 or 227mm long, 60mm wide and 8 to 15mm thick. Tile-setting operations required tile-workers high dexterity. The number of experienced tile setters has been and is decreasing, in spite of increasing the demand for tile-setting operations on the exterior wall of building in Japan.

To improve the quality and productivity of tile-setting operations, tile-setting robots for exterior walls has been and is being developed by Hazama Corporation, the Japan Ceramic Tile Association and Komatsu Ltd.. The process of the development of the tile-setting robots is outlined as follows:

Stage 1: design the concept of tile-setting robot and specify its indispensable functions ;
Stage 2: develop a specialized tile-setting robot and evaluate its availability ;
Stage 3: evolve further refinements of the tile-setting robot.

Tomaru,H., Haino,H. and Ishikawa,S. did research on the development of a tile-setting robot at the stages 1 and 2, and reported the findings of the research[1]. The tile-setting robot has been further improved at the stage 3.

First, this paper presents the improved tile-setting robot. Second, this paper reports the observations in the indoor and outdoor experiments to evaluate the operability of the improved tile-setting robot. Finally, we conclude with a summary of this study and discuss further research.

2. Evolving Further Refinements

Further refinements of the tile-setting robot have been achieved at the stage 3, focusing on the improvement of mobility and operability. The distinctive refinements are summarized below.

(1) Get smaller 40% in its size and lose 80% in its weight.

(2) Increase 50% in its productivity more than before.

(3) Simplify its operations so that unskilled workers can easily operate it just by pushing the ON/OFF switches. The previous version of tile-setting robot required tile-worker some computer works which they were unfamiliar with.

(4) The guide rails for a horizontal carriage of the tile-setting robot are mounted on scaffolding instead of being fastened by anchor bolts on the exterior wall. The disappearance of these anchor bolts enables to avoid manual works for tile-setting on the areas for anchor bolts of exterior wall.

The configurations of the improved tile-setting robot are shown in Photo 1. The specifications of the improved tile-setting robot are summarized in Table 1. Figure 1 shows the system components of the improved tile-setting robot.

Photo 1. The improved tile-setting robot

Table 1
Specifications of the improved tile-setting robot

Items		Specifications
Body of the Robot	Types of Joints	Lengthwise Horizontal Orientation Square Joint and Cross Joint
	Joint Width	Vertical Joint : 8±2mm Horizontal Joint: 9±2mm Expansion Joint: As possible as avoid
	Unit Execution Area	Approx. 0.4m^2 (for 1 setting)
	Method of Move-ment	(Horizontal) Guide Rails (Vertical) Chain Block, etc.
	Power Source	AC100V
	Established Place	Space between building and scaffold-ing (width 30-33cm)
	Outer Frame Dimen-sions	height 2.8 x width 0.8 x depth 0.26m
	Weight	Approx. 260kgf
Equipments	Guide Rails	height 60 x width 30 x length 1800mm
	Air Compressor	Above 2.2kw class(AC200V)
Appropriate Materials	Setting Mortar	Pre-mixed Mortar
	Tiles	227 x 60 mm tiles (thickness 8-15mm) Flat Surface Smoothness

A tile setter supplies the mortar hopper with mortar. The capacity of the mortar hopper is approximately 5 liters of which the volume is sufficient to process about 50 tiles (227mm x 60mm). The mortar trowel places the correct dose of mortar on the reverse side of the tile. A tile setter successively supplies the tile-holding unit with tiles.

As the tile-holding unit is carrying tiles, the tile-setting unit automatically and accurately places the tiles one next to the other to attain a continuous surface on the exterior wall. The tile-setting unit has an electric powered vibrator to firmly place tiles on the exterior wall, and has an air actuator with potentiometer that measures the finishing thickness of mortar required to accurately place tiles on the exterior wall.

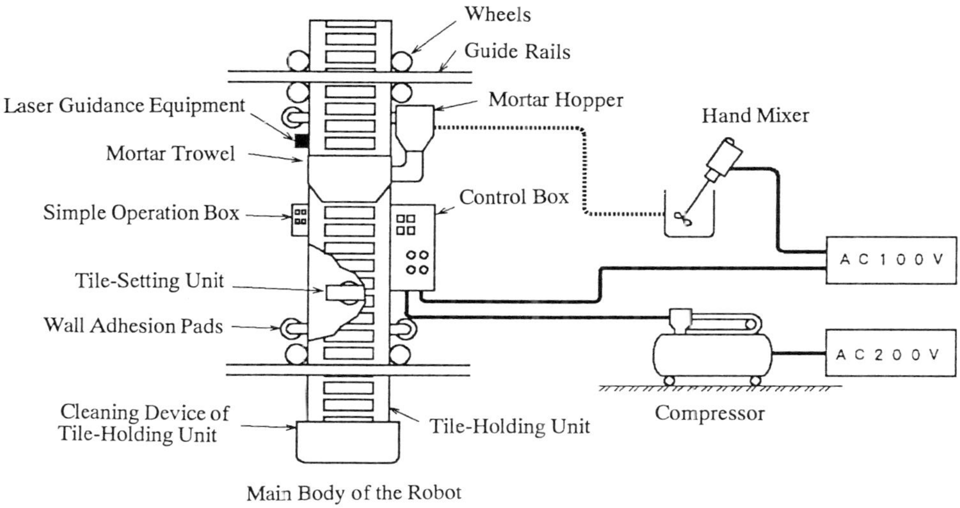

Figure 1. The system components of the improved tile-setting robot

This tile-setting routine is programmed off-line into the control box. The conditions of tile-setting, for example, the number of tiles in a line, joint width, a type of joint, location of expansion joint, were instructed on the control box by tile setter. Besides, the control box has a self-inspecting function to detect existing failure modes, to stop the operation of the improved tile-setting robot and to inform the tile setter of the potential causes as shown in Table 2.

Table 2
Failure modes and the potential causes

Item	Failure Mode	Potential Cause
Tile-setting unit	Motion as unintended - Not being in compliance with preprogrammed instruction. - Although a tile is being pushed toward the exterior wall, the tile can not be reached to the exterior wall.	- the snapping of wire - fail to supply the tile-setting unit with the correct dose of mortar - get mixed with alien substances - the distance between the tile-setting unit and the exterior wall may be longer than expected
Tile-holding unit	The sensor of the tile-holding unit can not fetch a tile.	forget to supply the tile-holding unit with a tile

3. Observations in Experiments

3.1. Indoor Experiment

Using a model wall, which is 7.5m long and 3.6m high, an indoor experiment of the operation of the improved tile-setting robot was carried out. The main items being tested were: (1) setting efficiency, (2) setting accuracy, (3) adhesive strength and (4) vertical movement time (i.e., between floors). The observations in the indoor experiment are summarized in Table 3.

Table 3
The efficiency of the improved tile-setting robot

Items	Robot operation	Reference Value
Setting efficiency	14 m^2/day (By an unskilled worker)	7m^2/day (Manual Operation by a skilled tile setter)
Setting Accuracy (Surface Unevenness)	Within ±1mm for 2m long span	Within a tolerance of ±2mm for 2m long span Specifications of JCTA*
Adhesive Strength (After working 28 days)	More than 6kgf/cm^2 Average 10kgf/cm^2	More than 4kgf/cm^2 JASS**
Vertical Movement Time	15 min. (Using a chain block)	-

legend: * Japan Ceramic Tile Association
** Japanese Architectural Standard Specification

3.2. Outdoor Experiment

To evaluate the operability of the improved tile-setting robot at an actual job-site, an outdoor experiment of its operation was carried out on the exterior wall, which is 9m long and 7.5m high, of Acoustics and Electric Wave Laboratory of the Technical Research Institute, Hazama. The numbers of setting tiles are about 2400 pieces. Photo 2 shows the existing conditions of the outdoor experiment. Photo 3 shows the exterior tiled wall, which the improved tile-setting robot produced.

Photo 2 The existing conditions of the outdoor experiment

Photo 3 The surface of the exterior tiled wall

It was seen from the outdoor experiment that the setting efficiency was more than 14m^2/day and that the adhesive strength (after working 28 days) was 17.2kgf/cm^2 in average and 2.1kgf/cm^2 in standard deviation.

From these observations, it can be seen that the improved tile-setting robot, which is operated by an unskilled worker, can realize the better quality and higher productivity of tile-setting operations than a skilled tile setter.

4. Conclusion

As can be seen from the observations in the indoor and outdoor experiments of the improved tile-setting robot, it could improve the quality and productivity of tile-setting operations and thus could overcome the shortage problem of skilled tile setters.

Further research problem is to evolve further refinements of the improved tile-setting robot so that it could also be available for tiles, 108mm x 60mm.

References

1. Tomaru,H., Haino,H. and Ishikawa,S.,"Tile Setting Robot for Exterior Wall", in Proc. of 8th. I.S.A.R.C., pp.929-936, 1991.

Automation and Robotics in Construction X
G.H. Watson, R.L. Tucker and J.K. Walters (Editors)

A System of Vertically Sliding and Installing Exterior Curtain Walls of a Building

Masaaki Iwamoto[a], Masayoshi Nakagawa[b], and Hidetoshi Oda[c]

[a]Building Construction Department, [b and c] Technical Department, Fujita Corporation, 2-1-16 Dojima, Kita-ku, Osaka 530, Japan

Abstract

This system, referred to as a "shuttle system," has been developed to enable the vertical sliding and subsequent installation of panels on the exterior walls of a building, starting from its uppermost story, while repeating a series of such operations as (1) sliding a carrier stage down to the ground, (2) on the stage, assembling panels into a panel unit of a size which entirely covers one story height, (3) sliding the panel unit, together with the carrier stage, up to the installation level, and (4) fixing the panel unit in place. Various sensors are provided to enable the panel unit to be installed accurately both in horizontal and vertical directions. The computer located in the field office is interlaced with these sensors. This paper provides a report of the case in which this system was successfully applied to the cladding of exterior walls of two mechanized parking towers.

1. INTRODUCTION

In Japan the construction industry has currently devoted itself entirely to technical innovation. At general contractors, special departments have been established to take on the challenge of the development of unmanned or mechanized construction systems.

As a step toward the industrialization of fully-automatic construction systems, the Fujita Corporation is the first to develop a "shuttle system." This is a system enabling the installation of panels to clad the exterior walls of a building, without the use of either a large crane or scaffolds. The development focussed on the aim of installing and automatically hoisting exterior-wall cladding materials aboveground. This development may take on the majority of the risks in the execution of construction, thereby eliminating the need of a labor force and providing improved productivity and safety. This paper presents an example of a certain construction which lent an impetus for us to develop the shuttle system. This article will proceed by first describing the history of the development of the system, then discussing the method overview and features and the effects of system application. Finally it will conclude with future development.

2. HISTORY OF THE DEVELOPMENT OF THE SHUTTLE SYSTEM

2.1. Construction overview of buildings to which the shuttle system was first applied

Figure 1 provides an outline of the buildings.

Construction : Steel frame
Building size : Two buildings, each being 31 meters in height
Building type : Parking towers of elevator type, each for 128 cars
Exterior wall finish : An enameled panel with approx. 600mm × 600mm over the entire wall surface

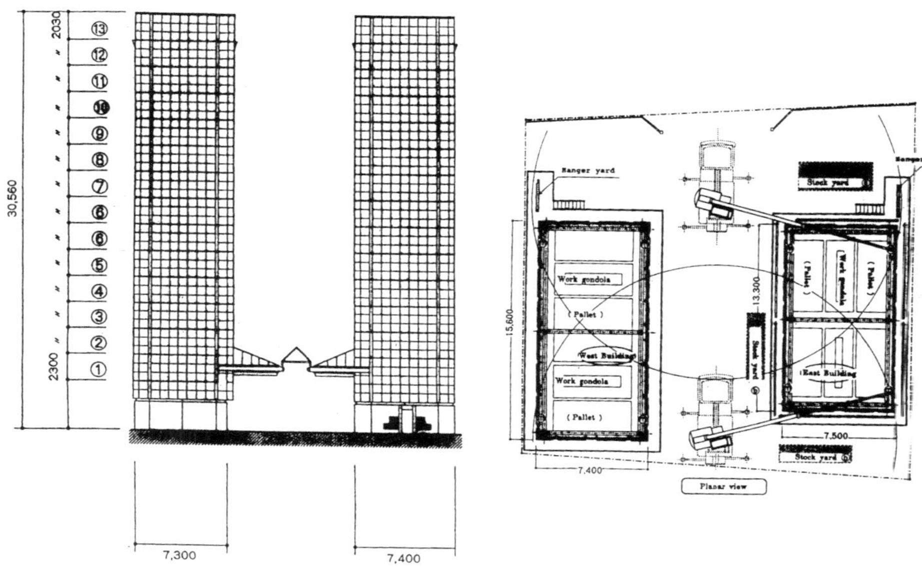

(a) Elevation (b) Planar view

Figure 1. Elevation and planar view.

2.2. Intent of system employment

The planning for the execution of the works by means of the shuttle system commenced with the assembly of 600mm × 600mm (some panels measuring 250m × 2000mm) enameled panels into the number of units required. Each unit was comprised of three vertical panels and four horizontal panels. The assembling operations were performed in workshop, for delivery to the construction site. This system obviously surpasses the conventional system in which all panels must be installed one by one on the construction sites.

As well, if the general method is employed to install exterior wall panels, scaffolding would have to be erected around and along the entire height of the buildings. Due to this procedure, we might have encountered the following limitations and problems;

① The existence of some closely located neighboring buildings did not allow for the installation of scaffolds around the buildings.

② Operations at elevated places without the use of scaffold might cause serious danger.

③ The installation of the panels would require a large crane, for example, with the hoisting capacity of 50 tons if no other effective method is available, which would increase the cost. Also, dangerous operations would have to be performed at this confined site.

We decided that the best solution available to these limitations and problems, was the development and implementation of the shuttle system, a method which is used to install exterior wall panels in a way which enables the installation by sliding them up the buildings.

3. SYSTEM OVERVIEW

The shuttle system provides for a system that repeats the operations ① through ⑧ shown in Figure 3: a set of four electric chain blocks are provided at the four corners of each building which hoist shop-assembled panel units successively, each time together with the stage. This is followed by their installation from the building's top down the wall toward the bottom. This system uses the term "shuttle" as a prefix because the stage moves up and down a building with panel units in it, reciprocating between the ground and the installation positions.

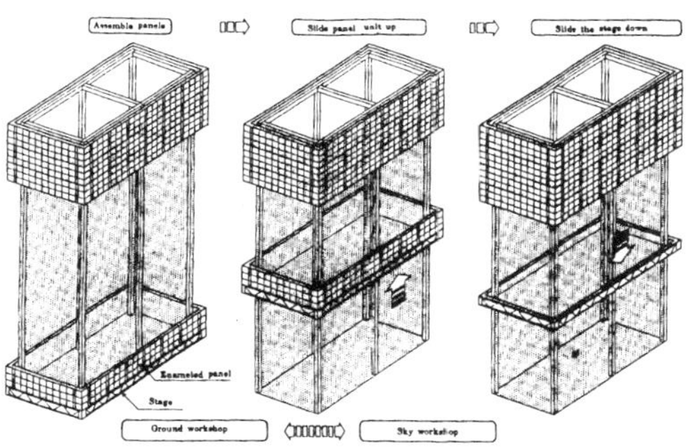

Figure 2. Conceptual shuttle system outline.

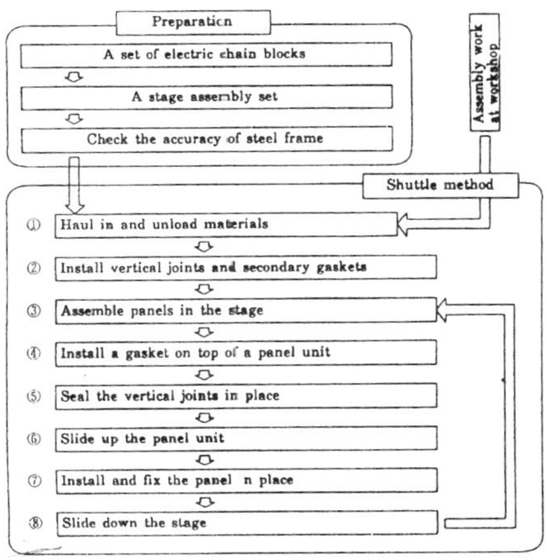

Figure 3. Flow chart.

To enable panel units to be hauled up to the exact installation positions, various sensors were used to automatically measure the horizontal and vertical positions while the panel unit is being slid up, so that the sliding device can be controlled. Sliding operations can be monitored on the display of the computer located in the operation room, in which it is integrated and computer control is performed.

4. SYSTEM APPLICATION EFFECTS

The incorporation of the shuttle system exceeded the conventional installation method in the creation of the following outstanding effects:
① Productivity was enhanced by about 20%.
② Approximately 40% labor savings could be achieved.

This section discusses the breakdown of the 40% labor savings which could be achieved through the mechanization and automation of panel installation operations. The two most important factors which led to the success in satisfactory completion of the buildings' exterior cladding are also discussed. Table 1 compares the shuttle system with the conventional method.

Table 1
A comparison between the shuttle system and the conventional method.

	Conventional method			Shuttle system		
Temporary works		10,330m^2	150 scaffolding men		Set	40 scaffolding men
General temporary works	A 50-ton crawler crane	3 months	75 operators	A 5-ton track crane	2 months	50 operators
	Others	Set	20 workers	Others		48 workers
Metal work	Installation of panels	2,400m^2	450 metal workers	Installation of panels	2,400m^2	225 metal workers
Water-proofing work	Sealing of joints between panels in place	8,700m	175 water-proofing men	Sealing of joints between panel units in place and installation of gaskets	7,680m 3,167m	150 water-proofing men
	Total		870	Total		513

Percentage of labor savings: 41%

4.1. System effects

The major effects which were achieved through the field application of the shuttle system are discussed below:

① Success in the achievement of the shuttle system, in which no scaffolds whatsoever are used, could assure an extremely high level of safety.
② Without involvement with operations performed on shaky-exterior scaffolds, on-the-ground assembly of panels could assure improved installation accuracy and quality.
③ The incorporation of various sensors into the measurement system permitted the operation of electric chain blocks through monitor displays in the operation room, until the installation of the last piece of panel unit. This also ensured reduced labor and improved comfort in the carrying out of the administration of the task.
④ The employment of silicon gaskets to seal the horizontal joints between the courses of panel units in place, eliminated the use of both sealing compound and gondolas.
⑤ The use of hanger type slide jigs wherever neighboring buildings were located close to the buildings enabled the entire operations to be performed by means of a 5-ton truck crane, but not by using any larger or heavier crane. This could cut the cost considerably. Other merits include the wider utilization of the confined site and improved operation efficiency.
⑥ A shortened construction period was achieved.
⑦ Since the shuttle system has generality, it can be applied to other similar constructions.

5. FUTURE APPLICATION

In the development stage, we encountered selective difficulties, and events which had never been anticipated. This section elaborates such information that may be useful in the employment of this type of method in the future. When choosing a single device from among many types of vertical carriers, it is necessary to scrutinize it from various angles by comparing its adequacy with that of other carriers. Table 2 compares an electric chain block with a hoist and a hydraulic jack.

Table 2.
A comparison in performance between an electric chain blck, a hoist and a hydraulic jack.

	Comparison item	Electric chain block	Hoist	Hydraulic jack	Comment
1	Work-performing ease	◎	△	×	≪ An electric chain block is of a simple construction and exhibits work-performance ease. ≫
2	Cost efficiency (general)	◎	△	×	≪ For an electric chain block, a variety of inexpensive, multi-use types are available on a lease basis, as well they are cost efficient. ≫

Table 2.
A comparison in performance between an electric chain blck, a hoist and a hydraulic jack.

3	Safety	◎	×	◎	≪ An electric chain block is of a simple construction offering a high level of safety.≫
4	Synchronized control accuracy	○	◎	◎	≪ An electric chain block will be able to perform synchronized control if using sensors.≫

Specifications of the electric chain blocks for use with the shuttle system	Lifting capacity	Lifting velocity	Quantity used	Gross weight	Stage	One panel unit
	2.8 tons/unit	1.7 m/min	4 units	7.5 tons	3.0 tons	4.5 tons

An electric chain block is used not only by the shuttle system but also by many other methods.

5.2. Horizontal and vertical accuracy control system

As compared with other industries, currently, few general contractors at their construction divisions have industrialized sensors of various types. With the shuttle system, the horizontal and vertical accuracy control system fulfills the role of the brains. The system effectively used various sensors as described on next page:

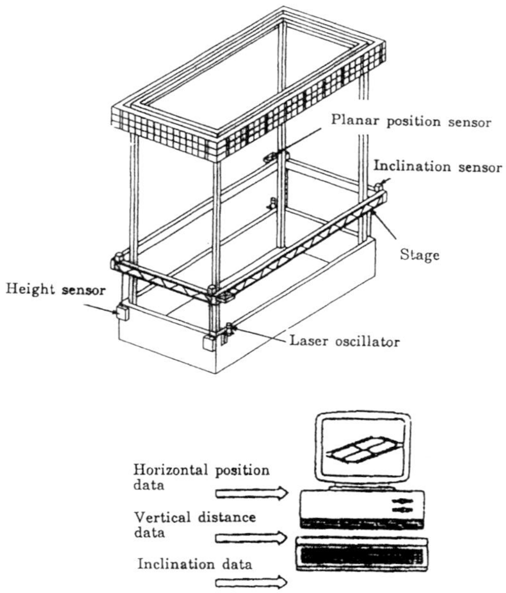

Figure 5. Conceptual illustration.

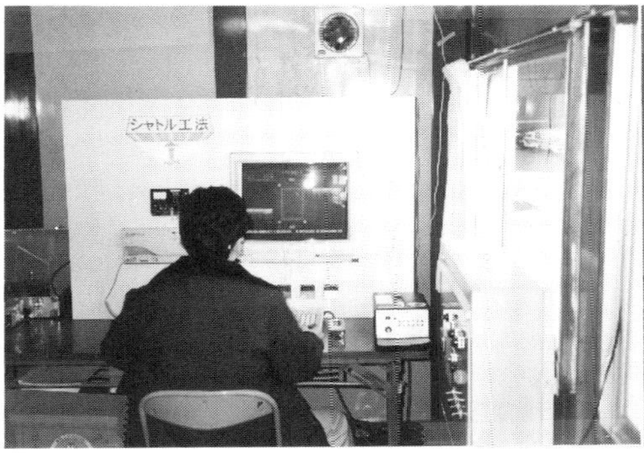

Photo 1. The state of the horizontal and vertical accuracy control system being in operation.

Photo 2. An inclination sensor as displayed on the computer screen.

- Four wire type height sensors were set at the four corners of the stage, for the detection of height.
- Two laser oscillator type planar position sensors were set at opposing points, for the measurement of planar deviation.
- An inclination sensor was provided at each side of the stage, for the measurement of panel distortion.

Measurement results transmitted from these sensors were processed and displayed at real time by the computer located in the field office, so that panel sliding-up and -down operations can be monitored. Up to 1mm of a panel position can be adjusted by remote-controlling each of the four electric chain blocks.

Using the following conceptual illustration and the photographs, the system is described in further detail.

5.3. A gasket to fill a horizontal joint between panel units

Since in principle the shuttle system uses no scaffold whatsoever, how to seal the horizontal joint between panel units gave rise to an issue to be resolved.

To cope with this issue, Fujita Corporation has developed a silicon string gasket. After panels have been assembled to form a panel unit, this silicon gasket is stretched along the top of the full length of the panel unit. Compression of the gasket will occur when another panel unit is butted to the gasket-applied top of the preceding unit, resulting in the creation of impermeability of the joint. Using this method, all impermeable joint strings could be installed before sliding up panel units, thus realizing an absolute no-scaffold method. Figure 6 illustrates the typical way in which an impermeable joint string is installed.

6. CONCLUSION

The shuttle system, we believe, can be applied to such steel-framed buildings as shopping centers, office buildings and sky scrapers. In the future, we will dedicate ourselves to improving the shuttle system by interlacing both an attitude control system and an automatic lifting system. The measurement results derived from the sensors and with an on-the-ground assembly workshop will be used, so that the unmanned installation of exterior-wall-cladding materials can be executed. Also, we will endeavor to create innovative technology toward the realization of a fully automated building construction system that can be generalized to the application to steel-framed buildings, an automatic up-down scaffolding system enabling bolting, welding, to the installation of exterior-wall cladding materials.

Photo 3. The sliding-up of a panel unit is proceeding.

Photo 4. The sliding-up of the stage is proceeding.

Automation and Robotics in Construction X
G.H. Watson, R.L. Tucker and J.K. Walters (Editors)

Configurable Control System of a Mobile Robot for On - Site Construction of Masonry

G. Pritschow, M. Dalacker and J. Kurz

Institute of Control Technology for Machine Tools and Manufacturing Units,
University of Stuttgart, Seidenstraße 36, W-7000 Stuttgart 1, Germany

Abstract

The dramatic shortage of skilled workers and a strong need of improving productivity and efficiency characterizes the situation of masonry construction in Germany today. Therefore, the construction enterprises affected must strengthen their efforts to develop and introduce automation technology for use on the construction site. Such concepts, however, can only be successful, if the national characteristics of masonry construction in Germany are considered, and if the construction industry's demands and needs in terms of flexibility, robustness and economy can be met. Taking these factors into account, the requirements of a mobile robot for on-site construction of masonry have been defined and a high-performance robot control system has been designed.

1. INTRODUCTION

Masonry, as a production technology, has had a long tradition in the German construction trade and is still of great economic importance. In 1991, artificial bricks and blocks at a total value of 1.6 billion German marks were produced in Germany /1/; masonry comprises 7 % of the total turnover in the construction trade and increases to 11 % in the field of building construction. The special importance of masonry for residential construction can be seen from the number of building permits given for the erection of new buildings. In 1991, the share of masonry in residential buildings was 90.0 %, in non-residential buildings as much as 51.8 % /2/. The market shares of commonly used bricks and blocks were as follows: 42.9 % bricks, 31.6 % sand lime bricks, 14.1 % aerated cement blocks and 11.4 % concrete blocks. Fig. 1 shows the mentioned values by diagrams.

The predominance of private clients and small workshop-like construction enterprises is a characteristic feature of the German bricklaying trade: Two thirds of the apartments erected are situated in buildings with only one or two apartments, and 80% of all construction enterprises employ less than 20 people /1/.

In spite of the masons' bad working conditions and high physical strain, bricklaying work is nowadays still mainly done by hand. This is due to the fact that there has been, up to now, no adequate technical support which meets the demands of the industry as far as economical factors and robustness are concerned. Due to the increasing shortage of skilled workers and the need of improved efficiency and productivity, automation technology is likely to play a major role in masonry construction in the near future.

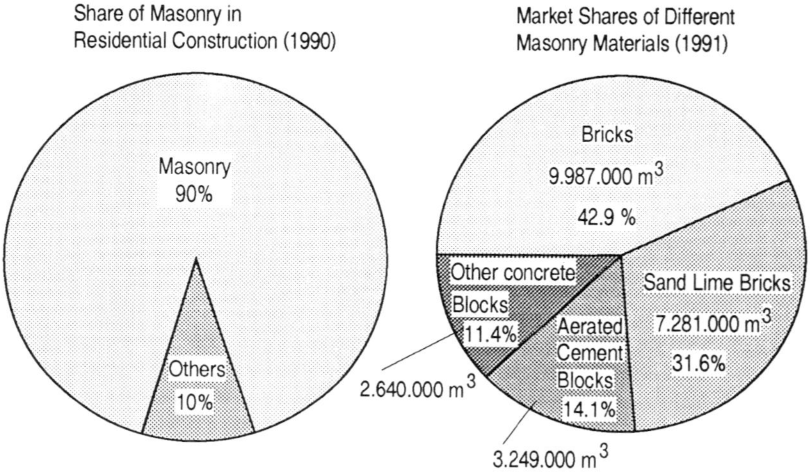

Fig. 1: Significance of masonry construction in Germany /1/

Mechanization and automation technologies for masonry construction presently available in industry are limited to the use of masonry cranes and semi-automatic bricklaying machines on construction sites /3/ and to the prefabrication of complete walls in production plants /4/. In Germany, these solutions are only of regional importance and, due to lacking flexibility and acceptance, have not been very successful on a large scale /5/.

International research activities concentrate upon the development of robot-oriented technologies for the use on construction sites. There are fundamental feasibility studies and realization proposals from USA /6/, Russia /7/, Finland /8/ and Germany /9/. Economic aspects of such approaches have been discussed in several papers /10,11/, with the results turning out promising. Current research activities in Israel and England are adopted to the respective national building technology: In Israel, gypsum blocks are used for dry wall construction by means of a mobile robot /12/. Research work in the United Kingdom focuses on traditional bricklaying technology using imprecise bricks and thick bonding material. Placing of the bricks is done by a stationary gantry type robot /13/. Within the European Community, 3 international R&D projects concerning automation of masonry construction are currently funded by EUREKA and ESPRIT programs /14,15,16/.

2. TECHNOLOGICAL REQUIREMENTS AND PERFORMANCE CHARACTERISTICS OF A BRICKLAYING ROBOT

The national characteristics of masonry construction in Germany described above show that a simple transfer of automation approaches from other countries to the German conditions will not be successful. The special middle-class structure of German construction enterprises, individual wishes of the clients concerning the ground plan of the buildings as well as the great variety of masonry materials used require an extremly flexible, economical and easy-to-use technology. From the construction industry's point of view, the following demands must be met:

- *On-site usage:* The technology must be available on the construction site for both new erection and rebuilding of houses.
- *Variety:* Erection of both exterior and interior walls by the robot, possibility of processing any type of brick or block which is in line with market requirements (especially bricks, sand lime bricks, aerated cement blocks and concrete blocks), application of related bonding technologies(masonry mortar, glue).
- *Flexibility:* Realization of any ground plan desired by the client.
- *Simplicity:* Trouble-free integration into the construction process, little stationary equipment, operation by semi-skilled workers.
- *Economy:* Increased productivity, lower cost.
- *Computer integration:* Transfer of CAD data generated off-line to the robot control system.

From these technological requirements mentioned above, the following performance characteristics of a bricklaying robot can be derived:

- *Mobility:* The working space of a manipulator performing the bricklaying task itself is limited by the reach of its arms. Putting the manipulator onto a mobile platform enables changing the working position, thus extending the working space to the whole floor.
- *Large reach:* The manipulator's reach should be no less than 2.5 m to reduce the number of changes of the working position.
- *Kinematic structure:* The working space of the robot is limited by construction site-specific bounds such as pallets of masonry material, building machinery or walls and ceilings which have already been completed. Such limitations must be considered when designing a suitable kinematic structure of the robot.
- *Small dimensions:* The overall dimensions of the mobile robot should be chosen so that passing through door openings is still possible.
- *Low weight:* The bearing capacity of the floor which the robot moves on is generally limited (a typical value is 500 kg/sqm). Consequently the weight of the robot is to be kept low .
- *High lifting capacity:* Economical construction favours the use of large-sized blocks with considerable weights. Therefore, the manipulator's lifting capacity should not fall below 50 kg.
- *Fast tool change:* Each type of masonry material may require a corresponding tool or handling device. Changing these tools must be possible in a fast and simple manner.
- *Safety:* The robot itself as well as the operating staff must be protected by

adequate safety precautions.

- *High performance:* Economy of the robot is largely influenced by the duration of the duty cycle. The application of intelligent control strategies results in minimum duty cycles.
- *Powerful control system:* Processing of sensor signals will be necessary for motion generation. That's why a high-performance robot control system with an open architecture and high functionality is required.
- *Overall robustness:* The bricklaying robot will be operated under rough environmental conditions (rain, heat, dust, vibrations). Fail-safe operation must be guaranteed by a robust construction.

A possible layout of a bricklaying robot with caterpillar tracks and a hydraulic manipulator is shown in fig. 2.

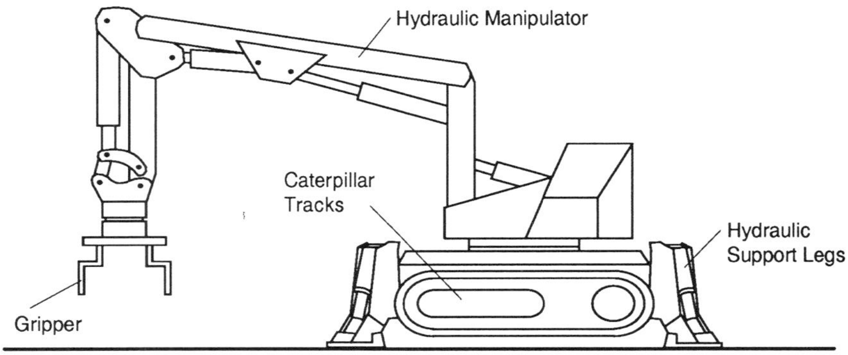

Fig. 2: Possible layout of a bricklaying robot

The hardware structure of the bricklaying robot will not be discussed in more detail. The next chapter focuses on the robot control system.

3. REQUIREMENTS OF THE ROBOT CONTROL SYSTEM

The purpose of the control system of the mobile bricklaying robot described above is the generation, execution and supervision of motion sequences for all locomotion, manipulator and handling axes involved in the bricklaying process. As the application is rather complex and a lot of axes will contribute to the motion of the end effector, powerful computer hardware and a realtime operating system are required. The functional profile of the robot control system results in the following requirements:

- *Integrated control system for mobile platform and manipulator:* The specific tasks of the mobile platform (changing the working position) and the manipulator (performing the bricklaying process itself) must be intgrated into the robot control system.

- *Programming system and CAD/RC - interface:* A easy-to-handle user interface must be provided for the generation and input of application programs. While program mode is on, CAD data which have been generated off-line should be directly transferred to the robot control system.
- *Sensor data processing:* Both positioning of the mobile platform and precise placing of bricks or blocks can not be done without support by proper sensors. The reason for this are inevitable tolerances of the robot and the masonry material. Therefore, adequate sensor interfaces must be available with the robot control system.
- *Motion generation:* Smooth reference paths have to be generated to avoid overload of the drive units. To achieve this, the dynamic limits of the drives must be taken into account during interpolation.
- *Coordinate transformation:* Programming of the robot by the user will in most cases be done in cartesian coordinates. For controlling the drive units, however, joint coordinates will be used. Transformation algorithms must be provided for the conversion of the reference values between the two coordinate systems.
- *Fast servo controllers:* From the control engineer's point of view, joints of manipulators and industrial robots are featuring substantial elasticities and little damping. In addition, strong non-linearities must be considered if hydraulic drives are used. Powerful servo controllers must be provided for the implementation of the high-level control concepts required.
- *Use of industrial standards:* Considering marketing requirements of the robot control system, the use of industrial standards is strongly recommended as far as the operating system, the control kernel interfaces, communication protocols and programming languages are concerned.

4. THE CONFIGURABLE ROBOT CONTROL SYSTEM ISWRC

The modern, freely configurable robot control system ISWRC has been developed at the Institute of Control Technology for Machine Tools and Manufacturing Units (German abbreviation: ISW) at the University of Stuttgart. The use of modular hardware components and an extremly open system architecture allow simple adaptation of the control system performance to the requirements of the respective application. In this chapter, the components of the control system hardware, the real-time operating system employed and the structure of the control software will be presented.

4.1 The structure of the control system hardware

A processor board equipped with a Motorola CPU 68040 is used as a control computer. Communication with several I/O devices, such as analog-to-digital converters, digital-to-analog converters, evaluation units for incremental encoders and binary I/O modules, is carried out via an industrial standard VMEbus system. A workstation is available for the development of the control system software which will be downloaded to the target via Ethernet coupling (fig.3).

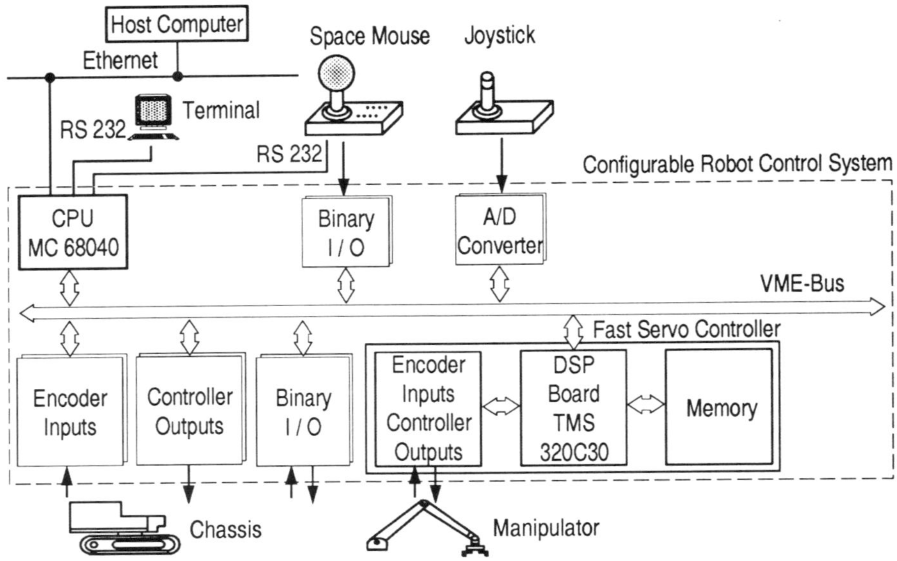

Fig. 3: Hardware structure of the Robot Control System ISWRC

The control system can be operated in either "manual mode" or "program mode". For manual operation, a joystick and a 6-degree-of-freedom space mouse have been provided: the joystick is used for controlling locomotion, whereas the space mouse allows motion generation for a 6-axes manipulator. Application programs for the program mode are entered via a video terminal and transferred to the control computer via a serial interface.

A fast servo control system based on the digital signal processor TMS 320C30 is used for position control of the robot joints. Sufficient computational power for the implementation of real-time high-level control algorithms has been provided /17/.

4.2. The real-time operating system

Organization and just-in-time execution of single tasks inside the control system such as motion generation or position control is performed by a powerful multitasking real-time operating system. It facilitates the integration of user-specific functions into the control system as well as future changes. The operating system is hardware-independent and, in addition, offers further advantages such as

- clear task-oriented structure of the control system software,
- short interrupt and task changing times,
- fast and efficient mechanisms for intertask communication and synchronization,
- preemptive scheduling,
- interfaces for data exchange with local and remote file systems,
- services for communication with peripheral devices and external computers.

4.3. The structure of the control system software

The software structure of the control system is shown in fig.4. A programming system forms an interface between the user and the robot control system /18/. In program mode, motion commands are entered as text using the robot programming language IRL (Industrial Robot Language) and compiled afterwards, in manual mode, respective commands are generated from the output signals of the manual control devices described in section 4.1. After a conversion of data formats, the reference values are transferred to the control kernel. Within the control kernel, preinterpolation and transformation functions are executed /19,20/. Sensor data may be on-line transmitted to the control kernel for further processing via several interfaces /21/. A servo controller interface has been provided for transmission of transformed joint reference values to the drive units.

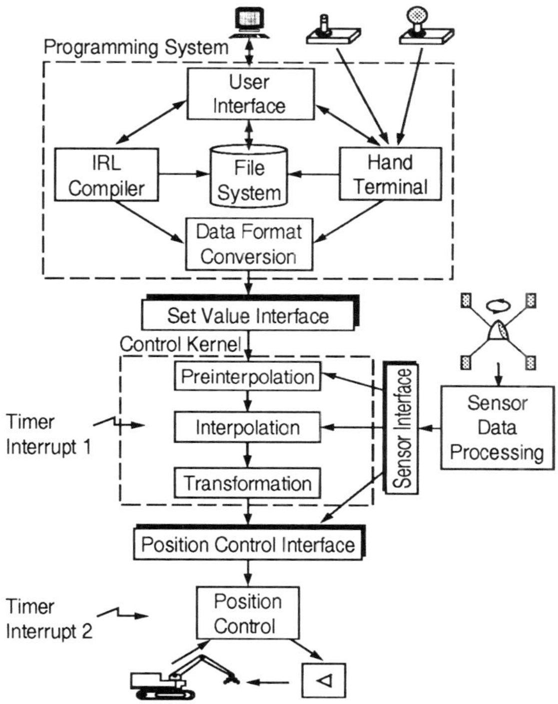

Fig.4: Software structure of the Robot Control System ISWRC

5. CONCLUSIONS

After a discussion of the national characteristics of masonry construction in Germany, requirements of an automated bricklaying robot and its control system have been defined. An open, configurable robot control system featuring modular hardware components , a hardware-independent real-time operating system, a task-oriented software structure and standard interfaces has been presented. The control system offers both manual and program mode programmable in either cartesian or joint coordinates, extensive sensor data processing capabilities and high-level control kernel functions. It may be easily adopted to any hardware configuration of a bricklaying robot.

6. REFERENCES

1 Statistisches Bundesamt, Fachserie 4, Reihe 3.1 und 5.1
2 Statistisches Jahrbuch 1991 für die Bundesrepublik Deutschland, Metzler-Poeschel, Stuttgart, 1991.
3 D.Böhm: The Mason's Elevator Handling Machine, Proc. 8th Int. Symp. on Automation and Robotics in Construction, Stuttgart, 1991.
4 Rimatem Fertig-Mauerwerk, product information by Rimatem, Dischingen.
5 J. Laukemper: Automation im Mauerwerksbau, PhD Thesis, Universität Stuttgart, 1991.
6 A. H. Slocum and B. Schena: Blockbot - A Robot to Automate Construction of Cement Block Walls, Robotics 4 (1988).
7 E. Malinovsky, E. A. Borschewski, E. A. Eler and V. M. Pogodin: A Robotic Complex for Brick-Laying Applications, Proc. 7th Int. Symp. on Aut. and Rob. in Constr., Bristol, 1990.
8 H. Lehtinen, E. Salo and H. Aalto: Outlines of Two Masonry Robot Systems, Proc. 6th Int. Symp. on Aut. and Rob. in Constr., San Francisco, 1989.
9 T. A. Bock: Concept of Building System for Robotization, Proc. 4th Int. Symp. on Rob. in Constr., Haifa, 1987.
10 G. Drees, J. Laukemper, G. Pritschow and M. Dalacker: Limits to Profitability of Automated Masonry, Proc. 8th Int. Symp. on Aut. and Rob. in Constr., Stuttgart, 1991.
11 Y. Rosenfeld, A. Warszawski and U. Zajicek: Economic Evaluation of Robotized Interior Finishing Works Based on Full-Scale Experiments, Proc. 9th Int. Symp. on Aut. and Rob. in Constr., Tokyo, 1992.
12 Y. Rosenfeld, A. Warszawski and U. Zajicek: Interior Finishing Building Robot "TAMIR", Proc. 8th Int. Symp. on Aut. and Rob. in Constr., Stuttgart, 1991.
13 D. Chamberlain, S. Ala, J. Watson and P. Speare: Masonry Construction by an Experimental Robot, Proc. 9th Int. Symp. on Aut. and Rob. in Constr., Tokyo, 1992.
14 EU 377 FAMOS BRICK - Highly Flexible Automated and Integrated Bricklaying System, EUREKA project (current).
15 EU 542- Mechanization of Bricklaying Technology on the Building Site, EUREKA project (current).
16 ROCCO - Robot Assembly System for Computer Integrated Construction, ESPRIT project No. 6450 (current).
17 W. Philipp and W. Scholich: Fließkomma-Signalprozessor-System für komplexe Regelungen, Elektronik Entwicklung 7 (1990).
18 G. Pritschow and O. Frager: Roboterzellen - Programmierung: Die Sprache IRL und der Zwischencode ICR, Robotersysteme 8 (1992).
19 M. Jantzer: Bahnverhalten und Regelung fahrerloser Transportsysteme ohne Spurbindung, Springer, 1990.
20 M. Bauder: Konfigurierbare Robotersteuerung mit allgemeiner Transformation, Springer, 1992.
21 G. Pritschow and A. Horn: Dynamik derzeitiger Sensorregelkreise für Industrieroboter, Robotersysteme 7 (1991).

Automation and Robotics in Construction X
G.H. Watson, R.L. Tucker and J.K. Walters (Editors)

GENERIC CRANE AND HOIST AUTOMATION WITH THE APPLICATION OF ADVANCED CONTROL ARCHITECTURES

N.A. Armstrong, P.R. Moore and R.H. Weston
Modular Systems Research Group
Department of Manufacturing Engineering
Loughborough University of Technology
Loughborough, Leicestershire
LE11 3TU, United Kingdom

Synopsis

Crane and hoist systems can potentially benefit from contemporary advances made in electronics based enabling technologies and control systems engineering. This paper describes the design and implementation of a modular distributed control system for crane and hoist automation in manufacturing and construction. The distributed control platform has been realised using the *FIP-Fieldbus* and the open machine control integration platform *UMC (Universal Machine Control)*. The technology is evaluated on a gantry crane 'demonstrator' which embodies control features such as; anti-swing, condition monitoring, tele-operation, automatic load coupling/decoupling, and automatic cycling. The project has successfully proven the applicability of modular distributed control systems for industrial automation and has substantiated the benefits of designing, implementing and maintaining control systems using an open machine control platform.

1.0 Introduction

A three year research programme entitled *MACHINE (Monitoring And Control of Hoists in INdustrial Environments*) started in September 1990 with the objective of applying innovative and cost effective improvements in the control, monitoring and mechanical design aspects of hoist and crane technologies. The desired outcome being semi-autonomous mechanical handling systems for diverse industrial sectors which demonstrate enhanced functional performance and improved efficiency and working conditions of operators. This collaborative research venture between six European partners was sponsored under the BRITE-EURAM initiative to a value of some 2 million ECU [1,2].

Principal aims of the research programme were to produce representative mathematical models of typical crane and hoist dynamics for subsequent use in designing improved crane control schemes; advance the performance of crane drive systems and hoist mechanisms; to establish a method of automatic or semi-automatic coupling of crane end-effectors to specified loads, and to demonstrate a reliability of coupling of up to 98%; to demonstrate guaranteed vertical lifting so as to eliminate swing at pick up by at least 90%, and to reduce the over swing of the load during traversing motions; develop a prototype performance record and condition monitoring system for cranes - useful for the production and maintenance planning; demonstrate operation with reduced personnel; provide real-time position data for operators, and demonstrate the efficient integration of the various sub-systems.

A control architecture was required which would integrate the control of crane axis drives and input/output hardware and support the collection of crane monitoring and collision avoidance data, provided a platform for swing control regimes, operator interface functions and partial autonomous operation. The control architecture needed to be sufficiently flexibility to enable such features to be developed independently and added, modified or removed; and allow the addition of, as yet unspecified, features which may be required in the future. For these reasons a distributed implementation of a control system based on the Universal Machine Control (UMC) architecture was chosen [3,4].

2.0 A Reference Architecture for Machine Control

The specification and design of Reference Architectures for machine and process control has been the subject of the ongoing research of the Modular Systems Group at Loughborough University of Technology. This research has produced the Universal Machine Control (UMC) system, which is an *open systems* methodology and software suite from which custom control systems can be designed and built from largely standard software modules; implementation-specific software is used to customise the application, but this is a significantly reduced amount compared with a specifically written software solution [5,6]. The current UMC run-time system is implemented on the Motorola 68XXX family of processors using the Microware OS9 real-time operating system. Run-time UMC machines typically use networked, diskless, rack based, single board computers with VME and/or G64 parallel backplanes. A machine's control system based on the UMC Reference Architecture (see Figure 1.) can be built quickly, simply and cost effectively from a range of re-usable software modules. Such machines can be simple or complex determined specifically by the application. The complexity of the control system reflects this and can be easily modified to accommodate updated hardware/software or for the system to be reconfigured or expanded to provide additional functionality as and when required.

Configuration tools allow the machine's structure to be defined in terms of the device hardware/software used and the user specific software which customises the control system for a particular application. Machine configuration data can also be used by associated modelling and simulation tools for offline development and testing of application code; these are useful for evaluating and emulating machine performance before it is used for a 'real-world' control system. An interface to higher level factory automation systems is also available through the CIM-BIOSYS *open systems* integration platform [7].

Run-time data associated with the machine and its hardware is stored in global data-areas which permits excellent data visibility for machine monitoring, diagnostics and error recovery strategies. Software modules (called handlers) provide a virtual interface through which the control system can access a range of devices. As a result of this uniformity, devices from different vendors can be seamlessly integrated in a single solution. The user's application software calls a set of 'standard' UMC functions which provide the interface, via handlers, to the machine hardware; as such the author of application software is freed from writing this often complex and hardware specific code and can concentrate on the task's primary purpose, namely, the user requirements. UMC architecture is hierarchically structured thus:

2.1 Machine Level

The machine level defines the overall format and structure of the control system. Utilities are provided for creating and modifying machine data (edit files which describe the machine in terms of number of axes, ports, tasks, events, device specific data etc) and location files. At run time a machine configuration utility creates the actual machine from these data files. Utilities exist for monitoring and dynamically modifying the functioning of a UMC machine during execution.

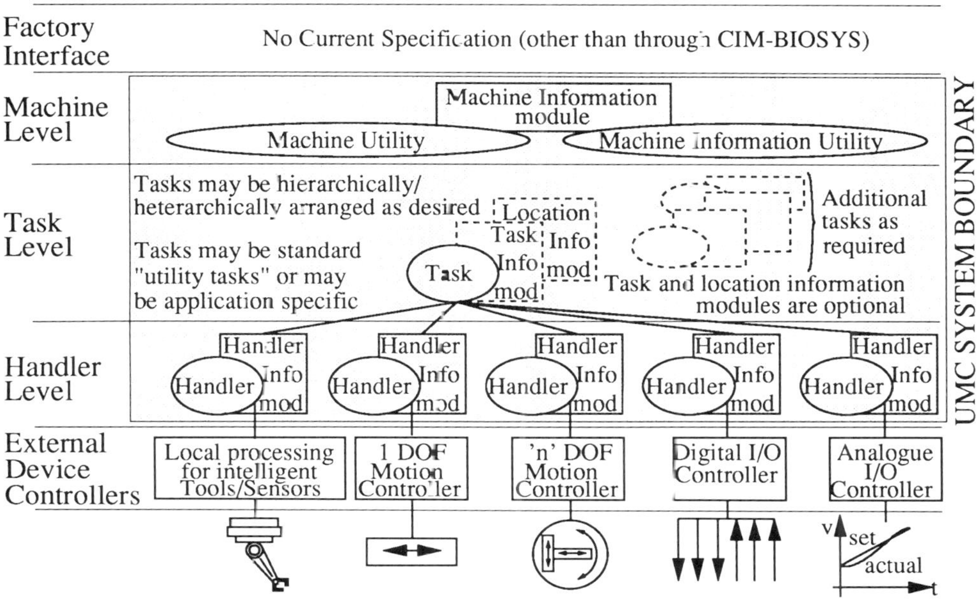

Figure 1. Universal Machine Control Hierarchy

2.2 User Tasks

Application specific programs (referred to as user tasks) are currently written using the C language. A library of 'standard' functions provides all the necessary links to the UMC control system. Users are then free to write whatever code they need, specific to their requirements, with access to all the UMC machine hardware and data that the task may require. Library functions include calls for reading input and writing output channels, single and multi-axis constant velocity and trapezoidal velocity moves, mapping and profiling of multiple axes, velocity control, teaching positions and moving to predefined locations. Library functions are also available for reading and setting parameters, general error recovery and status reporting, inter-task synchronisation and communication [8].

A machine will typically be made up of groups of devices, ie, the associated axes constituting a robot arm. Axis positions can be defined off-line using a dedicated editor or on-line using a specially written task. These locations are stored, at run-time, in a data module and can be saved as edit files (on disk) for re-use each time the machine is run. A user task might specifically control one group of axes with different tasks supervising other groups or performing other duties of the machine. A natural decomposition of the application is easily supported. Occasions may arise where a device is associated with two or more groups or user tasks, ie, an axis passing parts between work stations, in which case both tasks can share access of a single device. User processes may monitor the status of a handler at any time, without being the controlling task, by reading data from the device's global data module.

The UMC architecture imposes no constraints on the content or structure of the application code, linking to the library is necessary only if access is required to UMC functions. Proprietary (non-UMC) software modules will run alongside UMC processes, functioning completely autonomously or communicating using some form of proprietary inter-process protocol with UMC tasks.

2.3 Virtual Devices

A handler provides a 'virtual' software interface between user programs and specific hardware devices. By using a generalised set of commands and responses passed to the handler, a task can control a device or groups of different devices without being concerned with their specific operating or communication mechanisms. In this way a user defined task process can communicate easily (uniformly) with many types of manufacturer specific, class specific and device specific hardware. Replacing or upgrading the hardware and device-software used by a machine is simple and straightforward as task code should not be effected.

The handler is a state machine which acts as a translator, converting standard data and commands to device specific calls, receiving device specific responses and converting them to a standard format available to any user-task. The handler also monitors device status, reporting to the user any error conditions and provides a mechanism for taking corrective action.

Handler processes are only active for short intermittent periods, monitoring and processing changes in the device or task instructions; they then 'sleep', giving CPU time for other processes to run. If a task requires some special service from a device, ie, one which is not supported by the general command set, it can accomplish this via the UMC function 'specific' call, so making full use of a particular devices' specialist attributes.

3.0 Industrial Distributed Control Networks

Around 1984, standards bodies and industrial collaborative groups, in America and Europe began work on the creation of open standard digital communications networks for low level industrial devices (Fieldbus). This resulted in three distinct variations: the Instrument Society of America (ISA) Fieldbus, FIP from France and Profibus from Germany. The International Electrotechnical Commission (IEC) decided that the ISA fieldbus would form the basis for the international standard, although elements from FIP, Profibus and other factions will also have been included in the final standard.

In an attempt to hasten the realisation of an open standard, groups such as the International Fieldbus Consortium (IFC) and Inter-operable Systems Project (ISP) have formed to provide early 'Fieldbus' solutions, however, it is unlikely that devices with IFC or ISP fieldbus will be available before 1994. Meanwhile, the FIP and Profibus groups have continued to form national standards, vendor and user bases; and now see a broad, and increasing, use of their respective 'fieldbuses'. Network systems from different industrial sectors have also been proposed as suitable for the requirements of process, manufacturing and construction industry automation, Eg. MIL-STD-1553B a military standard and ARINC 629 an aviation standard; SERCOS from the machine-tool industry and CANBUS from the automotive industry. Proprietary networks from major PLC, motor-drive and instrumentation manufacturers and specialist networking companies have also become prominent, including Intel's Bitbus, Rosemount's Hart, Echelon Lonworks, and Pheonix contact's Interbus-S. These lists are by no means comprehensive, but serve to indicate the interest shown and potential market for such systems.

From the criteria for a simple digital replacement of an analogue standard, the draft Fieldbus standard now contains many (sometimes disparate) options in an attempt to provide a solution satisfactory for use in many diverse industrial application domains. It is hardly surprising therefore, that there would be a certain amount of procrastination and, at times, acrimony between the major players and that the formation of the IEC standard has been overdue. It is expected that the Physical-layer standard will be finalised very soon, and that the Datalink-layer is expected to become a draft standard by the middle of the year. There seems to be little progress towards an Application-layer specification which is acceptable to all the members of the associated working group [9]. At various times a (real-time) subset of Manufacturing Automation Protocol (MAP)'s Manufacturing Message Specification (MMS) services [10], global-object services similar to those offered by FIP and an application layer based on Profibus services have been suggested.

Potentially of greatest interest to a user of fieldbus is the 'User-layer'; a proposed additional tier in the hierarchy, although not part of the original ISO model [11]. The User-layer essentially 'hides' all networking functionality from the user by providing a set of virtual-device services which appear to directly interface the user's software with the remote device. For example, virtual-device function blocks would exist for instigating one of a standard set of PID loops on a remote device, what the fieldbus actually does to achieved this is of no concern to the user, other than wether is has been successful or not. As with the other layers, various classes of conformance, from completely defined (genuinely Open) to manufacturer/device specific have had to be included. NB: the UMC user-task library is analogous to the proposed IEC User-layer in providing a set of functions by which a user can control a specific device in a virtual manner.

3.1 Selection of A Distributed Control Network

An international standard fieldbus would have been desirable for this project, but would not have been available within it's life span. However, a system which offers features close to those likely in the IEC standard is appealing as it would then offer an easy migration path for future developments or applications arising from the project.

A 'distributed real-time database' was recognised to suit the global data-sharing aspect of a distributed version of the UMC architecture. The only *Open System* currently offering guaranteed real-time distributed-object services, for which multi-vendor hardware and software is available is the FIP fieldbus contender [12]. The implementation of FIP used also offers an acyclic messaging (and file transfer) facility; multiple physical media, speed and intrinsic safety options; (potentially) some MMS like services; ready availability of hardware from various sources; and an Open (French) national standard.

4.0 Distributed Control Implementation of UMC

The approach adopted for a modular distributed control system employs many of the concepts and attributes of the 'centralised' form of UMC. The users application code is defined within task programs which access the system via a library of standard functions. To convert a task written for a centralised UMC machine to one suitable for a distributed application should require little more than relinking it with the appropriate library. Task and Handler processes communicate in a consistent manner, whether they are situated on the same or different processors. Machine and device global data is accessible to any process that requires it, via the 'distributed data-base' maintained by the network. Distributed processes use a software interrupt mechanism for processing network service requests, this gives a fast response to aperiodic requests and allows the normal program-flow of tasks and handlers to be maintained.

Where centralised UMC uses a system of OS9 events, signals and shared data areas, the distributed form has network functions which globally mimic or locally reproduce these interprocess communication mechanisms. Attempts have been made to optimise the performance of distributed-system calls in-so-much as frequently used data is stored locally rather than repetitively requesting it from the network. However, some of the 'internal' functionality is not directly transferable between the two approaches, Eg. storing device data module addresses (pointers) in the machine data module of a centralised system is simple and efficient but not applicable in a distributed environment; in this instance the data module addresses will be different at each node and are therefore stored locally by each process.

The distributed implementation of UMC uses both the FIP synchronous object-variable 'data-base' and asynchronous message transmission services. Fip variables are used for maintaining a global copy of the real-time machine and device data. Messaging is used for the remote setting of OS9 events and signals and consumer to producer variable update requests. A combination of FIP variables and messaging is used for a global event mechanism.

Special distributed control utilities have been written to support local configuration of the data-modules (which define the access of networked global data), loading of task location modules, creation of application tasks, handler and network processes, and monitoring and resetting of global event and data-values.

The FIP implementation of distributed Universal Machine Control was first publicly demonstrated on a five axis modular robotic machine at a United Kingdom Department of Trade and Industry 'Fieldbus Open Day' at Loughborough in November 1992.

5.0 Implementation of the Demonstrator Crane Control System

5.1 Hardware Architecture

The demonstrator uses a suitably designed overhead-travelling (gantry) crane. The control-hardware architecture uses three FIP network nodes linked by a shielded twisted pair cable operating at a raw data-rate of 1Mbit/s. One node is mounted on the trolley (crab), this controls two variable-speed vector drives (for the cross-travel and hoist motors) via closed-loop motion controllers, plus an I/O module which supports digital and analogue channels. One node is mounted on the bridge, this controls the long-travel via a motion controller and vector drives and an associated I/O module; it also provides a 'gateway' between the FIP and proprietary-radio networks. The radio link is used to connect a data-logging node (for condition-monitoring), tele-operator and a collision avoidance system. The third node is mounted on the crane grab and provides the various I/O functions. Figure 2 depicts the control system hardware architecture.

5.2 Software Architecture

UMC is used to provide a control architecture for the fundamental crane control system and set of basic networking functions for communication between this and the proprietary, ie, non-UMC applications - data-collection, tele-operator and grab-control software (Figure 3). A UMC single-axis handler is used to control each drive, along with a UMC binary-analogue handler for each block of I/O.

The application code is split into three tasks. The first task (task1) provides the interface between the control system 'core' and the peripheral systems. It passes crane position data to the radio nodes and provides the man-machine-interface - importing commands via keyboard and tele-operator; these are processed and executed directly for grab I/O commands or exported as gross crane demand-position data / move-commands to the other tasks. These tasks use the gross commands and data to drive the crane hardware via UMC function calls to the handlers. The second task (task2) uses an anti-swing control algorithm to process the 'move' commands, and drives the handler/hardware in a low-level-UMC velocity control mode. The third task (task3) processes the same commands using high-level-UMC, multi-axis trapezoidal move functions. By switching between task2 and task3, the same gross command can be executed with or without the anti-swing control algorithm.

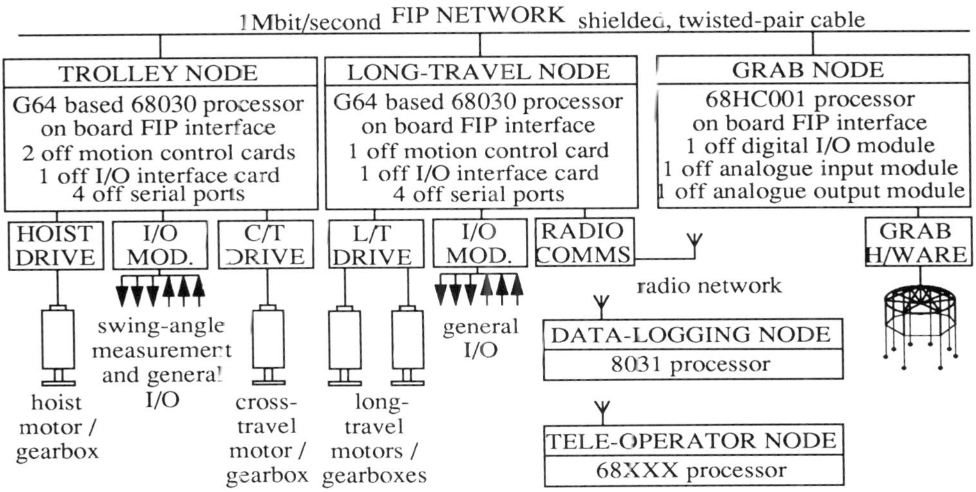

Figure 2. Demonstrator Hardware Architecture

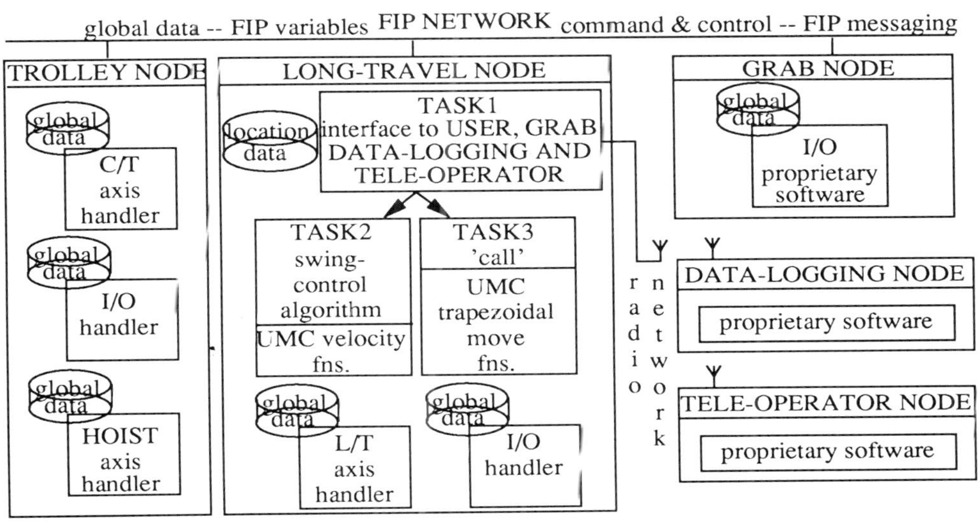

Figure 3. Demonstrator Software Architecture

By splitting the control requirements into discrete elements, each part can be written, debugged and tested in isolation. Furthermore, piecemeal changes can be made without forcing a major re-write of the software. ie, changing the drive hardware might require a different handler but the rest of the software would remain unchanged. Likewise, changing the anti-swing control algorithm would require modification of task2, but nothing else. This approach also applies to the integration of proprietary systems which can be installed, modified or removed in a similar fashion.

6.0 Conclusions

A consistent machine control reference architecture and associated modular software can offers cost-savings and other benefits in the design, implementation, service, modification and re-use of a wide range of machines, including those found in the construction industry. The ongoing MACHINE project will demonstrate how diverse crane-hoist technologies can be integrated successfully using modern, modular control methods to provide improved mechanical handling systems.

7.0 Acknowledgements

The authors gratefully acknowledge the contributions of other members of the Modular Systems Group and in particular those of Robert Harrison and Alan Booth in the foundation UMC research.

8.0 References

1 Anon. (1991), 'MACHINE: Monitoring and Control of Hoist Technology in Industrial Environments', BRITE EURAM programme synopses of current projects 1990-91 second edition, CEC. p238.

2 Dodman K (1991), 'Crane Automation', Cranes Today, March, p41-44.

3 Weston R H, Harrison R, Booth A H and Moore P R (1989), 'Universal Machine Control System Primitives for Modular Distributed Manipulator Systems'. International Journal of Production Research, vol27, No3, p395-410.

4 Weston R H, Harrison R, Booth A H and Moore P R (1989), 'A New Approach to Machine Control'. Computer-Aided Engineering Journal, February, p27-32.

5 Harrison R, Wright C D, Carrott A J, Booth A H and Armstrong N A (1992), 'UMC Conceptual Specification', dept. Manufacturing Engineering, LUT, Loughborough, UK.

6. Glad A S (1987), 'Software Engineering Guide for Real-Time Control Systems Development', Sixth Control Engineering Conference, Rosemount, IL. USA, May, p188-199.

7 Coutts I A, Weston R H, Murgatroyd I S and Gascoigne J D (1992), 'Open Applications Within Soft Integrated Manufacturing Systems'. International Conference on Manufacturing Automation, University of Hong Kong, August, p800-805.

8 Booth A H, Carrott A J (1992), 'UMC Version 2.0 Implementation Specification', dept. Manufacturing Engineering, LUT, Loughborough, UK.

9 Goodwin B (1993), 'Communication discord', The Engineer, 28 January, p18-19.

10 Anon. (c1992), 'MMS Open Systems in Manufacturing', dept. Trade and Industry, London, UK.

11 Lasher D (1992), 'A User's Standard - Why Is There An Eighth Layer?', Fieldbus '92, Institute of Measurement and Control and ISA International - England Section, London, November, (no page numbers.).

12 Anon. (c1991), 'FIP technical description', Club Fip, 3 bis, Rue de la Salpetriere, Nancy, France.

Automation and Robotics in Construction X
G.H. Watson, R.L. Tucker and J.K. Walters (Editors)

Intelligent control for robotic rebar bending

P. S. Dunston[a] and L. E. Bernold[b]

[a]Department of Civil Engineering, North Carolina State University, Raleigh, NC 27695-7908

[b]Department of Civil Engineering, North Carolina State University, Raleigh, NC 27695-7908

Abstract

Computers are becoming essential tools for integrating design, planning, fabrication, and construction. Computer-controlled construction equipment are poised to receive information from data generated in the design phase. The robotic execution of construction operations, however, depends on complex models for autonomous and intelligent control. This paper discusses the implementation of a strategy for the robotic bending of straight steel reinforcing bars (rebar) and presents the results of experimental work with an actual testing facility. Cold forming principles and the mechanical behavior of rebar are examined to develop rules for a real-time controller. Focus is placed upon exploring the dynamics between the bender and the rebar. The concepts of impedance, admittance, and compliance of both the machine and the environment are discussed in the context of control strategy development.

1. INTRODUCTION

The ultimate goal of Computer Integrated Construction (CIC) is the full linkage of all phases of a construction project, namely design, planning, fabrication, and construction. The flow of information through the offices of designers, fabricators, and constructors should continue to both the fabrication shop and the field where such information can be utilized to enhance productivity, quality, and safety. This information may then be employed in the real-time automated control of construction processes.

One area of construction that has been targeted for comprehensive research by the Construction Automation & Robotics Laboratory (CARL) at North Carolina State University is that of the design, delivery, and placement of steel reinforcement bars (rebar). A component of this research deals with the development of a computer-controlled bender for straight bars with real-time adaptive control capabilities. The current stage of research deals with understanding the behavior of the rebar and developing a model for controlling the operation in order to achieve consistently accurate bent bars.

Machine manufacturers have already developed automatic rebar benders that bend the small diameter coil rebar, and this technology is currently in use by rebar fabricators worldwide. However, these small size rebar only make up a fraction of the rebar typically used in concrete construction. The technology for automatically bending straight bars can dramatically change the traditional rebar fabrication process by increasing the quality of bending as well as just-in-time production and delivery, thus increasing the efficiency of rebar placement.

2. MECHANICS OF BENDING AND SPRINGBACK

The bending of rebar is a cold forming process that is affected by two physical aspects of the steel bar: (1) the moment of inertia of the rebar cross section and (2) the elastic-plastic properties of the reinforcing bar. The effect of the moment of inertia can be seen in Figure 1 below. Longitudinal ribs, a result of the hot rolling process and common to all rebar deformation patterns, add to the moment of inertia about the axis of bending when these ribs are oriented perpendicular (horizontal) to the bending axis. A collinear (vertical) orientation of the ribs relative to the bending plane results in a smaller moment of inertia and therefore requires less bending effort.

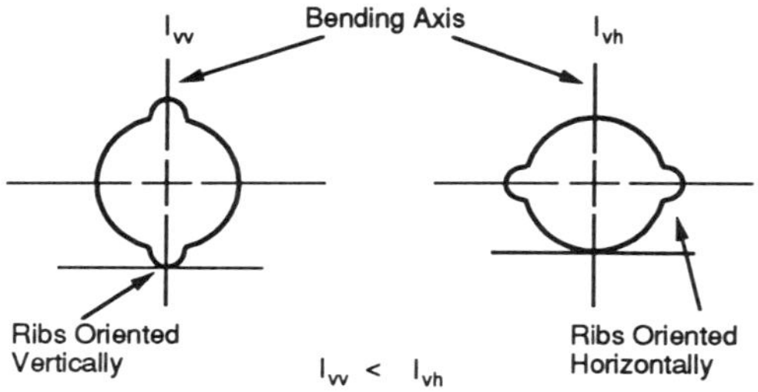

Figure 1. Effect of rebar orientation on moment of inertia.

The elastic-plastic properties of the reinforcing steel manifest themselves in the development of stresses during the bending process. The resolution of these stresses results in what is called springback, the major obstacle to accurate bending of rebar [1]. In the literature on the cold forming of metals, mainly sheet metal, the topic of springback has received the most attention [2, 3]. The analysis of the material behavior is based upon the conventional cantilevered beam theory for bending. The same general principles that apply to bending sheet metal also apply to the bending of bars.

Springback is governed by the variation of stresses that develop during bending. The largest tensile stress develops initially at the outer surface of the rebar and decreases toward the neutral axis. As bending continues, the zone exceeding the yield stress moves inward towards the neutral axis. Thus, the plastic region increases while the elastic region around the neutral axis decreases. When the final bend angle is reached and the bar is released, the elastic band attempts to return to a straight orientation, but the permanently deformed plastic band resists this motion. The resulting stress equilibrium is observed as a springback. The variables known to effect springback are elastic limit (or yield strength), bend radius, bend angle, and thickness (or diameter) of the metal workpiece. The only feasible method for overcoming springback in rebar is overbending, the traditional approach employed by the skilled rebar bender operator who relies upon his judgement to compensate for springback.

3. EXPERIMENTAL FACILITY

A prototype rebar bending system, named ARB (Automated Rebar Bender), for conducting empirical studies and testing of control strategies has been built at CARL. An illustration of the hardware used in the computer integrated system is shown in Figure 2. A 386 PC is used to operate a stepper motor via a Centroid controller (model CNC-3). Commands are sent from the PC to the motor controller through a communications port (COM1). The stepper motor drives the rotating table of the rebar bender through two chain-and-sprocket sets. The bending peg maintains a fixed point of contact as it wraps the bar around the center pin, but a roller bearing on the holdback allows free translation of the bar.

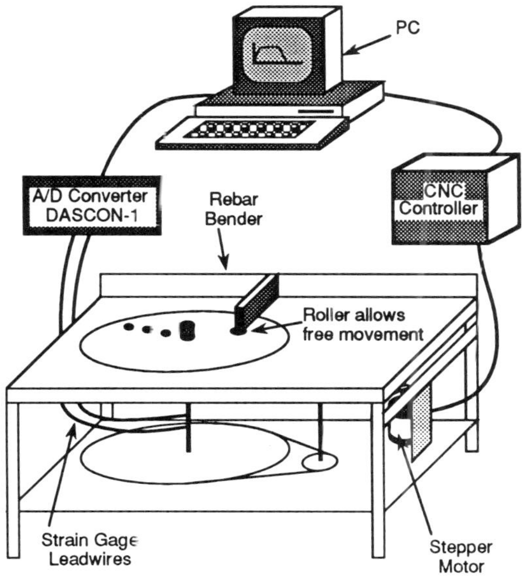

Figure 2. Schematic of ARB Hardware

The series of chains, sprockets, and axles make up the power transmission system for the rebar bender. One chain connects the motor shaft sprocket to a larger sprocket with the gear ratio of 9:112 which in turn drives a smaller sprocket. A second chain then allows this sprocket to drive the bending axle with a 10:112 gear ratio and an angular speed that is 0.7% of the commanded motor speed.

Strain gages on the main shaft of the bender act as force sensors for measuring the torque applied during bending. A DASCON-1 analog/digital converter board, housed inside the computer, converts the analog signal from the strain gages into a digital signal readable by the computer. A control program for the system is written in Microsoft QuickBASIC Version 4.5.

4. PATTERN ANALYSIS FOR CONTROL

The success of a real-time controller for bending straight deformed bars lies in its ability to use sensory data to accurately predict and compensate for the springback. The value of the sensor data as an input for control lies in the pattern information that can be extracted from it. Since the bending of rebar requires a manipulator which is able to provide rotational displacement, only the bending torque seems to provide useful sensory data. An example of bending torque data collected with ARB is shown in Figure 3.

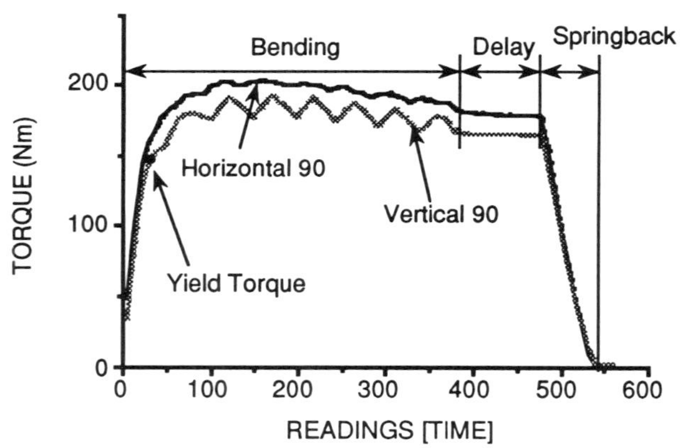

Figure 3. Fundamental features of the bending torque curve

Figure 3 depicts bending torque curves from two 90 degree bends made on one sample of rebar. One bend was made with the longitudinal ribs oriented vertically and the other horizontally (see Figure 1). This curve can be analyzed using the decision-theoretic approach to pattern recognition, wherein characteristic measurements or features are extracted from the input for classification and decision making [4]. Signature features of the bending torque curve can reveal valuable information about a specific rebar which can be used with rules for adaptive control.

Both example curves exhibit a linear increase in the start-up of the bending phase. After reaching a *yield torque*, the curve begins to level off with a gradually decreasing slope resembling the shape of a stress-strain curve. The bending phase is followed by a delay period during which the bender stops before returning to the start position. The final segment is the springback phase where a sharp decline in torque occurs. The amount of overbend necessary to compensate for the springback is related to the length and slope of the springback portion of the curve. The controller, monitoring and commanding the bender, must have a deep understanding of the factors and functions linking material properties with measurable operational data (e.g., torque) and the final outcome of the operation (e.g., bend angle).

Further examination of the bending torque curves in Figure 3 reveals valuable information about the rebar sample. It is clearly demonstrated that the amount of applied torque necessary to bend a bar with horizontally oriented longitudinal ribs (i.e., higher moment of inertia) is greater than the torque required for the same bar with a vertical orientation. The most distinct difference, however, is the smoothness of the horizontal curve relative to the jaggedness of the vertical curve. The peaks and valleys observed on the vertical curve are the result of encounters with the crosswise deformations on the surface of the rebar as it moves during bending. The longitudinal ribs provide a smoother translation of the rebar when it is oriented horizontally.

Figure 4 illustrates the types of features that can be analyzed from the bending torque curve data in addition to the rebar orientation. The features that may be used for predicting springback include an elastic torque modulus or slope from the purely elastic region of bending, the yield torque that can be identified using an offset criterion, and the maximum torque (average maximum torque in the case of a vertical orientation). Information that may be used in a feedback loop for recalibration or adjustment of control laws includes the range of the springback portion of the curve or the slope(s) of this portion of the curve. This portion of the curve has been divided into segments for slope calculations because it has been observed that the slope of the springback segment is not constant. It is possible that the changing slope may be due in part to a relaxation in the mechanical system of the rebar bender.

In order to establish links between the measurable features of the bending torque curve and the springback, torque data from numerous bend tests were collected and springback measurements were recorded. The torque data were fed into a data analysis program that was designed to process the data and extract values for the features of the bending torque curve. These feature values were then correlated with the corresponding springback measurements in order that the relationships between the feature values and the springback of the rebar may be studied. A scatter plot of one such correlation is illustrated in Figure 5.

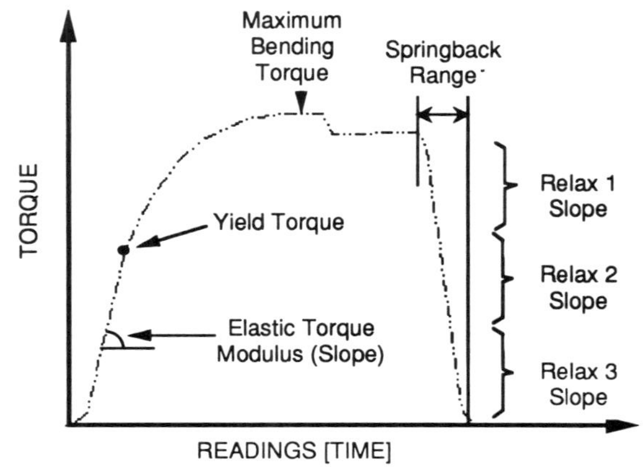

Figure 4. Feature based data points for predicting springback

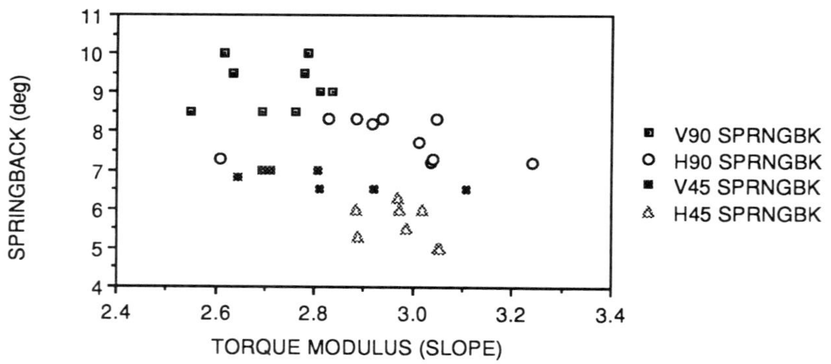

Figure 5. Sample data from bending tests

Figure 5 shows a plot of the elastic torque modulus (slope) versus springback for one batch of rebar. The data sets are described by combinations of the bend angle (45 or 90 degrees) and the orientation of the longitudinal ribs (horizontal or vertical) as illustrated above (see Figure 1). It is apparent that the degree of the bend angle has a significance since the springback for the 90 degree bend is consistently larger than for the 45 degree bend. Also, at any bend angle, an horizontal orientation of the longitudinal ribs results in a lower springback which corresponds to a higher elastic torque modulus. Finally, the grouping of the data by bend angle seems to reveal a trend with a downward slope. Such data sets as this are now being analyzed statistically for multiple batches of steel so that true patterns may be established. These patterns will subsequently be used to formulate control laws for a real-time controller.

5. CONTROL MODEL CONCEPTS

A primary task in creating a scheme for automating the fabrication process is the modeling of both the machine (manipulator) and the material (object). An exact model of the interaction between the manipulator and its environment is necessary for achieving precision in commanding accurately the manipulator's motions. The concepts of admittance, impedance, and impedance control [5, 6, 7] have been used in applications with similarly complex models where a manipulator has mechanical contact with the workpiece.

As explained by Hogan [5], a manipulator when coupled with the object to be manipulated assumes the behavior of an impedance, accepting motion inputs and yielding force outputs. The object (or environment) takes on the form of an admittance, accepting force inputs and yielding motion outputs. In this concept, the controller uses relative sensory measures to constantly readjust the control parameters of the manipulator, the impedance. The approach of controlling the use of such a manipulator by providing it with not only a motion command but also with a response to any disturbances that cause deviations in the execution of that command is termed impedance control [5]. Based on these definitions the rebar bender acts as the impedance while the rebar itself can be modelled as an admittance.

Rebar bending is a task requiring physical contact in which both the bender and the rebar exhibit a measure of compliance (yielding). The interaction of a compliant manipulator and a compliant object involves dynamic parameters and contact conditions which complicate the task of control [8]. The rebar bender compliance may be viewed as a mass-spring-dashpot system [7], that is, having inertia, damping, and stiffness parameters. In order to develop the necessary control laws, the relationships between motion and deviations in observable measure have to be clearly understood. The two critical measures in rebar bending are the bending torque and the springback.

6. SUMMARY

The robotic bending of straight rebar has the potential to dramatically impact the production and placement of steel reinforcing bars through accurate, just-in-time fabrication. Real-time adaptive control for rebar bending relies upon the establishment of control laws based upon models that accurately describe the rebar bender and the rebar. This paper has presented the type of data used for the development of a model for the rebar and has described the prototype automated rebar bender that is being used to obtain this data and to conduct experiments to test control laws developed from this data. Initial analysis of pattern information reveals a strong potential for establishing the control laws needed to have intelligent control for accurate bending through springback compensation.

7. REFERENCES

1. Bernold, L. E., "Process Driven Automated Rebar Bending," 8th International Symposium on Automation and Robotics, Vol. 2, June 1991, Stuttgart, Germany.
2. Eary, D. F., and E. A. Reed, *Techniques of Pressworking Sheet Metal*, 2nd ed., Prentice-Hall, Inc., Englewood Cliffs, New Jersey, 1974.
3. Mielnik, E. M., *Metalworking Science and Engineering*, McGraw-Hill, Inc., 1991.
4. *Digital Pattern Recognition*, K. S. Fu (ed.), 2nd ed., Springer-Verlag, New York, 1980.
5. Hogan, N., "Impedance Control: An Approach to Manipulation," *Journal of Dynamic Systems, Measurement, and Control*, Vol. 107, March 1985, pp. 1-24.
6. Lasky, T. A. and T. C. Hsia, "On Force-Tracking Impedance Control of Robot Manipulator," *Proceedings of the 1991 IEEE International Conference on Robotics and Automation*, Sacramento, California, Vol. 1, April 1991, pp. 274-280.
7. Liu, G. J. and A. A. Goldenberg, "Robust Hybrid Impedance Control of Robot Manipulators," *Proceedings of the 1991 IEEE International Conference on Robotics and Automation*, Sacramento, CA, Vol. 1, April 1991, pp. 287-292.
8. Annaswamy, A. M. and D. Seto, "Manipulation of Compliant Objects with Compliant Fingerpads in the Presence of Nonlinear Dynamics," *Proceedings of the IEEE Conference on Decision and Control*, IEEE, Brighton, England, December 1991, pp. 2124-2129.

Automation and Robotics in Construction X
G.H. Watson, R.L. Tucker and J.K. Walters (Editors)

SAFETY CRITICAL SYSTEMS FOR ROBOT EXCAVATION

LiLi Zhang, D.A. Bradley, D.W Seward

Engineering Department, Lancaster University, Lancaster LA1 4YR, U.K.
United Kingdom

Abstract

A number of mathematical and analytical tools are available to assist in risk assessment for industrial and other robots. This paper is concerned with the use of a hazard and operability study (HAZOP) methodology to analyse the range of hazardous conditions and to assess the degree of risk associated with construction robots. In this context, the paper sets out to identify the various hazard conditions that occur in relation to the sequence of operations of an automated and robotic excavator and to provide an assessment of the levels of the risks that result in accordance with a classification of the working volume of the excavator.

1. INTRODUCTION

The deployment of robots to carry out a range of operations once performed by humans such as may be associated with firefighting, space systems, nuclear inspection, maintenance and decommissioning and construction affords the opportunity to remove humans from a hazardous environment. However, construction robots themselves are a source of potential hazard and risk which could result in death or injury, for instance as a result of collisions with site personnel or site structures. In order to ensure that the dangers inherent to the deployment and use of construction robots does not offset their usefulness, they must be accompanied by appropriate safety related systems to guarantee their operation is accompanied by an acceptable level of risk. This, in turn, requires that the operation of the safety system must be fully understood and analysed as part of the design process of any automated and robotic system.

The development and deployment of automation and robotic plant in the construction industry has been slow compared to other industries. A variety of factors contribute to this slow transfer including the highly variable operation environments. In order to ensure effective and safe operation, construction robots have to be highly adaptive, responding rapidly to changes in their local and overall environment in relation to their operation and capable of making a series of linked strategic and tactical decisions about their actions. An AI based control system for automated and robotic excavation has been implemented by the Lancaster University Computerised Intelligent Excavator (LUCIE) project [1]. LUCIE is capable of autonomously digging a trench to a controlled depth in a variety of ground types and conditions. Safety systems for this automated and robotic excavator are now being studied.

To date, there have been developed a number of International Standards on the safety considerations for the design, construction, programming, operation, use, repair and maintenance of industrial robots [2]. The International Standard "Manipulating Industrial Robots---Safety" [3] was established in spring 1992. The International Electrotechnical

Commission document "Functional Safety of Programmable Electronic Systems" [4] and presents the hazard analysis, risk assessment and safety integrity requirements associated with this standard.

A number of mathematical and analytical tools are also available to assist in risk assessment for industrial robots including "Guidelines for Hazard Evaluation Procedures" produced by the Battelle Columbus Division [5]. The work presented in the current paper is concerned with using a hazard and operability study (HAZOP) method to analyse hazardous conditions and to assess the level of risk for automatic and robotic construction machinery with particular reference to excavation. In this context, the paper identifies various hazardous situations in relation to the sequence of operations of a robotic excavator and assesses the level of the associated risks in accordance with a classification of the working volume. With regard to the analysis of the possible range of hazard situations, the safety requirements can be obtained. Finally, safety strategies, which minimize the risks to an acceptable level, are considered.

Operational and control software in construction robots can also be a source of danger when it malfunctions. Obviously, software on its own cannot pose a threat to anyone, but some hardware failures may be the ultimate result of software malfunction. Safety-critical software is however the subject of a parallel study and is not covered in this paper.

2. FUNCTION AND STATE OF AUTOMATED AND ROBOTIC EXCAVATOR IN OPERATION

The analysis of safety requirements must be on the basis of overall operational requirement. It is, therefore, necessary that the operating modes of an automated and robotic excavator and its operating procedure are properly understood.

In operation, an automated and robotic excavator has essentially two operating modes, namely static and dynamic. Static mode implies that the base is held in a fixed position while the digging task is being carried out while the dynamic mode is associated with the movement of the excavator within working area.

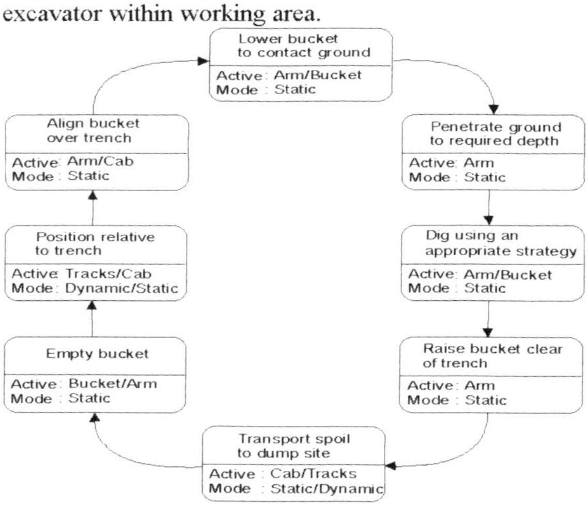

Figure 1 : An operational cycle for an excavator

When digging, the automated and robotic excavator is in static mode and the main active components are the bucket and arm of the excavator. The primary sources of hazards in this mode are collisions between arm or bucket of excavator and site personnel or site structures. When moving, the automated and robotic excavator is in dynamic mode. The highest level of risk is again associated with collision with site personnel or site structures. For spoil transport, the automated and robotic excavator can be either static or dynamic mode. Figure 1 sets out the main sequence of the operations and the corresponding modes of operation associated with the digging of a trench.

3. SAFETY REQUIREMENTS AND ANALYSIS OF OPERATION

The automated and robotic excavator is structurally essentially the same as a manually controlled excavator, working in an unstructured environment. However, its automated operating mode means that the excavator control system can itself define and implement operation in accordance with a specified task structure. As such, it has an enormous potential for initiating hazardous actions.

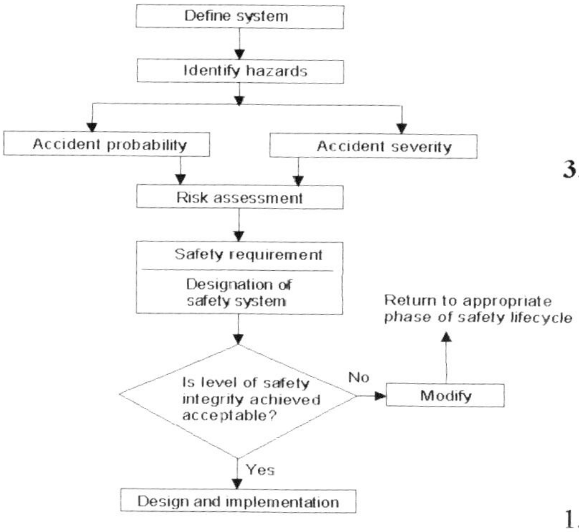

Figure 2 : Basic safety lifecycle model

3.1 The safety lifecycle of automated and robotic excavator

A safety lifecycle study provides a framework for the management of a safety system. The methods of development for safety systems must be fitted into this framework. Figure 2 presents a basic safety lifecycle model for an automated and robotic excavator.

The safety lifecycle may also be expressed as follows:

1. Define the required tasks for the foreseeable applications.
2. Identify the sources of hazard associated with any particular operation.
3. Evaluate and assess the associated risks.
4. Consider the safety strategies necessary to minimize the risks by reducing them to an acceptable level.
5. Assess the levels of safety integrity achieved for the system and ensure that these levels are acceptable.
6. If the safety integrity level achieved is not acceptable then repeat the appropriate steps.
7. If the safety integrity level achieved is lower than assessed tolerable risk level, design and implementation shall be carried out.

A number of analytical techniques can assist in the identification and structuring possible events leading to hazardous situations including:

- Fault Tree Analysis (FTA);
- Failure Modes, Effect and Criticality Analysis (FMECA);
- Hazard and Operability (HAZOP) Studies;
- Relative Ranking Techniques (DOW and MOND hazard indices)
- Preliminary hazard analysis;
- "What if" analysis [6].

The procedure associated with a 'Hazard and Operability (HAZOP)' study is specifically designed to identify any hazards and operability problems associated with the process. The possible deviations and their corresponding cause and consequences in relation to the execution of the process can then be analysed.

3.2 Preliminary hazard analysis

The definition of a hazard is a situation in which there is actual or potential harm to human life or limb, or to the environment. [7] These hazards and failures can be classified in one instance as, "Random Hardware Failure" or "Systematic Failures", [8] and in another as, "Direct Hazard" or "Indirect Hazard" [2]. In particular:

Random hardware failures are - "Those which result from various unspecified breakdown mechanisms that occur at unpredictable times."

Systematic failures are - Those which occur as a result of errors that have been made at some stage in the specification, design, construction, operation or maintenance of a system."

Direct hazards are - "Those which arise from human contact with the robot or with the tools and objects which it handles."

Indirect or secondary hazrads are - "Those which may arise from the robot acting in such a way that it is the cause of hazardous events in other parts of the overall system."

For the automated and robotic excavator, the primary sources of direct hazard are collisions with objects in the working volume, including intruders. This is, therefore, the basic concern for the design of the associated safety systems. The possibility of collision exists between the excavator and the following types and classes of site objects:

- Persons — Operator, other site personnel.
- Vehicles — Site vehicles such as dumpers, lorries and cranes.
- Site structures temporary — Buildings, pipes, columns, supports, scaffolding and associated structures.
- Site excavations — Trenches and holes.
- Site storage — Stocks of bricks, steel, and other material left on site.
- Site debris — General site debris, such as spoil heaps.

3.3 Hazard and Operability analysis

Unlike many conventional automated machines the spatial envelope within which the excavator can act is not immediately obvious to the casual observer. Hazardous situations can arise not only in the immediate locality of the automated excavator but also in a large surrounding volume. The different regions within this workspace then give rise to differing hazard conditions. To eliminate the main direct hazards, the classification of zones or levels of protection around the excavator should be considered. The definitions of the zones in the Figure 3 refer to those given in ISO 10218: 1992 "Industrial robots" [2], "Fault- Tree Analysis of Hazards Created by Robots" [9] and "A Robot Safety and Collision Avoidance Controller" [10].

Zone 1 — Maximum danger zone. A small volume surrounding the excavator arm.

Zone 2 — Restricted zone. The area bounded by the maximum reach of the automated and robotic excavator when stationary.

Zone 3 Dynamic danger zone. The region is outside the maximum reach of the automated and robotic excavator but within the emergency stop range.
Zone 4 Safety zone. The region within the working area of the excavator but outside the emergency stop range of the moving excavator.
Zone 5 Maximum detection zone. The region within the working area of the excavator but outside the minimum manoeuvring distance of the moving excavator.
Zone 6 Maximum working perimeter of the excavator.

As indicated, Zone 1 is a small volume surrounding the arm of the automated and robotic excavator. The arm and bucket perform their actions within this volume during a trenching operation. If the arm of the excavator impacts with obstacles in this zone, then severe damage is likely to be unavoidable. Maximum direct hazards will therefore occur in this zone. However, for the purpose of digging, impact with ground is also necessary and, as such, is not a hazard. Sometimes the surface of the ground is uneven and this unevenness must not be interpreted as obstacles. Thus the automated and robotic excavator must be able to distinguish likely hazards from necessary ground impact associated with the digging process.

The highest level of risk is associated with Zone 1. Collision between the excavator arm and site personnel or site structures could result in the death of site personnel and/or severe damage to site structures and possibly to the excavator.

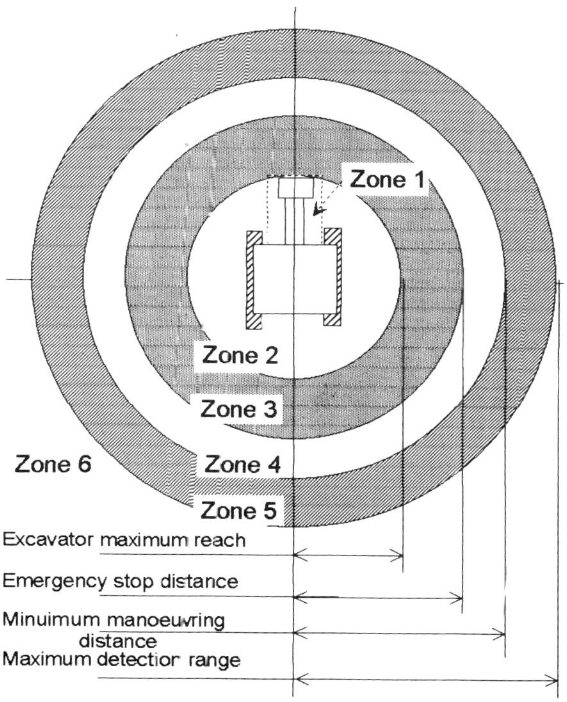

Figure 3 : The layout of the safety zones for an automated and robotic excavator

Zone 2 is a restricted zone relative to the excavator. It surrounds the excavator with the maximum reach of the excavator as its boundary. As for zone 1, any objects within this zone can be subject to a maximum hazard situation. Zone 2 therefore also encompasses the highest level of risk and hence should be kept clear of personnel and all fixed objects in the zone properly located and identified prior to operation. The risk evaluation in Zone 2 is therefore the same as that for Zone 1.

Zone 3 covers the volume outside the maximum reach of the excavator but inside the emergency stop distance of the moving excavator. Zone 4 is located between the emergency stop distance and the minimum manoeuvring distance of the excavator. Normally, there is no direct hazard situation associated with these zones when the excavator is operating in static mode. However, if the load was accidentally ejected from the bucket as a result of some action this could constitute a risk for any personnel or structures in these zone. When obstacles are themselves dynamic, they may progress from Zones 3 and 4 to Zone 2 in a relatively short time and this could result in a maximum risk condition if the surveillance system fails to detect their presence.

When the automatic and robotic excavator is operating in dynamic mode, the distance between the obstacles in Zone 3 and the excavator is always less than emergency stop distance of the dynamic excavator. Even if the excavator is subject to an emergency stop condition, collision is still unavoidable. Therefore, any object that exists in Zone 3 constitutes a direct hazard when the excavator is operating in dynamic mode.

For obstacles in Zone 4, if the automated and robotic excavator is operating in dynamic mode, or obstacles are themselves moving, then collision can be avoided by an emergency stop. Although the excavator can be subject to a stop condition for any object or personnel in Zone 4, the reliability of the surveillance system must be high otherwise the possibility of collision increases. A limit on the performance of the surveillance system is therefore set by the maximum relative velocity of any object within zone 4 as this dictates the time available for detection.

Zone 5 defines the maximum detection range of the system sensors. Using an appropriate strategy excavator can manoeuvre to avoid collision with any static object detected this outer zone. Static objects which exist in this zone, for instance a buildings, piles of bricks or site material, do not of themselves have the opportunity to come into contact with the excavator and cannot therefore be considered as direct hazards when the automated excavator is operating in static mode. However, when the excavator is operating in dynamic mode, then should the surveillance provision fail, direct hazards will be present.

Hazards in Zone 5 may in general be considered as indirect hazards. The level of inherent risk is therefore low. However, should the surveillance system fail to detect objects, the level of risk associated with this zone will increase.

Zone 6 defines a maximum working boundary within which the automated and robotic excavator is allowed to operate. As such, it defines the perimeter of the working area and thus presents the first line of defence against hazards. Any intruder entering into this outer zone may give rise to a possible hazard situation. Protection at this outer zone is often provided by the use of wire fencing, a light or infra-red beam barrier, pressure sensitive mats, or camera surveillance, to detect the entry of unauthorised personnel and vehicles. The automated and robotic excavator should respond appropriately if any individual or object which is not registered enters via the gate or crosses the light-beam barrier. From above, the risks can be summarized as follows:

- Risks resulting from the arm or bucket colliding with personnel or on-site structures during controlled motion or as a result of hardware breakdown or the dropping of the load.
- Risks resulting from a collision between the excavator while slewing and personnel or dynamic obstacles.
- Risks of collision between the excavator while subject to an emergency stop condition and obstacles that exist or move to within the emergency stop distance.
- Risks of collision between the excavator while not subject to an emergency stop condition and personnel or dynamic obstacles in safety zone.
- Risks of collision between the excavator while manoeuvring and personnel or dynamic obstacles within the maximum detection zone.

The designation, construction, and implementation of the safety systems should consider the operational characteristics of the excavator and all possible hazards. There are two fundamental principles for the selection and designation of the appropriate safety related system:

- The exclusion of personnel from the safeguarded space during automatic operation.
- The elimination of hazards or at least their reduction by detecting possible onset as early as possible.

For Zone 6, the selected safety system should be able to detect the entry of unauthorized personnel or vehicles to the safeguarded space. Methods used could be a mechanical guard such as a perimeter fence or barrier which completely encloses the operation space or an optical or photoelectric "curtain" guard comprising a series of transmitters and detectors, activated when a beam is intercepted. In terms of the cost of the equipment, the mechanical perimeter guard is likely to be cheaper. However, the guard must be readily movable as the operating volume of the excavator is continually changing.

For Zone 5, the main aim of the safety system is to detect objects entering the zone and to determine their precise positional and dimensional data so that the control system can modify its strategy in order to avoid collision. Because of this, a laser rangefinder installed on the top of the excavator is suggested as a possible safety-related sensor to guard Zone 5. Its rotation period and associated processing time must be less than the shortest anticipated time required for a dynamic object to move through this zone.

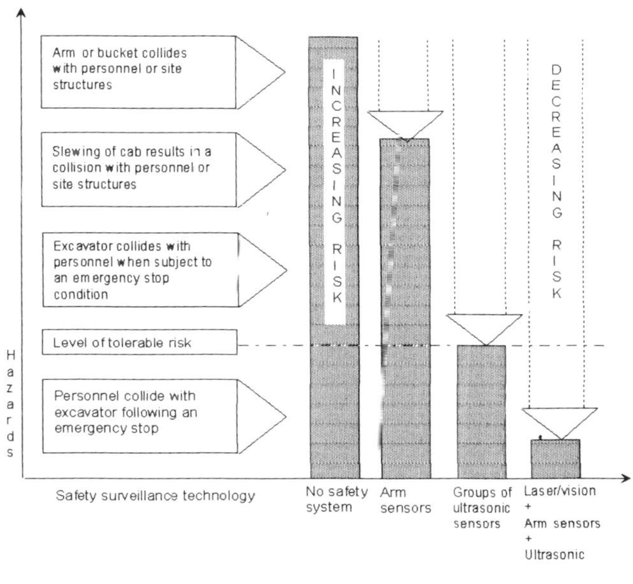

Figure 4 : Risk reduction with improvements in sensor technology and fit

For Zones 4, 3 and 2, the primary characteristic of the safety system is the detection of obstacles as early as possible. Their position and dimension are not necessary of major concern, but it must be known accurately whether an obstacle exists. Referring to the safety system requirements, a group of ultrasonic sensors installed around the cab of the excavator is suggested as a possible means of accomplishing this task. As infrared sensors are sensitive to the human body these could also be utilised for the detection of personnel.

The safety system is intended to ensure that the risks which exist in the operating environment are kept as low as possible, and certainly lower than the assessed tolerable risk. The safety integrity measure is intended to assess the degree of reduction of risks and to ensure that adequate precautions are taken against all possible hazards. Figure 4 shows the effect in reducing the level of risk after the adoption of the safety-related functions.

4 CONCLUSIONS

With the help of the Hazard and Operability Analysis method, we have presented a preliminary analysis of the safety requirements for an automated and robotic excavator. An initial prototype safety system for automated and robotic excavator has been proposed. However, to achieve implementation of the proposed the safety system, further work on the relevant criteria and integrity levels necessary for a more detailed assessment of risk are required. In conjunction with the design and implementation process safety verification and validation for the safety system as designed must be carried out.

It is, therefore, suggested that what is required is an Artificial Intelligent system for capable of simulating the operation of the safety system in order to verification and validation a proposed safety system. By answering various "what-if" questions the simulation process can quantitatively analysis possible deviations from expected behaviour. The results of simulation could then used to determine all possible states of the safety system to help to find and identify the potential hazards.

REFERENCES

1 Seward, D.W., Bradley, D.A., Mann, J. & Goodwin, M., 1992, 'Controlling an Intelligent Excavator for Autonomous Digging in Difficult Ground', 9th ISARC, Tokyo, Japan, Vol 2, pp 743-750

2 Schofield, M., 1992, 'Safety and Standards for Advanced Robots: A First Exposition'

3 ISO 10218, (BS 7228), 'Manipulating Industrial Robots Safely'

4 IEC Draft Standard, 'Functional Safety of Programmable Electronic Systems: Generic Aspects; Part 1', IEC Reference 65A (Secretariate 96)

5 Battelle Columbus Division, 1985, 'Guidelines for Hazard Evaluation Procedures', American Institute of Chemical Engineers

6 Kavianian, H.R., Wentz, C.A. & Peters, R.W., 1990, 'Safety System Management for the Design of Hazardous Processes', Professional Safety, Vol 35, March, pp 31-34

7 IEE, 1992, 'Safety Related Systems', IEE Professional Brief, January, Department of Trade and Industry

8 Bell, R. & Smith, S., 1990, 'An Overview of IEC Draft Standard: Functional Safety of Programmable Electronic Systems', Safety Related ACOS Workshop, 8-10 March, London

9 Sugimoto, N. & Kawaguchi, K., 1985, 'Fault-Tree Analysis of Hazards created by Robots', Robot Safety, International Trends in Manufacturing Technology Vol 4, Springer-Verlag, pp 83-98

10 Derby, S., Graham, J. & Meagher, J., 1985, 'A Robot Safety and Collision Avoidance Controller', Robot Safety, International Trends in Manufacturing Technology Vol 4, Springer-Verlag, pp 237-246

Automation and Robotics in Construction X
G.H. Watson, R.L. Tucker and J.K. Walters (Editors)

Evaluation of the position accuracy of the excavator tool

Fiorenzo Malaguti

CEMOTER-CNR, v. CANAL BIANCO 28, 44044 CASSANA (FE), ITALY

Abstract

The automation and robotics for earth moving machines were introduced to reduce labor cost and increase machines performances, the increment of position accuracy of the excavator bucket obtained by laser and/or automatic control systems meets both objects.

Experimental tests and simulations have shown excavating accuracy to be about two centimeters, the excavating accuracy depends on position accuracy of excavator boom prevalently, also a simulation of the boom dynamic system has shown it, in fact, during the precision excavation the arm is near the vertical position.

The excavator, or the basic machine, is used also as loader, forklift, pipe-layer, drill and more, therefore we consider right to value the position accuracy of all excavator linkages and complete equipment, and the design problems of the control system.

The accuracy evaluation of the complete arm is made by simulation, the evaluated system includes the dynamics of the linkages, hydraulic and power system (hydraulic cylinders, valves, pumps, engine) and the operator.

An important task is made by the operator , he is at the same time the feedback and controller of the machine, the use of the human model in the simulation doesn't prejudice the generality of the problem, because the excavator should became a telerobot probably, while it is less probable that it could become a complete autonomous robot.

The simulation is made using the dimensions and measure values of a microexcavator used as scale model.

1. Introduction

Introduction of the new laser systems in digging operations has allowed the operator to obtain a good performance in precision digging, and to reduce labour cost. These systems, making the necessary external reference, are the main sensors to automate the excavator so as the other earth moving and construction machines.

The present state of automation needs human operator again, he does almost all the controller and supervisor tasks of the machine operations,because he can fit many conditions, therefore it thinks that also in the future he will continue to control the heavy construction robot to fit it at the variable work condition.

In parallel to the research about automation and robotics for construction machines we have thought, therefore, to investigate about operator-machine iteration.

The digging precision depends on the boom position control mainly, but excavator is a multifunction machine by possibility to change the final tool, we analyze the positioning precision of the two main links of excavator, since the precision of the final tools depends on it.

For this analysis we have simulated the dynamics of the boom, arm, hydraulic system and controller model of the operator; putting in evidence the problems cause by the asymmetry of actuator. A dual bang-bang hierarchical model has been used to simulate the operator, this type of model was preferred on the continuous model, because it is more similar real operator behaviour.

The simulation is related to the commercial microexcavator fitted for robotics research.

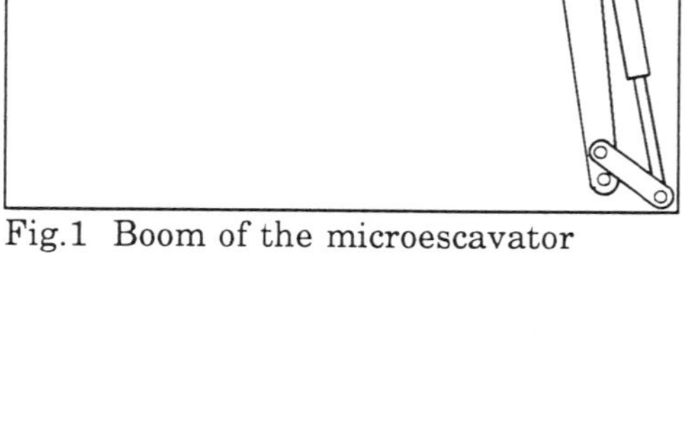

Fig.1 Boom of the microescavator

2. Dynamics of the boom

The motion equations of the boom and arm are given by the following Lagrange equations, the symbols of these equations are explained by fig. 2 .

$$T_1=(m_1h_1^2+m_1l_1^2)\ddot{\theta}_1+m_2l_1h_2\ddot{\theta}_2\cos(\theta sub2)-m_1gh_1\sin(\theta_1)$$
$$-m_2gl_1\sin\theta_1-m_2gh_2\sin(\theta_1+\theta_2)$$

$$T_2=m_2h_2^2\ddot{\theta}_2+m_2l_1h_2\ddot{\theta}_1\cos(\theta_1)+m_2l_1h_2\dot{\theta}_1\dot{\theta}_2^2\sin(\theta_2)$$
$$-m_2gh_2\sin(\theta_1+\theta_2)$$

The torques **T** are given by asymmetrical hydraulic cylinders, of which the forces are given by

$$F=a_1p_1-a_2p_2$$

where **a** and **p** are the area and pressures of two sides of piston.

Inverse kinematics of the link is carried out by simple relationships of a generic triangle made by two links and the distance between the origin and the final point.

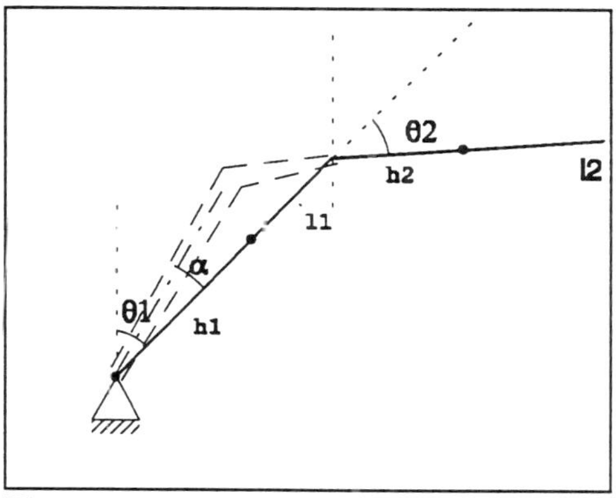

Fig. 2

3. Hydraulic system

The cylinder continuity equations have the usual form

$$\dot{p}=\frac{\beta}{V_0}(Q-av)$$

where β is the bulk modulus, **V** is the volume of a cylinder chamber, **a** is its area and **v** is piston speed.

The flows controlled by every servovalve are given by the following relationships

$$q_1=k_1 i\sqrt{p_s-p_1}$$
$$q_2=k_2 i\sqrt{p_2}$$

where **I** is the servovalve control current.

These relationships can be changed in linear form by

$$q=k_v i+k_p p$$

The servovalve can be represented by first or second order dynamic models, we use the first order model, in frequency domain it's

$$\frac{K}{1+s\tau}$$

while in time domain it's,

$$\tau\dot{q}+q=Ki$$

the valve parameters have been carried out by commercial specification.
A gear pump, driven by a.c. electrical motor supplies hydraulic power, we have neglected the dynamics of electric motor-pump-relief valve system, because electric motor has small r.p.m. change, while the supply pressure is supposed constant; moreover, we suppose that the flow, supplied by power system, is the necessary flow.
In fig. 3 is shown the block diagram of hydraulic system.

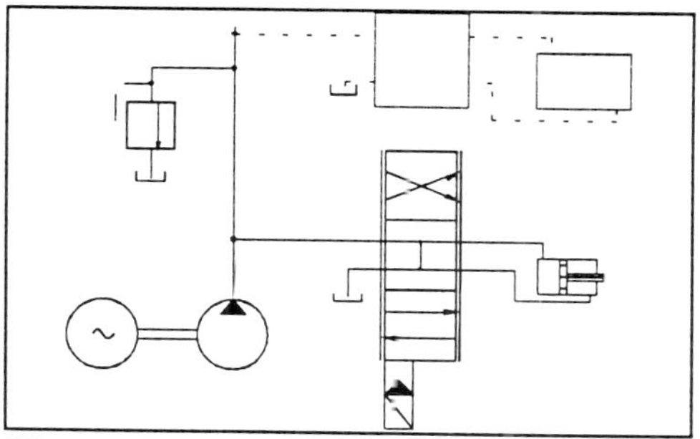

Fig. 3 Hydraulic system

4. Operator model

The operator is a multi impt-multi output controller, but is very difficult to find a MIMO models and they are linear usually, while the human behaviour is high not linear also.

We have used a model developed by Sutton and Towill for a crane, this is dual-mode hierarchical model, where for large control it uses bang-bang mode, while it uses tracking mode for small variations: we have modified tracking mode by a small bang-bang control.

A sample-and-hold and a delay complete the model.

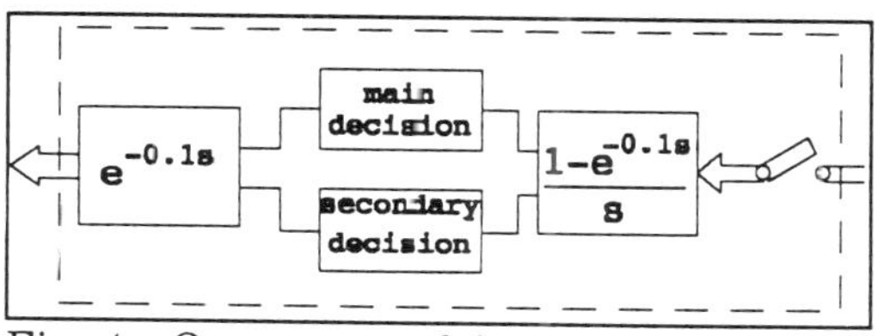

Fig. 4 Operator model

5. Simulation

The simulation has been made using a variable step Runge Kutta method used for linear system usually, but non-stiff methods to integrate differential equations systems are used in machines dynamics widely. As Kuczmarski and C. suggest, we have divided the integration time in three seconds long periods, so that the pressure transients should end, and using a RK method have con-

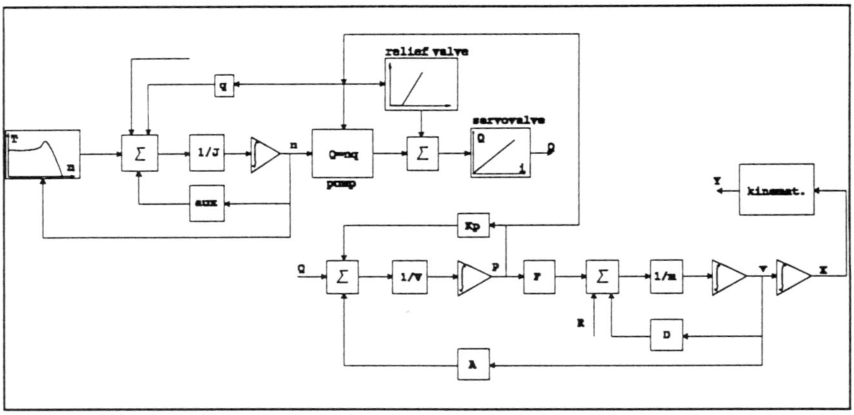

Fig. 5 Block diagram of a part of the system

tained the bang-bang control width.

The first aim, to verify a model of human operator, has been achieved, obtaining a positioning precision about 0.005 rad. for every link, which gives an error about 9mm for the boom and near one mm for the piston positioning: these results meet the results carried-out by Kuczmarski and C.

Also second aim has been obtained, to find possible problems, for exemple it isn't possible to use the continuity equation of ciylinder, when the valve return pressure is near zero, because the return pressure becomes negative by bang-bang control action, then we have replaced the continuity equation by the following

$$\frac{dp}{dt}=\frac{Ah_1\ddot{\theta}_1\cos(\theta+\phi)}{k_p}$$

where we have the piston area multiplied by piston acceleration, carried out from link acceleration, divided by flow-pressure valve coefficient (the return part of the valve).

6. References

F. Kuczmarski , T. Przychodzien, Accuracy of earthwork carried out by single buchet excavator, 3rd ISTVS European Conference, 15-17 September 1986

R. Sutton, Modeling Human Operators in Control System Design, RSP-John Wiley Inc., 1990

B. Paul, Kinematic and Dynamics of Planar Machinery, Prentice-Hall Inc.

H.E. Merritt, Hydraulic control System, John Wiley & Song Inc. 1967

Automation and Robotics in Construction X
G.H. Watson, R.L. Tucker and J.K. Walters (Editors)

Technology Strategy for Construction Automation

Keith D. Hampson[a] and C.B. Tatum[b]

[a] School of Construction Management, Queensland University of Technology, Brisbane, AUSTRALIA

[b] Department of Civil Engineering, Stanford University, Palo Alto, California, USA

Abstract

Automation technology can provide construction firms with a number of competitive advantages. Technology strategy guides a firm's approach to all technology, including automation. A more formal approach to the concept of technology strategy can benefit the construction manager in his efforts to remain competitive in increasingly hostile markets. This paper recommends consideration of five specific dimensions of technology strategy within the overall parameters of market conditions, firm capabilities and stage of technology evolution. Examples of the application of this framework in the formulation of technology strategy are provided for CAD applications, coordinated positioning technology and advanced falsework and formwork mechanization to support construction field operations. Results from this continuing line of research can assist managers in making complex and difficult decisions regarding their approach to new construction technology and benefit future researchers by providing new tools for analysis.

1. INTRODUCTION

The increasing technical feasibility of automation for construction brings major opportunities and requirements to develop approaches for the most beneficial use of this technology in firms. Automation technology will compete with human resources and other technologies for investment of scarce resources to continuously improve performance in construction firms. What approaches will bring the most improvement in the competitive performance of individual firms?

Whether explicitly stated or implicitly defined by managers' decisions and actions, the technology strategy for a firm guides the approach to all technology, including automation. Similar to human resource, finance, or marketing strategy, technology strategy is one of the functional strategies that support the competitive and corporate strategies of a firm. The purpose of this paper is to describe the elements of technology strategy for construction firms and apply these elements to the consideration of automation.

It is based on an ongoing investigation of technology strategy and competitive performance in infrastructure construction. This research will increase understanding of technology strategy and its relationship to competitive performance in construction, and support actions by construction researchers and professionals to improve the performance of construction firms through more effective use of technology. The following specific objectives structure this investigation: provide elements and measures to classify technology strategy in the construction industry, increase understanding of how technology strategy relates to competitive performance for infrastructure construction, and provide recommendations and techniques for analysis of technological alternatives and use of advanced technology to improve the competitive performance of construction firms.

The classification for technology strategy used in this research consists of five key dimensions: competitive positioning, sourcing of technology, technological scope, technical depth, and organizational fit. *Competitive positioning* is the relative command of specific technologies and monitoring alternative technologies potentially available from other sources. *Sourcing* is the acquisition of explicit (hardware) and implicit (knowledge) technologies. The *scope* dimension identifies the core and peripheral technologies for the firm. *Depth* highlights relative level of investment and extent of technical capabilities. *Organizational fit* includes the match of reward systems with technological objectives, the flow of technical information, and the matching of capabilities to project requirements. Each of these dimensions guides the construction manager's approach to the acquisition and implementation of technology.

This paper describes the application of this classification for technology strategy to construction automation. Since the concepts of technology strategy and automation technology are new to construction firms, this paper describes the process of formulating technology strategy and possible approaches for each of the dimensions related to construction automation and mechanization. This analysis of technology strategy for automation includes examples of CAD application to support construction and positioning technology and mechanization to support field operations. The conclusions highlight the steps and options in formulating technology strategy, and the benefits for competitive performance of construction firms.

2. TECHNOLOGY STRATEGY

2.1. Technology and Strategy

For this research we define *construction technology* as the combination of *resources, processes and conditions* that produce a constructed product. A resource can be either materials and permanent equipment or a temporary addition by construction. Construction processes are the methods and tasks needed to build a constructed product. Project requirements and site characteristics are the major conditions of construction technology (Tatum, 1988).

A *strategy* is a unified, comprehensive and integrated plan relating the strategic advantages of the firm to the challenges of the environment. It is designed to achieve the basic objectives of the enterprise (Glueck, 1980). Chandler (1962, p13) defined strategy as "the determination of the basic long-term goals and objectives of the enterprise and the adoption of courses of action and the allocation of resources necessary for carrying out these goals". To define *technology strategy,* Adler (1989) adapted Andrew's (1980) definition of business strategy: *technology strategy* is a pattern of decisions that sets the technological goals and the principle technological means for achieving both those technological goals and the business goals of the organization.

2.2. Elements of Technology Strategy

A range of determinants for technology strategy provided by others (Ansoff and Stewart, 1967; Freeman, 1982; Miles and Snow, 1978; Maidique and Patch, 1988; Burgelman and Rosenbloom, 1989; Malekzadeh et al, 1989) has been reviewed with the objective of selecting one set of parameters that ensures conceptual completeness, but avoids unnecessary duplication. The following technology strategy dimensions synthesizes the background:

- ***competitive positioning*** refers to the role that technology plays in establishing generic competitive advantage or how product technology, process technology and technical support capabilities are used to achieve cost leadership or differentiation (Porter, 1985). Maidique and Patch (1988, p241) raise the managerial question of "what new technologies to add to the firm's core set of technological skills?" Three factors that guide strategy in an important technology are: sustainability of the technology lead, first-mover advantages, and first-mover disadvantages (Porter, 1985).
- ***sourcing of technology*** incorporates the dual (but overlapping) concepts of Burgelman and Rosenbloom's "Technology and the Value Chain" and the Malekzadeh et al "Sources of Technological Capability". To remain competitive a firm must structure ways to acquire certain value-creating technologies, i.e.. the develop or buy argument. This decision to acquire new technology is, as described by Rosenberg et al (1990, p22), "inescapably an investment decision".
- ***scope of technology*** strategy describes the set of product and process technologies in which the business unit has invested combined with an assessment of the maturity of the technologies (Malekzadeh et al, 1989). No firm can hope to operate on the frontiers of all technologies relevant to its operations. The set of technologies that can have a material impact on its competitive advantage are *core* technology; all others are *peripheral.* Of course in a dynamic world, peripheral technologies can become tomorrow's core technologies and vice versa (Burgelman and Rosenbloom, 1989).
- ***depth of technology strategy*** encompasses the number of technological options the firm has available. The depth of a firm's technology strategy is determined to a significant extent by the intensity of its resource expenditures. Investment in technological resources is positively related to technical output (Mansfield, 1981). Expenditure on internal R&D has a strong relationship to increased profitability (Capon et al, 1990).
- ***organizational fit*** encompasses administrative and cultural factors of the firm. Patterns of technical communication throughout the construction organization are important determinants to the quality of organizational fit. From the Burgelman and Rosenbloom (1989) evolutionary point of view, organizational context serves as an internal selection mechanism affecting the strategic management of the firm. Burgelman and Sayles (1986) work suggests that reward systems may provide potential measures of one aspect of the organizational context of technology strategy.

2.3. Link Between Technology Strategy and Performance

The background literature contains examples of relationships between technology-related inputs and competitive performance. The Miles and Snow (1978) typologies include a technology-based component. The four basic patterns of strategy that emerged however, are principally based on a descriptive analysis of how an organization changes its products or markets. While each of these cases is of interest in their own right, no linkages are apparent corresponding to the recently unified concept of technology strategy which comprises a diverse range of inputs and competitive performance.

3. FORMULATING TECHNOLOGY STRATEGY FOR AUTOMATION

Selecting a firm's approach to a specific technology requires consideration of each element: positioning, sourcing, scope, depth, and organizational fit. This section describes a process of formulating technology strategy, primarily using examples of CAD applications in construction, coordinate positioning technology and highly mechanized falsework/formwork systems. By considering each of the elements for several candidate technologies, managers can formulate a technology strategy that best fits the capabilities and culture of a firm and creates substantial competitive advantages.

3.1. Competitive Positioning

Automation technology can provide several types of competitive advantages. Examples include: offering new products in new markets, developing distinctive or unique competence, serving a niche market, reducing project durations, and decreasing life cycle cost (Tatum, 1989a). Analysis of customer needs, types of contracts, and external influences on markets provides a basis to select a type of competitive advantage for a firm. This analysis can identify the competitive advantages that are desirable for the market conditions and feasible for the specific firm, along with the types of technologies that can provide this advantage.

Using CAD technology at the construction site to increase the effectiveness of construction engineering activities can create competitive advantages based on decreased project duration and decreased construction cost. For example, Construction Engineers can use CAD at the site to automate existing processes or to create new processes and products that allow more effective support of field engineering operations (Mahoney et al, 1990). Examples of using CAD to automate existing processes include planning survey control and layout, planning construction sequences and methods, and coordinating subcontractors. In each case, CAD drawings allow more complete analysis and improved communication of plans. Examples of using CAD for new processes and products include visualizing complex configurations, analyzing requirements for construction equipment, and detailing portions of the structure to allow the use of specialized construction methods. Each of these applications results in more complete technical information for field operations.

Automated positioning technology can provide advantages based on offering new products in new markets or developing distinctive competencies. For example, applications of automated positioning in the grading and paving industry could allow the accurate preparation of pavement subgrade and base, and permit placing the asphalt or concrete wearing surface to comply with more stringent tolerances for pavement location and surface finish. The high cost of concrete grinding and set-up labor for traditional equipment encourages use of this type of positioning technology through to decrease total construction cost.

Unique or distinctive competence can also be achieved within the concrete highway bridge construction sector through the development of mechanized falsework and formwork systems that address the industry drivers of labor cost and operational safety. For example, a series of fabricated steel trusses that efficiently combine the traditional functions of falsework and formwork and allow construction of a new double-deck freeway level to proceed over an existing busy highway provides clear competitive positioning advantages for its owner - especially when the investment in this more advanced construction mechanization is carried out during the inception of this type of continuing work. The entire system could also include prefabricated internal formwork and incorporate concrete placing and finishing equipment - all enclosed by debris panels to allow construction to proceed safely during normal working hours. When future project phases of a similar nature are bid, the strategic early investment in labor-saving construction automation/mechanization can provide unmistakably clear benefits.

3.2. Sourcing of Technology

Several options are available to acquire new technology for automation. These include: interact with a lead developer, conduct internal development, improve processes for repeated operations, drive technology development on specific projects, and interact with owners and operators (Tatum, 1989a). These approaches involve plans and commitment of resources to acquire a new technology and effectively use it. The most critical resources are not always financial - commitment of time by operations managers or field staff may be more important for success than funds.

Several different approaches for acquiring CAD to support construction illustrate the breadth of options available (Mahoney et al, 1990). The president of one firm bought hardware and software for CAD, put it on the site of a major projects, and instructed the staff to find productive uses. Illustrating a completely opposite approach, a field engineer in another firm used his personal computer and software to demonstrate beneficial applications of CAD in preparing lift drawings and planning concrete placements. In other firms, the home office engineering staff or consultants took the actions needed to acquire the technology, implement it, and demonstrate the advantages.

Throughout the development of the mechanized falsework system for double-decking existing freeways, the principal of one bridge construction firm maintained close personal involvement. (His employees jokingly referred to him as their "chief falsework designer".) Following the initial conceptual design carried out in-house, a number of external specialists were called upon to complete the detailed engineering and fabrication functions.

Co-operative interaction with the public facility owner through a value engineering scheme proved central in encouraging innovation from the construction community in developing alternative designs. The magnitude of one $52 million project encouraged the development of advanced falsework and formwork systems with almost 50 complete repetitions of the span-by-span construction of the 67' (20.4m) wide deck. The motivation to improve the processes for repeated operations enabled continual incremental improvements. This also placed the contractor in a commanding position for future similar contracts.

In an example of sourcing implicit technology (represented by knowledge or skills embodied in people), another construction firm took over the key staff from a retiring specialist subcontractor when they made the strategic decision to merge piledriving into their own existing structural concrete construction capabilities. Experienced (and committed) skilled labor was therefore smoothly integrated into their own organization, with the benefit of having witnessed the skills and work attitudes of the potential employees over a decade of close industry involvement.

3.3. Scope of Technology Strategy

Applications of construction automation provide examples of variations in the scope of construction technology. Although there are many examples of beneficial applications of CAD to support construction engineering, it remains a peripheral technology compared with more traditional equipment and tools used in construction operations. Computer applications to support project planning, materials management, and performance monitoring are additional examples of peripheral technologies. However, several types of construction firms illustrate the use of advanced equipment and early construction automation as core technologies. These include laser positioning for fine grading (Tatum and Funke, 1988), partially automated devices for underwater bolting (Stewart and Tatum, 1988), specialized equipment for transporting and setting heavy equipment (Tatum, 1983), and automated welding machines for large plate work and high pressure piping. For each of these technologies, several firms have developed a core competence and applied the technology to perform work that others cannot perform or to gain substantial schedule and cost advantages.

Another example reflects the potential benefits gained through the scope of abilities embodied in the firm's people or through strategic alliances with external professionals. Strong in-house engineering skills, for instance, can reveal opportunities available through optimizing falsework materials, and through redesigns under the province of value engineering.

One highway construction firm has demonstrated a successful integration of site project control through the on-site application of computer project programming and job costing. The progress of job sectors or individual operations can be tracked almost daily to allow potential problems to be highlighted and corrections applied, before they significantly impact project success.

3.4. Depth of Technology Strategy

Construction applications of CAD illustrated variations in depth of technology strategy. The commitment and level of resources ranged from a major investment in multiple high-end workstations for one project, to purchase of software compatible with the low-end personal computers designated for administrative use at the site (Mahoney et al, 1990). The core technologies identified above illustrate the increased depth necessary to achieve distinctive competence.

Infrastructure construction firms benefit through their ability to provide prompt and effective technical response to site problems. Two differing approaches characterize successful performance in this area:

- technically qualified and experienced supervision operating independently at the construction site, and;
- production-oriented and trade experienced supervision based on site - with technical expertise, e.g. engineering - based in home office, but allocated to specific projects and readily accessible by phone, fax, or radio (potential for video communication for future). In this manner, the depth of professional skills based at head office can be effectively spread over a number of projects - thereby lowering job overheads.

Expenditure on R&D in the formal sense is minimal in the highway infrastructure firms surveyed. However, in a similar way that Nam and Tatum (1992, p522) have reported in their evaluation of the source of key technical ideas, a substantial number of innovations and enhanced project automation or mechanization comes from informal R&D activities described "as a clever collection, transfer, or improvement of already-known technologies". The innovative investment in the trussed falsework system for double decking freeways is an example of more advanced construction process hardware developed in this fashion.

3.5. Organizational Fit

Several examples of construction innovation from heavy and building construction illustrate the importance of the organization in creating the environment necessary for technical change (Tatum, 1989b, Nam et al, 1991). Key actions in organizing for innovation include establishing supportive policies and priorities, maintaining flexibility in grouping and providing technical resources, facilitating intra- and interorganizational coordination, and staffing to satisfy specific requirements for key positions (Tatum, 1989b). Further, linking personal reward structures to the technological objectives of the firm serves as a meaningful signal to employees (Burgelman and Sayles, 1986).

Examples of CAD applications on construction sites illustrate the types of organizational environments that foster the adoption of new technology. These included strong support by senior managers, established technical excellence in the firm, and willingness to allow experimentation with new approaches. These innovative organizations lacked rigid structure and position descriptions, fostered individual initiative and experimentation, and included several types of champions (Mahoney et al, 1990).

More successful infrastructure construction companies demonstrate a number of consistent organizational characteristics including: strong personal involvement by principals, an established core of technical competence in the firm (either at the home office or on site), personal and group reward structures that reflect productivity and safety improvements, and personal communication networks unimpeded by either project location or organizational responsibility.

Since labor is the largest single cost component of highway bridge construction, a technology strategy to maximize productive output per employee through construction automation and mechanization is of paramount importance to enhance company performance in this competitive environment.

4. CONCLUSIONS

Technology strategy guides the construction firm's approach to the consideration and implementation of technologies within the context of the overall competitive and business goals of the organization. This paper has identified and summarized elements of technology strategy synthesized from the background literature. It then provided examples of its application to the integration of CAD, co-ordinate positioning, and site mechanization in construction activities.

The five key dimensions to be considered in formulating an appropriate technology strategy are: competitive positioning, sourcing of technology, scope of technology, depth of technology strategy and organizational fit. Through selecting a strategy that best suits the existing capabilities of a firm, and addressing the construction sector market forces, managers can be better prepared to face the increasingly competitive environment in which they operate. Formulating a technology strategy cannot guarantee market success. But the diligent application and implementation of the technology strategy concept can provide the foundation for the creation of substantial competitive advantage.

5. REFERENCES

- Adler, Paul S, "Technology Strategy: A Guide to the Literatures", in *Research on Technological Innovation, Management and Policy*, Rosenbloom, Richard S. and Burgelman, Robert A. (eds), Vol 4, JAI Press, Greenwich, CT, 1989, pp25-151.
- Ansoff, H.Igor and Stewart, John M., "Strategies for a Technology-Based Business" in *Harvard Business Review*, Nov/Dec, 1967, pp71-80.
- Burgelman, Robert A. and Rosenbloom, Richard S., "Technology Strategy: An Evolutionary

Process Perspective" in Rosenbloom, R.S. and Burgelman, R.A., (eds), *Research on Technological Innovation, Management and Policy*, Vol 4, Greenwich, CT, JAI Press Inc, 1989.

- Burgelman, Robert A. and Sayles, Leonard R., *Inside Corporate Innovation: Strategy, Structure, and Management Skills*, The Free Press, New York, NY, 1986.
- Chandler, A.D., *Strategy and Structure*, MIT Press, Cambridge, MA, 1962.
- Freeman, Christopher, *The Economics of Industrial Innovation*, 2nd ed., MIT Press, Cambridge MA, 1982.
- Glueck, W.F., *Strategic Management and Business Policy*, McGraw-Hill, New York, 1980.
- Mahoney, John J., C. B. Tatum, and Kenji Kishi, "Construction Site Applications of CAD," *Technical Report No. 36, Center for Integrated Facility Engineering*, Stanford University, Stanford, California, October, 1990.
- Maidique, M.A. and Patch, P., "Corporate Strategy and Technological Policy", in Tushman, M.L. and Moore, W.L. (eds), *Readings in the Management of Innovation*, 2nd ed, Ballinger, 1988, pp236-248.
- Malekzadeh, Ali R; Bickford, Deborah J. and Spital, Francis C., "Integrating Environment, Competitive Strategy, and Structure with Technology Strategy: The Strategic Configurations", *Academy of Management 49th Annual Meeting*, Washington DC, August, 1989.
- Mansfield, Edwin, "Composition of R&D Expenditures: Relationship to Size of Firm, Concentration and Innovative Output", *The Review of Economics and Statistics*, Vol 18, Iss 4, November 1981, pp610-615.
- Miles, Raymond E. and Snow, Charles C., *Organizational Strategy, Structure and Process*, McGraw-Hill, New York, 1978.
- Nam, C. H., J. G. Gasiorowski, and C. B. Tatum, "Microlevel Study of Integration in High-Strength Concrete Innovation," *Journal of Construction Engineering and Management*, ASCE, Vol. 117, No. 2, June, 1991, pp294-309.
- Nam, C. H., and C. B. Tatum, "Strategies for Technology Push: Lessons from Construction Innovations," *Journal of Construction Engineering and Management*, ASCE, Vol. 118, No. 3, September, 1992, pp507-524.
- Porter, Michael E., *Competitive Advantage: Creating and Sustaining Superior Performance*, Free Press, New York, NY, 1985.
- Rosenberg, N.; Ince, P., Skog, K. and Plantinga, A., "Understanding the Adoption of New Technology in the Forest Products Industry, *Forest Products Journal*, Vol 40, Iss 10, October, 1990, pp15-22.
- Stewart, William S., and C. B. Tatum, "Segmental Placement of the Renton Outfall: A Construction Innovation," *Journal of Construction Engineering and Management*, ASCE, Vol. 114, No. 3, September, 1988, pp390-407.
- Tatum, C. B., "Heavy Rigging for Nuclear Projects," *Journal of Construction Engineering and Management*, ASCE, Vol. 109, No. 1, March, 1983, pp48-60.
- Tatum, C. B., "Classification System for Construction Technology," *Journal of Construction Engineering and Management*, ASCE, Vol. 114, No. 3, September, 1988, pp344-363.
- Tatum, C. B., "Design and Construction Automation: Competitive Advantages and Management Challenges," *Proceedings of the 6th International Symposium on Automation and Robotics in Construction*, San Francisco, June 6-8, 1989a, pp332-339.
- Tatum, C. B., "Organizing to Increase Innovation in the Construction Firm, *Journal of Construction Engineering and Management*, ASCE, Vol. 115, No. 4, December, 1989b, pp602-617.
- Tatum, C. B. and A. T. Funke, "Partially-Automated Grading: A Construction Process Innovation," *Journal of Construction Engineering and Management*, ASCE, Vol. 114, No. 1, March, 1988, pp19-35.

Automation and Robotics in Construction X
G.H. Watson, R.L. Tucker and J.K. Walters (Editors)
1993 Elsevier Science Publishers B.V.

Robot Implementation Decisions in the Australian Construction Industry

Cecily Neil, Gustav Salomonsson and Miroslaw Skibniewski

CSIRO Division of Building, Construction and Engineering
P.O. Box 56, Highett, Victoria 3195, Australia.

Abstract

This paper presents a model for managers to use to assess the feasibility of robotising particular tasks on a construction site. A case study illustrates the application of the model with respect to the use of concreting robots on a construction project. This case study highlights the importance of taking into account the overall work process on a site when considering the use of robots.

1. INTRODUCTION

This paper presents a model for appraising the feasibility of robotising particular tasks on specific construction projects. The model was developed from a series of case studies, the results of one of which are presented here. The robots chosen for the prior-use evaluation in the latter study were concrete screeding and finishing robots.

2. PRIOR-USE EVALUATION OF THE AUTOMATED EQUIPMENT

From the work of Blackler and Brown[1], it is possible to define three essential components in prior-use evaluation of new automated construction technologies:

- the direct costs or benefits affecting the immediate project application, as defined by evaluation of both cost-substitution and value-added benefits;
- the evaluation of the wider effects of the new technology on the organisation adopting it; and
- the evaluation of what is likely to be required to ensure a smooth implementation process within the company.

2.1. Direct cost/benefits

Calculating the cost-substitution component of the evaluation of a new piece of automated equipment will involve comparison of the relative costs of:

- the labour, using the two types of equipment;
- the capital involved in the purchase or leasing of the two types of equipment;
- the operating costs for the two types of equipment;
- material waste that can be attributed to the use of the two forms of equipment; and
- any rework that may be influenced by the use of either form of equipment.

Comparison of the relative labour costs will depend firstly on the ratio of the number of person hours required to carry out the task by conventional methods, to the number of person hours required to carry out the task using the new equipment. However, for most tasks, the use of automated equipment would not exclude all manual work: finishing around columns and in corners with concrete screeding robots, for example, would require some manual input. Thus, the design characteristics of a particular project will be significant in determining the savings to be gained from reduced labour requirements. Labour use logistics will also be particularly important in determining relative costs: any savings will be contingent on effective use of all labour required on the site for the initial preparatory or finishing-off work associated with the tasks to be undertaken by the new equipment.

If less labour is required when using a new technology, other benefits may also arise, such as savings from reduced site infrastructure requirements or from a reduced number of stoppages affecting the overall construction schedule – especially if the completion of tasks to be undertaken using the new technology is critical to the production schedule, or if there is a shortage of labour with the appropriate skills. Similarly, the use of robots on sites with a restricted local labour market can reduce the cost of recruiting, transporting and housing labour.

At the same time, however, any additional costs involved in acquiring labour with the necessary skills to operate the new technology will also need to be taken into account in calculating cost-substitution savings. This may be particularly relevant in those companies which are encouraging multiskilling, and pay workmen at the highest rate for which they have trained, regardless of the rate for the particular task they might be undertaking.

The value-added component of the evaluation will depend on:

- the relative match between quality desired and quality that can be achieved using the two forms of equipment; and
- any changes that may occur in customer service (eg improved schedules, better predicability of productivity and completion dates) as a result of using the new technology.

2.2. The effects on the organisation of the new technology

In addition to potential cost-substitution and value-added benefits, the introduction of a new technology may have a number of other organisational impacts, although, for the most part, these will be consequences which cannot be directly quantified. They may include:

- an impact on the health and safety of workers;
- the potential for upgrading an organisation's technological image, which may translate into an improved relationship with construction owners and increasing competitiveness;
- the potential for business expansion into new areas and ventures;
- the impact of any necessary job redesign on the job satisfaction of end-users; and
- any potential impact on the balance of power among different groups in a company.

2.3. Organisational factors influencing the implementation process

Whatever advantages might potentially accrue from using a new technology, the effective realisation of these advantages is likely to be contingent on its smooth introduction.

Labour organisation will be particularly important in this respect. Where labour is organised on a craft basis, if there is any uncertainty over which union's members should use the new equipment, the costs of preliminary negotiations with multiple unions would need to be considered in any evaluation. Australian construction labour is represented by craft unions, which, in the past, have exercised rigid work demarcation. However, recent movements have been made towards multiskilling. It is now proposed that all structural work, for example, will ultimately be carried out by the same workers, with no formal demarcation. If successfully implemented, multiskilling could resolve most of the potential demarcation problems likely to be associated with the use of robots. It would also assist the effective use of labour that might be released from the task that has been automated. Multiskilling may also ensure that training in the use of a piece of automated machinery is not confined to one section of the workforce.

Different styles of management can also be ordered in terms of how effectively they lend themselves to the initial phases of introducing new technologies. The consultative style, for example, is likely to result in the least problems in actual implementation.

A number of other organisational factors are also likely to influence the chances of a smooth implementation process. The policy of using innovative techniques wherever appropriate is likely to facilitate the introduction of a new technology, especially if a company already has an effective procedure in operation for evaluating new technologies. Companies with existing on-site training programs will have a clear advantage if these training programs can be readily modified to incorporate modules relating to the use of the new equipment. Unless its introduction has reached the point where its use is accepted as part of the normal routine, the willingness of employees to use a new technology on-site will also depend at least in part on their willingness and capacity to retrain. Thus, workers' experience of, and response to, using automated equipment, and their computer literacy and attitudes to using computers, will all be relevant. Any language barriers may also present problems either for training or retraining, or for ensuring that guidelines for the safe use of a new technology are fully understood. Finally, the company's equipment acquisition policy, and its cash flow model for the use of equipment, will help determine whether the company can afford to lease or purchase new equipment.

Ultimately, the feasibility of introducing a robot may depend on the extent to which the appropriate characteristics for its smooth introduction already exist in a company, or, if not, whether certain organisational changes are warranted in order to enable this to take place.

3. A CASE STUDY APPLICATION OF THE MODEL

The case study chosen to illustrate the above prior-use evaluation involved examination of the concreting work in a $30 million project consisting of the renovation and addition of five floors to an existing building. The site was located in the central business district of a major city. Although access was limited, this would not have hindered the use of smaller robots.

The contract was based on a fixed price, and the time allowed to tender for the contract was not sufficient to permit a non-conforming bid. Thus, there was no opportunity for the contractor to modify the design to suit any technology he might wish to use. Most tasks on this project, including the concrete work, were subcontracted to specialist subcontractors.

3.1. Basic characteristics of the concreting task

The construction techniques employed on this project were conventional. The structure to be built was reinforced concrete, and involved a total slab area of 22 500m^2. The concrete was placed using the combination of a concrete pump and a crane with bucket. The average work cycle to complete a floor level was 1.5 weeks, with 2 pours per week. The surface finish was obtained using walk-behind power trowels, although the numerous column and bracing wall sections in the layout meant that approximately 25 per cent of the slab area was inaccessible to the power trowels, and had to be finished by hand trowelling. The average size of a pour was 1000 m^2. There were some 23 pours completed by a team of 11-12 men.

3.2. Potential for using robots

A prior-use evaluation was carried out for a concrete levelling and screeding robot and a finishing robot. The machine considered for the levelling and screeding was a vehicle-mounted screeder developed by Takenaka. The machine deemed most applicable for the concrete finishing on this site is one manufactured by Schimizu. Although there are well-established travel path algorithms to cover even the most complex areas[2] that would have enabled programmable robots to have been used even on this particular job, the machine selected is remotely controlled by an operator. On jobs such as this, the concrete slab areas are frequently of varying depths. The uneven exposure to the elements due to the surrounding buildings can cause an uneven initial curing resulting, as it did on this job, in large wet patches located in the areas otherwise ready for trowelling. A remote-controlled machine would have provided an opportunity for the operator to detect any such patches of concrete not properly cured, and respond accordingly.

3.3. Prior-use evaluation from the perspective of the subcontractor

The subcontractor awarded the tender for the structural work was a family firm which, six years prior to this project, had been set up to undertake full structural packages. At the time of the project, the subcontractor had undertaken projects ranging in value upwards to $18 million for many of the local leading firms in the construction industry. Five separate projects were being carried out by the subcontractor at the same time as this contract, although the amount of work on hand at any point of time was variable. The largest contracts undertaken had involved 50 to 60 of the subcontractor's employees; the smallest, around 7 to 8.

The subcontractor maintained a permanent workforce, numbering around 140 to 160. The average length of service of the subcontractor's employees was around three to six years.

3.3.1. Direct cost/benefits: cost-substitution savings

Overall labour requirements: The concrete pour was usually commenced early in the morning. Levelling, screeding and finishing started after a suitable time lag, and ultimately involved the whole team that had initially worked on the pour. Work would usually finish in the late evening. While not all men necessarily worked an equal number of hours on any one pour, each team member was paid according to the total number of hours the pour took. Hourly cost for concrete labour was approximately $30, including all overheads.

The use of a concrete screeding robot on this project would have required an operator plus one person to assist. Concrete finishing robots require at least one skilled worker to operate the machine, set up and dismantle the robot on a site, and ensure that proper transport procedures are strictly followed. The cost of operator labour may be somewhat higher than that of a traditional concrete labourer, given the specialised training required, but should not exceed $45 per hour. However, it would be necessary to train more that one person in a team to operate each robot. Given the subcontractor's policy of paying his men at the rate appropriate for the highest level of skill they have attained, this could to some extent inflate the general labour costs among those not operating a robot at any given time.

It was estimated that due to the complex floor layout, approximately 25 per cent of the work would still need to be done manually, and that two people would be required for this work.

Labour logistics: The use of concrete screeding and finishing robots would lead to some reduction in labour costs through enabling the entire team to complete the concrete finishing in fewer hours than required using conventional methods. However, any additional labour cost savings would depend on efficient labour utilisation. The logistics of efficient labour usage in conjunction with the use of the robots for screeding and finishing seemed to present the greatest barrier to labour cost savings through the use of the concrete robots on this site. Using screeding and finishing robots would reduce the number of men required after the concrete placement had been carried out. However, given the way the work was organised with respect to the concreting team, labour cost benefits could only accrue from using a robot if:

- work could be found for the men responsible for pouring concrete who were surplus after the pour had taken place; or
- the number of people required for the initial concrete placement work could be reduced.

The subcontractor felt that it might have been possible to reduce his team size on this project by 2, and still place the concrete efficiently. However, factors such as concrete truck delivery delays and intermittent site inaccessibility due to traffic jams would have made making any substantial reduction in team size particularly difficult.

The alternative of finding work elsewhere for surplus team members would have proved much more difficult. It would have necessitated:

- the use of a multiskilled work force, which could be used on alternative skilled work;
- the use of unskilled labour on the concrete pour, with the possibility of using this labour for unskilled tasks when the pour was finished; or
- the company having a sufficiently large portfolio of projects with pours on the same day, so that a crew could finish pouring on one site, and then move to another for a second pour.

Given that the subcontractor was contracted to undertake all the structural work on this project, the first of these two options appears to have been a viable long-term option, although labour was not yet sufficiently multiskilled for this to have occurred on this project. The second option was also theoretically viable, but the deskilling of the work and breaking up of the team organisation would have been unpopular. The logistics of the third option made it unviable. Thus, on this particular site, productivity savings could not have been fully realised, and potential labour cost savings were, therefore, substantially reduced.

Equipment cost: The type of equipment costs to be compared included those associated with the robots on one hand, and those of manual screeders and engine-powered helicopters on the other. The purchase cost of the types of robot discussed here, when commercially available, was estimated to be likely to be around $50 000. This figure would have been financially viable for the subcontractor, and would not have had to be offset against a single project in order to avoid cash flow problems. The cost of owning and operating each of these machines was estimated to be likely to be around $400 per day, but will vary with the supply and demand.

The cost of the dumping levellers, straight-edge boards and vibrators is very low, and was therefore not included in this comparison. The purchase price for the most commonly used helicopter trowels is around $1500, rising to $2500-3000. These types of machines usually remain operational for many years with a minimum of maintenance, and are very economical to run. Discussions with concrete contractors and hire companies indicate that an average cost of owning this equipment, including fuel, would not exceed $50/day.

Rework: On some sites, the use of concrete screeding and finishing robots may reduce the total amount of rework necessary, thus reducing cost and minimising schedule delays. However, most of the concrete rework that had to be done on this site was patching necessary as a result of rain that had occurred after pours had taken place. According to the subcontractor, none of the rework that had been necessary could have been avoided by the use of robots.

Material waste: Material waste in the concrete work on this project had been around 0.2%, and again, would have been unlikely to have been reduced by the use of robots.

Table 1
Summary of comparative assessment

Cost category	Traditional work ($)	Cost category	Robot assisted ($)
Concrete Pump (8 hour)	1 920	Concrete Pump (8 hours)	1 920
Crane (1 day)	400	Crane (1 day)	400
Crane Crew (9 hours)	1 260	Crane Crew (9 hours)	1 260
Traditional labour (12 men/11 hours/$30)	3 960	Traditional labour (8 men/9 hours/$30)	2 160
		Robot operators (2 men/9 hours/$45)	810
		Screeding robot (1 day level & screed 750 m^2)	400
Helicopter Trowels (2)	100	Trowelling robot (1 day trowel 750 m^2)	400
Total Traditional	7 640	Total Robots	7 350
Traditional/unit slab	7.64/m^2	Robotic/unit slab	7.35/m^2
Job total - traditional (22500 m^2)	171 900	Job total - robotic (22500 m^2)	165 375
Difference			- 6525

Summary cost-substitution savings: Overall, cost-substitution evaluation suggests that the total direct substitution benefit in this case would have been around $6525 (Table 1). The savings in these initial calculations are those likely to be accrued after the employees have reached the top of the learning curve for the use of the particular robot. Since this is not likely to be achieved the first few times the machines are used, the calculations to some extent overstate the labour savings likely to have accrued on this particular project.

3.3.2. Direct cost/benefits: value-added benefits

Schedule savings: Since concrete placement operations are usually on the critical path in construction, shorter performance times may translate into a cash bonus if it means project delivery ahead of schedule. However, the size of the pours possible on this site was a major determining factor in the scheduling of concrete finishing, and it was unlikely that even if finishing times could be reduced, it would have been possible to increase the frequency of pours. In some situations, the shortened performance time for finishing might also help in project resource allocation should shortages of skilled labour take place in the course of the project. However, this was not a problem on this site.

Stoppages and delays: The subcontractor felt that this particular project had been one of the fastest jobs he had undertaken for the design type. There had been no industrial stoppages affecting the concrete work, and the only serious delays that had occurred had resulted from the weather. Thus, the availability of the robots would not have substantially reduced time lost.

Improved work quality: Robotic concrete finishing may sometimes translate into a better quality concrete surface by avoiding the human error inherent in the traditional performance of these tasks. According to both the contractor and the subcontractor, however, there had been no problems achieving the desired quality of concrete finish on this project.

3.3.3. Organisational effects of introducing robots into the company

Improved health and safety of workers: There were no major accidents on this site. However, concrete levelling and screeding are physically demanding tasks, often requiring brute force to disperse large volumes of concrete mix on the floor surface. Cements used in concrete mixtures may contain varying doses and types of fly ash, some of which are slightly radioactive. In addition, cases of human bone and ligament disorders caused by handling work tools subject to low frequency vibrations have been reported; concrete mix vibrators may, therefore, create a potential health hazard to the workers using them. Hearing and back problems are also prevalent among concreters. The robotic performance of concrete tasks may, therefore, lead to a long-term reduction in various health problems and hearing problems.

Impact on labour supply: By reducing the number of men necessary to do the concreting, the use of robots can reduce any problems that might exist with respect to the supply of skilled labour. According to a director of the subcontracting firm, there has been a problem in the supply of skilled concreters over the last couple of years, reflected in the fact that concreters are among the oldest of the company's employees. He foresaw an increasing problem in recruiting sufficient skilled concreters in the next ten years - although it has been suggested that this may change with the growth of multiskilling. The introduction of screeding and finishing robots could, however, help to forestall any possible impending shortage of skilled labour.

Potential for business expansion: Potential for business expansion as a result of a company's introduction of robots onto this construction site could come from several sources, including:

- a greater capacity to win contracts as a result of an improved technological image; and
- the acquisition of new workforce skills that could either improve competitiveness, or make it possible to tender for work not previously undertaken;

Since the subcontractor was already winning tenders with a number of major contractors, the chances were that any experience gained through the use of sophisticated automated equipment on one project could readily be turned to an advantage on other projects. However, the firm already used some of the most up-to-date equipment available, and the general consensus in the industry was that while technological sophistication might impress a client, in the current economic climate it was not a factor likely to be particularly significant in the winning of tenders: the only factors the owners consider important are tender price and the contractor's reputation as a solvent business with a record reflecting an ability to deliver the required services. Further, in the current economic climate, the scope for moving into new areas of construction are limited. Nonetheless, the company's acquisition of skills relating to the use of robots, if appropriately integrated into the work organisation, could significantly improve overall productivity, and thereby be a worthwhile asset with respect to obtaining subsequent contracts. The subcontractor's labour retention policy ensured that skills acquired by his employers would be retained in the company.

Impact of any necessary job redesign on the job satisfaction of end-users: Since the robot operators would still be very much part of the team work, they were unlikely to become isolated in the work context, as can happen with the use of robots in mining, for example. However, the subcontractor's use of robots on this site would have influenced the job design of concreters who became robot operators for this project in several ways:

- the work would have been less strenuous and dirty;
- the impact of most robot operators' personal skills on any one project would have not been so much on the quality of the finished product, but on their skill in keeping the job on schedule through efficient programming and set-up and dismantling;
- there could have been a potential reduction of individual initiative and decision-making, depending on whether the robot operator was employed predominantly as a machine minder, or trained to program and do minor maintenance work;
- the training to use robots would almost certainly have resulted in pay increases; and
- the introduction of robots could have introduced the opportunities for personal advancement, and continued work opportunities for older workers or workers who had sustained previous injuries that restricted their ability to undertake strenuous work activities.

The actual changes to the job design that would have come about for concreters working on this and future projects utilising robots, had concrete screeding and finishing robots been introduced, would have depended on the way in which work arrangements were reorganised, the number of men trained as robot operators, and the frequency with which trained operators were likely subsequently to be using robots, rather than carrying out other structural work. These factors, together with the current sources of job satisfaction and work values of operators, would have, in turn, influenced the impact of using robots on job satisfaction.

The decisions regarding the reorganisation of concreters' jobs would have needed to have been carefully made if this company were to have used robots. In particular, the potential disadvantages of training only a few men would have had to be weighed against the costs of upgrading the skills of the majority of the workforce, both in terms of training and subsequent rates of pay. In making these decisions, it would have been necessary to have had, *inter alia*, a clear understanding of sources of job satisfaction among the men involved in using the machines, in order to achieve a job redesign that would minimise any detrimental changes in those areas important to their job satisfaction.

Interviews with the concreters working on this job suggested that ensuring that robot operators remained part of the work team would have been important as well as feasible. The concreters usually worked in teams made up of approximately the same group of people. All regarded working with people whom they found friendly and easy to work with very important. The *status quo* appeared to reflect preferences with respect to team structure. Thus, if robots had been introduced, it would have been important that the majority of men did not find themselves regularly working on their own.

The importance that all the concreters interviewed attached to being able to make good use of their skills could have also presented some problem if only a few people were trained to work continuously as robot operators. However, in other respects, the interviews suggested that the possible changes to work patterns that could have come about as a result of job changes induced by the use of robots would have been unlikely to have had any serious impact on job satisfaction for the majority of concreters. Only a minority of those interviewed, for example, regarded being able to use initiative or being able to make decisions about how they did the job as very important. Similarly, the scope for increased personal advancement offered little prospect for significantly improved job satisfaction for the majority of the workers interviewed - the latter were equally divided as to whether having prospects of some sort of future career path was very important, important, or unimportant. Nonetheless, given that the workforce is an aging one, the introduction of a technology that reduced the physical effort involved in concrete work could be an important factor in prolonging the ability of employees to continue in concreting work, and in relieving the pressures on younger workers obliged to carry a disproportionate load of the most strenuous work.

Potential impact on the balance of power among different groups within the organisation: While the use of robots might improve productivity, the possibility of doing the work by conventional methods would not be excluded by the introduction of robots into a company. At the same time, if robots were introduced, there would never be a process of total deskilling among concreters, although the traditional skills of the tradesman might be less important to a project than previously. Indeed, the training required for robot operators would involve a process of reskilling rather than deskilling. Thus, the introduction of robots would have been unlikely to have had a significant impact on the pre-existing balance of power between management and workers in the company. In any case, the management-worker relationship in the company was such that this was unlikely to have been a significant issue.

3.3.4. Organisational factors influencing a smooth introduction

Managerial technological knowledge and approach to evaluating new technologies: The equipment the subcontractor used tended to be the most up-to-date in the field for the work being carried out: thus, there was no resistance to change in the company's management structure. There was, however, no formal procedure for evaluating new technologies.

Nature of industrial relations: The director of the subcontracting company prided himself on the quality of the company's industrial relations. The firm sought to be supportive to employees in all matters, and industrial disputation was said to be low.

Company training programs: There were no formal training for concreters in the company. Some changes would have been needed in this respect if robots had been introduced, to train both operators in the use of the machines, and the remainder of the team in safety guidelines.

Language barriers: Permanent employees had a wide range of different nationalities. However, language was not considered to be a problem with respect to team work, passing on instructions or training.

Workers' experience of, and response to, using automated equipment: The closest that any of the concreters interviewed had come to using semi-automated or automated equipment was the use of helicopters for concreting. Only one of the men had ever used a computer, and there were mixed reactions among the remainder with respect to learning to do so.

Equipment acquisition policy and cash flow model for use of equipment: The subcontractor was a well-established firm in a position to be able to afford to either lease or purchase a medium-priced robot. Thus, it would not have been dependent on a possibly reluctant plant hire firm for the supply of concreting robots.

4. CONCLUSION

Analysis suggested that there were no major impediments to the smooth introduction of robots on this site. It also showed that there could be some definite advantages for the organisation in their introduction. However, the estimated cost-substitution savings on this project were only $6525, due primarily to some limited construction labour savings. It is unlikely that any company would be prepared to introduce a new technology for such limited cost savings. However, it is possible that larger substitution benefits could have been achieved on other projects concurrently in the contractor's portfolio, provided that more suitable conditions for the robotic performance of concrete screeding and levelling existed. The capacity to use robots on a number of sites simultaneously would have further increased their attractiveness to the company. Nonetheless, on this project alone, the introduction of concreting robots would have been basically uneconomical.

REFERENCES

1 F. Blackler and C. Brown, Current British Practice in the Evaluation of the New Information Technologies, In F. Debuse and H.W. Schroiff (eds.), Psychology of Work and Organisation, Elsevier Science Publisher B.V., North Holland, 1986.

2 R.A. Jarvis and J.C. Byrne, An automated guided vehicle with map building and path finding capabilities, invited paper, 4th International Symposium of Robotics Research, University of California at Santa Cruz, pp.155-162, 9-14 August 1987.

Automation and Robotics in Construction X
G.H. Watson, R.L. Tucker and J.K. Walters (Editors)

SIMBASE : A Decision Support System for Economic Justification of Automated Construction Technology

Jae-Jeung Rho[a], Deborah J. Fisher[b], and Michael W. O'Neill[c]

[a]Ph.D. Candidate, Department of Industrial Engineering, University of Houston, Houston, Texas, 77204-4812, (713) 743-4186,

[b]Assistant Professor, Department of Industrial Engineering, University of Houston, Houston, Texas, 77204-4812, (713) 743-4195,

[c]Professor and Chair, Department of Civil and Environmental Engineering, University of Houston, Houston, Texas, 77204-4791, (713) 743-4252.

Abstract

Methodology for the economic justification of automated technology has already been developed in the manufacturing arena. This paper discusses the modifications required for application of this methodology to the construction industry. Modifications are necessary due to the industry's conservative nature in adapting new technology, the fact that the construction process resembles more a job shop production system rather than a flow shop (more difficult to automate), and due to the fact that the majority of construction contractors, relatively speaking, are small in size and capital poor. An economic justification methodology appropriate for the construction industry is presented with four modules: Cost and Performance Information Module (CPIM), Cost and Performance Evaluation Module (CPEM), Project Bids Estimation Module (PBEM), and Total Expected Profit Module (TEPM).

1. INTRODUCTION

Since the early 1980's, the construction industry has been trying to adapt automated technologies, including robotics, in order to improve its decreasing productivity. It has been reported, in the manufacturing industry, that successfully applied automated technology can improve productivity by 20% to 30% [1]. However, the construction industry has been slow to adapt automated technology for the following reasons: a harsh working environment, dependence on a high degree of manual skill, minimal standardization, and frequent reconfiguration of the operation. Many researchers have broken the problem down systematically by working on the identification of tasks which have the most potential for automation, instead of trying to automate the entire construction process at once. Kangari and Halpin [2], Tucker, *et al.*[3],

Fisher and O'Connor [4], and Rho *et al.* [5] have addressed issues of task identification for different construction processes using different criteria.

Unfortunately, none of the above models include an economic justification methodology appropriate for the construction industry. Contrary to the manufacturing industry's unified structure of large corporations, the construction industry is composed of several smaller participants (*i.e.*, owners, architects and engineers, and contractors). It is the contractors, among all the industry participants, that will introduce automated technologies into construction operations. Since contractors are, relatively speaking, small in size and poor in capital for research and development compared to their manufacturing counterparts, they are reluctant to adapt automated technology. Moreover, contractors in construction make profits based on each project's bids (*e.g.*, job shop production) unlike manufacturing which bases its profit on the number of products in the mass production system (*e.g.*, flow shop production).

Therefore, an economic justification methodology, the total expected profit (TEP) method, is developed in order to support a construction contractor's decision on incorporating and/or adapting new automated technology. This methodology is used in the development of software called SIMBASE. SIMBASE integrates digital simulation and a database system in order to estimate total expected profits accumulated from all projects over the user-defined time period.

2. TOTAL EXPECTED PROFIT (TEP) METHOD

The economic justification of replacing a conventional construction method with an automated method is becoming more complex as technology advances. Most of these automated technologies are expensive when considering only initial capital investments. Traditionally, automated technologies have been justified by applying a general expected savings analysis [6]. This analysis may be an appropriate method for the mass production system of the manufacturing industry. However, this expected savings methodology can not be directly adapted to the construction industry for the following reasons:

(1) The construction industry, in general, has been conservative in accepting new technologies. Therefore, automation in construction should be evolutionary, rather than revolutionary [2]. That is, it has to yield solid total expected profits over the initial investment cost and the life cycle operating cost.

(2) The construction process is characterized more as a job shop production in the sense that it is engaged in the production of low

quantities of various kinds of specialized products on a project basis [7]. In a job shop production system, it is more important to improve the probabilities of winning future project bids through improved credibility of the contractor by completing projects with reduced project duration and/or reduced rework than merely reducing the production cost per number of products. This is true even moreso today with the increased use of partnering arrangements for repeat business. With this strategy, contractors can increase the total expected profit and reduce the costs caused by being idle between projects.

(3) From the contractor's viewpoint, profits can be determined by the profit margin of the project bids. Since project bids include separate cost items for material, labor, and equipment, cost savings due to replaced laborers by the automated technology can not be considered as a direct benefit to the contractor even though it constitutes a major benefit in the economic justification for automation in the manufacturing industry.

For the above reasons, there needs to be an economic justification methodology which represents the job shop characteristics of the construction industry and focuses on the contractor's viewpoint. The total expected profit (TEP) method, modified from the net present value (NPV) method, was developed and implemented to calculate the total expected profit. TEP can be defined as the difference between the present value of total project bids and the present value of total costs, which represents the actual cash-flow of the contractor related to the technology being evaluated. TEP can be determined by the following equation;

$$TEP = \sum_{t=o}^{n} B_t (1+i)^{-t} - \left\{ \sum_{t=o}^{n} C_t (1+i)^{-t} + I.C \right\} \tag{1}$$

Total project bids is the summation of the yearly project bids, B_t, over the user-defined period, n, which has to be set equal to several economic life cycles of the evaluated technology. Total costs related to the automation technology can be composed of initial investment cost to adapt the technology, I.C., yearly operating costs, C_t, which include maintenance costs and labor costs required to operate the technology. Yearly costs and yearly project bids can be evaluated at the end of each year through digital simulation, and then adjusted to the present value with the user-defined minimum attractive rate of return, i.

Since the tôtal expected profit of each method is simulated under the same condition, the method with the greater total expected profit should be selected as the economical method. If the total expected profit of the automated method is greater than that of conventional method, the evaluated automated construction technology can be justified economically.

3. PRINCIPAL MODULES OF THE SYSTEM

The benefits for adapting automated technology are generally grouped into two categories, quantitative and qualitative benefits. Quantitative benefits are readily provable and quantifiable in the traditional economic justification methodologies, such as rate of return (ROR) analysis or cost-benefit analysis. Savings in labor cost, savings in project completion time, and tax benefits are some examples of the quantitative benefits. Other benefits areas are usually considered to be qualitative benefits and are frequently excluded from the formal economic justification calculations. Quality improvement, elimination of overtime charges, reduction in accident rates, and flexibility increases are some examples of qualitative benefits. Although these benefits are difficult to express in numerical values, the pervasive effect of these benefits are much greater than the quantitative benefits, in the sense that they can increase the contractor's credibility to customers for repeat business.

In order to express the qualitative benefits by numerical values as accurately as possible, SIMBASE integrates digital simulation and a database system with the economic justification methodology. Contractor's credibility can be represented by the probabilities of winning project bids, the probabilities of doing rework, and the probabilities of an accident. These probabilities are expressed as a factor and incorporated into the simulation procedure by storing these statistical parameters into the database for each construction method analyzed. These values will then be transferred to the simulation module when a project event arises. Therefore, automated technology will produce more expected profit than conventional methods due to more projects, higher utilization rate for equipment, less rework, and fewer accident claims. SIMBASE is composed of four modules; Cost and Performance Information Module (CPIM), Project Bids Estimation Module (PBEM), Cost and Performance Evaluation Module (CPEM), and Total Expected Profit Module (TEPM). The structure of SIMBASE is shown in Figure 1. Its functions and interaction between its modules are described in the following sections.

3.1 Cost and Performance Information Module (CPIM)

This module is a database file management program whose functions are as follows:

(1) To provide cost and performance information on the automation technology for the simulation process.
(2) To provide project type information to the PBEM.
(3) To store project bids and project operating costs, which will be provided to the TEPM.

Database files managed in this module are listed in Table 1.

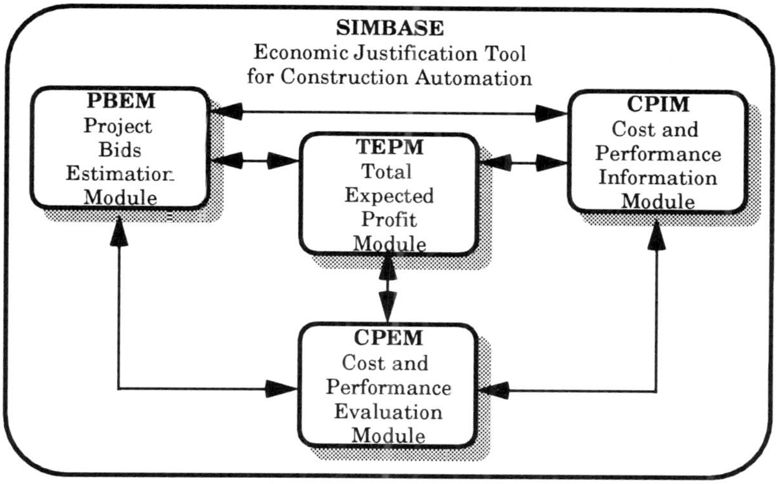

Figure 1. Overall structure of SIMBASE

The database system developed in this module can be used in many other applications later. First, all the files pertinent to the operating cost can be utilized in replacement analysis. Second, input of the project/equipment information can be automated by transfer from this database. Third, when combined with the design and engineering information, this will be the first step to computer integrated construction (CIC).

Table 1.
List of main file names and their interaction between modules.

File Name	Comments	Direction
ProjInfo	All information related to project type available in the simulation, such as project size, configuration, complexity, *etc*.	To PBEC To CPEM
ProjBids	Estimated project bids based on the project information and the technology performance if the bid is successful.	From PBEC To TEPM
PerfInfo	All information related to the performance of technology such as production rate, such as credibility factor, rework ratio, accident rate, *etc*.	To PBEC To CPEM
OpPara	Statistical parameters that each type of operation cost event will be occur during simulation process.	To CPEM
OpCost	Costs for each type of operating cost item, such as repair, downtime, maintenance, *etc*.	To CPEM

3.2 Project Bids Estimation Module (PBEM)

The functions of this module are as follows:

(1) To calculate unit cost per hour for owning and operating construction technology being evaluated.

(2) To calculate material cost and labor cost from the information on the given project type.

(3) To estimate the project bids with the calculated separate cost items and the predefined overhead percentage and profit margin.

Information on the performance of the construction technology being evaluated and information on the project type generated in the simulation are extracted from the database files in the CPIM, respectively. Project bids estimated in this module will be accumulated until the end of each year. This amount represents yearly income of the contractor and will be stored in the project bid database file in the CPIM if the bid is successful.

3.3 Cost and Performance Evaluation Module (CPEM)

This module is a digital simulation program, and its functions are as follows:

(1) To calculate project duration based on performance parameters of the technology for the given project information.

(2) To generate detailed annual project cost data for each project over the user-defined time periods.

(3) To convert contractor's credibility improvement from the qualitative benefit into numerical values by using statistical parameters.

CPEM performs simulation over the user-defined period that has to be set at several economic service lives of the automation technology in order to calculate its life cycle cost. CPEM will first schedule all the projects with the starting date and its type over the time period. Information related to each project type will be extracted from the project information file in the CPIM. (See Table 1.). Then PBEM will generate the project bids using information on the project and the technology. Finally, CPEM will decide whether the contractor will win the project or not by comparing the randomly generated number and the credibility factor of the technology. If its credibility factor is greater than the generated number, the contractor will win the project bids, and the calculated project bids will be stored in the project bids file in the CPIM. Project duration can be determined from the project information and the production rate of the technology. After completing the project, contractors will bid the next available project. The above procedures will be repeated until the end of the user-defined time period. During the simulation, CPEM will generate all events related to operating costs, such as repair, maintenance, downtime, rework, and accidents, based on their statistical parameters from the performance information file. Those operating costs will be accumulated to form yearly costs and then will be stored in the CPIM. CPEM will perform another simulation under the same project information for conventional methods. Due to the differences of the statistical parameters, total project bids and total costs will be different. The difference will be calculated in the TEPM.

Integrating digital simulation into the economic analysis can provide several advantages, one of which is the availability of sensitivity analysis. Sensitivity analysis involves repeated computations with different decision factors to compare results obtained from these substitutions with results from the original data. The sensitivity of any of the uncertain decision factors used in the economic justification calculation can be checked. Minimum attractive rate of return (MARR), time period for simulation, credibility factors, rework ratio, and accident rates are some examples of the decision factors in this module. Analysis can be displayed on graphs that show the effects of percentage variation for key factors.

3.4 Total Expected Profit Module (TEPM)

Yearly project bids and yearly operating costs, as well as the initial investment cost, will be provided to the user in tabular form, in order to indicate expected profit for each year. This table will show the above items for the conventional method and the automated method in two columns for the comparison purposes. The total expected profit can be calculated from equation (1), described earlier. If the total expected profit of the automated method is greater than that of conventional method, the evaluated automated construction technology can be justified economically.

4. CONCLUSION

This paper has provided an overview of the decision support system, SIMBASE, developed for the economic justification of automated construction technology, from the contractor's view point. Total expected profit (TEP) methodology has been developed by modifying the economic justification method for the manufacturing industry. The TEP method is appropriate for the project-oriented, capital poor contractors operating in the construction industry.

SIMBASE is composed of four modules. The Cost and Performance Information Module (CPIM) manages all the database files necessary for the other modules. The Project Bids Estimation Module (PBEM) estimates project bids based on the project information and the performance of the technology.

The Cost and Performance Evaluation Module (CPEM) simulates projects over the economic life cycle of the technology and generate project operating costs, which represents qualitative benefits of the technology. The Total Expected Profit Module (TEPM) calculates the present value of total expected profit and compares the results with those of conventional methods.

SIMBASE will be integrated into the software called "Drilled Shaft Decision Support" (DS^2), which is under development by the authors. DS^2, which is integrated software of an expert system, digital simulation, and database management system, can assist in decision making for construction of drilled foundations for bridges and other structures.

5. ACKNOWLEDGMENTS

The research reported in this paper was supported by a grant from the Texas Advanced Technology Program. The authors are grateful to the assistance provided by ADSC: The International Association of Foundation Drilling.

6. REFERENCES

1 The Business Roundtable. *More Construction for the Money*. Summary Report of the Construction Industry Cost Effectiveness Project, 1983.

2 R. Kangari and D.W. Halpin. Potential Robotics Utilization in Construction. *J. Constr. Engng. and Mgmt., ASCE*, Vol. 115 No. 1, March 1989, pp. 126-143.

3 R.L. Tucker, P. Claire, J. Meyer, and T. Simonson. *Needs Assessment for Construction Automation*. Publication No. 16-1. The Construction Industry Institute, 1992.

4 D.J. Fisher and J.T. O'Connor. Constructability for Piping Automation: Field Operations. *J. Constr. Engng. and Mgmt., ASCE,* Vol. 117, No. 3, September 1991, pp. 468-485.

5 J.J. Rho, D.J. Fisher, and M.W. O'Neill. A Systems Approach to the Integration of Automation Technology in Drilled Shaft Construction *Presented to 72nd Transportation Research Board Meeting*. Washington D.C., January 1993.

6 J.L. Riggs and T.M. West. *Engineering Economics, Third Edition*. McGraw-Hill, New York, New York, 1986.

7 M.P. Groover. *Automation, Production Systems, and Computer Integrated Manufacturing*. Prentice-Hall, Englewood Cliffs, New Jersey, 1987.

8 C. Hendrickson and T. Au. *Project Management for Construction*. Prentice-Hall, Englewood Cliffs, New Jersey, 1989.

Automation and Robotics in Construction X
G.H. Watson, R.L. Tucker and J.K. Walters (Editors)

Robotic Systems for the Processing of Residual Soil from Construction Projects

K. Saito and S. Obayashi

Japan Industrial Robot Association

Abstract

In recent years there has been an increase in the amount of soil removed from construction sites. Factors contributing to this trend include a steady expansion in construction demand, larger structures, and greater use of underground spaces. In urban areas such as Greater Tokyo, processing of the residual soil from construction work has become a significant social problem and a serious impediment to the construction industry. Given the need to build more infrastructure to support what are increasingly sophisticated and diverse lifestyles, disposal of residual soil is a problem that must be solved and an issue of great concern to the people of Japan.

This document contains the results from the first year of activity by the Technical Committee on Development Strategies for Robotic Systems for the Processing of Residual Soil from Construction Projects. Committee members include many engineers who bring with them vast on-site experience in construction work.

We began by making a detailed view of how residual soil is currently processed, collating previous studies, and elucidating problem areas. From this base, we develop three new viewpoints that we use to propose feasible robotic systems able to dispose of residual soil. While there are problems to be overcome in many of the component technologies before the robotic systems proposed in this paper can be realized, the proposals are nonetheless valuable as an indication of where the construction industry will be headed in the future.

1. INTRODUCTION

Although some of the soil produced on construction sites is used on-site as backfill, most of it must be removed from the site and disposed of at inland receiving areas or waterfront reclamation projects, or else used as fill, landfill, or backfill material at other construction sites. Significant quantities are also being dumped into the ocean. This paper uses the generic term "residual soil" to refer to soil that is produced at and must be removed from construction sites. In recent years the quantity of residual soil has been increasing steadily, fueled by a steady increase in construction demand, larger structures, and greater use of underground spaces, a result of the increasing concentration of population in urban areas. In Greater Tokyo, for example, the amount of residual soil produced grew from 42.40 million m^3 in 1975 to 56.70 million m^3 in 1987 and 69.80 million m^3 in 1990, with the increase during the three years from 1987 to 1990 equivalent to the combined increase for the 15 previous years.[1]

At the same time, there has emerged an absolute shortage in the number of sites available to receive the residual soil. Contributing factors include a reduction in vacant land due to continuing urbanization, land use restrictions enacted to protect the environment, and a reduction in the number of ocean reclamation projects. This is particularly true of metropolitan areas, where the shortage of receiving areas is in direct proportion to the growth in residual soil volume. The end result has been the exacerbation of a variety of problems, including longer distances to the receiving sites, higher costs for transportation and receiving, greater traffic congestion from the trucks carrying the soil, and the environmental harm caused by illegal disposal. In other words, the problem of residual soil processing is beginning to have a major impact both on construction design and work methods and on construction costs.

According to forecasts of the production and utilization of residual soil from construction projects, an estimated 847.00 million m^3 of residual soil will be produced in Greater Tokyo during the 13 years from 1988 to 2000.[1] In contrast, the maximum possible utilization is estimated at 663.60 million m^3, even assuming the active use of residual soil in inland housing development projects and seaside reclamation projects. This means that in the near future roughly 20 percent of all residual soil produced at construction sites in Tokyo will be left unused. The fact that construction foremen are now handling this soil on their own paints an uneasy picture for the future.

The processing of residual soil is one of the most urgent and most serious issues facing the construction industry. At the same time, we are fast approaching a time when it will no longer be possible to process this soil using traditional methods, such as diverting it to reclamation projects or other construction sites. Thus, new ideas offering a fundamental solution to the problems associated with residual soil are called for.

2. PURPOSE AND SCOPE OF RESEARCH

Recently, Japan has seen wide-ranging surveys and research on the topic of residual soil from construction projects. Because the creation of measures to deal with residual soil is an important administrative issue, surveys and research are often carried out at the national or local government level for use in drafting government policy. These studies began around 1975, when the issue of residual soil processing first received attention as a problem for society.

Most of the work up to now, however, has offered analyses of existing problems or dealt with the organization and management of related information. The focus, therefore, has been on resolving current problems, and as yet there have been no concrete proposals for the future based on new concepts.

At the same time, together with the worldwide swell in concern about environmental issues, we see a questioning of traditional values about the surrounding environment and of the ideal form which construction should take. Specifically, there have been vocal calls for the elimination of the traditional priority accorded to construction productivity in favor of a new system that gives priority to production activities suited to the natural environment.

This paper offers a proposal, from a new standpoint and unbound by traditional concepts, concerning the use of automated or robotic systems to eliminate the problems posed by residual soil from construction projects. Specifically, it presents a "desirable" method for the processing of residual soil from the standpoint of the construction industry and based on the assumption that it will see actual use by construction foremen.

Our basic philosophy concerning the effective processing of residual soil can be summarized in two points: 1) minimization of the amount of residual soil that is produced; and 2) preparation of receiving sites in accordance with the amount of residual soil that is produced. Also essential when considering residual soil countermeasures are environmental measures such as the following: 1) the effective utilization of limited resources; 2) protection of the natural environment in the soil receiving site and prevention of damage caused by the dumped soil; 3) protection of the natural environment by limiting the taking of further resources (i.e., soil); and 4) elimination of traffic-related problems caused by the vehicles that remove the residual soil from construction sites.

Taking these factors into consideration, we decided to carry out surveys and analysis and submit proposals for the following:

(1) A system that minimizes residual soil production
(2) A system that utilizes residual soil as a new resource
(3) A new system for the removal and processing of residual soil

3. PAST RESEARCH ON RESIDUAL SOIL FROM CONSTRUCTION PROJECTS

When past Japanese research on residual soil is classified and collated, it can be grouped into the 12 categories listed below. <1> through <3> focus on systems for minimizing residual soil output, while <4> through <12> offer systems for the reuse of residual soil as a new resource. Figures in parentheses indicate the number of studies reviewed.

<1> Minimization of residual soil output through proper earth works planning (6)
<2> Minimization of residual soil output by improving earth work methods (11)
<3> Minimization of residual soil output by the on-site reuse of soil (11)
<4> Development of standards for the utilization of residual soil (>23)
<5> Collation of residual soil-related information; creation and utilization of an information network (10)
<6> Appropriation of residual soil over wide areas (4)
<7> Transport of residual soil (9)
<8> Construction of stockyards (7)
<9> Construction of soil improvement plants (21)
<10> Residual soil improvement technologies (48)
<11> More construction using residual soil (16)
<12> Effectiveness of using residual soil (6)

Based on these past studies, specific technical and system-related issues for the future can be divided into the two groups listed below.

A) Technical Issues
<1> Technologies to improve the accuracy of residual soil-related information
<2> Development of methods to evaluate the effect of reusing residual soil from the standpoints of economics and the environment
<3> Establishment of quality standards for improved soil (based on the premise of

active reuse) and construction management standards
<4> Testing and research on the changes over time in improved soil that has been reused (surface properties, reexcavation, etc.)
<5> Effect on trees of using improved soil for backfill
<6> Improvements in plant equipment and operating methods for the efficient improvement of viscous soil
<7> Economical methods of improving residual soil with high moisture content
<8> Clarification of the definitions of "soil" and "sludge" from engineering, administrative, and legal perspectives
<9> Examination of construction methods for minimizing the amount of residual soil which is created or which must be removed from the site
<10> Gathering of past data on the transport and utilization of residual soil
<11> Development of new technologies to reduce residual soil output and allow improvements in soil quality at construction sites
<12> Examination of technologies to confirm the stability of planned disposal site layouts
<13> Examination of methods to store, transport and check the quality of residual soil received from construction projects
<14> Numerous issues remain in the area of slurry shield works:

- The development of more compact slurry treatment plants that are free from noise and vibration
- Research and development of non-polluting solidified materials for use in slurry treatment plants
- Development of more advanced detection systems for slurry and solidified materials at slurry treatment plants
- Development of more advanced system technologies for investigating subbase properties
- Development of shield technologies that eliminate the need for processing and treatment of transported soil.

<15> Planning for new soil disposal sites and development of environmental technologies
<16> Establishment of earth work indicators capable of actively promoting the reuse of residual soil

B) System-related Issues
<1> Establishment of a system that will improve awareness of the problem of residual soil and actively promote the utilization thereof
<2> Establishment of specialized organizations for the integrated processing of related information and the coordination of supply and demand
<3> Promotion of planned layouts for soil improvement facilities and stockyards and assistance measures for the construction of facilities
<4> Establishment of programs to provide economic assistance, financing, and special depreciation for projects that use residual soil from construction projects
<5> Enhancement of systems for cooperative research in the academic and business fields
<6> Promotion of new projects allowing the large-scale utilization of residual soil

As can be seen from the results of past studies, a great deal of research has been conducted on systems that minimize residual soil output and systems that utilize residual soil as a new resource. As yet, however, there have been no concrete proposals concerning the transport and processing of residual soil.

4. A NEW PROPOSAL FOR THE PROCESSING OF RESIDUAL SOIL

In line with the three stated objectives of this paper, we carried out diverse investigations into "desirable methods of processing residual soil" for the future. Because of the wide range of areas involved, our studies involved repeated trial and error based on numerous assumptions to narrow the search down to several possible visions for the future. Fig. 1 shows the areas for investigation and their interrelationships. The systems proposed during the investigation process are outlined below. Needless to say, functions deemed necessary from the analysis of past studies have been included in each system. Detailed descriptions will be omitted for lack of space.

<1> Development of a shield machine that minimizes cross-sectional area
—Multicircular, oval, and rectangular cross-sectional shield machines to reduce unnecessary cross section
<2> Proposal of new shield machines which minimize or do not use additives
<3> Development of instantaneous soil drying vehicles and dump trucks equipped with soil compacting functions
—Facilitate removal by reducing the weight and volume of residual soil
<4> Robotic system for the compacting of excavated soil
—A system comprising a number of individual robots, each performing a separate process on residual soil
<5> Robotic system for the molding of granulated soil
—Form the granulated soil on-site using excess slurry (including the hardening agent for cement) removed from subbase improvement works
<6> Robot using slurry to make turf
—The inorganic sludge created during slurry shield works and other basic construction work is separated into solid and liquid components using a liquid cyclone, and the resulting substance is dewatered using electroosmosis. The dewatered sludge is then mixed with water-absorbing resins, grass seeds, and nutrients, pressurized, and sprayed over green belts. The mixture can also be compacted and formed to create sod.
<7> Robotic system for the processing of sludge
—A robotic system that performs on-site solidification of sludge (i.e., industrial waste) and uses the resulting material to form construction materials such as blocks and aggregates.
<8> System for the building of man-made islands
—A soil transport robotic system designed to build man-made islands using residual soil from construction projects. Based on a self-advancing robotic transport system equipped with a global positioning system (GPS). Designed to operate only on man-made islands.
<9> Mud recycling system
—An unmanned system designed to separate mud into soil and slurry on-site, transport it to a processing and treatment center, and recycle it.
<10> Grading and grade adjustment system
—A robotic system that automatically grades all residual soil and adjusts grade and pH
<11> Robotic system for the reuse and recycling of residual soil
—A robotic system for the reuse of residual soil according to the desired quality and

Fig. 1
Areas of Investigation for a Residual Soil Processing Robotic System

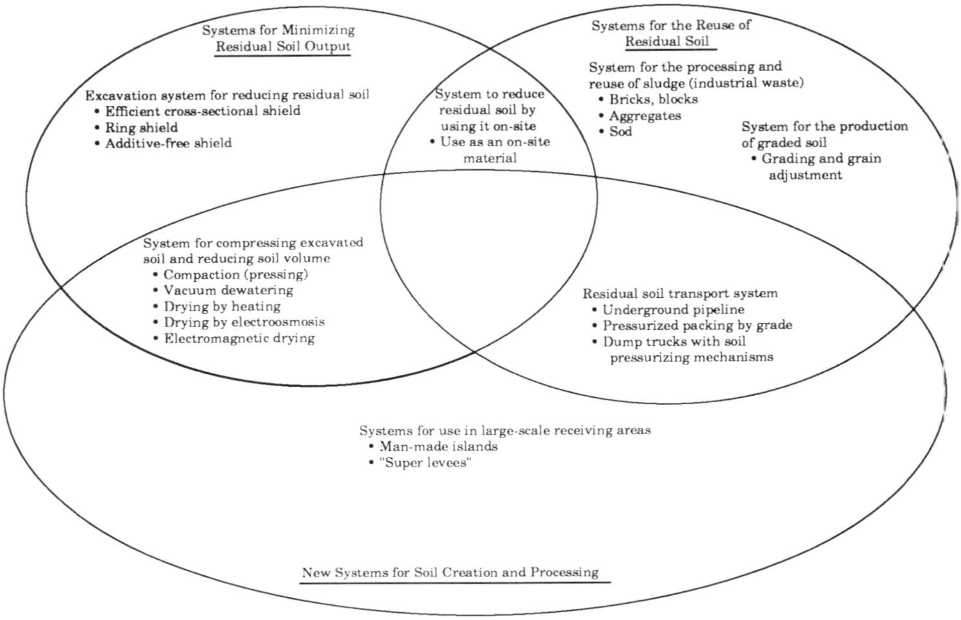

recycling as a secondary processed product
<12> Robots for sea shipment and processing of residual soil
—A large-scale residual soil transport system designed for the construction of man-made islands
<13> System for the separation of residual soil by soil quality and transport
—An unmanned system that automatically separates residual soil on-site according to soil quality and transports it to stockyards

The systems proposed here are robotic systems designed (or assumed) to operate with specific functions. In this paper, we would like to propose a "new comprehensive robotic system for the processing of residual soil," to be formed by integrating various systems having single functions or performing partial operations. The comprehensive robotic system is designed to build "super levees" based on adequate consideration of existing residual soil transport volumes, transport distances, and the type of use for the transported soil. In a separate sense, the super levee is an issue of major concern from the standpoint of the overcrowding of urban areas, with their extremely high concentrations of population and assets, and the prevention of disasters in these areas. Plans for these super levees are now being implemented in surrounding areas, and in the future such areas will be able to accept large quantities of residual soil from construction projects.

The necessary conditions, functions, and component technologies for the comprehensive soil processing system are listed below. The overall image of the comprehensive system will be omitted here from lack of space.

A) Necessary Conditions

(a) Consideration of the view from the construction site (natural view, urban view). Should not generate ill will among neighboring residents (i.e., due to dust, noise, vibration, pollution of the water supply, etc.)
(b) Safe operation and management
(c) Long-distance transport of large quantities of soil
(d) Construction of super levees
(e) Handling of all varieties of residual soil
(f) Transport speed not a primary consideration
(g) Low transport costs

B) Necessary Functions

(a) Functions for soil discrimination and screening by soil quality
—The system must have functions for discrimination of construction sludge, organic soil, fine-grained soil, sandy soil, etc. and functions for separating the soil according to the results of this discrimination.
(b) Functions for dewatering and drying of residual soil (primary processing functions)
—The system must have functions for the reduction of soil weight and volume and to facilitate handling of the soil.
(c) Functions for processing of the residual soil into products (secondary processing functions)
—The system must have functions for the processing of residual soil into bricks, aggregates, fuel, fill materials, etc. as needed.
(d) Functions for transporting the residual soil
—The system must have functions for the transport of the soil from the site where it was produced to the processing areas for discrimination and grading, and the transport of the graded soil and primary or secondary processed products to their destinations.
(e) Functions for unmanned transport and unloading
—The system must have functions for the unmanned loading, transport, and unloading of soil.
(f) Automation of earth work operations within the super levee areas
—Unmanned roller robots for packing of the soil surface

C) Necessary Component Technologies

(a) Soil discrimination techniques incorporating the latest technologies
(b) Energy-efficient dewatering and drying technologies
(c) Efficient (i.e., low-cost) product development technologies
(d) Soil improvement technologies (stable processing technologies for conversion into high-quality soil)
(e) Efficient transport technologies (land, sea, and underwater)

(f) Technologies for unmanned transport, loading, unloading, storage, operation and management

5. CONCLUSION

This paper has offered a direct approach to "the processing of residual soil" from construction projects, which represents one of the most difficult problems facing the construction industry today. Following a careful analysis of numerous previous studies, we collated current technology- and system-related issues to propose a vision for the future. Although we have made some departures from current legal regulations and from engineering verification in the strict sense, we decided to give priority to the hopes and dreams of the engineers who are struggling to resolve the issues presented by the processing of residual soil. Research and development on the numerous topics branching off from our study will be continued long into the future. Nonetheless, we believe that this paper is valuable as a summary of the ideas for the near future held by young engineers working in the construction field.

The following engineers participated in the deliberations for this paper:

Makoto ENOMOTO	Kajima Corporation
Yoshinari HANYUDA	Obayashi Corporation
Yukichi HASHIMOTO	Maeda Corporation
Masaharu KATSUMATA	Maeda Corporation
Kazuya KIKAWADA	Hazama Corporation
Akira NAGATA	Kumagai Gumi Co., Ltd.
Hiroaki NAKAMURA	Chuo Kaihatsu Corp.
Yoshiyuki OHARA	Shimizu Corporation
Hiroshi TOIDA	Taisei Corporation

REFERENCES

1 Ministry of Construction, Committee for the Study of Comprehensive Measures for Residual Soil, Report on Comprehensive Measures for Residual Soil from Construction Projects (1990).

Automation and Robotics in Construction X
G.H. Watson, R.L. Tucker and J.K. Walters (Editors)

Construction Cladding Joining Methods: A Guide to Design for Automatic Assembly

B. L. Atkin[a], P. Atkinson[b], C. Bridgewater[c], B.L. Davies[d] and R. D. Wing[c]

[a]Department of Construction Management & Engineering, University of Reading, Whiteknights, Reading, RG6 2AZ, United Kingdom

[b]Department of Engineering, University of Reading, Whiteknights, Reading, RG6 2AX, United Kingdom

[c]Department of Civil Engineering, Imperial College of Science, Technology and Medicine, South Kensington, London, SW7 2BU, United Kingdom

[d]Department of Mechanical Engineering, Imperial College of Science, Technology and Medicine, South Kensington, London, SW7 2BX, United Kingdom

Abstract

One of the fundamental lessons learnt by manufacturing industry was that assembly tasks must be redesigned to suit the strengths and weaknesses of machines vis a vis human workers. The basic principles of *Design for Automation* and *Design for Assembly* can be applied to construction tasks in order to improve productivity either by simplifying existing tasks or by integrating machines with processes in a more structured construction system. This paper outlines some of the design principles which have been used successfully in manufacturing industry and applies them to the different processes which are found in the majority of industrial joining methods.

1. INTRODUCTION

Previous research into the use of automation systems for the cladding of buildings led to the design and simulation of a prototype erection device[1]. The research highlighted the necessity for components and processes used in the assembly of the facades to be designed for automation systems[2][3]. The research also showed that the methods used for joining were primitive and likely to prove troublesome for machines. Alternative methods for joining are therefore being investigated with the aim of identifying useful technologies for construction automation[4][5].

Other research has shown that construction automation requires a detailed analysis of both the task to be performed and the context in which it will be used[6]. This implies that the elements of the construction task will need modification to suit the peculiar needs and requirements of automation[2]. The principles which are embodied in "Design for Automation" closely resemble what is known as "Buildability" in the UK construction industry. Designers are encouraged to avoid difficult or complex details to improve the accuracy and speed of construction[7]. Experience has shown that applying "Design for Automation" to existing assemblies leads to improved productivity and fewer errors even when automation systems are not used[8].

Many of the techniques used in construction for joining parts on site have remained unchanged for decades and are labour intensive[9]. The use of construction automation will lead to the introduction of new techniques and methods which place more emphasis on machinery. While many of the joining methods which might be used by automated systems on construction sites are used routinely in manufacturing industry, conditions on site are different to those which are found in factories. It follows, therefore, that many of the joining techniques which may be considered will be unsuitable for use on construction sites[10].

The principal aim of this paper is to give designers of cladding fixing systems a systematic framework which will help them assess the difficulty of using different joining systems for automated cladding. To this end, the underlying principles of "Design for Automation" will be discussed in the context of cladding fixing systems. After this, the stages which are common to all joining methods will be reviewed before the guidelines for assessing the utility of particular methods are presented.

2. PRINCIPLES OF DESIGN FOR AUTOMATION

Previous research into high volume manufacturing has already defined the basic principles of "Design for Automation"[11]. These principles will be discussed in the following sections with a particular emphasis on cladding.

2.1 Combine Functionality of Components

Fewer assembly operations are needed if the functionality of several items are combined into one or two components. However, problems may arise with fabricating the new components as some materials may be less ductile or malleable than others.

2.2 Modular Design Philosophies

The use of a modular design philosophy such as Group Technology[12], can widen the applicability of automated fixings to many different cladding types. This will increase the attractiveness of automated solutions since the range of cladding types that can be erected is extended.

2.3 Reduce the Number of Fasteners

A distinction can be made between site fixings and factory fastenings as the latter are simpler and more reliable. As outlined in Section 2.1, parts can be designed to incorporate many of the functions of fasteners so that they can be pushed together without the need for special joining devices and operations.

2.4 One Operation at a Time

The complexity of the site automation systems are reduced if the need to handle several components simultaneously is eliminated. In general, assembly plans which are sequential require simpler equipment and can be carried out more quickly.

2.5 Simplify the Part Motions

The robot motions necessary to place and fix components should be as simple as possible. If a component can be inserted into an assembly along one axis with no extra rotations or sideways movements, then the robot moving the part can be simpler than one which is capable of performing complex motions within the assembly.

2.6 Improve Quality Control

Strict quality control must be enforced so that assembly errors do not arise from defects and tolerance problems. Component tolerances must be checked to ensure that suppliers comply with agreed limits, as any delays caused by incorrect manufacture will have implications for erection times and hence profitability.

2.7 Design Specific Delivery Mechanisms

In factories, feeders are used to ensure that parts are presented with the correct location and orientation. They also ensure that the parts are waiting for the robot while an assembly is in progress, not vice versa. Such mechanisms should also protect components so that they are less likely to be damaged.

2.8 Reduce Ambiguity in the Assembly

The assembly should be designed to ensure that components are assembled correctly on site. Splines or keys can be added to make it difficult for parts to be forced into the assembly. Clear and concise instructions are also necessary to enable site operatives to check that the automation systems have functioned correctly.

3. CHARACTERISTICS OF JOINING METHODS

In order to assess the complexity of different joining methods, an analysis must first be made of their constituent processes. The principles of "Design for Automation" can then be applied to the subtasks in order to automate the joining method as a whole. Some typical joining methods are given in Table 1, which also includes an assessment of the relative importance of each method in different sectors of industry. It is apparent that there are some methods which are used in manufacturing industry but which are not used in construction. This will be due partly to conservatism in the industry towards new materials and methods, but will also be related to the necessity for changing current working practices to suit the new construction technology and methods.

Table 1
Usage of Fixing Types by Sector of Industry

Fixing Type	Ae	Au	Co	Mn	Sh
bolts	C	C	C	C	C
rivets	C	C	C	C	C
screws and plugs	C	C	C	C	R
"push fit" connectors	C	C	R	C	R
deformation fixings	C	S	R	C	R
ties and chains	N	N	S	N	R
clamps	R	C	S	C	R
stitching and staples	N	N	R	S	R
welding	C	C	C	C	C
pressure welding	R	R	N	R	S
brazing and soldering	R	S	N	R	N
adhesion	C	C	S	R	N
grouting and in situ concrete	N	N	C	N	S
freezing	N	N	S	N	N

Key: **Au**: Automotive **Ae**: Aerospace **Co**: Construction
Mn: Manufacturing **Sh**: Shipbuilding and Offshore

C: Common **S**: Special **R**: Rare **N**: Never

3.1 Basic Operations in Fixing and Joining

From a further analysis of the joining methods given in Table 1, it emerges that the joining of construction materials together on site is essentially a threefold process of positioning, fixing and protecting[5]. These three areas further decompose into six discrete stages which are always present in any joining method. However, some methods may combine several stages into a single process. What is important is the ease with which the different stages can be performed and the degree of accuracy that can be obtained in aggressive and inhospitable site environments. Each of the six stages in the joining process that have been identified will be discussed in the following sections.

3.1.1 Preparation of Components

Some joining methods require little or no preparatory work, whereas others require high degrees of cleanliness between the faces which are to be joined. For example, adhesives and welds will often not "take" in the presence of rust, water or oil.

3.1.2 Assembly of Components

After preparation, the components can be transported to the assembly point. In a factory, conveyor belts and jigs ensure that components arrive at the assembly point in the correct position and orientation, but on a construction site there is much less control. An allied problem is that of close alignment and positioning of the components relative to the building. Uncertainty in the global positioning of components is one of the fundamental problems of automating cladding assembly tasks on site. This is due to the fact that construction tolerances are much larger than those allowed in factory production methods. For example, tolerances of between 15 and 25 mm are common in structural frames of insitu concrete, whereas the cladding components are typically manufactured to tolerances of 5 mm[13]. Overcoming positional and dimensional uncertainty either involves ingenuity in altering the assembly methodology, or it requires the use of expensive sensing systems to compensate for the lack of firm knowledge about the environment.

3.1.3 Mating of Assembly to Fixing Device (or vice versa)

During the course of the fixing process, the components and machines must be brought together and then prepared for use. In a factory setting, the assembly travels to the fixing device or to a known point in space to which the fixing device can be moved. By contrast, on a construction site, all of the positions and orientations are defined relative to features in the partially constructed building.

3.1.4 Initiation of Fixing Process

Once the fixing device has been positioned correctly, some sensing may be required to ensure that the fixing process progresses satisfactorily and that there are no problems which might cause the fixing to be aborted. A specialised fixing unit can be attached to the robot in advance, which adds to the complexity and cost of the robot but which eliminates some local sensing.

3.1.5 Withdrawal of Machine After Fixing

This is a simple operation but may become complicated if the fixing has to be aborted. It is unlikely that a simple reversal of the fixing operation will lead to a proper removal of the components, since the fixings may be damaged in the process. If the fixing is completed successfully and needs no further attention, the fixing device may be moved to the next location.

3.1.6 Post fixing Operations

Some of the processes used in manufacture require extensive testing and finishing before they are considered to be complete. Welding and adhesives are obvious examples, but checking of correct bolt torque is also important. Other operations such as painting or rustproofing may be necessary as well as nondestructive tests.

4. GUIDELINES FOR DESIGNERS

Procedures exist for the selection of appropriate fixing systems given different materials, structural requirements, and so on. One set of criteria was produced for the Royal Institute of British Architects, and has been summarised in Table 2[14].

Table 2
Summary of Primary Selection Procedure

Criterion	Subsidiary Factors
Structural	Loadings, vibration, vandalism, material characteristics, ductility, ranges.
Movement	Required movement, tolerances, rigidity.
Installation	Effects on rest of building, finishes, inhabitants, performance, removability.
Durability	Treatments, longevity of material, cost.
Adjustability	Design life, reversible, necessity for periodic adjustment, temporary fixings.
Appearance	Visibility, compatibility with other fixings and finishes, cost.
Accessibility	Neighbouring components, location, insulation, condensation, finishes.
Responsibility	Supply, delivery, installation, checking of integrity.
Programme	Schedule, cost, performance, information.
Other	Protection from weather, interaction with other building components.

Most selection procedures of this type, however, assume existing fixing systems and do not assume the use of any kind of automation, be it factory based or automation on site. But experience in other sectors of industry suggests that the principles of "Design for Automation" will have considerable impact on the six stages of joining which were discussed in Section 3. The criteria which derive from a consideration of "Design for Automation" have been summarised in Table 3 and should be used in conjunction with the sections dealing with site installation and inspection in Table 2. Not all of the principles apply at every stage of the fixing process, but some dominate at different times.

The need to reduce the number of components and fixings is important in the earlier stages, whereas the need to simplify or eliminate complex robot motions, particularly "peg in hole" insertions, is more critical in the middle stages. Quality control is vital in the preparation and post fixing stages but becomes less important once the fixing is actually underway. Other factors, such as the need for a modular design of joining component, can be assessed in a similar fashion.

Table 3
Considerations of Design for Automation

Stage	Subsidiary Factors
Preparation	Modular design, quality control.
Assembly	Combine functionality, modular design, sequential operations, simplify motions, special feeders, reduce ambiguity.
Mating	Combine functionality, simplify motions.
Initiation	Combine functionality, modular, design, reduce number of fasteners, sequential operations, reduce ambiguity.
Withdrawal	Simplify motions.
Post fixing	Quality control, inspection, testing.

5. CONCLUSION

This paper has discussed the principles of "Design for Automation" and has presented guidelines which will help designers understand the difficulties associated with using automatic joining methods in cladding applications. The principles are based on research into joining methods which has sought to apply the experience gained by manufacturing industry to construction problems.

Many of the fixing methods used extensively in other industries are not widely used for construction purposes partly because of conservatism in construction sector with regard to the scalability and durability of new materials and methods. Previous experiments with industrialised building systems in the 1960's were certainly unsatisfactory in this respect. Considerable standardisation of fixing methods has been applied in manufacturing industry, which can enjoy the benefit of higher levels of repetition than are usually to be found on building sites. Therefore, it can be seen that considerable changes to current working practice will be needed to utilise site automation and factory based production technology in a general construction setting.

One of the wider aims of the research programme is to continue to develop and refine methods of comparing joining technologies from different sectors of industry. Preliminary results suggest that many of the joining techniques which are widespread in construction are unsuitable for automation when compared with those which are used in other sectors of industry, for example adhesives and push-fit connectors

6. ACKNOWLEDGEMENT

The work described in this report has been carried out by the authors as part of SERC funded grant: GR/H 46374 "Robotic Fixing Systems for Cladding".

7. REFERENCES

1. Ibanez Guzman J., Atkin B. and Atkinson P., Simulation of a Panel Positioning Robot, in **Proc 9th Int Symp on Aut and Rob in Cons**, **1**, Japan Society of Civil Engineers, Tokyo, 1992, pp 451-458.

2. Atkin B., Atkinson P., Bridgewater C. and Ibanez Guzman J., A New Approach to Automating Construction, in **Proc 6th Int Symp on Aut and Rob in Cons**, Construction Industry Institute, San Francisco, 1989, pp 118-126.

3. Bridgewater C., Atkin B., Atkinson P. and Ibanez Guzman J., **Cladding Components in the Parts Set**, Occasional Paper 23, Department of Construction Management & Engineering, University of Reading, UK, 1990.

4. Atkin B., Adapting Building Technology to Facilitate Robotisation, in: **Proc 5th Int Symp on Rob in Cons**, **1**, Japan Industrial Robot Association, Tokyo, 1988, pp 221-228.

5. Wing R., Connections for Construction: A Robotic Approach, in: **Proc 7th Int Symp on Aut and Rob in Cons**, **1**, Conseil International du Batiments, Bristol, 1990, pp 145-150.

6. Slocum A. and Schena B., BlockBot: A Robot to Automate Construction of Cement Block Walls, **Robotics**, **4**, 1988, pp 111-129.

7. Ferguson I., **Buildability in Practice**, Mitchell, London, 1989.

8. Boothroyd G., Poli C. and Murch L., **Automatic Assembly**, Marcel Dekker Inc., New York, 1988.

9. Martin B., **Joints in Buildings**, George Godwin Ltd, London, 1977.

10. Vallings M., **Mechanisation in Building**, 2nd Edition, Applied Science Publishers Ltd, London, 1975.

11. Myrup Andreasen M., Kahler S. and Lund T., **Design for Assembly**, IFS (Publications) Ltd, London, 1983.

12. Groover M., **Automation, Production Systems and Computer Aided Manufacture**, Prentice Hall Inc, New Jersey, 1980.

13. Brookes A., **Cladding of Buildings**, Longman Scientific & Technical, London, 1990.

14. Launchbury W., **Handbook of Fixings and Fastenings**, The Architectural Press, Royal Institution of British Architects, London, 1971.

Automation and Robotics in Construction X
G.H. Watson, R.L. Tucker and J.K. Walters (Editors)

Robotized system of GRC panels for construction industry

C.Balaguer, F.J.Rodríguez, J.M.Pastor, R.Aracil, A.Barrientos, E.A.Puente
J. Flórez*, M. Esteban*, J.A. Comas*, E. Márquez*

Div. Ingeniería de Sistemas y Automática (DISAM), Universidad Politécnica de Madrid, c/José Gutierrez Abascal, 2, Madrid 28006, Spain.

*Dragados y Construcciones, S.A., c/Orense, 81, Madrid 28020, Spain.

Abstract

This paper presents a robotized system that produces high quality GRC panels in an automatic way. The system is installed in a Dragados y Construcciones factory near Zaragoza. The input of the system are the 3D AUTOCAD drawings of the panels. The output of the system are the robots paths, which are generated automatically, and real-time optimum sequences of the system processes: robot movements, spray gun working, panels transportation and mortar preparation.

1. INTRODUCTION

The GRC (Glass Reinforced Cement) technology was developed through the last 30 years. The GRC panels can be design as wall units, windows wall units, columns, etc. [1]. The lightness of GRC material permits to perform very big panels (4x3m) easy to assembly in the site. However the GRC panels fabrication process is performed nowadays in a artisanal and manual way. Using this manual method adequate quality is a limiting factor in the production capability of GRC panel factories.

The manual process is performed using a conventional mortar spraying gun equipped with a glass fiber cutter. Two simultaneous shots, one of mortar and one of glass fiber, are directed on the panels. These shots are usually not concentric. Required thickness of the panel is obtained by progressive spraying and forming of 2-5mm layers. Compactation by roller or trowel is necessary after each spraying. In this way several types of panels will be performed by the developed system: shell with and without isolation, sandwich and stud frame. Its minimum thickness is about 8mm.

The automatization of the spraying process will permit to increase the quality and uniformity of the panels, save an appreciate quantity of materials, and organize in a more efficient way the factory production. This is why the developed system is a very important step towards fully automated prefabricated system, like in the manufacturing industry [2].

2. SYSTEMS STRUCTURE

The scheme of the developed system is presented in Figure 1. It is formed by four sub-systems: a) spraying cell based on the ASEAs 6DOF IRb-3200 robot, b) compacting cell based on four manual workplaces (in the future will be automatic), c) roller based transportation system for panels, and d) mortar preparation cell. Each of the sub-systems has its own control system which are coordinated by the main computer.

The spraying cell is the main sub-system. It is presented in the Figure 2 and is formed by the robot, the spraying workplace, and the electronic pump spray station. The position of the robot is selected in order to spray panels of maximum dimensions: 3x4 m. The robot is equipped in its tip with the concentrical spraying gun. The maximum and minimum productivity of the gun are 36 and 18 kg/min respectively. The mortar and the glass fiber (three separate wires sets) reach the robots tip in such way that no interferences take place during its movement. This restriction strongly influences in the joints path planning.

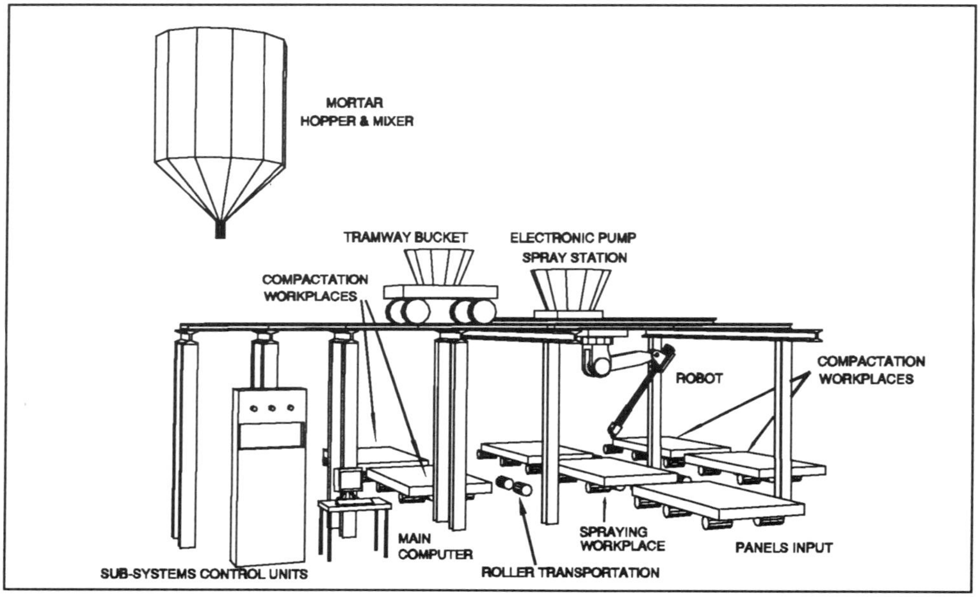

Figure 1. Systems structure scheme.

The sparying workplace is omnidirectional. The panels reach this workplace from every other workplace of the system using roller transportation. The level of the mortar in the electronic pump and the possibility of being the fiber unexpectedly cut are controlled automatically. If this happenes the robot automatically stops and moves to a previous programmed position in order to repair this abnormal situation.

The compactation cell is formed by four workplaces which are manual ones in this stage of the project. But in the near future they will be robotized in the same way that the spraying workplace is. In order to optimize the productivity of the system main computer plans in real-time the most adequate place for panel transportation. In order to know when and where the human operators are to start and had finish the concurrent compactation processes, several indicators and switches are situated near the compactation workplaces.

The input and output of the panels are performed in a manual way using the conventional crane. The instant and the type of panels to input or output are assisted by the main computer using display.

The transportation of the mortar from the hopper&mixer to the electronic pump spray station is performed in an automatic way using a tramway bucket.

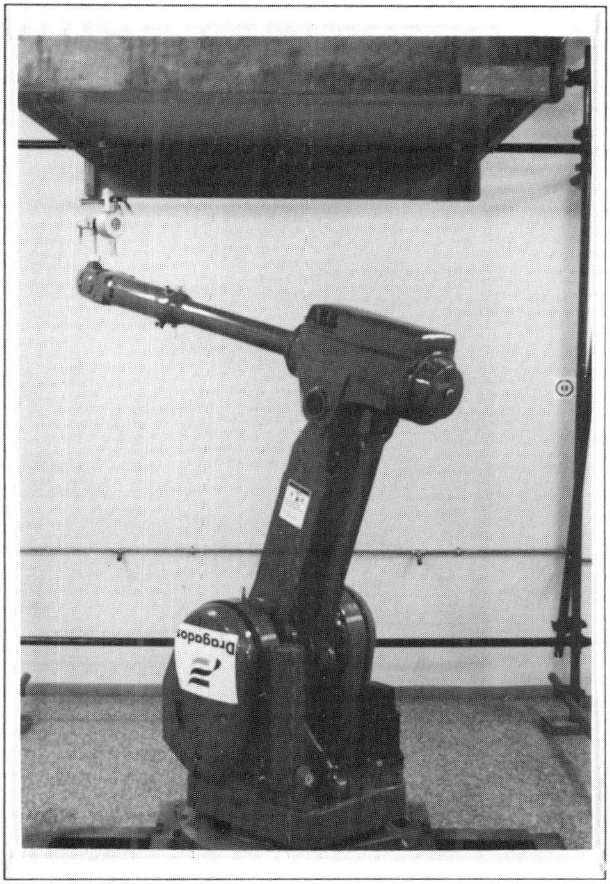

Figure 2. Spraying cell.

3. CONTROL SYSTEM

The control system is based on the distributed and hierarchical concept. There are three main levels in the control system. The main computer, which is situated in the second level, coordinates all the sub-systems control units and performs the man-machine interface. The structure of the control system is presented in the Figure 3. The high level control is for off-line programming. It is based on a powerful PC where AUTOCAD version 12 and specially developed software run. The input of this level are the drawings of the several layers of the panels, type of materials and clamps, position of isolation materials, etc. The output is the robots path.

The second level is formed by the real-time control PC. Its main mission is to coordinate the sub-systems control units. The input is the order list of panels to be performed during the day and their robots paths. The output is the commands to the processes: robot, transportation system, spraying gun, hopper&mixer, and the sincronization between them.

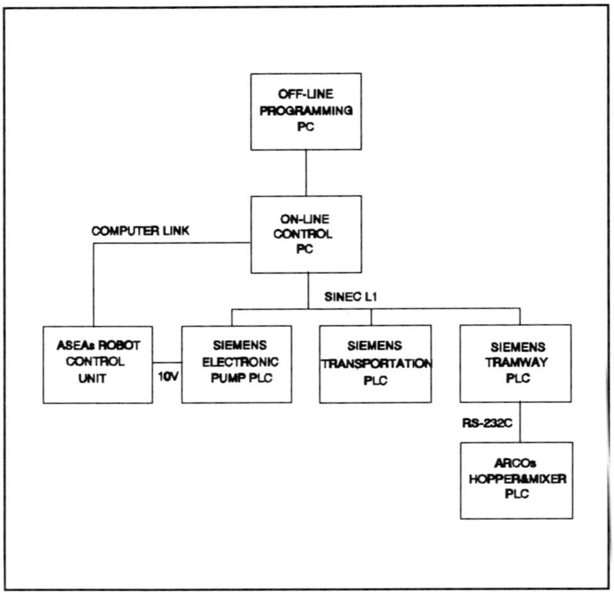

Figure 3. Systems control scheme.

The main difficulty of this level is the necessity to use MS-DOS due the fact that computer link between robot and PC is performed using commercially available ASEAs software Communication Tools [3]. It is well known that this operating system is not for real-time applications. But its advantage is the hardware low cost. Our system has two main input/output channels: robot serial line and SIEMENS network SINEC L1. For the network communications special a hardware with a big buffers is used. In this order the simultaneous communications don´t cause any problems to the systems.

Other important channel (keyboard, mouse and display) is the man-machine interface. It is designed in order to have a very easy data and commands exchange between the system and the human operator. An example of the man-machine interface during the real-time control is presented in the Figure 4. Other important feature is the very fast response, specially during emergency stop of the system. The sample period of the control loop is in average about 200 msec.

The on-line control PC has others secondary off-line missions: one day panels production planning, optimum ordering of the planned panels, data login, maintenance, manual (step-by-step) control of the system, etc.

The third level of the control system is formed by the robot control unit and four PLCs. Each of them is responsible for one of the system processes. There are no connections between PLC, except the connection between robot and pump. This is an +-10V analog channel. It is used to coordinate the robots velocity, and the guns mortar and fiber flows. Without this space and time sincronization the high quality of the GRC spraying layer is not possible.

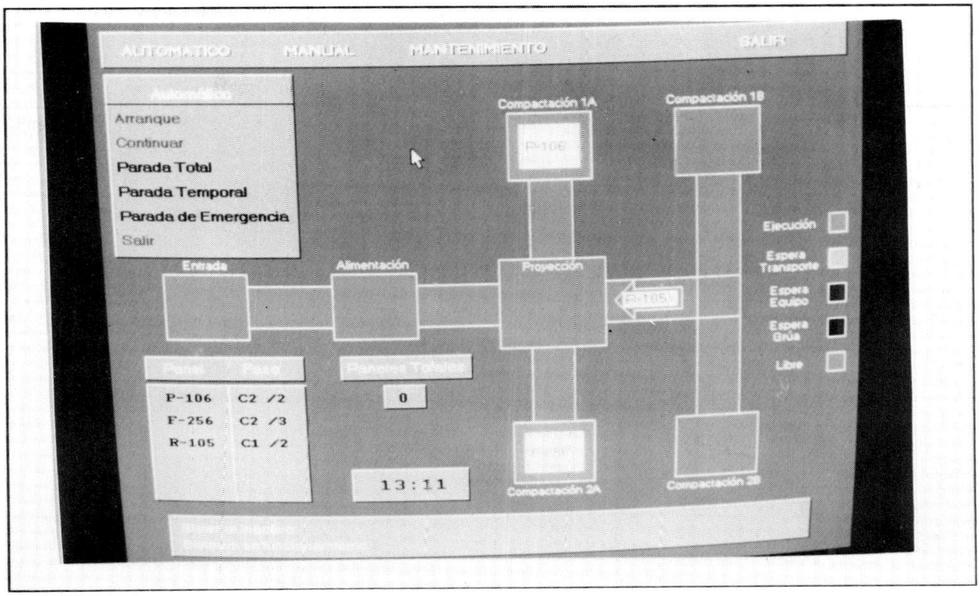

Figure 4. Man-machine interface.

4. PANELS DEFINITION AND ROBOTS PATH PLANNING

One of the best advantages of the system is the automatic robot path planning from 3D AUTOCAD v12 panels drawings. The 3D AME v2.0 software [4] has been used and linked with specially developed C software.

The drawings of the panels are performed through an GRC menu which helps the human operator to draw the panel layer by layer and introduce the following information for each layer (see Figure 5):

- type of the layer (bottom or side),
- type of the layers material (with or without fiber, mortar type, etc.),
- thickness of the layer,
- type and position of isolators and clamps,
- cone of the spraying gun,
- etc.

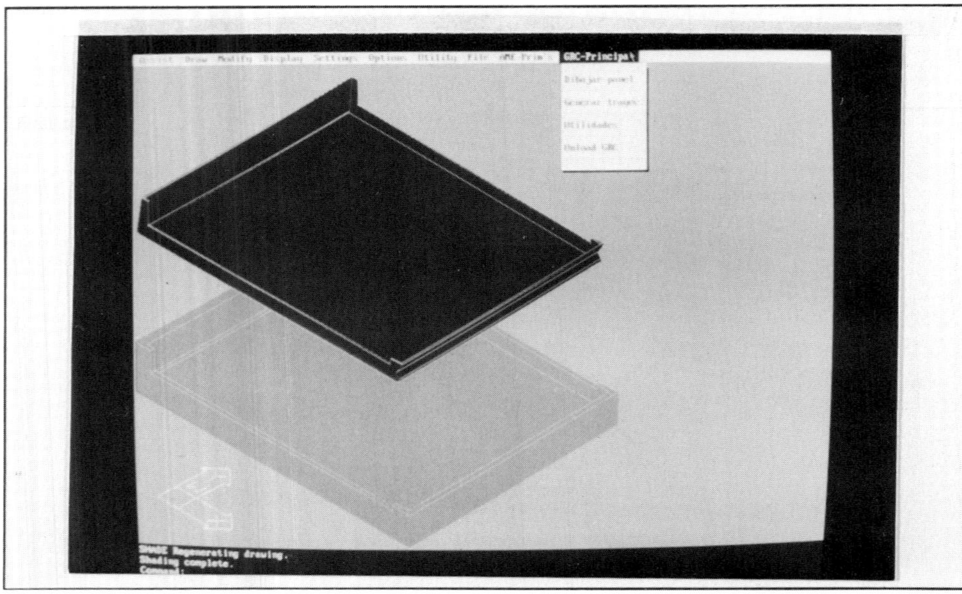

Figure 5. Panel definition through CAD system.

Other parameters like distance of projection, and very specially orientation of the spraying gun and robots path are generated automatically. There are four steps:

1) Computation of the 3D convex volumes and its conectivities which represent all "free" space between this layer and the previous one. For this purpose we extrusioned the previous layer and cut it by planes or by normalized revolution surfaces of the borders. This is why only panels with any number of planes and revolute permits are possible. Developed algorithm is similar to Dobkins one [5].

2) Computation of the robots position and orientation in such way that no collisions occurs between the robot with the spraying gun and the panels, and the robot and the gun. For this purpose several free-collision path planning algorithms have been used, like [6].

3) Computation of the robots path in order to obtaine continuous paths without stops or with minimum number of stops. This is a necessary condition in order to optimize the path. Used algorithm is based on the Hansens idea [7].

4) Modification of the robots path keeping in mind the robots joints limits (specially for the first axis), the limitation of the joints rotations during the movements in order to do not roll up the mortar and fiber tubes, the changes in the robots configuration to obtained smooth or spline paths, etc.

6. CONCLUSIONS

The developed system presents a new step toward fully automatic prefabricated manufacturing. The quality of GRC panels and factory productivity are increased. But the most important advantage is the flexibility of the system in order to introduce new types of panels. This procedure has been done in very fast and easy way. It is assisted by the standard CAD system and is structured in order to be independent from systems lay-out and robots type.

7. ACKNOWLEDGEMENTS

This work is sponsored by construction company Dragados y Construcciones, S.A. and the Spanish Ministry of Industry. Special thanks to E. Pinto and L.F. Peñin for their developments. The subcontracted companies of this project are: ARCO Electrónica, S.A., ODEMSA, S.A. and SIEMENS, S.A. The assistance of MULTICAD is very appreciated.

8. REFERENCES

1 E.S.Knowles (editor), Recommended Practice for Glass Fiber Reinforced Concrete Panels, *Prestressed Concrete Institute*, 1987.

2 A.Barrientos, H.Amaris, A.Yela, R.Aracil, C.Balaguer, Control System for an Automatic and Intelligent Packing Cell, *International Symposium on Robotics, Mechatronics and Manufacturing Systems (IMACS/SICE RM²S)*, Kobe (Japan), 1992.

3 Communication Tools User´s Guide, *ABB Robotics*, 1991.

4 AUTOCAD AME 2.1 and API Reference Manual, *Autodesk, Inc*, 1992.

5 D.Dobkin, L.Guibas, J.Hershberger, J.Snoeyink, An Efficient Algorithm for Finding the CGS Representation of a Simple Polygon, *Computer Graphics*, vol.22, nº4, 1988.

6 C.Balaguer, F.J.Rodrigues, J.M.Pastor, E.Gambao, R.Aracil, A.Barrientos, A Rule-base Selection of C-space representation for Rapid 3D Robot Path Planning, *7th IFAC/IFIP Symposium on Information Control Problems in Manufacturing (INCOM)*, Toronto (Canada), 1992.

7 A.Hansen, F.Arbab, An Algorithm for Generating NC Tools Paths for Arbitrary Shaped Pockets with Islands, *ACM Transactions on Graphics*, vol.2, nº2, 1992.

Automation and Robotics in Construction X
G.H. Watson, R.L. Tucker and J.K. Walters (Editors)

An Experiment in Semi-automated Spray Fireproofing

Iwao UCHIZAKI[a] and Yasuaki UCHIDA[b]

a. Construction Equipment Unit, Technical Research Laboratory, TAKENAKA Corporation, 2-5-14, Minamisuna, Koto-ku, Tokyo 136, JAPAN

b. Development Department, ASK Corporation, 2-5-5, Turumichuou, Turumi-ku, Yokohama 230, JAPAN

Abstract

Fireproofing material spraying is one of the labor intensive construction works under inferior working environment. Toward improvement of the working surroundings as well as the work efficiency, a semi-automatic fireproof material spraying robot has been experimentally constructed. It features a double-nozzle design and a man-machine cooperative system. Small-scale spraying tests were carried out with the prototype. The tests indicated that the following ideas can provide a guide to the development of construction robots:

1) Complex mechanisms and advanced control functions can be eliminated by adopting a man-machine cooperative system; this makes it possible to reduce the cost of the robot.
2) A robot can operate under heavy loads that would be difficult for humans, making it possible to raise work efficiency.

1. Introduction

In general, the fire resistance of structural steel members in buildings is ensured by applying fire-resistant material to them. About 30,000,000 m^2 of fireproofing, in which the share of labor cost is 50%, is carried out annually in Japan. In performing this work, workers manually operates nozzles which spray refractory material. While spraying, rock wool scatters and cementitious liquid splashes, creating an unhealthy work environment. This, coupled with the recent shortage of young or skilled laborers, has accelerated the introduction of automation technology into such a labor intensive task as fireproofing which has otherwise changed little over years.

Against this background, we have focused on a "wet" application method using factory prepared pre-packed material, mixed with water on site and spray through a slurry pump, nominal density of 500 kg/m^3. The reason is that the demand of this type of fireproofing work has been increasing with considerably high rate and will likely continue to increase. Although some spraying robots for the work have been developed using current automation technology, they have not yet found wide application.[1)]

2. Basic outline of development

The following two drawbacks have generally been pointed out when replacing manual construction work with robot methods:

1) Higher cost
2) Lower work efficiency

To ensure that this would not be a problem in our case, we decided to improve the idea of simply replacing workers with robots. Put more precisely, we decided that in order to simplify the robots, they should not be designed to works suitable for humans(e.g. nozzle positioning or direction control), but only to do works that machines are good at (e.g. works that require power or speed). The conclusion drawn from discussions centering on the application of the philosophy to the automation of "wet"-type fireproofing spray work was that the following systems are the most important, most feasible and most cost-effective for installation in a robot system:

1) A man-machine cooperative system based on a mechanical balancer
2) A multi-nozzle sweep system

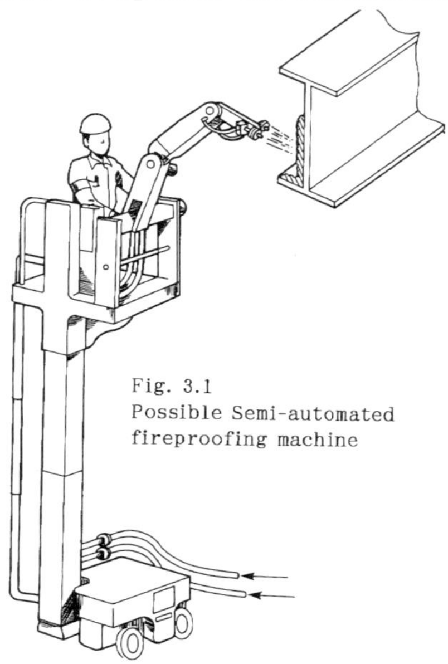

Fig. 3.1
Possible Semi-automated fireproofing machine

3. Element performance test

Figure 3.1 shows a prototype fireproofing machine designed at the initial stage of development. A multi-nozzle sweep mechanism and a mechanical balancer mounted on a vehicle capable of high lift work are conceptually presented. To obtain appropriate design information for the prototype machine, a series of performance tests are conducted.

3.1 Experiments on nozzle system

To investigate various designs for the multi-nozzle system, tests using a single nozzle were carried out. As a control, tests of manual spraying by workers were also performed.(See Photos 3.1 and 3.2.) The results of these tests can be summarized as follows.

1) A highly complicated mechanism is required to replace manually operated spraying nozzles.
2) Using a multi-nozzle design may increase the work efficiency, but it makes it difficult to form a uniform thickness layer. (In particular, the overlaps in batch spraying operations cause problems.)
3) A function which corrects for variations in the flow rate at a spray gun is required.

Photo 3.1 Manual Spraying

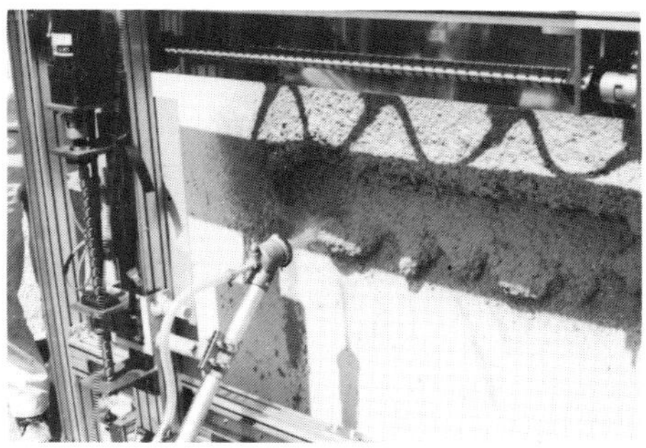

Photo 3.2 Mechanical Spraying

These results led us to select 1) a double-nozzle design and 2) a mechanism swinging parallel to the surface to be sprayed.

3.2 Experiments on vehicle for high lift work

The actual use of the machine was assumed to be limited to the following two methods:

1) A method in which the spraying position changes while the vehicle moves; that is, the worker's main task is to maneuver the vehicle.
2) A method in which the vehicle stops in a particular position and

the spray nozzle is controlled manually with the aid of the balancer.

Method 1) above clearly requires a vehicle which can travel parallel to a structural beam at a constant speed with great accuracy. Both of the vehicle performance were tested using a commercially available multi-directional aerial work platform. Figure 3.2 and 3.3 show the test results. The linear travelling performance of the selected platform was within 25mm per 5m and the speed variation was less than 10% as shown in the figures, offering a performance suitable for our application.

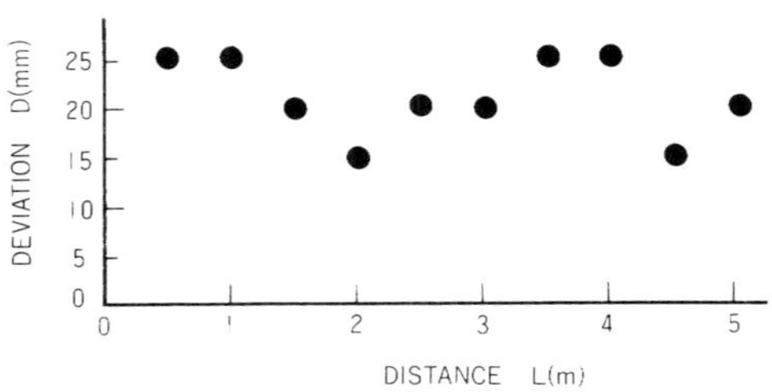

Fig.3.2 DEVIATION FROM A STRAIGHT COURSE

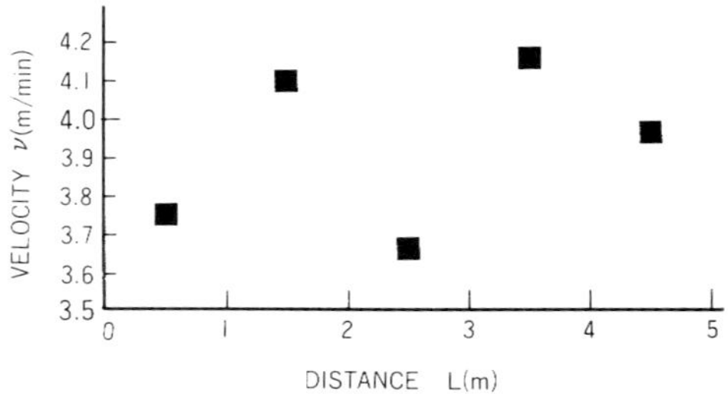

Fig.3.3 UNIFORMITY OF RUNNING SPEED

4. Prototype machine

Figure 4.1 and Photo 4.1 show the prototype machine and Table 4.1 gives its specifications. The main mechanical features of this prototype are the followings:

1) The machine has a double nozzle capable of sweeping and spraying over a width of about 40 to 50 cm at a time.
2) The double-nozzle is supported by a mechanical balancer and can be positioned and controlled by the operator with little effort.
3) The whole mechanism is mounted on a multi-directional aerial work platform.

This mechanism makes it possible to carry out the two types of spraying described above, spraying while the vehicle moving straight, and semi-automated spraying by a worker with the aid of the balancer while the vehicle stays put. The expectant area that can be sprayed per unit time is significantly increased by incorporating the double-nozzle design.

To maintain a constant distance between the nozzle and the surface being sprayed, a telescopic mechanism controlled by an ultrasonic distance sensor is also fitted.

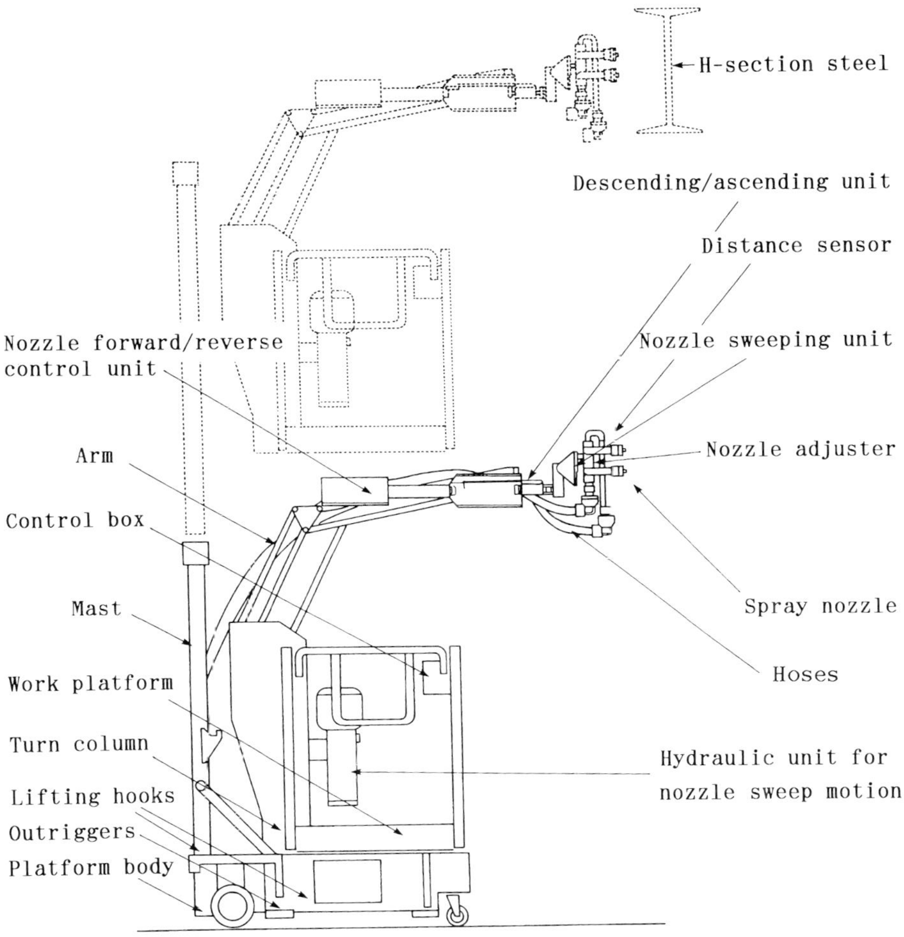

Fig. 4.1 Prototype Machine

Photo 4.1 Prototype Machine

Table 4.1 Specifications

Platform specifications	
Platform	KOMATSU LS100
Length	2,150 mm
Width	1,425 mm (outriggers extended), 780 mm
Height	2,270 mm
Maximum payload	100 kg
Weight	1,050 kg
Power source	Battery DC 24V
Drive type	Hydraulic diagonal two-wheel drive
Safety devices	
	Overload alarm
	Emergency descent valve
	Emergency stop unit
	Inclination alarm
	Travel limiters
Working performance	
Forward/backward distance	440 mm (manually)
Travel speed	0 to 6.7 m/min.
Descending/ascending stroke	4,650 mm, including descending stroke of work platform
Nozzle forward/backward stroke	280 mm (automatic) with ultrasonic distance sensor
Nozzle forward/backward speed	870 mm/min.
Nozzle forward/backward drive type	Electric motor, AC 100V
Nozzle inching rate	0 to 150 cycles/min.
Nozzle inching drive type	Hydraulic motor, AC 100V
Nozzle sweeping	90 degs. (manually)

5. Spray tests with the prototype

Photo 5.1 Spraying Test

A simple spraying test was carried out on an actual construction site in which fireproofing materials were applied to a wall using the prototype machine(See Photo 5.1). The purpose of the test was to investigate how much the spraying efficiency is improved by the double-nozzle design, as well as to acquire a qualitative comprehension of the advantages and disadvantages with regard to the semi-automated man-machine cooperative system.

The test results can be summarized as follows.

<u>1) Test under motion</u>

From the running test, which involved spraying while the vehicle was in motion, an instantaneous spray efficiency was obtained. This value represents the area sprayed per unit time, excluding all preparatory work. The value obtained was 1.1m^2/min., almost double the record for manual spraying, which is 0.6m^2/min. Figure 5.1 shows the thickness of the layer of fireproofing materials sprayed in the test.

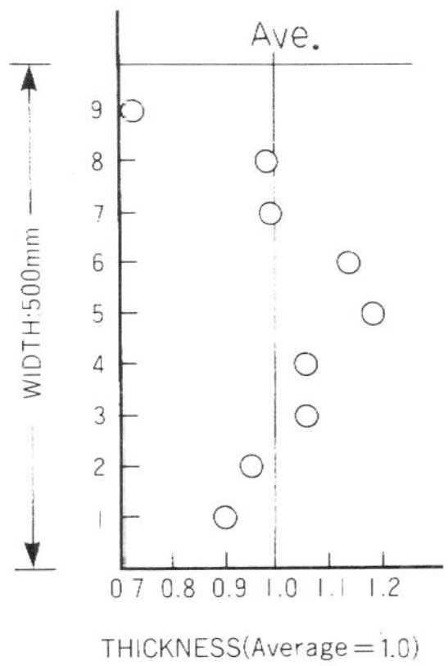

Fig.5.1 THICKNESS OF SPRAYED COMPOSITION

<u>2) Fixed-vehicle-position test</u>

Spraying tests were carried out in which the double-nozzle was positioned by a worker while the vehicle was held in one position. The direction and position of the nozzle were controlled manually through a combination of the platform's descending motion and zero-weight positioning with the mechanical balancer. It is obvious that operator's workload was considerably lightened, and they got less tired than with conventional spraying work. Observations of the test indicate that even those deficient in physical strength could achieve intensive works with the aid of this automated unit. An uneven spray thickness resulted when the operator lacked skill, so we are now planning to investigate how this can be improved in the next step.

6 Conclusions

A semi-automatic fireproof material spraying robot has been experimentally constructed. It features a double-nozzle design and a man-machine cooperative system. Small-scale spraying tests were carried out with the prototype. The tests elucidated that the following ideas, conceived at the initial stage, can provide a guide to the development of construction robots;

1) Complex mechanisms and advanced control functions can be eliminated by adopting a man-machine cooperative system; this makes it possible to reduce the cost of the robot.
2) A robot can operate under heavy loads that would be difficult for humans, making it possible to raise work efficiency.

References

[1] H.Miyamoto, et al., "Development of System Capable of Spraying Fireproof Covering Materials," in Proceedings of the 9th International Symposium on Automation and Robotics in Construction, June 3-5, 1992.

Automation and Robotics in Construction X
G.H. Watson, R.L. Tucker and J.K. Walters (Editors)

Robotics and Automation in the Construction of the Sliding Domes of King Fahd's Extension of the Prophet's Holy Mosque in Madinah, Kingdom of Saudi Arabia

Khalid A. H. Taher[a], Bakr M. Binladin[b], Fuad A. Rihani[c], and Mahmud Rasch[d]

[a]Civil Engineering Department, King Saud University, P.O. Box 88971, Riyadh 11672, Saudi Arabia
[b]Saudi Binladin Group, P.O. Box 12995, Jeddah 21483, Saudi Arabia
[c]Saudi Binladin Group, P.O. Box 12995, Jeddah 21483, Saudi Arabia
[d]Postfach 801012, 7000 Stuttgart 80, Germany

Abstract

The King Fahd's extension of the Prophet's Holy Mosque in Madinah, Saudi Arabia, is the largest and most complex addition in the Mosque history. One of the most innovative features in this landmark project remains to be the unique twenty seven sliding domes, 18m x 18m each, covering its twenty seven interior open courtyards. With complex geometric and quality requirements, the design and construction of most components and segments of these domes had to rely on computer-driven robots. This paper describes the sliding domes and their major elements, materials and features. It also describes in detail the process used in their production, with emphasis on the role of robotics. The paper elaborates on how the use of robot-intensive industrial processes are utilized in the construction of such a major civil project. The successful completion of these magnificent domes witnesses to the validity and practicality of the use of robotics in the construction industry.

1. INTRODUCTION

Traditionally, the quality of a building comes from its design, the arts and crafts developed over the centuries and the special techniques used in its construction. Industrial production processes have gradually replaced the traditional ways of making objects. Until recently the design and construction processes have oscillated widely between excessive commercialization of ornamentation, typical of late nineteenth century, and its absence in the "modern" movement.

The advent of a wider application of computers and robots in the design and construction of civil projects has ushered a new era into the world of architecture and engineering where these new tools have received increasing attention by both researchers and practitioners around the world [1]. One of their significant contributions is reflected in how they have allowed a very refined development and control of the desired forms that can serve human aspirations and spirit rather than the law of the machine. This is enhanced by the many other advantages of the new applications. It is estimated that computer-integrated design and construction can result in approximately 10 to 15 percent increase in the overall productivity with accompanying significant improvements in cost, safety and quality [2].

On the other hand, robotics in construction have faced numerous challenges, some shared with the industrial processes, that have required extensive research and investment. These challenges have been grouped into three general areas: the development and control of the robot itself; its application into a production process; and the formation and delivery of the final product. In the area of their development and application, robotics have dealt with several significant issues which included the level of programmability, flexibility and intelligence of the robot itself, the scale and size of the product, degree of repetitiveness in the operations, number of product elements, economics of automation, design optimization and the reconciliation of traditional crafts and materials with their modern counterparts [3-5].

The design and construction of the Prophet's Holy Mosque in Madinah utilized numerous sophisticated processes, such as the automated production of artificial stones, marble and granite and other materials. However, the design and construction of the sliding domes remain to be the most innovative feature of this project where computer-controlled robots have been used extensively. Following is a description of the sliding domes and the use of computers and robotics in their design and construction.

2. THE PROPHET'S HOLY MOSQUE AND THE SLIDING DOMES

The Prophet's Holy Mosque in Madinah, the second holiest site in Islam, consists of the existing old part, about 16,500 square meters in area, and the new King Fahd's Extension of 82,000 square meters. The new extension includes twenty seven inner open courtyards, 18m x 18m each, which required a mobile cover that could be opened and closed to control and support the air-conditioned environment of the Mosque. This resulted in the design and construction of twenty seven sliding domes, Figure (1).

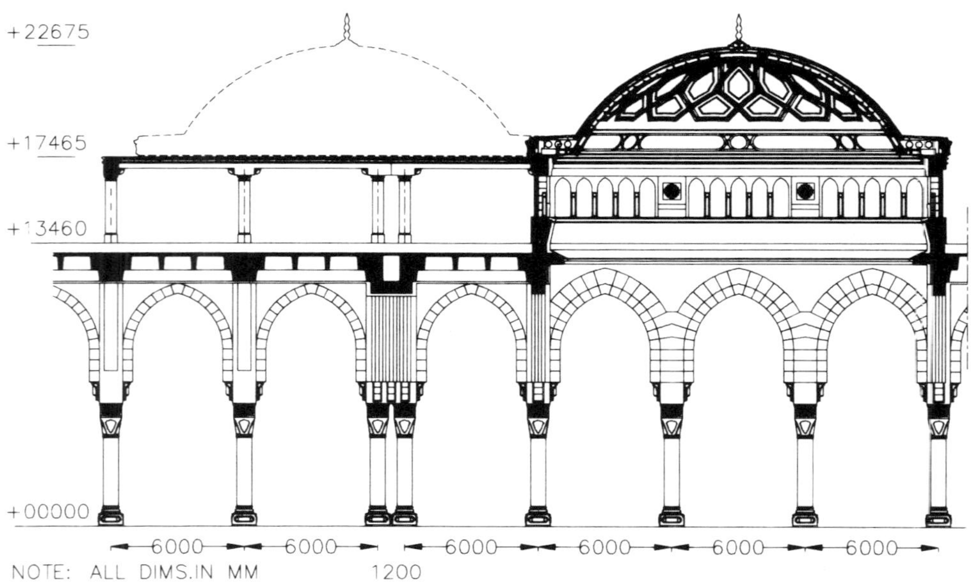

Figure 1: Cross-Section of a sliding dome showing the open and closed positions.

The sliding dome consists of a spherical portion and a support frame riding on four steel wheels on steel rails. Each wheel is driven by a direct-drive electric motor with a manual over-ride. The spherical portion of the dome consists of the following elements(Figure 2):

1. A steel structure composed of 24 meridians and 3 ring beams. The structure supports the inner and outer cladding in addition to an internal insulation and waterproofing systems.
2. An inner cladding of wood-laminate reinforced by wood-laminate flanges and supporting the inside intricate decorative elements.
3. An outer cladding of 50 mm hexagonal ceramic tiles glued with epoxy resin to carbonfiber laminate which in turn is reinforced with carbonfibre flanges.

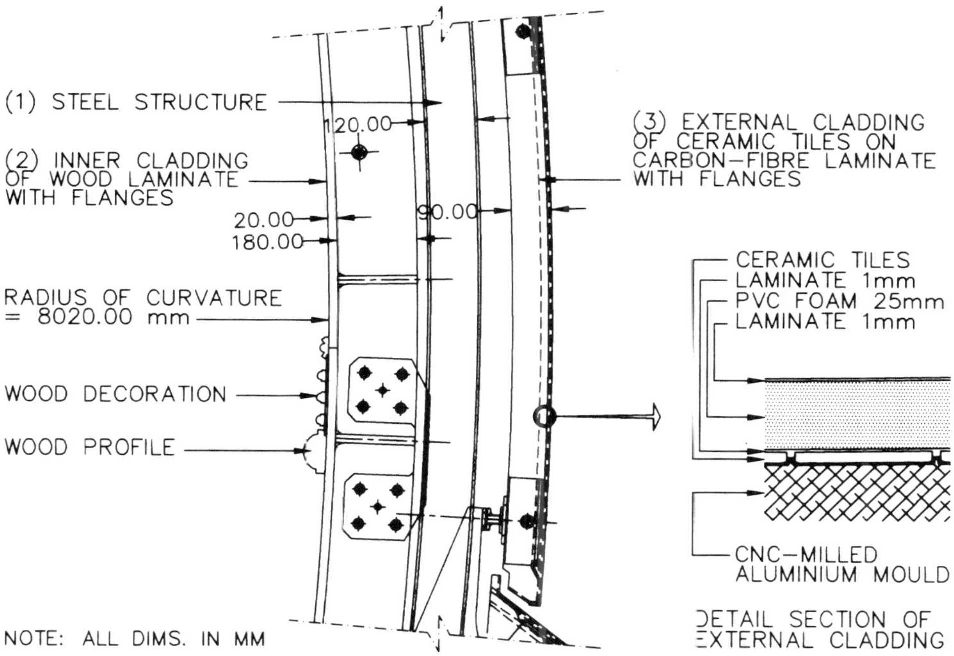

Figure 2: Cross-Section of the Spherical Portion of the Sliding Dome.

3. DESIGN PROCESS OF THE SLIDING DOMES

At the start of the project, initial thoughts were directed to selection of known traditional materials, maximum use of off-the-shelf components and reliance on in-situ construction. The uniqueness of the Mosque and its sliding domes and the demanding aesthetic requirements of Islamic design revealed rapidly that a wide-range evaluation of all possible designs, materials, crafts and technologies was warranted.

The adopted process involved the following steps:

(a) Preliminary designs for all feasible options;
(b) Working scale models;
(c) Review, evaluation and choice of preferred design and materials;
(d) Final design; and
(e) Construction of a full-scale working prototype.

It was through the last two steps that construction methodologies were evaluated and the use of CAD and robotics in the construction of the prototype was decided. Economic analysis of robotic production of critical elements in the prototype predicted equivalent cost to those of traditional methods but with far superior quality and precision. This was an early solid proof that application of robotics to the total project would be economically competitive.

4. DESIGN AND CONSTRUCTION OF THE SLIDING DOMES

Technological adaptation of successful concepts from other industries to construction offers an innovative progress [6]. This is true inspite of the common belief that the construction industry has been viewed as a prototype-oriented industry and as such unable to apply production methods developed for the manufacturing industry [7]. Whittaker, commenting on the problem of adaptability and flexibility, stated that "it is common to mistake or overestimate chaos in a task environment simply because form and understanding are not apparent. There is a great prospect for structuring the apparently unstructured either by discovering structure or by imposing it" [8]. One characteristic of a well-structured and efficient production system is repetitiveness. Construction projects are considered unique products and therefore the use of the repetitive processes is difficult. However, Halpin and Woodhead proposed that if one organizes the functions used to create a product hierarchically, and evaluated the repetitiveness in each level, a more diversified analysis will become apparent [9].

This understanding was translated into a special process for the design and construction of the sliding domes (Figure 3). In this process the dome was segmented and divided into different components with different levels of repetitiveness, and the production methodology, including the use of robotics, was easily determined for each.

The first level of the process involved the use of Computer-Aided Design, CAD, where very complex designs were created, optimized and brought into reality. In this process, design tasks such as the finite-element structural analysis and design optimization were integrated through the CAD tool. CAD has allowed designs to be processed at an incredible degree of sophistication and proved that knowledge and advanced technology gained from hi-tech manufacturing industries could be adapted and applied to this construction project.

The next level involved the direct utilization of Computer-Aided Machining, CAM, in the production process where incredible advantages could be realized (Figure 4). Many forms that could have only been approximated were now made with great precision. Errors, that came from transfer of information from one medium (i.e.design drawing) into another (i.e.shop drawing) were completely avoided. Tolerances could be reduced to minimal ranges and unparalleled high quality standards in technical and artistic details were reached.

The design of the sliding domes was made from the beginning with a three dimensional CAD program on PCs. The spherical shape of the dome dictated the use of a design process that could handle the complex mathematical formulation that described some of the inner and outer decorative components.

The inner cladding was planned to be a composite wood laminate consisting of several layers of wood veneer bonded with epoxy resin. The entire shell was segmented into single

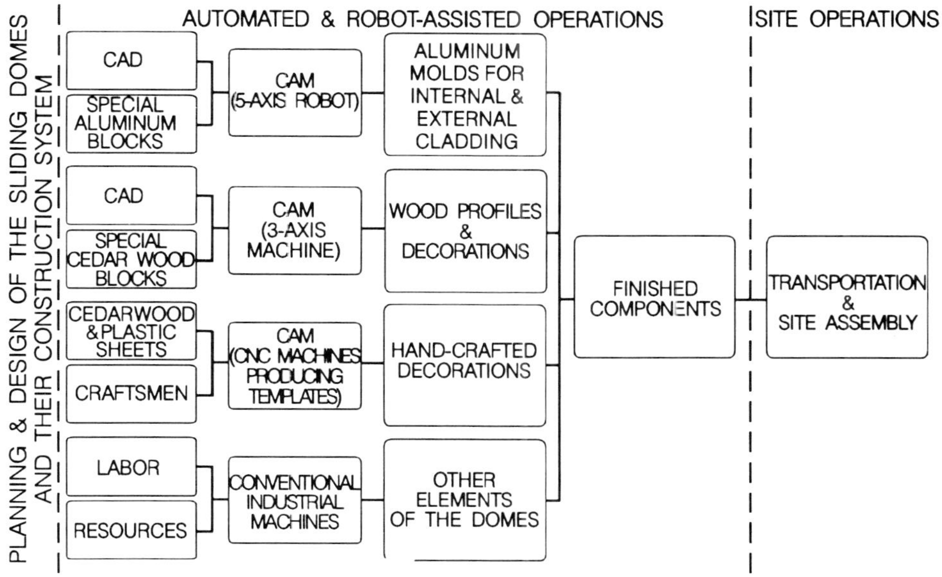

Figure 3: Robotics and Automation in the Design and Construction of the Sliding Domes.

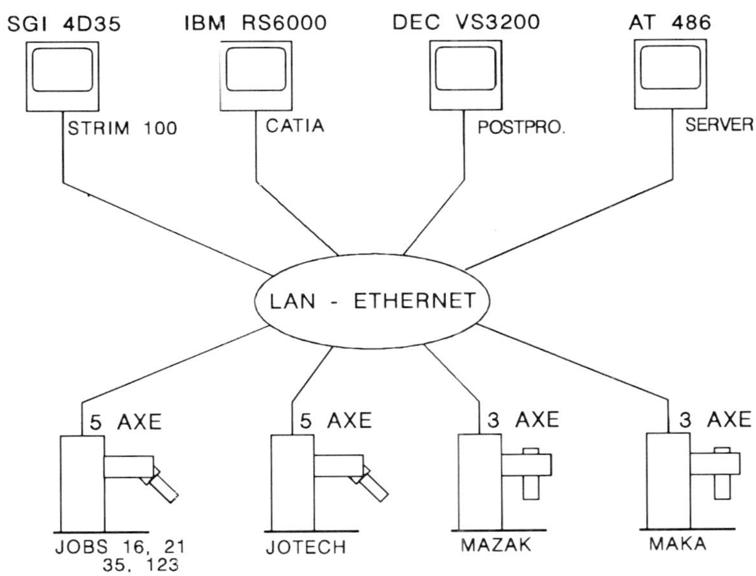

Figure 4: Interaction of CAD and CAM systems in the Sliding Domes Project.

parts that took advantage of repetitive operations and accommodated shipping requirements (Figure 5). Applied on these panels were decorative elements made of cedar wood on the inside and stiffeners and flanges on the outer side. The floral ornaments were designed to be traditionally carved floral pieces framed by a bold geometrical pattern of double-curved cedar wood profiles. All of this required great precision, especially where the joints were exposed. To meet such demanding tolerances, all parts were made on CNC machines. Some of the surfaces were guilded, and semi-precious stones were fitted through gold-plated brass frames. It was clear that without the help of these CNC machines the integration of traditionally hand-crafted objects into the industrially produced panels would have been rather impossible.

The outside panels were also subdivided into segments that would ship easily (Figure 5). But unlike the inner panels, designed to be a background for artisan work, these panels were industrially produced in every aspect. The design required flat and spherical panels covered with multi-color pattern of ceramic tiles. This design used a traditional Islamic theme: a band of turquoise tiles placed in a white background and separated by a golden contour. In order to achieve the necessary aesthetic quality, the pattern was designed in CAD detailing the correct placement for every single tile. A program was developed to drive a very large 5-axis routing robot by which the various aluminium molds where milled. The molds had a pocket for each one of the 50 mm hexagonal tiles, bands and contours.

The resulting three dimensional aluminium molds were used directly in the production process. In every pocket the respective tile was laid upside down with part-tiles precut according to the CAD files. The cutting of tiles was done on a CNC water-jet cutting machine. On the back of these tiles the first layer of epoxy-carbonfibre laminate was cast directly. Then a PVC layer was placed and the final layer of epoxy-carbonfibre laminate was cast to complete the sandwich. The entire package was then covered with a foil and pressed in vacuum under high temperature. After the resin was hardened, the panels were taken out of the molds ready for cleaning and shipment. The resulting panels were of great precision, stability and beauty. Without the application of such CAD/CAM processes, using one of the largest milling robots in Europe, these panels would not have been feasible in the quality required for this very special Mosque.

The steel structure, waterproofing, insulation, driving system and assembly tools and materials were all manufactured according to established industrial processes. All components of the Sliding domes were shipped to site for assembly where the knowledge and experience gained from building the original prototype proved to be of great value.

CONCLUSIONS

The design and construction of the Sliding Domes have been completed in a successful, economic and timely manner. The project represents a real-life example of a new era in architectural design and engineering work, where complex designs and construction requirements were brought together in an optimum fashion, and where the utilization of programmable and flexible robots in construction was achieved. The application of computers, robotics and automation, normally associated only with hi-tech industries, demonstrated the successful integration of new technologies with traditional designs, crafts and materials.

The Sliding Domes project has also proven that the adverse effect of physical distances between the construction site and the locations of production operations is losing importance rapidly. This was demostrated by how resources as well as design and construction processes in four continents were integrated and brought together to build a fitting and functioning part of this magnificant edifice.

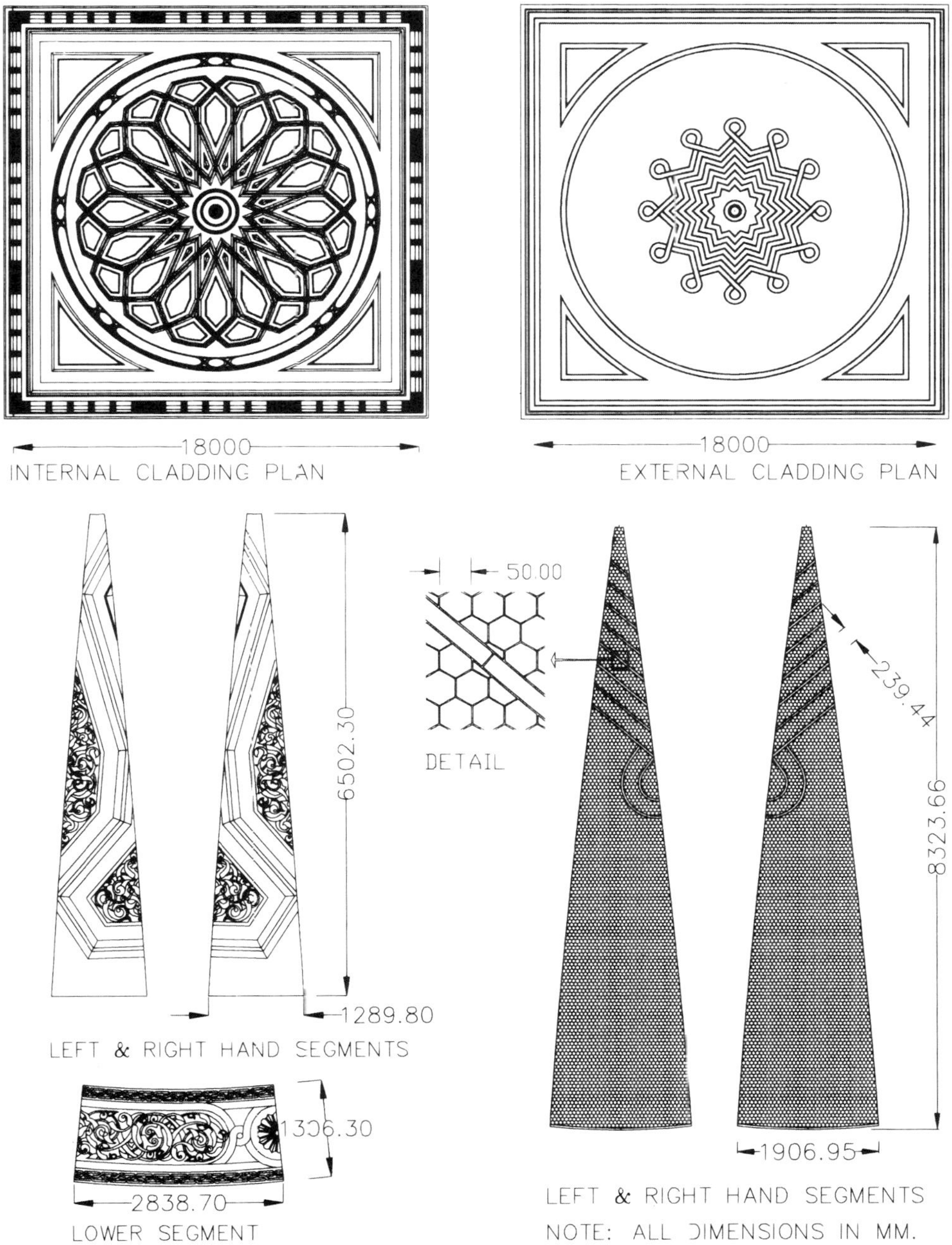

Figure 5: Patterns and Segmentation of Internal and External Cladding of the Sliding Domes.

Finally, the project reflects the high level of sophistication in the construction industry in Saudi Arabia, as represented in the positive use of a new advanced technologies, and in the maturing of a new breed of national firms capable of handling such demanding and complex projects. The experience and knowledge gained from the sliding domes project point to a promising future where the use of robotics and automation in construction is expected to receive wider acceptance resulting in the fulfillment of its envisioned role.

REFERENCES

1. Demsetz, L. January 1990. Automated Equipment and Robotics for Construction. *Construction Specifier*, 84-93.
2. Skibnsewski, M. and Nof, S. 1989. A Framework for Programmable and Flexible Construction Systems. *Robotics and Autonomous Systems*, no.5 Elsevier Science Publishers.
3. Singh, A. April 1991. Construction and Robotics: Problems and Solutions. *Construction Congress '91*, 234-239, ASCE.
4. Moseihi, O. and Hason, S. October 1989. Robotics in Construction: Implementation and Economic Evaluation. *Canadian Journal of Civil Engineering*, 678-683.
5. Skibnsewski, M. and Russell L. June 1989. Robotic Applications to Construction. *Cost Engineering*, 10-18.
6. Groover, M. et al. 1989. Survey of Robotics Technology in Construction. *SME Technical Paper* (series), Lehigh University, Pennsylvania, USA.
7. Bernold, L. and Abraham, D. April 1990. FMS Approach to Construction Automation. *Journal of Aerospace Engineering,* 108-121, ASCE.
8. Whittaker, W. 1986. Construction Robotics: A perspective. *Special Report*, Robotics Institute, Carnegie-Mellon University, Pittsburgh, Pennsylvania, USA.
9. Halpin, D. and Woodhead, R. 1976. Design of Construction and Process Operations. John Wiley and Sons, Inc., New York, N.Y., USA.

Automation and Robotics in Construction X
G.H. Watson, R.L. Tucker and J.K. Walters (Editors)

Propeller type wall-climbing robot for inspection use

A. NISHI and H. MIYAGI

Faculty of Engineering, Miyazaki University
Gakuen-Kibanadai, Miyazaki 889-21, JAPAN

Abstract

Use of a wall-climbing robot for inspection or maintenance of wall surfaces has been anticipated for a long time. Three quite different models have been developed in our laboratory. The present model can move on a wall by using the thrust force of a propeller and can fly whenever it is required. Its mechanism and control system are discussed.

1. INTRODUCTION

A robot capable of moving on a vertical wall has been looked forward to a long time. It could be used for wall inspection, rescue and fire-fighting in high-rise buildings. Three quite different types of wall-climbing robot models have been developed in our laboratory over the last 20 years. The first model, shown in Fig.1, has a large sucker and crawlers as a moving

Figure 1. A large sucker model.

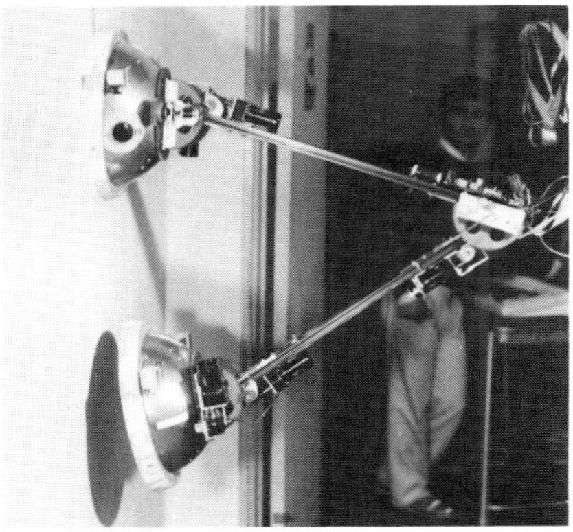

Figure 2. A biped walking model.

mechanism.[1] Recently, many varieties of this type model have been developed for wall inspection in Japan. The second type, shown in Fig.2, is a biped walking model, which has a small sucker on each foot.[2][3] As this can be applied to almost all irregular wall surfaces, it has a wider applicability than the former.

In general, the walking motion is not very quick so that the walking model takes much time to climb up to higher places on the wall. Unfortunately, a robot would be required to climb up to higher places on buildings in a short time for the inspection or emergency purposes. The third model, shown in Fig.3, aims at these purposes.[4]-[8] It has propellers, and their thrust forces are inclined a little toward the wall to make use of the frictional force between the wheels and the wall surface, as well as to support the robot itself. Sometimes unexpected strong winds occur on the walls of high-rise buildings. In such cases, the control system to compensate for the wind load is important to avoid having the robot falling from the surface. This situation has been previously discussed in detail in Refs.(6)-(8).

On the other hand, often there are many obstacles at the lower part of buildings, such as trees, eaves, entrances, etc.. In such cases, it would be useful if the robot could fly over these obstacles and can reach the upper wall surface. Moreover, if the robot falls from a higher place on the wall surface by accident, it needs to make a soft landing to avoid endangering itself or its surroundings. These maneuvers can be done by employing a mechanism and control system enabling it to fly. As the robot has an enough thrust force to sustain itself in the air, it can fly or land. This is the fourth type model and its control mechanism is discussed in this paper.

Figure 3. A propeller type wall-climbing robot model.

2. MECHANISM OF THE ROBOT

2.1 Wall-climbing mechanisms of insects and small animals

The wall-climbing mechanisms of insects and small animals are of interest. A black fly, a grasshopper, a small frog and many other small animals can walk on vertical window glass. They have their own spacial feet mechanisms. Most of these small insects and animals have microscopic mechanisms utilizing adhesive cilia or body skin.

On the other hand, a small bee, like the one shown in Fig.4, is shown flying along a glass surface. It can sustain its weight by the reaction of the air flow produced by its wings. This, then, is an another mechanism for moving on vertical surfaces, and is the model for the present fourth type of robot mechanism.

2.2 Thrustor of the wall-climbing robot

An example of a conceptual model is shown in Fig.5. A conventional thrustor is a combination of an engine and a propeller. Although a small jet engine and rocket motor can be made available for emergency use, they are expensive and noisy compared with the conventional engine and propeller. There are two types of propellers : those with larger diameters similar to helicopter rotors, and smaller ones similar to light plane propellers. The former can produce a larger thrust force for the same engine power. However, the wind load is larger, and it is difficult to design a wall-climbing robot using this type. Therefore, the latter has been used for the model concerned here.

Figure 4. A small bee flying along a glass.

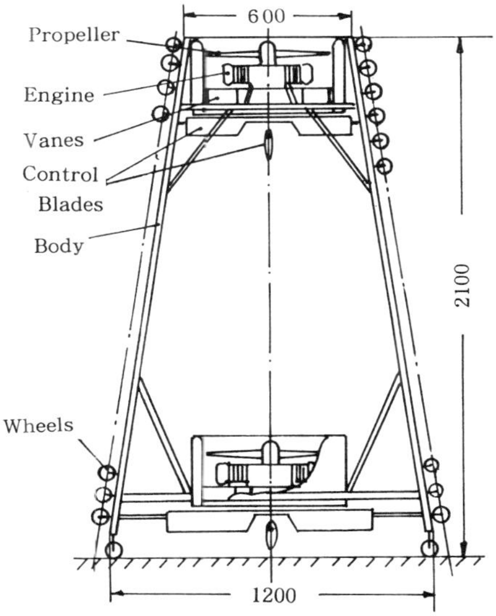

Figure 5. Schematic of a conceptual model.

2.3 Mechanism of the flight model

To use a robot for the inspection of a wall surface, first, it is necessary for the robot to reach and attach itself to the wall surface at the lower part of the building. However, often there are obstacles such as trees, eaves and entrances at the lower part of the building. In such cases, it may be difficult for the robot attach itself to the wall directly. As the robot has enough thrust force to lift itself up in the air, it can fly over these obstacles employing a suitable mechanism and control system.

Almost the same situation will occur at higher parts of the buildings, where there are long eaves or steps on the wall surface.

Moreover, there is danger of the robot falling from the wall surface by an unexpected strong wind or accident. In such cases, the robot must make a soft landing, or the fall would be very dangerous for the robot itself and its surroundings. These situations can be avoided by designing the robot so that it is able to fly or land.

However, its major purpose is not flying like a helicopter, but moving on a wall surface using its wheels, since moving on a wall is safer. Therefore, it has a long body with many wheels and its thrust force has a component toward the wall to produce frictional force between the wheels and the wall surface. The robot was designed for this requirment. Two propellers have been installed at the top and bottom of the trapezoidal body. Many light weight wheels have been arranged on the outside of the body. So that, any outer plane can touch the wall surface. When the robot touches the wall, the thrust force axis is inclined a little toward the wall, and the frictional force of the wheels is produced automatically.

2.4 Control system of the flight model

There are two ways to produce the control force in the air. One is a method of inclining the thrust force directly. The other is that the propeller slip stream by the vanes. The former can produce larger force, however, its mechanism is heavier, so that it has a larger time lag of control than the latter. On the contrary, the control mechanism is easy to construct and regulate for the latter.

In general, as the wall-inspection robot is used in a calm or weak wind conditions on the wall, the large force does not need for the control. Therefore, the latter method can be employed to the flight model.

The slip stream of the propeller rotates and thus producing reaction torque on the body. Its magnitude is variable for changing the thrust force depending on the up and down motion of the robot. The regulating stream vanes are installed in the slip stream, and a couple of the vanes are controlled to eliminate the reaction torque. This motion is sensed by a gyroscope. Such a gyroscope unit has been already developed for the control of model helicopters. It can be applied to this model with a small adjustment.

The up and down motion of the robot is produced by controlling the thrust force, i.e., the engine revolutional speed.

Flight control is accomplished by control blades installed in the slip stream of the propeller. These are set in the down stream of the regulating vanes. Two blades are set closswise. Each of them are controlled independently to produce a side force in any direction.

Both the regulating vanes and the control blades are employed in each engine slip stream, as shown in Fig.5.

As the body is symmetrical along the thrust axis, a righting force is not produced by the thrust force, when the robot is inclined in the air. Therefore, the control blades in each engine slip stream should be adjusted independently to produce a righting force around the center of gravity. This control occurs automatically by the onboard computer.

The flight control can be summarized as follows :

(1) avoidance of the rotational motion produced by the reaction torque is accomplished by the automatic control of regulating vanes ;

(2) the up and down motion of the robot is controlled manually by the engine revolutional speed ;

(3) the side movement of the robot is developed by the side force produced by the control blades ; and

(4) the control of inclined angle of the body is made by the independent control of the upper and lower blades.

3. MECHANISM AND TESTING OF THE LABORATORY MODEL

3.1 Mechanism and control system of the laboratory model.

Before the construction of a flight model, an investigation was carried out by using a laboratory model, shown in Fig.6. It had a mechanism and control system similar to the flight model. It was designed to obtain experimental data easily. The mechanisms and communication system were as follows :

(1) the rotational mechanism,

(2) the up and down motion,

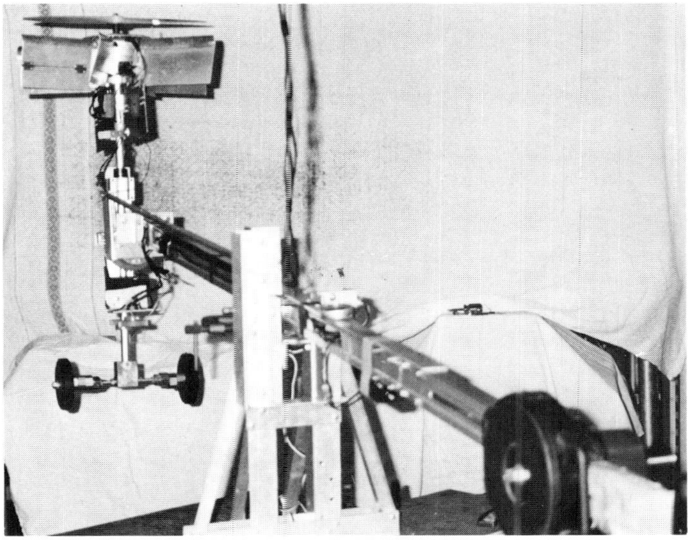

Figure 6. A laboratory model.

(3) the side movement, and
(4) wireless communication between the computer and the model.

A model which had a propeller driven by a motor and a simply-shaped body was set at the end of a bar. At the other end of the bar a counter-weight was installed. When the propeller was rotating, the weight was almost balanced. The bar was supported with a hinge and a pivot, so that it could move both up and down, and to the sides. The model was also supported by a similar mechanism at the end of the bar. It could rotate about the motor axis when the reaction torque of the propeller was produced. The laboratory model had five degrees of freedom.

The rotating angle produced by the reaction torque was sensed by a potentiometer. It was adjusted by regulating vanes with feedback signals from a computer. The movements up and down and to the sides were controlled by regulating the motor revolutional speed and the angle of the control blades, respectively. As each control was independent, the equations of motion were simple. The angles and angular velocities were taken as the feedback veriables in each direction.

3.2 Wireless communication system and its preliminary testing

Wireless communication is required to control the prototype in the air or at higher places on a wall. The radio controller of a model plane was employed to simulate the control system for the laboratory model. As shown in Fig.7, the sensor signals from the model were given by wire to the computer. The feedback signals to the model were given by radio signals. Two channels were used. These were the control signals to the servo-motors of the regulating vanes and control blades. The preliminary experiment demonstrated that this method was an effective and easy way to control the robot. This can be developed for the mutual communication between the flight model in the air and the computer on the ground.

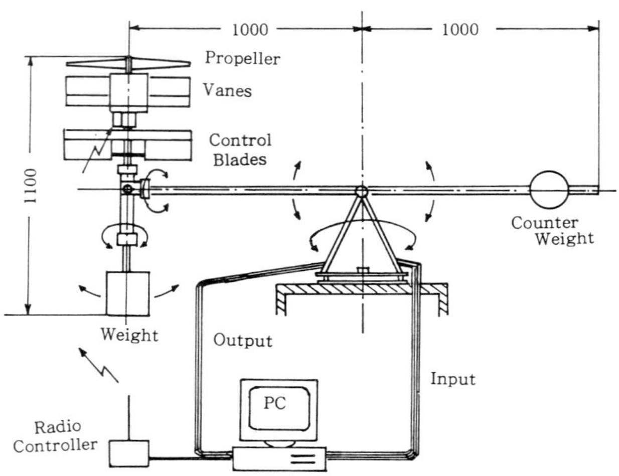

Figure 7. Schematic of a control system.

3.3 Experimental results

The experiment was carried out in three stages. First, the rotation of the model by the reaction torque could be controlled easily by the vanes, and the model could be fixed at any position of rotation. For the flight model, this control will be accomplished by using a gyroscope unit, which has been widely used for model helicopter control.

The second control was an up and down motion of the model. This was sensed by a potentiometer as the inclination angle of the bar at the support point. The model could be fixed at a given height by controlling the motor revolutional speed. For the flight model, the height control would be made manually.

The third stage was the movement of the model to the sides. This was accomplished by changing the control blade angle. Strictly speaking, the magnitude of the thrust force might be changed by controlling the blade angle, since the drag force of the blade decreased the thrust force. Therefore, the revolutional speed of the motor had to be changed to maintain a constant height. However, as the drag force was small compared with the side force, this effect did not appear in the model. In this case, a constant side force was acting on the model, and the position could be adjusted by feedback control of the rotational angle and the angular velocity of the bar.

4. CONCLUSION

Three different types of wall-climbing robots have been developed in our laboratory. The first model had a large sucker and could be utilized on flat and/or wide wall surfaces. The second was a biped walking model, which had a small sucker on each foot. This could be applied to almost all irregular wall surfaces. The third model could move on a vertical wall at high speeds. It had propellers and the thrust force supplied both the forces to lift the robot itself up and produce the frictional force between the wheels and wall surface to be safe on a wall.

When the third model is employed to inspect a wall surface, it is necessary for it to reach and attach itself to the wall surface, at first. Often there are obstacles at the lower parts of the buildings, such as eaves, trees, entrances, etc.. Therefore, it would be very useful, if the robot could fly over these obstacles and reach the upper wall surface. Moreover, when the robot falls from the wall surface by accident, a soft-landing control has to be employed to avoid danger to the robot and the surroundings. For these purposes a flight model has been considered. The mechanism and control systems for such maneuvering have been discussed. A laboratory model was constructed and tested. As a result, it was demonstrated that the laboratory model could be controlled easily, and similar control methods can be applied to the flight model. The flight model is now in preparation.

REFERENCES

(1) Nishi A., Wakasugi Y. and Watanabe K., Design of a Robot Capable of Moving on a Vertical Wall, Advanced Robotics, Vol.1, No.1 (1986),33.

(2) Nishi A., Biped Walking Robot Capable of Moving on a Vertical Wall for Inspection Use, Proc. 5th Int. Sym. Robotics in Construction (1988), 535.

(3) Ikeda K., Nozaki T. and Shimada S., Development of Wall-Climbing Robot, Proc. 1st Int. Sym. on Construction Robotics in Japan (in Japanese) (1990), 107.
(4) Nishi A. and Miyagi H., A Wall-Climbing Robot Using Thrust Force of Propeller, 1st Report, Trans. Jpn. Soc. Mech. Eng. (in Japanese), Vol.57, No.543C (1991), 176.
(5) Nishi A. and Miyagi H., A Wall-Climbing Robot Using Thrust Force of Propeller, 2nd Report, Trans. Jpn. Soc. Mech. Eng. (in Japanese), Vol.58, No. 547C (1992), 163.
(6) Nishi A., A Wall-Climbing Robot for Inspection Use, Proc. 8th Int. Sym. on Automation and Robotics in Construction, (1991), 267.
(7) Nishi A., A Wall-Climbing Robot Using Propulsive Force of Propeller, Proc. 5th Int. Conf. Advanced Robotics, (1991), 320.
(8) Nishi A., Control of a Wall-Climbing Robot Using Propulsive Force of Propeller, Proc. IEEE/RSJ Int. Workshop on Intelligent Robots and Systems IROS '91, (1991), 1561.

Automation and Robotics in Construction X
G.H. Watson, R.L. Tucker and J.K. Walters (Editors)

Autonomous Learning in Vertical Environments

Professor Matthew Cusack[a] and Dr. Terence Fogarty[b]

[a]Faculty of the Built Environment, University of the West of England, Coldharbour Lane, Frenchay, Bristol BS16 1QY, Great Britain

[b]Bristol Transputer Centre, Faculty of Computer Studies and Mathematics, University of the West of England, Coldharbour Lane, Frenchay, Bristol BS16 1QY, Great Britain

Abstract

This paper describes the further development of earlier work on wall climbing robots and recent research on the development of control systems, to enable the robots to learn the best strategies for climbing and carrying out various inspection and maintenance tasks on civil engineering and building structures.

There are three main strands to this research: a rule-based learning system, the wall climbing robots, and the integration of the two into an autonomous, adaptive system.

1. INTRODUCTION

The aim of this research is to design a rule-based learning system to support the various kinds of adaptation useful to autonomous systems in general and climbing robots in particular, i.e. evolution, learning and programming. Rule-based learning systems have been built that can achieve one or two of these kinds of adaptation on a particular simulated task. To date, no one has built a system to achieve all of these and they have not as yet been used on real systems.

The rule-based learning system to control a climbing robot will be programmed with rules enabling it to:-

i. recognise the various surfaces it will encounter via its sensors,

ii. constrain its actions to those which are both safe and sensible,

iii. have a goal towards which it can strive,

iv. measure how well it is doing.

Thus an interface between rules at the programming level and rules at the learning level will be provided. It will also be capable of discovering new rules and presenting these to a human being.

2. RULE-BASED LEARNING SYSTEMS

Rule-based learning systems [1] lie midway between neural networks and symbolic processing systems and combine the benefits of both. They are massively parallel, message passing, rule-based systems which use genetic algorithms [2] to discover new rules as well as providing for reinforcement learning and programming. It has been proposed that a suitable application of genetic algorithms is to evolve robots [3]. The most suitable way to use genetic algorithms to evolve the control systems for real robots is within the rich framework provided by rule-based learning systems.

Autonomous control systems need to be able to adapt to circumstances, i.e. to weigh up a number of strategies based on how well they are working and to choose the best amongst them with high probability. So appropriate reinforcement algorithms are being used. The standard reinforcement learning algorithm for rule-based learning systems is the bucket brigade [4] but the interaction between this and the genetic algorithm is complicated and has not been shown to work reliably on any system. It is proposed that the genetic algorithm be decoupled from the reinforcement learning algorithm and that a number of other algorithms be looked at to provide the reinforcement component of the rule-based learning system. These will include back-propagation and temporal difference methods.

This process of adaptation must work within the constraints of programmed and learned rules. It is this interaction between evolution, learning and programming that must be resolved to allow the genetic algorithms to be used on real systems.

At a recent workshop on learning systems [5] a number of groups presented successful applications of the genetic algorithm to control problems. However, one cannot evolve complex systems with a simple genetic algorithm nor is it wise or safe to start from scratch in real applications where programmed knowledge can provide constraints for the genetic algorithm to work within it [6]. If the genetic algorithm is to be used to evolve systems for industrial [7] or commercial [8] applications this is best done in the framework provided by rule-based learning systems. These:-

i. can be programmed,

ii. can learn rules from examples,

iii. can weight rules by reinforcement such as provided by reinforcement learning algorithms,

iv. can discover new rules using parallel genetic algorithms which allow both co-evolution and growth to take place.

3. WALL CLIMBING ROBOTS

Prototype wall climbing robots that can be used as test beds for a range of inspection and maintenance end effectors are being designed and constructed. Each robot will:-

i. be able to travel quickly to the points requiring inspection,

ii. be able to move in smaller steps when using measuring equipment,

iii. turn, incrementally, on its own axis,

iv. have as high a power to weight ratio as possible so that reasonable payloads can be carried and to ensure that the robot is as stable as possible when using the tools and end effectors fitted to it,

v. enable the tools and measuring equipment to be easily fitted and interchanged,

vi. be able to check the feet for vacuum.

The robots will be able to move over the surfaces of buildings and be able to perform tests on the structure and carry out remedial work. The design will aim at the following target group of structures:-

i. a typical motorway bridge

ii. a Yorkon type office building

iii. a multi-storey residential building

It will be possible to transfer the basic concept to other robots in this field. The cost of designing a robot for a specific building will be small compared to the capital cost of the building itself. The main thrust of the project will be to develop the enabling technology which can be easily transferred to a variety of structures. Sites at Bristol and London have already been identified for test purposes. Initially, a small-scale prototype will be built to:-

i. test various feet configurations and develop climbing strategies,

ii. test a new concept in vacuum feet which will conform to most of the surfaces in use in construction,

iii. have a high degree of mobility, i.e. be able to move up and down surfaces by rotating at either 5° or 90° to traverse building structures including overhanging projections,

iv. act as a test bed to further the work already carried out at the University on testing building structures,

v. develop the technology to carry out the remedial work required on the test structures.

When the small-scale trials are in an advanced stage, a full-scale robot will be built and tested on the selected, previously referred to, test sites. The following test methods will be used:-

i. visual

ii. rebar

iii. carbonation

iv. chlorination

v. rebar corrosive

vi. dalamation

vii. voids.

Two wall climbing robots have been built at Bristol Polytechnic. The first -Sasquatch [12] - created considerable interest at the May 1991 Automan Exhibition at the National Exhibition Centre.

The second - Sysiphus [13] - was designed to carry instrumentation for measuring defects in structures and tools for carrying out remedial work. (Figure 1)

Figure 1. Sysiphus

This robot is controlled by software written in C++ on a PC and downloaded to the target system using Sourceview™. The target system is based on a 80C188 microprocessor running on an STE bus. The design is based on a sliding vacuum foot attached to a deck which can be raised and lowered relative to the main frame using pneumatically operated cams. The target processor then controls the robot movement through its parallel input/output (PIO) interface. The PIO in turn opens and closes the 12v solenoid valves allowing control of the pneumatic system and ejectors used to create vacuum for the feet.

The main frame contains the cam mechanism and has small feet attached to it, thus raising or lowering the cams, either a big foot or small feet are in contact with the surface to be traversed. The robot is capable of forward, reverse motion in a large step mode, and inching forward motion in small step mode, rotation in either clockwise or anticlockwise direction by 5 degrees per step. The platform may be parked with all twelve feet adhering to the surface, which gives a very stable platform from which to operate monitoring equipment.

4. INSPECTION AND TESTING

Some preliminary work has already been undertaken to identify appropriate testing techniques. A lightweight mechanism, operated by a pneumatic piston mounted on a supplementary frame parallel to the surface being measured, has been developed. This has been used in tests undertaken using a standard cover meter (Figure 2). Preliminary trials indicate that this method can produce reliable information on rebar cover and concrete cover depth.

Figure 2. Cover Meter

A feasibility study [9] undertaken for the DTI by CIRIA - the Construction Industry Research and Information Association - between May 1986 and July 1987 identified the inspection of civil engineering and building structures as having the greatest potential for developments in automation and robotics. Providing access has been identified as the major component in the total cost of inspection [10]. Safety statistics [11] for the period 1981 to 1985 showed that maintenance activity accounted for between 34% and 50% of the total number of fatal accidents in construction and that of these 30% were caused by falls from scaffolds with 20% as a result of falls from ladders. Apart from the human factor, accidents of this nature have substantial financial consequences for the industry.

5. CONCLUSIONS

The successful development of prototype wall climbing robots is now at an advanced stage. Initial research into the potential use of neural networks and more specifically rule-based learning systems control looks extremely promising.

Autonomous control systems need to be able to adapt to circumstances, i.e. to weigh up a number of strategies based on how well they are working and to choose the best amongst them with high probability, so appropriate reinforcement algorithms are being used. It is proposed that the genetic algorithm be decoupled from the reinforcement learning algorithm and a number of other algorithms be looked at to provide the reinforcement component of the rule-based learning system. It is the interaction between evolution, learning and programming that must be resolved to allow the genetic algorithms to be used on real systems.

6. REFERENCES

[1] Holland J H & Reitman J S (1978) edited Waterman D A & Hayes-Roth F Cognitive systems based on adaptive algorithms. Pattern directed inference systems

[2] Holland J H (1975) Adaptation Natural and artificial systems

[3] Brooks R A (1992) (eds) Varela F J & Bourgine P Artificial Life and Real Robots. Towards a Practice of Autonomous Systems p3-10

[4] Holland J H (1987) eds. Michalski R S, Carbonell J G & Mitchell T M Escaping Brittleness: The Possibilities of General Purpose Learning Algorithms Applied to Parallel Rule-Based Systems. Machine Learning: An Artificial Intelligence Approach, Vol.2, pp593-623

[5] Niven J (1992) The application of learning systems to electro mechanical engineering - workshop report. C&I Committee Sec., SERC

[6] Fogarty T C (1989) edited by Rzevski G Learning new rules and adapting old ones with the genetic algorithm. Artificial Intelligence in Manufacturing, pp275-290

[7] Fogarty T C (1990) edited by Gottlob G & Nejdl W Adaptive rule-based optimisation of combustion in multiple burner installations. Expert Systems in Engineering: Principles and Applications, pp241-8

[8] Fogarty T C, Ireson N S & Battle S A (1992) Developing rule-based systems for credit card applications from data with the genetic algorithm. IMA Journal of Mathematics Applied Business and Industry, Vol.4, No.1, pp53-9

[9] CIRIA (1987) Feasibility Study on the Application of Advanced Robotics to Civil Engineering and Construction. Report to the DTI

[10] Radivisky R A & Garas F K (1988) Advanced Robotics in the Field of Survey/Inspection Maintenance and Repair of Buildings and Structures. 5th International Symposium on Automation and Robotics in Construction, Tokyo, Japan

[11] HMSO (1988) Black Spot Construction: A Study of Five Years Fatal Accidents in the Building and Civil Engineering Industries. Health and Safety Executive

[12] Cusack M M & Thomas J G (1991) Exploration and Innovation in Robotics. Advance, Issue 2, Vol.1, Autumn

[13] Cusack M M & Thomas J G (1992) Wall Climbing Robots for the Inspection and Maintenance of Buildings. 9th International Symposium on Automation and Robotics in Construction, Tokyo, Japan

Automation and Robotics in Construction X
G.H. Watson, R.L. Tucker and J.K. Walters (Editors)

A Sensor based autonomous climbing vehicle for unstructured environments.

N.R.Bevan[a]BEng(Hons), Prof.A.A.Collie[b],
B.L.Luk[b]PhD MSc BSc(Hons), T.S.White[a]BEng(Hons).

[a]Portsmouth Technology Consultants (Portech) Ltd., Unit 12, Airport Service Road, Anchorage Park, Portsmouth, Hampshire, PO3 5RQ U.K.

[b]University of Portsmouth, Portech Mobile Robotics Laboratory, School of Systems Engineering, Anglesea Building, Anglesea Rd., Portsmouth, PO1 3DJ, U.K.

Abstract

This paper describes the latest developments to Robug IIs and in particular the modifications and additions to its sensor system and how this has helped to increase its autonomy in unstructured environments. The configuration of Robug IIs' local intelligence is described along with the system for control of each leg. The paper also describes Robug IIs' ability to perform automatic floor-to-wall transitions along with its alignment and positioning to the wall or obstacle by means of ultrasonic ranging. The final section of the paper describes the future enhancements being considered with a view to further increasing the autonomy.

1. Introduction

Apart from transporting components around the workplace mobile robots can also be used to carry tools to perform various remote operations. Such robots can provide a cost effective alternative for applications such as inspection or maintenance of large buildings, chemical plants or nuclear reactors etc. Such applications are either hazardous to human beings or involve erecting costly scaffolding.

Robug IIs has been developed at the *Portech Mobile Robotics Laboratory* of the *University of Portsmouth* and is being used to address the problems of unstructured environments. Robug IIs is primarily controlled by a teleoperator control system consisting of a video display (displaying views from the on-board camera system) and computer interface for user interaction. The user controls the robot by use of high level commands which cause the robot to carry out coordinated motion subject to pre-programmed instructions.

The robot is able to cope with a relatively disordered environment but the control becomes more difficult when entering an unstructured environment where unforeseen obstacles are encountered. It is for this reason that we are decreasing the need for operator interaction and increasing the autonomy of Robug IIs.

Robug IIs is able to manoeuvre on horizontal and vertical surfaces. It also has the ability to perform autonomous floor-to-wall transitions to overcome obstacles such as walls and other large objects and has the intelligence to seek and verify foot holds.

This paper describes the sensor system developed for Robug IIs and describes how it uses this information to avoid obstacles and perform floor to wall transitions.

Its sensor system incorporates an ultrasonic ranging system controlled by a specially designed microcontroller board. The system is designed to aid with alignment and ranging during automatic movement. The sensor system also incorporates ;

(i) Attitude sensors
(ii) Vision
(iii) Tactile sensing

The paper will describe these systems in detail and will discuss their use in performing tasks in relatively unstructured environments. Other sensors in the system are potentiometers for recording leg positions and pressure transducers for determining leg forces. These parts of the system will also be described in some detail.

2. Microcontroller configuration

2.1 Computer control system

Control of Robug IIs is based on a hierarchial control system. At the top of the system hierarchy is a PC which serves as a supervisory computer. It performs path planning, on-line user interaction and the functions necessary for coordinated motion.

A master-slave network configuration via a RS232 serial link is used for communication between the supervisory computer and five individual microcontrollers *see figure 1*. The microcontroller card is based on the NEC 78310a microprocessor. Each card is dedicated to a specific part of the mechanical structure. Cards 1 to 4 are dedicated to control of an individual leg and the 5th card is dedicated to external sensor control.

All the microcontrollers listen to the network and each microcontroller has its own identification code, the board becomes active when it receives a board selection signal with its own identity code. The microcontroller then executes all the subsequent commands until a different microcontroller is selected. Upon receiving a command the microcontroller echoes the received command back to the supervisory computer to confirm the command. This arrangement simplifies the communication protocol and still achieves a high reliability of communication. Communication between the individual boards is carried out via the supervisory computer.

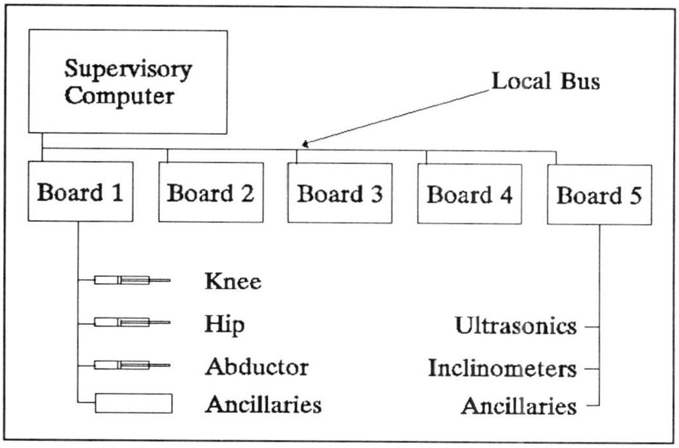

Figure 1. Configuration of hierarchial control structure.

2.2 Leg control

At the lower level are the individual limb controllers, each of which is a dedicated microcontroller board. The limb microcontroller controls all three joints of a single limb. To do this it uses the information from the potentiometers and pressure transducers to calculate leg positions and leg forces.

Each leg is controlled by nine electrically operated valves (3 for each cylinder). In order to provide the capability for Robug IIs to negotiate and climb over obstacles, each joint can be controlled in position control mode or compliant control mode. In the first mode the microcontroller uses position feedback to control the movement of the leg. This mode does not take into account the force feedback and hence is unable to detect obstacles in unstructured environments.

In the compliant control mode the individual microcontrollers monitor both position and force feedback. The demand pressure of the cylinders is calculated by multiplying the position error with a stiffness coefficient plus the force demand *see equation 1.*

By adjusting the stiffness coefficient and the force demand, compliant control can be achieved. In this mode obstacles can be detected and the robot can feel for foot holds on the surface e.g. if the leg is obstructed the pressures will build.

$$\boldsymbol{\Psi} = \mathbf{K.e} + \mathbf{F_D} \qquad (1)$$

$\boldsymbol{\Psi}$ is the demanded pressure, **e** is the position error, **K** is stiffness coefficient, $\mathbf{F_D}$ is the force demanded

2.3 Ultrasonic control

Again at the lower level is the sensor microcontroller card. This card is used to control the ultrasonic sensors. The ultrasonics are controlled via the digital I/O and the time stored relates to the range of the obstacles.

Two primary components comprise the ultrasonic ranging unit : an acoustical transducer (40mm , 50Khz Polaroid transducer) and a ranging circuit board (Polaroid). Together these components can detect the presence and distance of objects within a range of approximately 150mm to 10.5m. The ranging unit currently comprises two ultrasonic transducers positioned on the front of the vehicle, *see figure 2*.

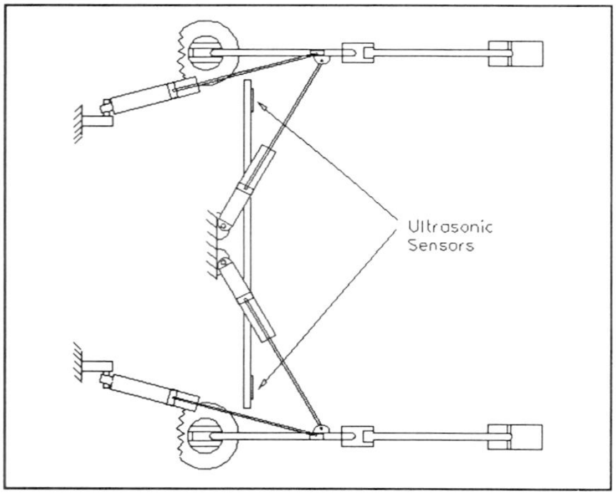

Figure 2. Front section of Robug IIs, showing position of Ultrasonic sensors.

These two transducers are activated independently by the microcontroller card. Once the unit is activated the micro controller card starts a counter and the transducer emits a sound pulse. The emitted pulse is a 50 kHz pulse lasting for about ½ millisecond and consisting of sixteen pulses. The unit then waits to receive the echo returning from the object.

Upon receiving the echo the unit signals the microcontroller card with an echo received signal and the microcontroller stops the counter. This time is passed to the high level controller where it is used for alignment and ranging purposes.

3. Floor-to-wall transition and obstacle avoidance.

Whilst Robug IIs is approaching a wall or obstacle it is very easy for the tele-operator to see obstacles or walls in front of the vehicle but it is difficult for the tele-operator to judge the distance to that obstacle or wall using the on-board vision carried by Robug IIs.

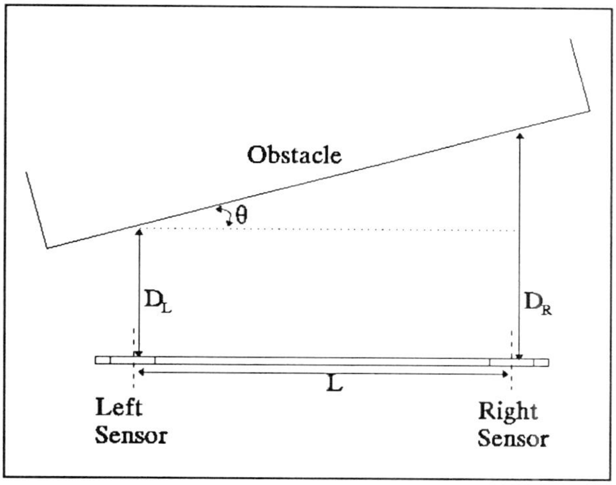

Figure 3. Configuration of ultrasonic sensors.

The information collected by the sensor board now becomes very useful. The ultrasonic sensors on Robug IIs are triggered independently by the supervisory computer via the serial link. The microcontroller card returns two independent times to the supervisory computer , one from the left and one from the right sensors. *Figure 3* shows the configuration of the sensors on the front of Robug IIs.

The high level software uses the information received from both the left and right sensors to calculate first of all the distance *see equation 2* and secondly the skew angle of the obstacle to the robot, *see equation 3*.

$$\mathbf{D_L = (C * \Gamma_L) / 2} \tag{2}$$

D_L is the distance from the left sensor to the obstacle, C is the speed of ultrasound in air and Γ_L is the returned time for the left sensor from the sensor microcontroller. It can be seen that a similar calculation can be done for the right sensor.

$$\theta = \mathbf{tan^{-1} [(D_L - D_R) / L]} \tag{3}$$

θ is the skew angle of the robot to the obstacle/wall, D_L is the distance from the left sensor to the obstacle/wall, D_R is the distance from the right sensor to the obstacle/wall, and L is the distance between the two sensors.

After calculating the two distances the supervisory computer categorises the distances into three levels :

(i) Obstacles/Walls over 2 metres away.
(ii) Obstacles/Walls in the range 2 metres - 1 metre.
(iii) Obstacles/Walls under 1 metre away.

Robug IIs behaves in different ways in these three ranges. In the first range when an obstacle/wall is detected at a distance greater than 2 metres, Robug IIs walks in its normal way, taking full strides with each step until it enters the second range.

In the second range the supervisory computer calculates the skew angle to the obstacle/wall it then alters its skew angle by turning on the spot until Robug IIs is square to the obstacle/wall. It then carries on making progress to the wall until it enters the third range. Whilst approaching the third range it is constantly correcting its skew angle.

Once it has entered the third range Robug IIs begins to take smaller steps, whilst gradually increasing the width of its stance and also maintaining its attitude to the wall. This action of increasing the width of stance enables the robot to get close to the wall without its feet hitting the wall. Robug IIs carries on moving closer to the obstacle/wall until it reaches a distance of about ½ metre. At this point Robug IIs is ready to perform a floor to wall transition to manoeuvre over the obstacle or climb onto the wall.

The spine of Robug IIs can be controlled to assist with a floor-to-wall transition. Once the robot is in position to carry out the floor-to-wall manoeuvre it begins by bending the spine so that the front section is at approximately 45° to the wall. Along with the movement of the legs this allows the front section of the robot to make the transition to the wall.

If however the wall is too far away for the front feet to make the transition the spine cylinder can be controlled and bend to a certain angle (specified by teleoperator interaction) to enable the front feet to make contact.

After the robot has begun its floor-to-wall transition it uses its attitude sensor to keep it in a vertical orientation during the transition. The attitude sensor used on Robug IIs is an Accustar® electronic clinometer. The sensor is a capacitance based device and has no moving parts, it has a range of ± 60° and a resolution of 0.001°. If Robug IIs begins to veer to the left or the right the Robug IIs supervisory computer adjusts the step length of the legs to compensate for this. The attitude sensor is also used to keep Robug IIs at a fixed angle on the wall, again adjusting step lengths etc. *Figure 4.* shows Robug IIs performing a floor to wall transition at the FRED test facility [1].

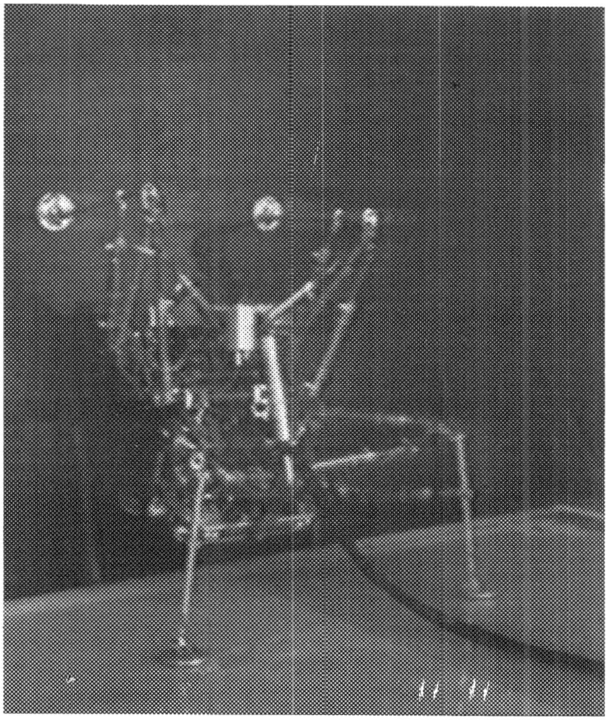

Figure 4. Robug IIs performing a floor to wall transition on FRED

4. Conclusions

Robug IIs has been fully tested both at the *Portech Mobile Robotics laboratory* at the *University of Portsmouth* and at the *FRED test facility* on a full scale reactor mock-up. Its performance has been compared with the performance of the *Nero* [2] vehicles developed by *Portech* for *Nuclear Electric Plc* which successfully completed its part of a 12 month inspection programme at the *Trawsfynydd Nuclear Power Station* [3-5]. Robug IIs has the advantages of floor to wall transfer over the Nero robot and was found to have a comparable ability once situated on the vessel. Following these trials further modifications are being carried out to further improve the autonomy of the system. Some of the enhancements being considered follow ;

Robug IIs is able to perform the alignment to the wall and the subsequent floor-to - wall transition but as yet is unable to perform the transition from wall-to-floor. It is not felt that a pair of ultrasonic sensors are required merely a single sensor on the rear of the robot to give us a distance to the floor. The alignment with the floor can be achieved using the attitude sensor found on Robug IIs.

Enhancement of the vision system would also lead to increased autonomy. Especially in the fields of automatic path planning/obstacle avoidance.

Whilst work on Robug IIs is being carried out another robot development which has recently begun is another version of the Robug robot, Robug III. This project is a two year Teleman scheme funded by the Commission of the European Communities.

5. References

1 FRED - Facility for Reactor Engineering Development, Nuclear Electric Plc.

2 T.S.White, A.A.Collie, J.Billingsley, B.L.Luk, and D.A.Moseley, "A Robot climbing vehicle and its Man-Machine interface for use in the nuclear industry", The 8th International Symposium on Automation and Robotics in Construction, 3-5 June 1991, Stuttgart, Germany, pp 851-859.

3 A.A.Collie, B.L.Luk, T.S.White and N.R.Bevan, "Experience of climbing robots on a nuclear reactor", IBC conference on Remote Technology for the Nuclear Ind.

4 A.Curry, M.S.Burrows and R.P.F.Clayton, "Nuclear reactor pressure vessel inspection climbing vehicles/robots", IBC conference on Remote Technology for the Nuclear Industry, 2-3 December 1991, London.

5 M.S.Burrows, A.A.Collie and A.Curry, "Inspection of a nuclear reactor pressure vessel by vacuum attached and magnetically attached mobile robots", 4th International Symposium on Offshore, Robotics and Artificial Intelligence - Telerobotics in Hostile Environments, 11-12 December 1991, Marseille, France.

Automation and Robotics in Construction X
G.H. Watson, R.L. Tucker and J.K. Walters (Editors)

Graphical Site Simulation Using an Object Oriented CAD Model

Toshihiko Aoki, Takahiko Kimura, Kazuhiro Momozaki, Aketo Suzuki

Construction Technology Development Department, Taisei Corporation, Sanken Building, 25-1, Hyakunin-cho 3-Chome, Shinjuku-ku, Tokyo 169, Japan

Abstract

Recently, CAD research has concentrated on shifting CAD systems from the traditional line based structure to a more intelligent and object oriented structure. Most work in this area, however, has concentrated on developing applications for design work, leaving construction phase applications largely ignored. The authors of this paper have begun development work on PRISM (Project Representing Intelligent Site Modeler), an intelligent, object oriented CAD tool that will be used during the construction phase as well during design. PRISM will consist of a 3D CAD tool linked to a knowledge base system. The knowledge base will hold various types of site information in the form of object attributes and rules and will be able to reason about the project while CAD will be used to graphically display the results of that reasoning. PRISM will be initially used at the site to represent and simulate retaining wall construction. Readings from inclinometers will be fed directly to PRISM which will graphically display these measurement results. Furthermore, the knowledge base will be able to carry out back analysis and reason about the measurement results, offering suggestions where possible.

1. INTRODUCTION

Early CAD systems were limited to drawing and drafting by the computer technology which was available at the time. Those CAD systems were essentially electronic drafting tools which were capable of producing only two dimensional drawings. While these graphic presentation techniques revolutionized civil design, they remain largely unused during the actual construction process, the reason being the complexity and vast amounts of information needed and generated at the site. Recent advances in computer technology and CAD technology in particular, however, have given computers the power to represent objects as three dimensional solids and have greatly improved computer response time. Presently, CAD systems are evolving from line based systems to more intelligent object oriented systems that can represent drawings as collections of drawing "objects" capable of holding attributes and behaviors. With advances in computer technology, CAD will expand from design and drafting to the realm of project management and site simulation.

2. Project Goals

The authors of this paper have begun development of a new computer tool to be used at the construction site. The tool, PRISM (**P**roject **R**epresenting **I**ntelligent **S**ite **M**odeler) is an object oriented intelligent 3D CAD application composed of a CAD system and a knowledge base system. While object oriented intelligent CAD development has been the topic of research for many years, efforts have concentrated on the design phase usage of such systems. Traditionally, CAD has not been utilized at the site because its only use was as a drawing aid.

PRISM, however, will be used at the site not to draw, but to display information from the knowledge base. Thus the aim of our research is to develop a comprehensive intelligent CAD system which can be used at the site as well as the design office. Our target is to develop a system that will be able to create a computer model of a project that will follow a project from design to construction. Engineers will add information to the model throughout its existence, making modifications when necessary. The project goals can be categorized into the following four areas:

1) Graphic site representation: The authors would like to present construction data in a graphical, easily understood format. This is especially true for "invisible" information such as rebar strain or soil pressure.

2) Information consolidation: Site information which is presently scattered through various computer applications and log books will be stored in a comprehensive site model.

3) Design and construction integration: Since the site model will be used throughout the life of the project, information from both design *and* construction will be held by the model. Data, rules, and behaviors pertinent to all construction phases will be recorded in PRISM giving its users a more general concept of the whole project.

4) Store construction knowledge: PRISM will accumulate information in the form of new attributes, rules, or elements as it is used and will act as a storehouse for construction knowledge.

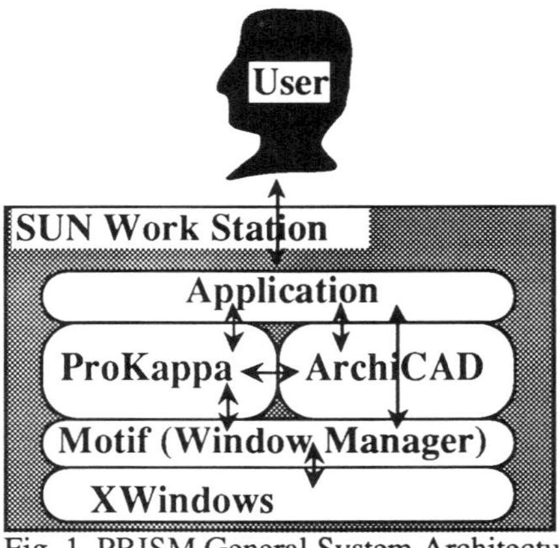

Fig. 1 PRISM General System Architecture

3. PRISM Architecture

The CAD system used in PRISM is called ArchiCAD. It is a 3D CAD system developed by Architectural Systems, a Japanese software company. (Please note that this CAD system is *not* the same ArchiCAD that was developed in the U.S.) The knowledge base system that will hold the project attributes is ProKappa by IntelliCorp from the U.S. The two applications will be connected by a two way dynamic link allowing changes made to the drawing to be registered in ProKappa and allowing reasoning output to be displayed by ArchiCAD. The reason for the two way dynamic link rather than a one way link or an indirect link stems from two characteristics of construction work. First, site engineers need immediate feedback to remain up to date on site conditions. This is especially true during emergencies. Second, site conditions are always changing unlike manufacturing where production takes place in a controlled environment.

The overall architecture of PRISM is shown in figure 1. The system runs on a Sun 4 engineering work station using the XWindows graphics platform. The Motif window manager will support ProKappa and ArchiCAD. The user will access the system through an application that provides a graphic user interface.

4. OBJECT ORIENTED CAD

Object oriented, when used with reference to CAD, means that lines and shapes are grouped together by function and named to form objects. These objects hold information about the actual object in the form of attributes and rules. The object information in PRISM will be held by the knowledge base. While many CAD packages allow the user to assign attributes to drawings, the attributes are usually limited to text. In object oriented CAD, attributes can be anything from alpha numeric data such as length, type, and material, to rules and behaviors such as the use of the formula $wL^2/8$ to determine the bending moment at the center of a simply supported beam with a concentrated load. Each object in the drawing is capable of acting independently since each holds its own attributes, rules, and behaviors.

Object oriented CAD offers many advantages over conventional CAD systems. Since drawing objects are capable of holding information, interpretation of the drawing can now be done by computer rather than by humans. Computer will be able to recognize long, "I"-shaped objects in the drawing as I-beams. The implications of this shift from line based CAD to object oriented CAD represents an important advance in CAD technology. CAD diagrams have always contained a wealth of information which was incomprehensible to computer applications without human interpretation. To the computer, a detailed drawing of a construction project would be nothing more that a long list of points and vectors. An engineer at the site would have to look at the drawing, interpret it as a group of objects constituting a construction site, and re-enter the information into various computer tools such as cost management and scheduling software. In the case of an object oriented CAD drawing, computer applications would be able to interpret the drawing and obtain all of the needed information directly. This combination of CAD and knowledge base system results in many benefits which neither application could offer alone.

5. Modeling Methodology

The general framework of PRISM is shown in figure 2. Objects are represented in both the knowledge base and CAD. Graphic data about the objects (i.e., shape, size, orientation, and position) are contained in the CAD drawing. Each drawing object also holds an ID number that acts as a pointer to a corresponding instance in the knowledge base. All text data and rules associated with the site objects are held by these instances. By holding non-graphic data in the knowledge base the amount of information carried by the CAD drawing is reduced, making the drawing easier to manipulate.

Element libraries will contain commonly used objects. There will be a number of element libraries, one for each type of construction. Each element in the element library will correspond to a subclass in the knowledge base. For example, if the library consists of anchors, walls, and struts, there will be three corresponding subclasses in the knowledge base called

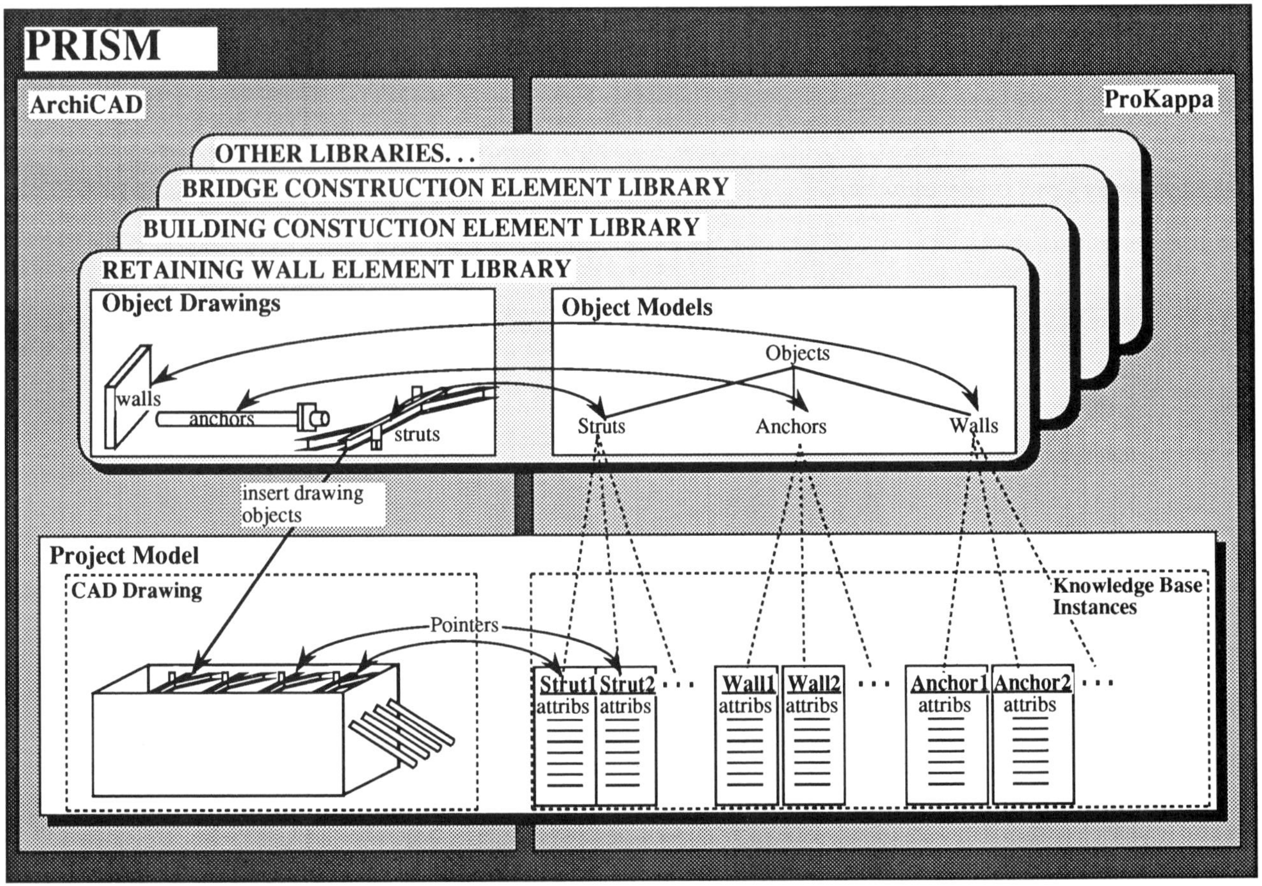

Fig. 2 General Framework of PRISM

"anchors", "walls", and "struts". During project design, the design engineer will select the desired element library from a library menu and load it into PRISM. To create a site model, a designer simply chooses objects from the menu and inserts them in the drawing. Input is mouse driven wherever possible except for attribute specification which is done by keyboard through dialog boxes. As the objects are placed in the drawing, messages are automatically sent to the knowledge base to create instances under the appropriate sub-classes. Every element inserted into the drawing from the element library will correspond to an instance in the knowledge base.

6. Retaining Wall Simulation

While PRISM can be used for a variety of site applications, the authors plan to use it at first to simulate retaining wall construction during excavation work. Usually, when the word simulation is used with reference to construction, it refers to the simulation of a construction process or task for the purpose of feasibility verification or process streamlining. When simulating retaining walls, however, PRISM will be used to simulate ground and wall movements during construction, allowing site engineers to monitor site conditions that cannot be easily visualized.

The typical information flow during retaining wall monitoring is shown in figure 3. Monitoring can be divided into three main areas: measurement, back analysis, and failure management.

Measurement is relatively simple. Presently, inclinometers and strain gauges at the site feed information to computers which can graphically display the incoming data. The output is usually in the form of two dimensional graphs as shown in figure 4. Although this lets the engineer discern the wall deflections at that point, the graph gives no context. The engineer must use his imagination to determine how the wall deflections relate to the surroundings.

Back analysis is the process of recalculating values assumed during design, such as soil friction angle or cohesiveness, from the actual measured deflections. In retaining wall construction, it is impossible to know all of the pre-existing conditions necessary to analyze wall behavior completely accurately. Assumptions are made concerning soil properties and wall properties during design to predict deflections. During actual construction, deflections are monitored and compared with design values. If excavation is carried out to a certain level and the difference between the design and actual deflection is too great, back analysis is performed. Using values recalculated during back analysis, wall deflections for subsequent excavation work can be predicted at a higher degree of accuracy, thus enhancing site safety. Presently, back analysis systems exist on the computer, although they are stand alone systems.

Failure management is at the core of retaining wall monitoring. In this phase, decisions are made concerning the condition of the site and what actions are required. This task has traditionally been the responsibility of the site engineer because of the reasoning abilities and experience needed.

With both reasoning capabilities provided by ProKappa and presentation capabilities provided by CAD, the authors of this paper hope to integrate these three phases of retaining wall monitoring through the use of PRISM. Measurement data can be collected in the PRISM site model and graphically displayed along with the surrounding area (figure 5). System users will be able to see wall deflections and how those deflections will effect surrounding buildings or even surrounding construction areas. In addition, walls, other supporting members, and

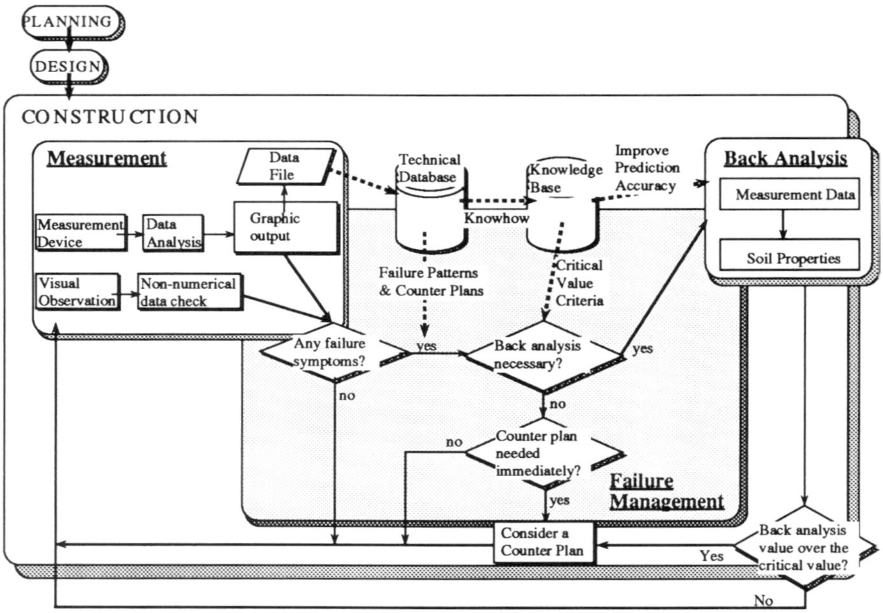

Fig 3 Observational Construction Flow

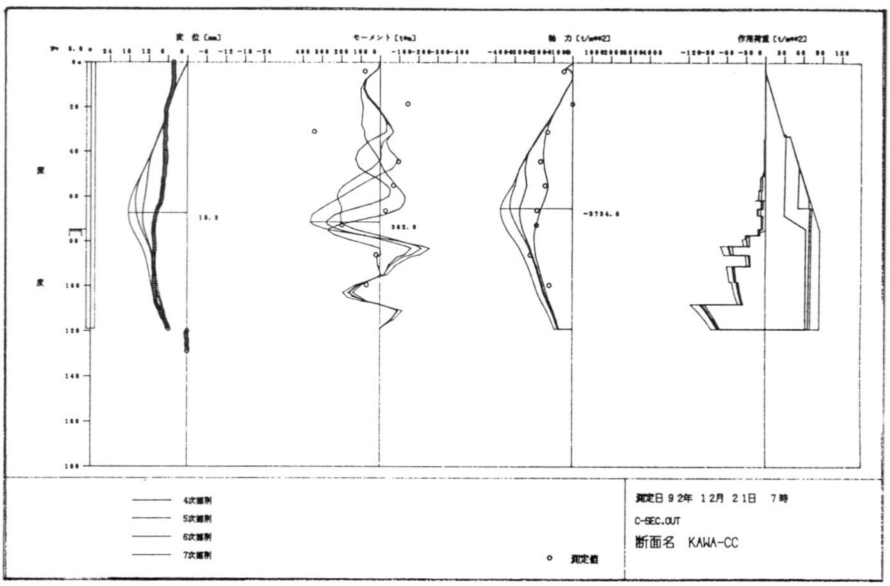

Fig. 4 Two Dimensional Measurement Data Output

Fig. 5 Three Dimensional Displacement Output

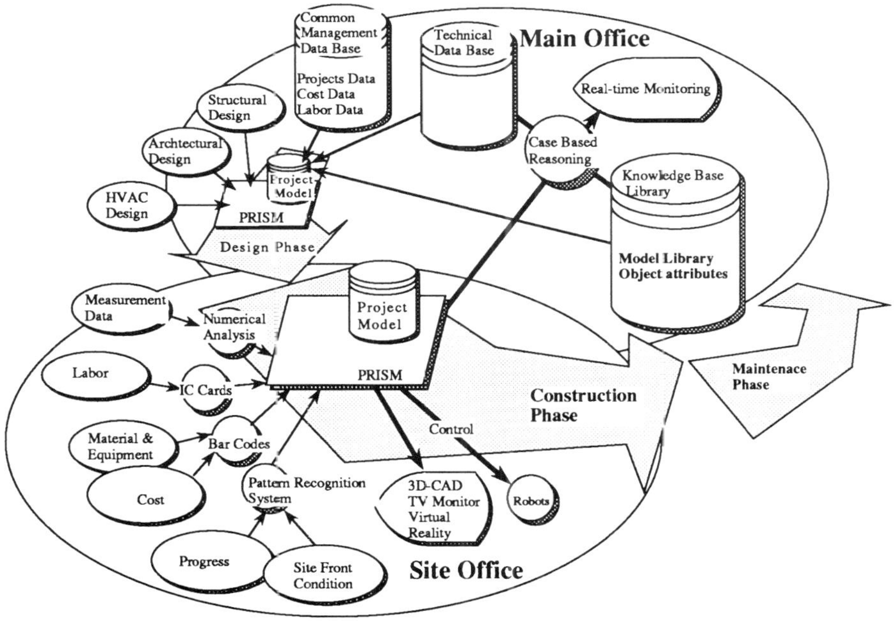

Fig 6 Project Model Life Cycle

soil layers can be displayed with the deflection plot, visually presenting the relationship between all of these elements.

PRISM will also be able to monitor deflection values and if there are any failure symptoms, the user will be warned of the potential danger. PRISM can, then, automatically feed the pertinent data to a back analysis system rather than having an engineer manually input the data into a back analysis model. As PRISM becomes more developed, it may also be able to provide suggestions to the engineer as to how problems of possible failure could be addressed.

7 CONCLUSION

Although PRISM is still in the development phase, the authors believe that it will prove to be a valuable tool at the construction site. When the modeling capabilities of PRISM are sufficiently developed, the authors hope to use it as a storehouse for site information. (Figure 6) Engineering knowledge can be stored in the system in the form of new rules, elements, and attributes. In addition, PRISM will act as a catalyst for integration. As stated at the beginning of this paper, site information is scattered at the site in various notebooks and computer applications. By creating one comprehensive model of the construction site, engineers will have a single information base reducing the need for redundant data. There are two major benefits of integrating separate computer applications. First, information and applications are easier to access when they are located in one area. What is more important, when information from a variety of functions is consolidated in one area, relationships between these functions become more apparent. The influence of structural design on project cost and the requirements of constructability are two examples. While this level of computer integrated construction is yet to be achieved, PRISM represents an early step toward this goal.

REFERENCES:

[Aoki 91] Aoki, T., "A Measurement Planning System for Information Integrated Construction", *Center for Integrated Facility Engineering (Stanford University) Technical Paper*, July 1991.

[Chang 91] Chang " An Object-Oriented Simulation System for Construction Process Planning ", *Proceedings on the American Society of Civil Engineers Construction Congress,* April 1991.

[Dym 91] Dym, C.L., and Levitt, R.E., *Knowledge-Based Systems in Engineering*,1991

[Froese 91] Froese, T. and Paulson, B.C. Jr., "An Object-Oriented Approach for Integrated Project Management Software", *Center for Integrated Facility Engineering (Stanford University) Technical Paper*, 1991.

[Goto 92] Goto, S. and Marui, I., "A Study on Application of Site Work Simulation System for Civil Construction Planning Using Object Oriented Three-Dimensional CAD/CAM System" *Proceedings from The 9th International Symposium on Automation and Robotics in Construction*, June 1992.

[Jain 88] Jain, D and Maher, M.L., "Combining Expert Systems and CAD Techniques", *Microcomputers in Civil Engineering*, vol.3, 1988.

[Kunz 89] Kunz, J.C., Stelzner and Williams " From Classic Expert Systems to Models: Introduction to a Methodology for Building Model-Based Systems ",1989

[Momozaki 91] Momozaki, K., "Information Integration in Retaining Wall Construction", *6th Meeting of the Hokkaido Soil Improvement Committee*, Feb. 1991.

[Levitt 87] Levitt, R.E.,"Expert Systems in Construction: State of the Art" *Expert Systems for Civil Engineers: Technology and application, Chapter 5*, 1987.

[Pohl 92] Pohl, J., Myers, L., Cotton J., Chapman, A., Snyder, J., Chauvet, H., Pohl, K.J., and La Porta, J., *Design Institute Report CADRU-06-92*, 1992.

Automation and Robotics in Construction X
G.H. Watson, R.L. Tucker and J.K. Walters (Editors)

A Knowledge Representation for Construction Operations

Naruo Kano

Department of Architecture, Waseda University,
3-4-1 Okubo, Shinjuku, Tokyo, JAPAN 169

Abstract

This paper proposes a semantic structure representing construction operations with advanced equipment and robotics and a methodology to implement it as a knowledgebase for planning and controlling the operations.

1. Introduction

Construction management, using a large amount of various information about operations at the site, enables the construction to be implemented smoothly. In the past, when management was done manually, information about construction operations was handled by people processing information with great flexibility. With the introduction of computers to construction management, a well-structured format has been required of information used for management.

In addition, it has become more inevitable to implement a knowledgebase to utilize the information in computer systems in order to perform construction operations efficiently using advanced construction equipment and robotics.

As the construction sites vary in organization, personnel, and equipment related to operations, and the site situations change in accordance with the progress of construction, it is difficult to standardize the format of information. This difficulty complicates information management at the site.

This paper proposes a semantic structure which enables the information to be organized in a well-structured mode, and a methodology to implement it as a knowledgebase for planning and controlling for the operations.

2. Information about Construction Operations

2.1. Problems in Processing the Information

The format to describe information at the site depends largely on each subcontractor of construction operations. In the case of a robotized construction site, the information format also varies according to developers of the software to control operations, manufacturers and owners of the robots. Therefore, even though each robot constantly provides adequate information about operations, this difference raises the problem that effective communication among operational organizations and among construction robotics is not achieved.

One of the resolutions to this problem is to standardize the information format. This method, however, would not be practical, because of the situations at construction sites.

2.2. Representation of Construction Operations

2.2.1. Objects Related to Construction Operations

This method of representing activities, substance, space and time in construction operations as objects was used in the author's previous papers to describe construction planning[1] and simulation[2]. This paper employs the description method with partial changes to make it adaptable to a wide range of information processing for construction management, aiming at formalizing objects which represent construction operations.

The contents of construction operations are clarified by attributes and relations among objects. The earliest start date of structural work can be specified by assigning the date as an attribute of the structural work. To represent the precedence relation between structural work and finishing work, the concept of precedence relation should be set between the objects of both works.

These attributes and relations are defined by terminologies which represent the meaning of the concepts. For the above examples, attribute "EarliestStartDate", and relation "PrecedenceRelation" might be used. The above information about works are represented in the following two sentences.

EarliestStartDate of StructuralWork is 20Jun93.

StructuralWork has PrecedenceRelation to FinishingWork.

2.2.2. Mutual Relations of Semantics

The representation format using attributes and relations would differ with the user who inputs information for a certain purpose; thus, it is difficult to follow the uniform input rule. Moreover, when information is used as a reference, the way of referring to attributes and relations differs widely depending on the purposes.

The following are examples of several representation formats in a relationship between structural work and finishing work:

1) Finishing work is a following work of structural work.
2) Structural work is a preceding work of finishing work.
3) Finish date of structural work is earlier than start date of finishing work.
4) Start date of finishing work is later than finish date of structural work.

In the construction industry, where various roles are involved and many people and subcontractors are associated, it is not easy to standardize the process of input and reference of information into one mode. Even if it could be standardized, unstandardized or exceptional items are sure to be included.

In order to prevent these communication obstacles, this paper has managed the inconsistency in the input and the reference of information by defining the mutual relationship among terminologies used in attributes or relations.

In the above case, the following mutual relationships are required.

1)"X is following work of Y" is equivalent to
"Y is preceding work of X"
2)" X is earlier than Y" is equivalent to
"Y is later than X"
3)if "X is following work of Y"

then "Start date of X is later than
finish date of Y"

3. Semantic Structure of Construction Operations

3.1. Semantic Network

This paper, based on a semantic network, proposes a structure for managing references to information with different formats from what have been used for input, by defining the mutual relationships among terminologies associated with object, attributes, relation, and modifier to be used.

A semantic network, known as a conceptual graph, was designed by R.F.Simmons & J.Slocum[3], R.F.Fikes. & G.Hendrix[4], N.V.Findler[5], J.F.Sowa[6].

3.2. Description of Construction Operations

3.2.1. Object and Alpha-numerical Data

Objects which represent the elements of construction operations are of two types: *class* and *instance*. A *class* represents an object which has a concept of substance or activity, whereas an *instance* represents an object which has a name of specific substance or specific activity. To describe an *instance*, the name of the place separated by "/" is added to a class name to which it belongs.

The following are examples of *class* and *instance* objects:
[Class]: StructuralWork, FinishingWork
[Instance]: StructuralWork/2F, FinishingWork/2F

When information about alpha-numerical data is referred to, information is described as data, not as an object. Numerical data must always have a clarified numerical unit.
example: (val 20 Days), (val "move 20,30,40')

3.2.2. Attribute and Relation

An object has attributes and relations which represent its own content as well as relationships to other objects. An attribute is described as follows.

[Attribute]: (Attribute1 of Object1 is Object2)
 example: (PrecedingWork of FinishingWork/2F is
 StructuralWork/2F)

Another attribute can be added to an attribute as follows:
[Attribute]: (Attribute1 of (Attribute2 of Object1) is Data1)
 example: (Variance of (Duration of FinishingWork/2F) is
 (val 5 Days))

The notation of attributes in a semantic network is shown in Fig.3.2.1 and Fig.3.2.2.

A relation is described as follows.
[Relation]: (Object1 has Relation1 (to Object2))
 (Object2 has Relation2 (from Object1))
 (Object1 has Relation3 (from Object1 to Object1))
 example: (StructuralWork/2F has PrecedenceRelation
 (to FinishingWork/2F))

The notation of relation in a semantic network is shown in Fig.3.2.3.

3.2.3. Modifier

In order to narrow a concept defined in a *class* to a more specific concept, a modifier is added to the *class*. A modifier of the *class* is added after the class name as shown below.

[Modified Class]:(Class Modifier)
example: (Crane (which is Large))

A modified class becomes a class by itself, so it is possible to have another modifier. The following is an example of this case.

example: ((Crane (which is Large))(which is Movable))

Because an instance is an object which indicates a specific item, it would not be modified or narrowed down to a more specific concept; the modifier with an instance represents a sentence which explains the additional content of instance.

Therefore, a modifier sentence in Eq.1 and Fig.3.2.4 can be divided into two sentences as in Eq.2.

[Modified Instance]: (Instance1 Modifier)
example: (FollowingWork of StructuralWork/2F is
(FinishingWork/2F (of which duration is
(val 20 Days))) ---(1)
(FollowingWork of StructuralWork/2F is
FinishingWork/2F)
(Duration of FinishingWork/2F is (val 20 Days)) ---(2)

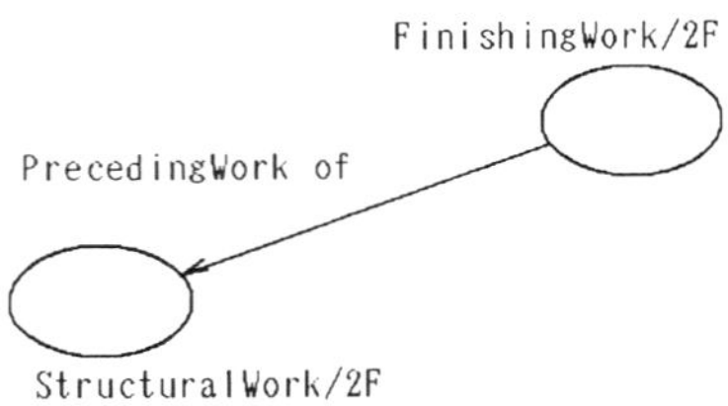

Fig. 3.2.1 Semantic Network for an Attribute

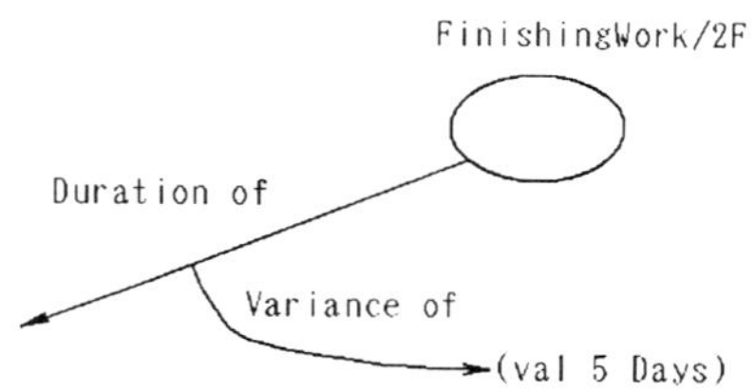

Fig. 3.2.2 Semantic Network for Two Attributes

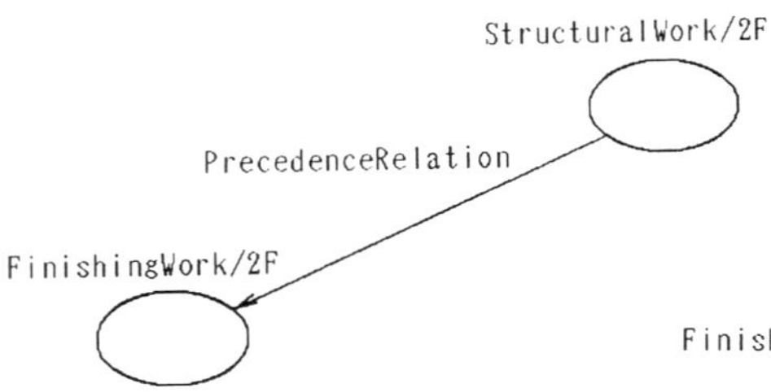

Fig. 3.2.3 Semantic Network for a relation

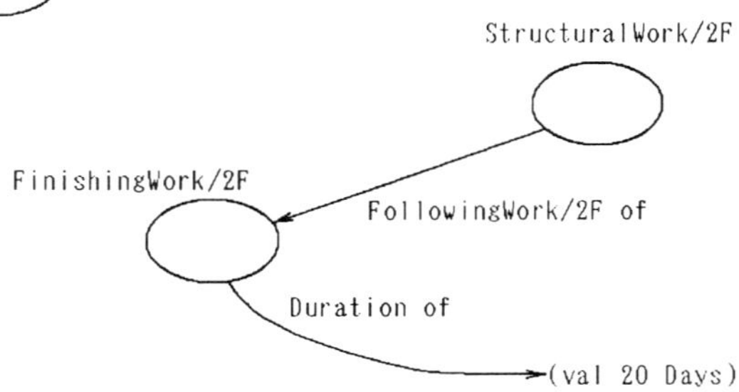

Fig. 3.2.4 Semantic Network for Instances

A specific relation can be described by adding objects to be associated with the relation as modifier phrases.

[Modified Relation]:

```
 (Relation1 (from Instance1 to Instance2))
example: (LeadTime of (PrecedenceRelation
          (from StructuralWork/2F to FinishingWork/2F))
          is (val 20 Days))
```

3.3. Organization of Information

3.3.1. Subgraph in a Semantic Network

One of the problems in semantic network methodology is that the definition of each information becomes unclear or each information is overly related, because information is linked continuously without being divided into layers. In order to alleviate this problem, the author divided information described in a semantic network into subnetworks, which can be integrated if necessary, as shown in Fig.3.3.1. This figure depicts a compound graph used for the following if-then clause.

[If-then clause]

```
(ifthen
 (if
  (Duration of FinishingWork/2F is (val 20 Days))
  (PrecedingWork of FinishingWork/2F is StructuralWork/2F))
 (then
  (LeadTime of (PrecedenceRelation
     (from StructuralWork/2F to
     FinishingWork/2F)) is (val 5 Days))))
```

3.3.2. Structural Flexibility

Information, if once input, must have a structure which provides an appropriate response to any references in various formats. Because the format of information is not standardized, if the format in the reference is different from what in the input, the structure of relationships among concepts defined in

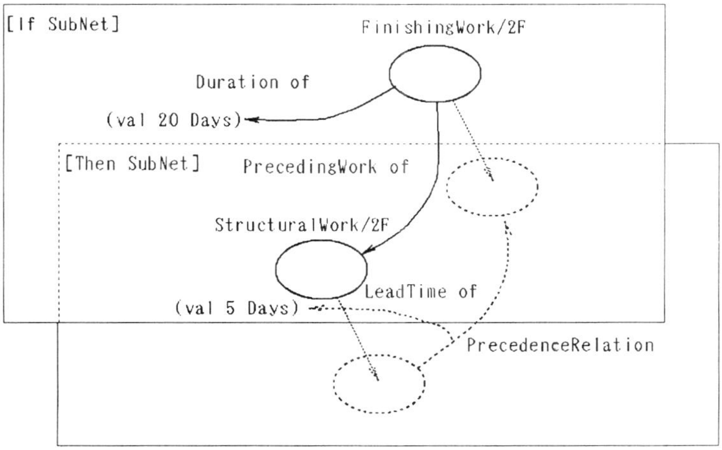

Fig. 3.3.1 If-then clause in compound graphs

a semantic network has the role of responding to it in a flexible manner.

When input or reference is made using a new format of information, the structure which clarifies mutual relationships between a new format and other formats of information should be established in advance to maintain consistency and flexibility in the knowledgebase.

To get an appropriate response to the reference using a different format, it is necessary to search for the existence of a possible alternative format which agrees with the referred content. For an instance in Fig.3.3.2, the following two sentences are in knowledgebase:

The earliest finish date of structural work
on the second floor is DT930104.

The earliest start date of finishing work
on the second floor is DT930201.

Then, the following question format is provided.

(?what is LagTime (from Date (which is EarliestFinishDate of
StructuralWork/2F)) to Date (which is EarliestStartDate of
FinishingWork/2F))

In the above reference, the difference in dates between DT930104 and DT930201 has not been calculated, and the terminology "Lag Time" is not defined between two date objects. Therefore, the system searches for the concept of "Lag Time" in the upper class of the dates; and if the concept exists as a relation, it is assumed that "Lag Time" exists between two dates as a relation. Then, supposing that the concept "Lag Time" exists between these two dates, the value is examined. If the value doesn't exist, a method which calculates the difference of dates is activated to obtain a value to the reference as follows.

(LagTime (from (Date (which is EariestFinishDate of
StructuralWork/2F))
to (Date (which is EarliestStartDate of FinishingWork/2F))
is (val 20 Days))

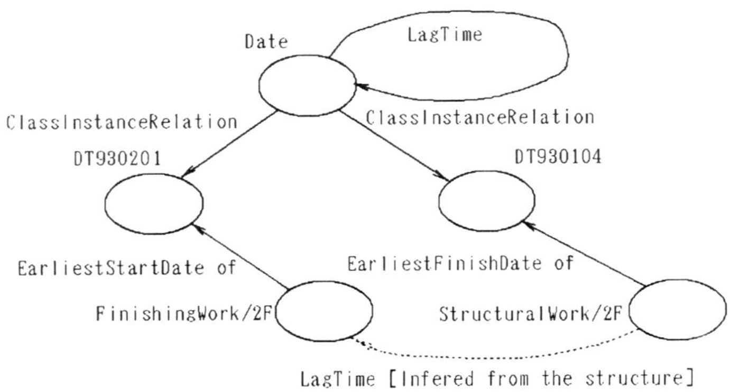

Fig. 3.3.2 Flexible Structure in Semantic Network

After the value is obtained, the knowledge remains in the structure. Therefore, the inferring process mentioned above is not necessary in the second reference, thus improving efficiency in referring.

3.4. Case Study for Robotized Construction Operations

The following is an example of information about current situations in construction operations done with robotics. A robot takes out the material from the elevator which arrives on the third floor where two stockyards of material are located; then the robot operates by itself to carry the material to the specified stockyard. In this process, the situations of the elevator, robotics and stockyards are interrelated closely.

The above operations are described in Fig. 3.4.1 as a semantic network.

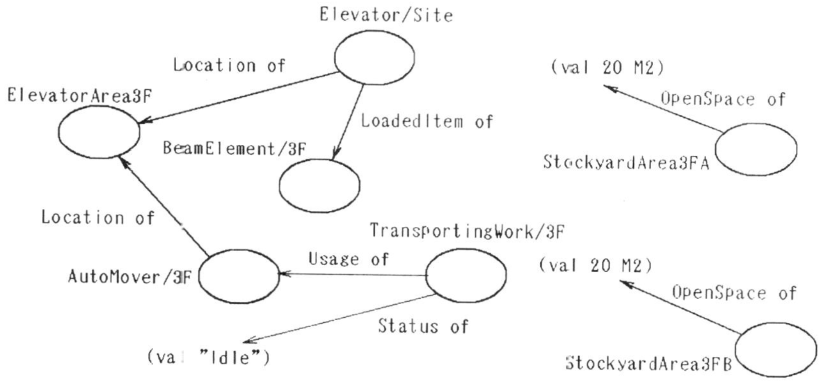

Fig .3.4.1 A Semantic Network for robotized Operations

4. System of Processing Information on Construction Operations

Based on the above-described methodology, the author proposed a common communication handling system, in which consistency is, maintained by organizing information in a well-structured manner, rather than by standardizing the information.

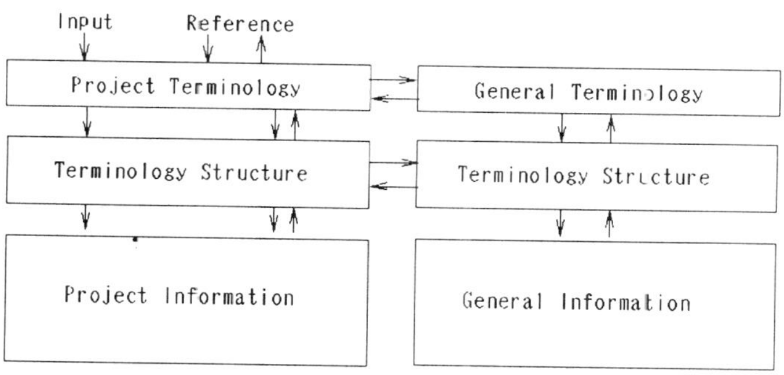

Fig.4.1.1 Concept of the system

This system organizes past records, present situations, estimated values and estimating logic for future situations in construction operations into common information. The concept of the system is shown in Fig. 4.1.1.

This system is named ConsText (Construction Context), developed on ProKappa. Fig.4.1.2 shows the user interface of ConsText for elements, relations and the breakdown structures of construction operations.

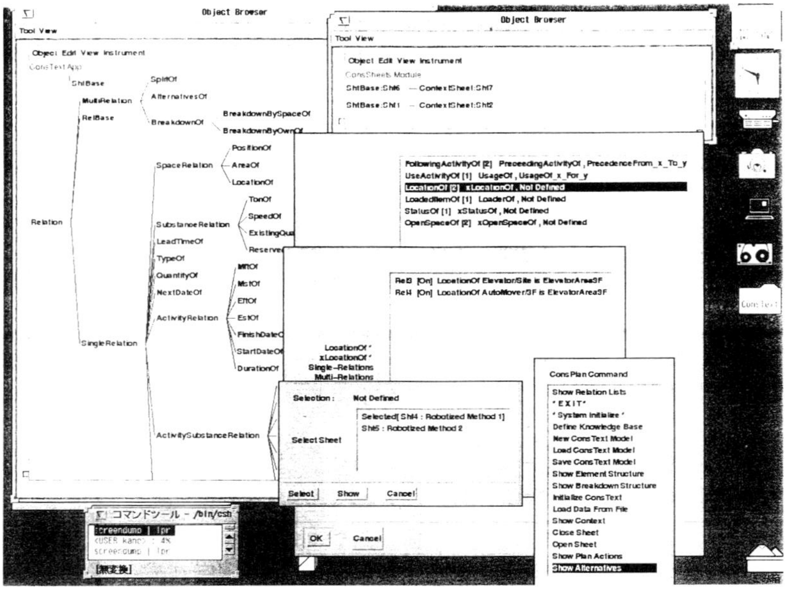

Fig. 4.1.2 User Interface of ConsText

5. Conclusion

For the effective use of information at the construction site, this paper proposes organizing the information in a semantic network. The development of the prototype system confirms that the exchange of information between different organizations, personnel, and equipment could be done flexibly without standardization, by means of storing the information in a semantic structure.

References

1)N.Kano:Proc. of Automation & Robotics in Construction, UK, 1991

2)N.Kano:Proc. of Robotics in Construction, Japan, 1992

3)R.F.Simmons,J.Slocum:"Generating English Discourse from Semantic Networks," CACM,Vol.15,pp891-905,1972

4)R.E.Fikes,G.Hendrix:"A Netowrk-based Knowledge Representation and its Natural Deduction System," IJCAI, Vol.5,pp235-246,1979

5)N.V.Findler:"Associative Networks, The Representation and Use of Knowledge by Computers," Academic Press,1979

6)J.F.Sowa:"Conceptual Structure, Information Processing in Mind and Machine," Addison-Wesley,1984

Automation and Robotics in Construction X
G.H. Watson, R.L. Tucker and J.K. Walters (Editors)

Path planning and sensing for an experimental masonry building robot

D. A. Chamberlain, S. Akrawi and J. Watson

Construction Robotics Unit, Structures Research Centre, Department of Civil Engineering, City University, London

Abstract
An experimental robot cell is being used to investigate the enabling technology for a masonry tasking robot. A CAD/CAM facility has been devised by which wall designs are translated into the robot's 'theoretical task'. However, this can not be directly implemented with out real-time adjustments derived from sensors. A further complication is that the form of a move influences the dynamic response of the robot structure. Rule-based intelligence has been applied to the solution of both these problems.

1. INTRODUCTION

Masonry is an important building element due largely to its high value and architectural significance. Considering the economics of automated masonry production in the UK, it has been shown [ref 1] that only 10% of blockwork construction need be amenable to this in order to justify the development of automation devices. This has provided the impetus for the reported research into the enabling technology for masonry production using an advanced robot [ref 2]. On the international front, substantial progress has been achieved in masonry automation research, each contribution reflecting the national character of masonry [ref 3].

The experimental robot cell described in pervious ISARC proceedings is being used to produce both dry wall and bonded wall constructions. Excluding overall mobility, all aspects of the automation are represented in the cell; delivery, survey and quality assessment of supply masonry units, navigation, mix dispensing and assembly. Apart from standard production masonry units, interlocking blocks and special bricks are under investigation . Interest in the later arises from the weight advantage that a robot can have over a human worker.

The CAD/CAM regime shown in fig. 1 has been adopted by which a CAD derived 'project description' is processed into a 'theoretical task'. This cannot be precisely implemented due to a number of factors. However, it serves to guide the closed loop, task-sensed operation of the cell. A particularly valuable aspect of the 'theoretical task' definition is the provision for optimised moves, which represent the reconciliation of the requirements for least cost and low vibration inducing motion. The determination of the robots theoretical task and the run-time implementation of this are the main subjects of this paper. This research is funded by the Science and Engineering Research Council.

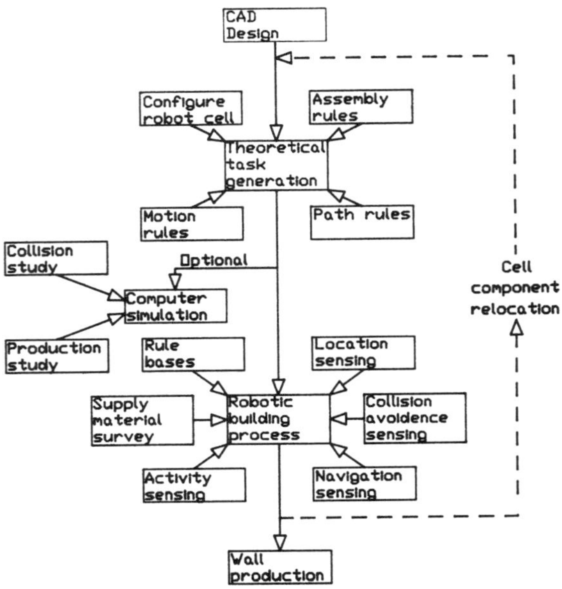

Figure 1 The CAD/CAM Regime

2. EXPERIMENTAL ROBOT CELL

The robot cell comprises a 5m long industrial conveyor, a vertical axis laser beacon that provides horizontal referencing, a mix dispensing station and a gantry type robot. The robot has a clamp type gripper with various sensors mounted on it. A working envelop of 5.0m x 2.5m plan and 2.0m elevation is available and masonry units up to 50kg weight can be manipulated on it's three prismatic and two revolute axes.

The robot's motion logic is implemented on three parallel SMCC's (Smart Motion Control Cards) which communicate with a master 80486 processor via a RS232 communication line. The master also communicates with distributed 80286 processors allocated to the conveyor and robot end effector. These credit size processors are combined with A/D (Analogue to Digital) convectors supporting the multiple sensing requirements. Further processors are provided for the navigation beacon and mix dispensing station.

3. PROJECT DESCRIPTION GENERATION

The dedicated CAD facility is coded in AutoLisp, the programming language of AutoCad. This enables projects to be solid modelled as single skin brickwork or blockwork or cavity construction comprising both bricks and blocks. The cavity width can be varied and standard window and door openings introduced. For cavity wall construction, rule-based reasoning avoids small block lengths and vertical joint alignment. On conclusion, a 'project description' is generated. This is essentially a list of N sets of entity attribute lists, N being the total number of bricks and blocks in the project. The attribute list comprises the 'XYZ' co-ordinates of the target location for the centre of a unit relative to a ground datum, 'Oa' the orientation index (0/1 according to alignment on the CAD grid), 'Bb' the brick or block type identifier, 'Sp' the special unit indicator (standard/non-standard) and 'Ul' the special unit length if appropriate. Figure 2 shows an example of a cavity wall project having openings and special blocks.

In the context of the conveyor supply, where units may be rejected, operator intervention is necessary for special units. Whilst outside of the scope of the current programme, a microprocessor controller block cutter could be mounted on the conveyor.

4. THEORETICAL TASK GENERATION

To derive the 'theoretical task' from the 'project description' it is necessary to consider (i)the assembly method, (ii) collision freeways and (iii) move optimisation. The first item concerns the order of assembly and the bond application requirements. Whilst a matter for further study, the units are currently laid in a clockwise sequence, switching between the inner to outer leaf construction in cavity wall projects. As far as possible, access is kept to one side of the project. However, it is observed that potential collision zones arise for the end effector under this restraint. This matter is currently being studied by computer simulation in order to arrive at a generalised rule-base. This will append an access parameter to each of the N entity lists and reorder them. A bond application parameter will also be appended which indicates the faces

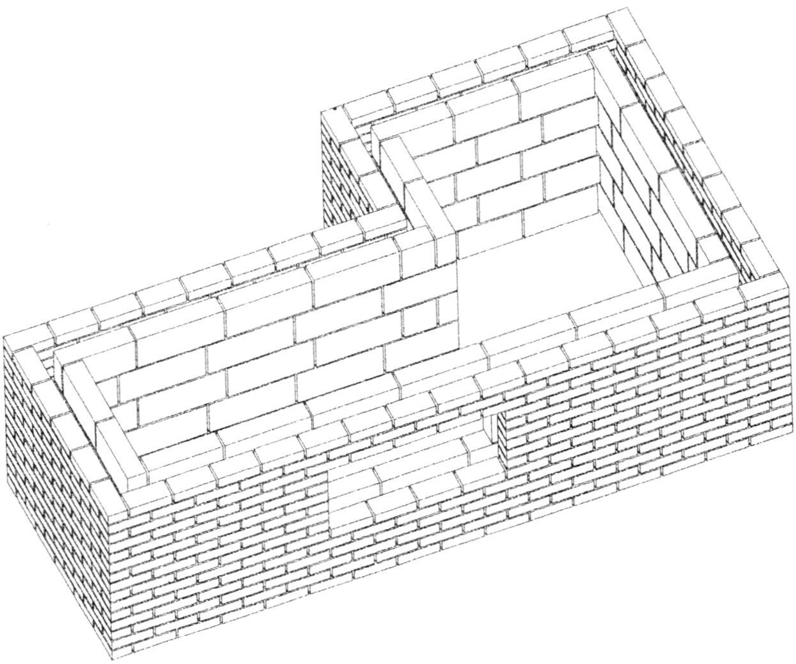

Figure 2. CAD Generated Project

requiring the bond mix.

Regarding the second item, this is complicated by the changing workspace topography. However, the problem has been substantially simplified by considering only the end effector. The adopted search algorithm is of the 'combinational' type having a finite number of solutions for each stage of construction and region. In this, freeways [4] are identified in which vertical planes can be analysed for clearance using the CAD model. A move matrix is calculated which allows sufficient leeway for the end effector to rotate about its roll and yaw axes. In the event of a cell component being relocated during the execution of a project it would be necessary to reprocess this matrix. Whilst such events are foreign to manufacturing environments they are relevant to construction sites where activities tend to compete for work space.

The last item addresses the problem of move optimisation that is complicated by the dynamic behaviour of the end effector. This arises from the partial compliance of the jaw axis about which the combined inertia of the roll axis motor, gripper and gripped block is substantial. This effect is most pronounced when the axis of the gripper is set perpendicular to the motion path and maximum acceleration achieved during a short move. In this case the settling time is very significant. To understand this further,

the effective move time (Effective Move Time = Move Time + Settling Time) has been investigated. Fortunately the SMCC's are able to support motion control to 'S' curve order under point-to-point and continuous path control. By these means, experiments have been conducted to determine the influence of the move order on settling time. An accelerometer located at the furthest radial point of the end effector enabled the combined translational and rotational accelerations to be measured. Three types of moves ere run each covering the same distance in equal time using (i) linear, (ii) parabolic and (iii) smoothed 'S' curve velocity variations. Figure 3 shows the acceleration plots corresponding to moves of seconds duration. These are near system limit moves made on a single linear axis. Whilst the 'S' curve provides the smoothest path, the high value of acceleration achieved leads to greater settling time than with the parabolic move. However, 'S' curves do produce the least 'effective move time' for longer moves where induced oscillations have time to decay during motion.
A rule-base is currently being prepared which will select the appropriate form of moves.
For this, the least time moves will be determined by fitting cubic splines clear of collision points. Apart from under continuous path control, where these can be precisely followed, the motion requirements for good dynamic behaviour result in some noticeable departures between collision clearance points.

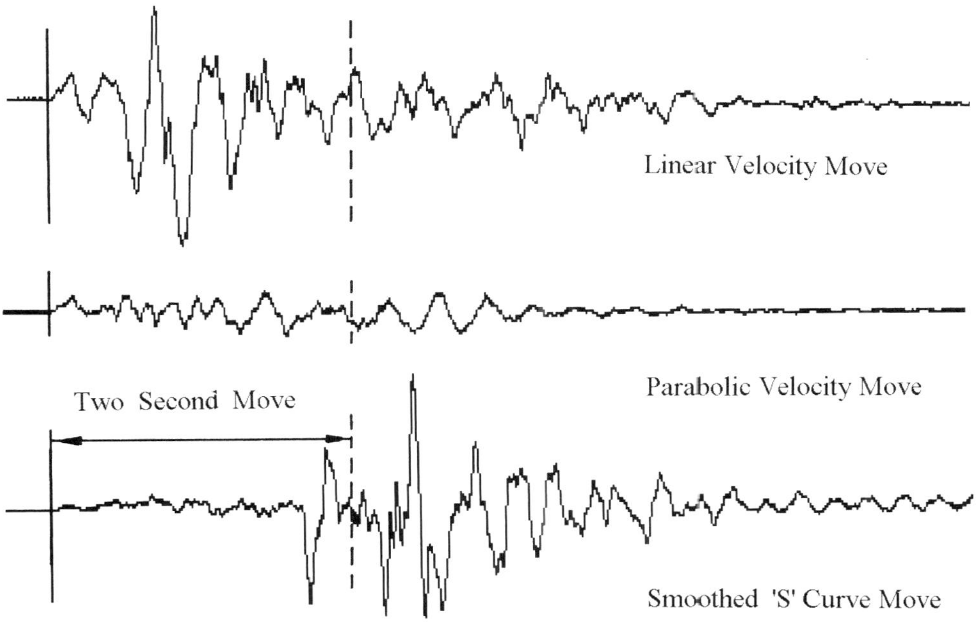

Figure 3. End Effector Dynamics

5. END EFFECTOR SENSORS

We confine our attention to the various sensors located on the end effector, these supporting the location, gripping and. placement of masonry units, survey of the assembly and collision avoidance. Of these, two highly focused near ranging Ultra-Sonic (700mm range) transducers and two displacement transducers are clearly visible in fig 4. The former are used to locate masonry units at the conveyor pickup point and avoid close obstacles during motion and the later to determine the angle and offset of gripped blocks relative to the gripper. Strain gauges are mounted on the side elements of the gripper that enable the contact forces to be monitored in relation to the offset information provided by the then vertical near ranging transducer. This combination avoids potentially damaging contact forces arising from misalignment of the gripper during pickup and also confirms the full or empty status of the gripper. Following pickup, the gripped unit is presented to a linear laser scanner that determines the transverse location of the masonry unit relative to the gripper. A pair of Polaroid ranging sensors are currently being investigated for overall collision avoidance, these operating on the 'bat's eyes' principle. For the purposes of work survey, a precise laser ranging unit is provided, this able to resolve the masonry features. Whilst not implemented on a real-time basis, image processing methods have also been successfully applied to the recognition of individual masonry units within an assembly. This is inherently a better approach to the survey problem as it does not rely on the accuracy of the robots absolute position.

Figure 4. Robot End Effector Sensors

6. SENSING INTELLIGENCE

The runtime sensing intelligence is cast in a hierarchy of rule-sets that operate acyclic graphs. In this, reasoning is applied to the filtered and compressed sensor data in order to determine the required high level instructions for the robot. The use of trained Neural Networks is currently being studied for use with rule-bases. Whilst incomplete, these provisions should allow the robot to safety and reliably execute the 'theoretical task' with the sensor determined adjustments.

In the robot's operation, two classes of moves are apparent, interference moves in which controlled contact is required between objects and clearance moves in which contact between objects is forbidden. For the masonry robot the former applies when picking units, presenting them for bond mix application and locating them in the assembly. Other than these, all moves are of the latter class. For collision avoidance, 'active' and 'semi-active' sensing regimes are being studied, the top level structure of the latter shown in fig. 5. The essential difference between these is that the 'semi-active' delivers the move requirements to the master processor whereas the 'active' delivers notification only when changes in prescribed states occur. The former is compatible with point-to-point adjustments based on relative positioning and the latter continuous motion relative to objects.

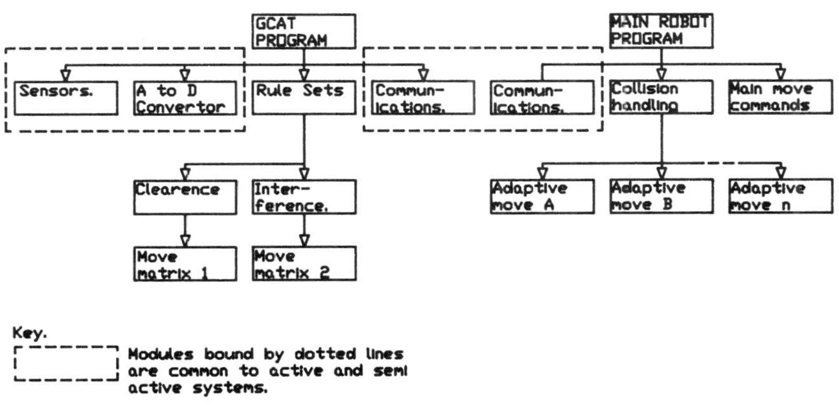

Figure 5. Semi-Active Sensing Regime

7. CONCLUSIONS

The CAD/CAM regime for an experimental robot has been described in which readily designed masonry projects are translated into robotic construction. In this, a theoretical task description is determined from the project design and implemented using closed loop sensing. Parallel processed rule-based intelligence is used in this. Concerning the robot's motion, it has been demonstrated that the form of move can substantially affect the end of move settling time.

8. ACKNOWLEDGEMENTS

The authors wish to acknowledge the financial support of the Science and Engineering Research Council without which the research would not be possible. A debt is also owed to Wiltshier PLC. and John Doyle Construction Ltd. for their interest and support.

9. REFERENCES

1. D.A. Chamberlain, Developments in Construction Robotics, Construction Robotics and Automation Symposium, Reading University, 21st. March, 1991

2. D.A. Chamberlain, G.A. West and P.R.S. Speare, A Masonry Tasking Robot, Journal of Mechatronics Systems Engineering, Kluwer publishers, December 1990

3. D.A. Chamberlain et al, Masonry Construction by an Experimental Robot, 9th. International Symposium on Automation and Robotics in Construction, Tokyo, 1992.

4. R.A. Brooks, Solving the Find-Path Problem by Good Representation of Free Space, IEEE Transactions on Systems, Man, and Cybernetics, SMC-13 (2), March 1983

Automation and Robotics in Construction X
G.H. Watson, R.L. Tucker and J.K. Walters (Editors)

Modeling of building construction activities using forms

Amr A. Oloufa[a]

[a]Department of Architectural Engineering, Penn State University, 104 Engineering Building "A", University Park, PA 16802, U.S.A.

Abstract

In the course of improving the efficiency of building construction activities, the application of discrete event simulation tools may be extremely beneficial. Simulation is used to evaluate construction scenarios and allows the construction manager to evaluate between alternatives. Selection of an optimum mix of equipment and crews may also be studied.

However, the practice of simulation in building construction has not been popular for a variety of reasons. Perhaps a major reason is the investment in time and cost needed to develop the simulation model and program the corresponding code in a simulation language. The situation is further aggravated by the lack of training in simulation in most undergraduate and graduate engineering curricula.

This paper reports on ongoing research aimed at simplifying the development of simulation models. The construction manager will select equipment and crews from a library of resources and then specify the interaction of these resources in an intuitive mode with no programming.

With the advent of object-oriented modeling and programming, simulation entities may be described in a format that encapsulates their characteristics and capabilities. The user will be capable of changing the number and/or mix of equipment with little or no change to the underlying simulation code.

In this paper we will develop a non-network based methodology for specifying the interaction between construction resources and explain an approach for mapping to an object-oriented simulation language. The utility of this approach will be demonstrated in a building construction application.

1. INTRODUCTION TO MODELING AND SIMULATION

Discrete event simulation is a tool for the study of operations. Shanon (1975) defines simulation as *"The process of designing a model of a real system and conducting experiments with this model for the purpose either of understanding the behavior of the system or of evaluating various strategies (within the limits imposed by a criterion or set of criteria) for the operation of the system".*

Simulation can be used to evaluate BETWEEN a set of competing alternatives. The simulation approach may be used to study almost any problem. Each simulation run becomes an experiment on the system that can be observed and controlled.

The simulation study encompasses a multitude of phases starting from the development of the system's model which is intended to serve as a surrogate of the system under consideration. The model should encompass the major system components that are likely to affect its behavior. The design and composition of the simulation model is always a tradeoff between its simplicity and how well it approximates the system considered. The incorporation of every possible feature of the system is time consuming and impractical. On the other hand, the omission of one or more items will render the model useless. For this reason, the success of modeling is largely a function of the correct choice of entities and parameters to be used in the simulation.

After the simulation model is developed, the model is coded in a simulation language. Two important concepts are involved namely validation and verification. Validation is concerned with how well the model truly represents the simulated system and is independent of simulation language issues. Verification on the other hand is concerned with computer language implementation issues, i.e. similar to debugging of programming languages. Validation and verification are iterative in nature and the typical simulation study should involve a complete and reliable investigation of both.

When the developers of the simulation model are confident of the accuracy of its logical (i.e. *validation*) and physical (i.e. *verification*) representations, the model is subjected to simulation "runs" under various operating conditions. Statistical data is collected during these runs and analyzed later for its significance. Conclusions about the model and the study can then be drawn.

It is obvious that a major ingredient for the success of a simulation study is related to the ease of development of the simulation model. Perhaps a major reason against widespread application of simulation in construction is related to the time and cost needed for developing and coding the simulation model. The author's ongoing research reported here strives to simplify the development and coding of the simulation model. The ultimate goal is the development of an environment where the simulation model can be created through an interactive dialogue with the construction project team. Since comprehensibility of the simulation model by its end users is a major step towards the credibility and success of the simulation study, this system will allow the individuals with the most intimate knowledge of the project to develop the simulation model .

2. OBJECT-ORIENTED MODELING & PROGRAMMING

Object-oriented approaches bridge the gap between the physical system and its computer representation. The object-oriented view is that the system is composed of interacting physical objects that are the central focus of the simulation. The first object-oriented programming language developed in the

1960's, named SIMULA, was a simulation language. This language laid the foundations of object-oriented programming despite of its commercial failure. Recently, the emergence of object-oriented simulation languages for manufacturing applications has received a lot of attention (Cammarata et al (1987), Knapp (1986, 1987), Roberts et al (1988), Rothenberg (1986) and Ulgen (1986)).

In describing a system, we define its components and how these components interact together (Booch 1991). We also declare the valid operations these components engage in, and how this engagement affects their states before, during and after these operations. It is important to distinguish between object-oriented modeling and object-oriented programming. Object-oriented modeling deals with the description of the system in terms of its resources. Each resource embodies in its definition its respective capabilities. Resources may inherit one or more capabilities from other resources and could therefore be defined as a subset of other resources. A conventional (*non object-oriented*) programming language may be used to translate the object-oriented model to its computer counterpart. However, what used to be understandable physical components will now be represented in terms of queues, methods, etc. in a simulation language. This is the strongest reason for the communications gap between users and programmers.

A better approach for the mapping of an object-oriented model to its computer counterpart is through the utilization of an object-oriented computer language. This approach establishes a direct mapping between the model and its coded description. However, it is worthwhile to emphasize that the utilization of object-oriented programming without a corresponding object-oriented modeling approach yields its benefits exclusively to the computer programmer, not the software user. It is estimated that companies spend 70-80% of their software budget on software update not writing software from scratch. Therefore it is extremely advantageous to develop a methodology that simplifies code development, modification and extension. The software user however, is not able to determine, and indeed has no interest in whether the application program was coded in any specific programming language.

3. A FORM-BASED APPROACH

Shu (1988) defines visual programming as "*The use of meaningful graphic representations in the process of programming*", and "*the application of graphical techniques and pointing devices to provide visual environments for program construction and execution; for information retrieval and presentation and for software design and understanding*". Several researchers (Gordon and McNair (1987), Davis *et al* (1988), Glicksman (1986), Thomasma and Ulgen (1988), Tseng *et al* (1988) and Ozden (1991)) and a few software packages such as XCELL+ (Conway and Maxwel 1991) and SIMFACTORY (Tumay 1987) have attempted to automate the model building process through visual programming. All these implementations however were designed specifically for manufacturing environments and are incapable of representing some common situations in construction simulation. In construction applications, Riggs (1986), Odeh *et al* (1991), and Ioannu (1992) used icons to create the simulation model. However, the user has to be familiar with the simulation language CYCLONE which is a non object-oriented language.

The author proposed the development of an interactive environment for simulation model generation that allows the decision maker to develop discrete event simulation models with no programming. Major requirements for such a system (adapted from Wales and Luker 1986) are as follows:

1. Allow users to define the system in terms familiar to them not in terms of the simulation language used.
2. Guide users in the steps required for model development.
3. Checks users responses for consistency and completeness.
4. Creation of new models by the adaptation of existing ones.
5. Provide a visual representation of the system in a form understandable by decision makers.
6. Generate the simulation model code automatically.

The proposed system will function as two layers between the user and the simulation language being used. The first layer will be language-independent. An additional layer will change the output from the first layer to the simulation language employed. The language independent module is composed of two submodules RESPEC and REACT (Oloufa 1992). While the RESPEC module remains identical to the one reported before, the module REACT has been extended. An example demonstrating its utility will be described in a later section.

3.1 RESOURCE SPECIFICATION MODULE [RESPEC]

The resource specification module (explained in Oloufa 1992) is used to define the various resources used in the construction project. This definition is a complete representation of the resource that includes its attributes, capabilities, type, and the number of units available. Resources represent equipment, material and labor. Equipment resources belong to families of specialized equipment types. For example, both trucks and loaders may belong to the resource type "Vehicle" which is capable of translational motion. However each type has its own attributes and specialized capabilities. The object-oriented approach offers a very elegant mechanism for the definition of resources due to the hierarchical nature of their description. Resources belonging to a company may be saved in a "Resource Library". The module REACT explained in the next section will access this library and retrieve information belonging to the definition, methods and attributes of various resources interacting in the model.

3.2 RESOURCE ACTION MODULE [REACT]

The Resource Action Module (Fig. 1) is used to define the interactions between the entities in the simulation. The data model for its implementation is shown in Fig. (2).

Resource Action Module										
Resource			Attribute			Capability				
#	Actor	Capture	Capability	Owner	Attribute	of	Seq.	Release	Create/Dis.	Next

Fig. 1

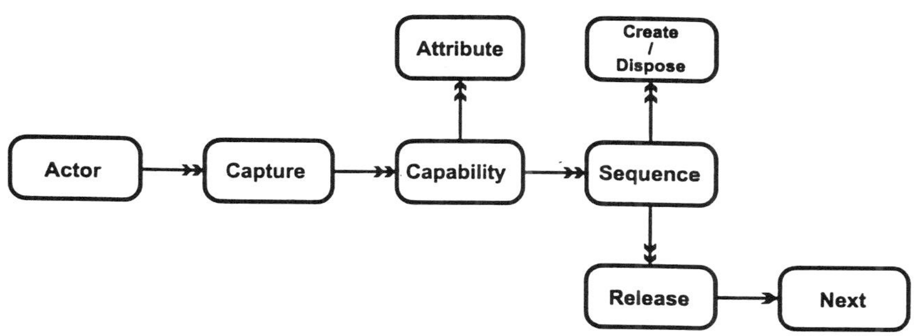

Fig. 2

An actor is an object that captures one or more resources and then proceeds to implement one or more of its own capabilities or capabilities belonging to "*captured*" or other resources. For every captured resource, the user can ask several capabilities to be invoked by entering the "*owner*" resource of these capabilities. Each of these capabilities may need input from the resources seized. This input is in the form of attributes belonging to the actor, captured resources or global variables. For example, a loader may need to inquire about the truck's capacity to determine how long it will need to load it.

Invoked capabilities may occur simultaneously or sequentially depending on the sequence number given in the Seq. column. At the completion of a capability, one or more resources may be created or disposed. (*Resources may also be created or disposed without the need of a capability invoking them*). At the completion of each capability, one or more resources may be released.

After the resource is released, it can have one of two options. It may move back and join the pool of idle resources, or it may become an actor and invoke other capabilities. A resource that is an actor and was not released continues to play that role. Each resource released may have only one or no entries in the *Next* column. If no entry is entered in the *Next* column when a resource is released, this means the resource will rejoin its pool. An entry is the *Next* column means that the resource will start capturing zero or more resources. A *Next* entry that has no corresponding resource released in the *Release* column, refers to an Actor that continues to be involved in the simulation. Parallel activities are shown as an identical number in the # column that corresponds to the number entered in the *Next* column. An entry of 0 in *Next* column refers to the first line in REACT (i.e. start of simulation). (*The proposed approach is still lacking in Start, Terminate and Branch mechanisms. These features will be added in future implementations of the system*).

4. EXAMPLE

To demonstrate the proposed approach, we selected an example from (Halpin and Riggs 1992). This example simulates the operation of a concrete precast plant. Resources are Batch Permit, Batch, Crew, Set Position, Forms, Crane, Cure Position, and Truck. After a Batch Permit is given and a Batch is available, and with the aid of a Crew and the availability of a Set Position, the batch is poured. It is left to set then the Crew pulls the forms and cleans them for the next batch. The element is loaded in a Cure Position, with the help of a Crane, for steam curing. After the cast has cured, a Truck transports it to storage. Fig. 3 shows the CYCLONE network and Fig. 4 shows the corresponding model using the approach proposed here.

5. CONCLUSION

This paper reported ongoing research aimed at simplifying the development of the simulation model. The research attempts to integrate the power of object-oriented modeling and programming to enhance the ease by which the model is built and coded in a simulation language. The interface proposed will be language independent. A "*translator*" will be needed to change this specification to a language dependent implementation.

The proposed approach is being applied in a variety of building construction scenarios to evaluate its expressive power. Future work will include mechanisms for specifying starting and stopping criteria of the simulation.

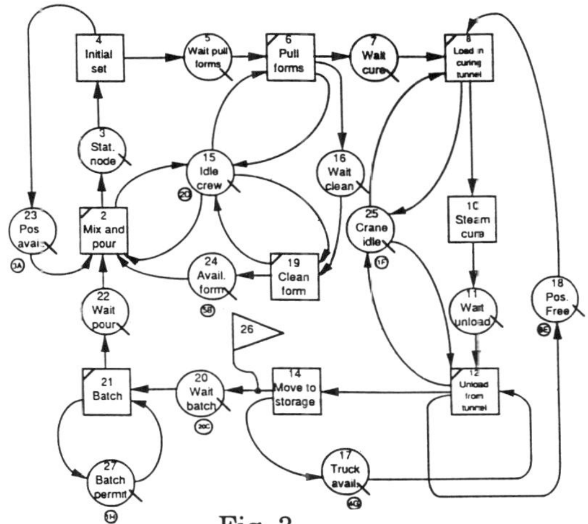

Fig. 3

Resource Action Module										
Resource			**Attribute**			**Capability**				
#	**Actor**	**Capture**	**Capability**	**Owner**	**Attribute**	**of**	**Seq.**	**Release**	**Create/Dis.**	**Next**
	Batch Permit	Batch	Batch	Batch Permit				Batch Permit		
										1
1	Batch	Crew	Mix & Pour	Crew			1	Crew		
		Form								
		Set Position	Initial Set	Batch			2	Set Position		
										2
2	Form	Crew	Pull Form	Crew				Crew	+ Cast	
										3
3	Form	Crew	Clean Form	Crew				Form		
								Crew		
3	Cast	Crane	Load	Crane			1	Crane		
		Cure Position	Steam Cure	Cure Position			2			
										4
4	Cast	Crane	Unload	Crane				Crane		
								Cure Position		
										5
5	Cast	Truck	Go Storage	Truck				Truck	- Cast	
								Batch		
										0

Fig. 4

6. REFERENCES

Booch, G.: "*Object Oriented Design With Applications*", The Benjamin/Cummings Publishing Company, Redwood City, CA, 1991.

Cammarata, S., Gates, B. and Rothenberg, J.: "*Dependencies and Graphical Interfaces In Object-oriented Simulation Languages*", Proceedings of the 1987 Winter Simulation Conference, Atlanta, GA.

Conway, R., Maxwell, W. L., and Worona, S. L.: "*User's Guide to Xcell Factory Modeling System*", Scientific Press, Palo Alto, CA, 1985.

Davis, C. K., Sheppard S. V. and Lively, W. M.: "*Automatic Development of Parallel Simulation Models In ADA*", Proceedings of the 1988 Winter Simulation Conference.

Glicksman, J.: "*A Simulator Environment For An Autonomous Land Vehicle*", Proceedings of the 1986 Winter Simulation Conference.

Gordon R.F. and McNair, E. A.: "*A Visual Programming Approach To Manufacturing Modeling*", Proceedings of the 1987 Winter Simulation Conference.

Halpin, D. W. and Riggs, L. S.: "*Planning and Analysis of Construction Operations*", John Wiley, 1992.

Knapp, V. : "*The Smalltalk Simulation Environment*", Proceedings of the 1986 Winter Simulation Conference, Washington, D.C.

Knapp, V. : "*The Smalltalk Simulation Environment, Part II*", Proceedings of the 1987 Winter Simulation Conference, Atlanta, GA.

Odeh, A. M., Tommelein, I. D. and Carr, R. I.: "Using Design Drawings And Project Plans To Construct Discrete-Event Simulation Networks", Proceedings of the 8th International Symposium on Automation and Robotics In Construction, Stuttgart, Germany, June 1991.

Oloufa, A. A.: "*Object-Oriented Simulation of Earthmoving Operations*", Proceedings of the 7th International Symposium on Automation and Robotics In Construction, Bristol, England, May 1990.

Oloufa, A. A.: "*Intuitive Simulation Modeling Using Object-Oriented Constructs*", Proceedings of the 8th International Symposium on Automation and Robotics In Construction, Stuttgart, Germany, June 1991.

Oloufa, A. A.: "*Visual Simulation Programming For Construction Operations*", Proceedings of the 9th International Symposium on Automation and Robotics In Construction, Tokyo, Japan, June 1992.

Ozden, M. H.: "*Graphical Programming of Simulation Models In an Object-Oriented Environment*", Simulation, February 1991.

Riggs, L. S.: "*Interactive Graphing of Simulation Networks*", ASCE Journal of Computing, Vol. 3, No. 2, April 1991.

Roberts, S. D. and Heim, J.: "*A perspective on Object-oriented Simulation*", Proceedings of the 1988 Winter Simulation Conference.

Rothenberg, J.: "*Object-Oriented Simulation: Where Do I Go From Here?*", Proceedings of the 1986 Winter Simulation Conference, Washington, D.C.

Shannon, R.E.: "*System Simulation: The Art and Science*", Prentice Hall, Englewood Cliffs, N.J. 1975.

Shu, N. C.: "*Visual Programming*", Van Nostrand Reinhold, New York, 1988.

Thomasma T. and Ulgen, O.: "*Hierarchical, Modular Simulation Modeling In Icon-Based Simulation Program Generators For manufacturing*", Proceedings of the 1988 Winter Simulation Conference.

Tseng, F. T., Zhang, S. X. and Wolfsberger, J. W.: "*Automatic Programming Assistant For Network Simulation Models*", Proceedings of the 1988 Winter Simulation Conference.

Tumay, K.: "*Factory Simulation With Animation: The no Programming Approach*", Proceedings of the 1987 Winter Simulation Conference, IEEE, Atlanta, GA.

Ulgen, O. J.: "*Simulation Modeling in an Object-oriented Environment using Smalltalk-80*", Proceedings of the 1986 Winter Simulation Conference, Washington, D.C.

Automation and Robotics in Construction X
G.H. Watson, R.L. Tucker and J.K. Walters (Editors)

Research on Components of Underground Excavation Robot

Keiko Homma[a], Tatsuo Arai[a], Hironori Adachi[a] and Tatsuya Nakamura[b]

[a]Autonomous Machinery Division, Robotics Department, Mechanical Engineering Laboratory, Namiki 1-2, Tsukuba, Ibaraki 305, Japan

[b]Department of Mechanical Engineering, Mie University, Kamihama 1515, Tsu, Mie 514, Japan

Abstract

This paper describes the basic research for an underground excavation robot. This robot consists of three system components: the manipulation system, the locomotion system, and the task planning system. In order to keep the operator away from the hazardous worksite, the task should be performed through teleoperation or performed autonomously. Two different teleoperation systems using parallel link manipulators have been built for this purpose. A locomotion system has been proposed, consisting of a mobile platform supported by a pair of crawlers and four legs. A hierarchical control system with one host computer and nine transputers has also been developed. For excavation task planning, a measuring method using tactile sensor data has been proposed. With this method the shape and stiffness of the target soil is determined from force and position data. These soil characteristics are passed to a cutting model of the soil and optimal parameters of excavation are estimated.

1. INTRODUCTION

This research is being conducted as part of a national research and development program on underground space development technology, which was started in 1989 by the Japanese Ministry of International Trade and Technology. Our research addresses the three systems that will comprise the underground excavation robot. In section 2, the manipulation system is described. In section 3, the locomotion system is described. In section 4, the task planning system is discussed. Results are summarized in section 5.

2. MANIPULATION SYSTEM

We have developed a parallel manipulator for excavation. The manipulator was designed to provide uniform force production capability based on a static mechanism analysis. For underground dome excavation the robot will operate either in teleoperation mode or in autonomous mode. While the robot may work autonomously for routine tasks, it should be controlled by a teleoperator in non-routine tasks, for example, when it confronts unexpected and emergent events, or in the process of searching and modeling its environment. The master-slave control technique is well known for refined teleoperation in which a human operator may command the robot simply and dexterously. If we, in addition, feed back reaction forces to the operator, this constitutes a bilateral system and the teleoperation task may be executed more smoothly. Sometimes it is preferable that the robot is capable of working in both teleoperation and autonomous modes, and is capable of switching between modes with no specific control command.

We have proposed and developed two kinds of teleoperation techniques for the excavation task. One is a bilateral master-slave system using a parallel link master. The other is joystick teleoperation enabling the robot to work in either or in both autonomous and teleoperation mode.

In the bilateral system, the master and the slave arms have the same configuration, but differ in size. This fact brings the problem that a translational motion and an orientational motion can not be preserved at the same time if the direct link servo method is introduced. For

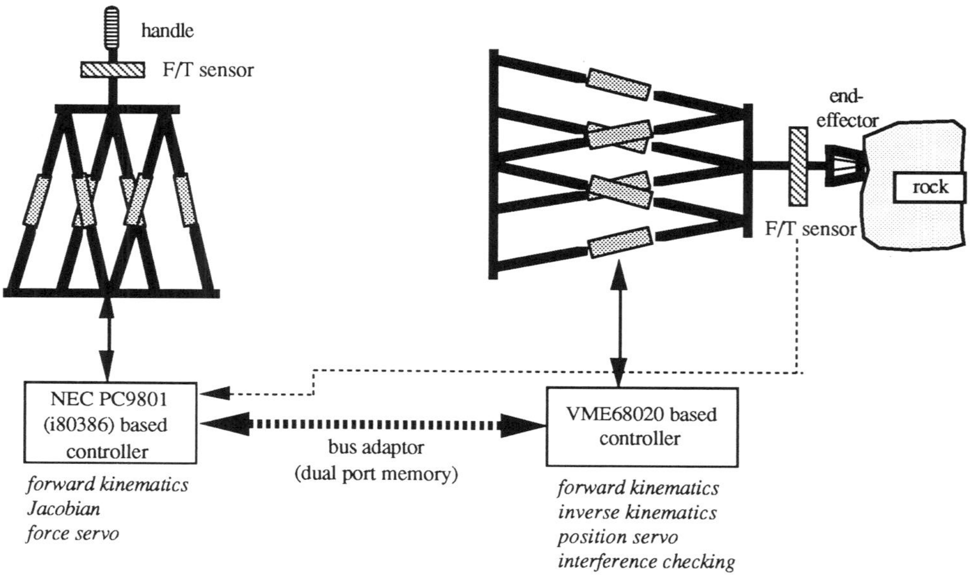

Figure 1. System configuration of bilateral teleoperation system.

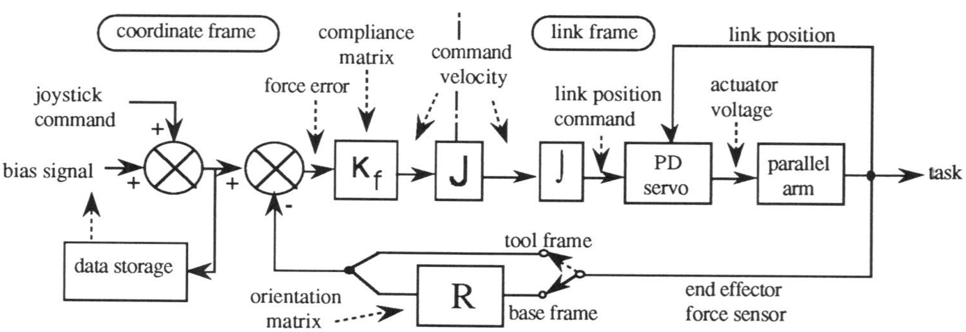

Figure 2. Block diagram of the joystick teleoperation system.

serial manipulators, kinematically dissimilar bilateral master-slave systems have been proposed to realize general bilateral master-slave systems that consists of any kind of master and slave manipulators. In order to solve the present problem, the same concepts can be applied to the parallel manipulator system. Figure 1 shows the configuration of the overall system.

The joystick teleoperation system is proposed based on the following design concept: (i) capable of working both in base and in end effector coordinates, (ii) capable of moving based on rate control without interference from the environment, (iii) capable of working based on hybrid control with interference, (iv) capable of commanding complete 6 DOF motion from the joystick, (v) capable of superimposing joystick commands onto the programmed commands. Figure 2 shows the block diagram of the proposed system. By using this system, the operator can easily intervene to teach the robot task or to modify the programmed robot motion.

3. LOCOMOTION SYSTEM

In order to construct a huge underground dome, it is necessary for the excavating machine to have locomotion capability. The locomotion system has to offer good mobility for efficient task execution. In addition, the locomotion system has the potential to assist the excavation operation because it is the contact point between the machine and the supporting land surface. At present, three types of locomotion have been considered for this machine: wheels, crawlers, and legs. Wheels offer fast and efficient locomotion on hard and flat terrain, but have difficulty on rough terrain. Crawlers have terrain adaptability to same extent, but it is difficult to execute the fine motion for assisting the excavation task. Although legs are capable of adapting to a variety of terrains and generating motions for the task, they are not capable of fast locomotion. Therefore, a hybrid locomotion system consisting of both legs and crawlers was adopted for the excavation machine. Figure 3 shows the experimental vehicle and the locomotion system for the excavation machine. The vehicle has four 3-DOF legs, and a pair of crawlers.

Figure 3. Apparatus of the experimental vehicle.

Figure 4 shows the control system and sensors for the vehicle. It consists of one host computer and nine transputers. Control software is developed on the host computer, and executed on the transputers. It has a hierarchical structure, with one transputer for higher level control and eight transputers to control the interface boards. The lower level transputers are divided into two groups. One group is for controlling the legs and crawlers, and the other group controls the attitude sensor in the body and the force sensors in the feet.

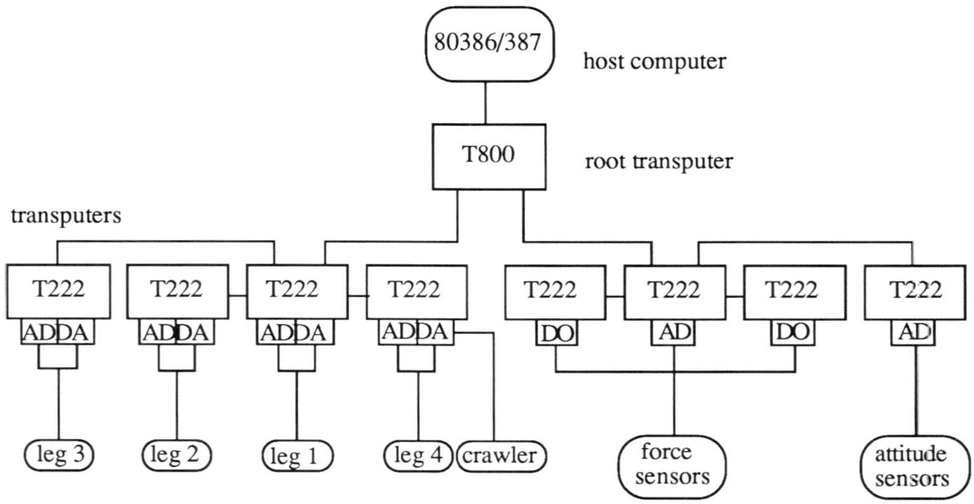

Figure 4. Control system of the experimental vehicle.

4. TASK PLANNING SYSTEM

In the case of underground excavation, an operator cannot operate the machine directly since the worksite is located deep underground and filled with underground water. Also they have difficulty in getting information about the target soil they want to excavate since the contaminated water prevents optical or ultrasonic sensors from measuring it. Therefore we propose a task planning method based on force and position data obtained from contact between the excavation manipulator and the soil.

As an example of motion planning with force data, Gocho et al.[1] use hydraulic pressure to control the motion of the bucket in an automatic wheel loader system. In their system the load is calculated from the measured pressure, and the bucket angle is determined according to the load.

4.1. Concept of the task planning method with data from contact

In the following discussion, we assume that the excavation manipulator is equipped with a force/torque sensor and probe. Figure 5 shows the flow of the task planning method we propose. First, the characteristics of the target soil are determined from the contact data. Then those parameters are used by the excavation model to estimate the reactive force of cutting. From the estimated force, we determine the optimal parameters for excavation.

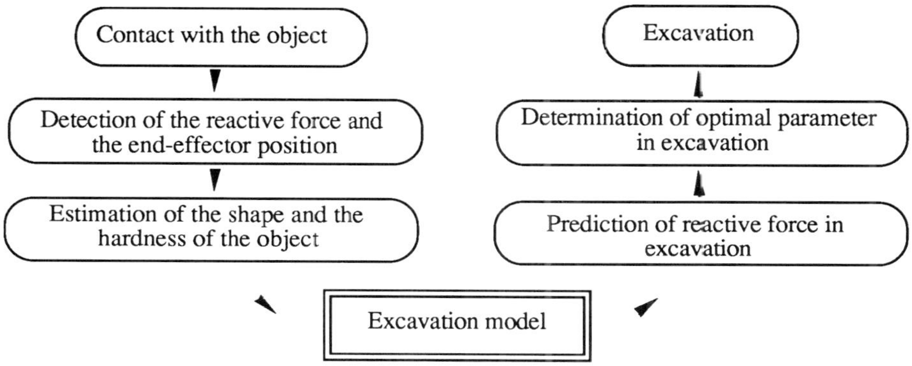

Figure 5. Task planning with contact data.

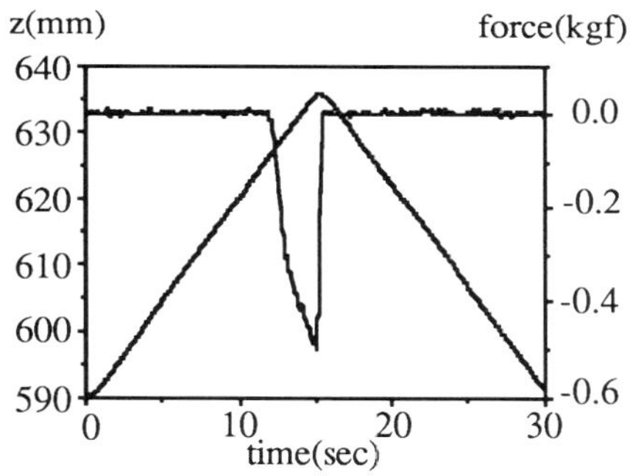

Figure 6. Contact data of soil.

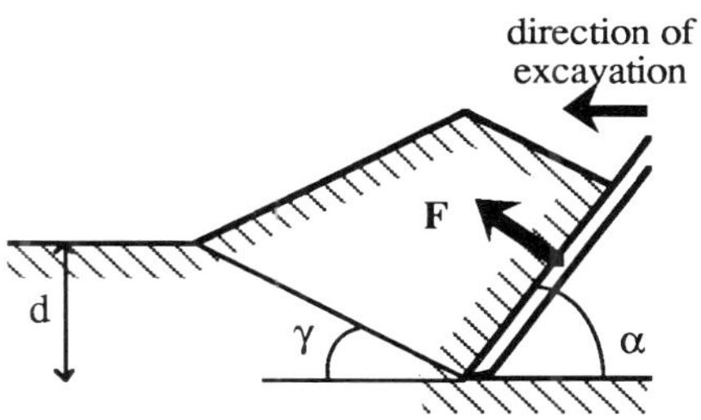

Figure 7. Two-dimensional cutting model.

4.2. Identification of shape and hardness of the soil

Measurement of shape and hardness by force sensor has the advantage of robustness and efficiency over measurement by non-contact sensors, such as optical sensor or ultrasonic sensor. That is, the shape and hardness of the object can be measured simultaneously.

To obtain the surface of the object, the robot is kept in motion and the measurement of force and position is repeated until a reactive force is obtained or the robot moves a certain distance. In the former case, the surface of the object is assumed to be at the position where the reactive force is measured. In the latter case, it is assumed that there is no object at least in the area where the robot passed.

To determine the stiffness, the probe is pushed into the object after the surface is determined. If the object is hard, the reactive force grows instantly while the position of contact between the robot and the object moves only slightly. If the object is soft, the position of contact keeps on moving a certain distance after contact occurred, and reactive force grows gradually.

Figure 6 shows the position of the end of the probe and the reactive force around the moment of contact.

4.3. Cutting model of sand

If we can obtain the characteristics of the target soil, then we should adapt them to the soil model and predict the reactive force in cutting. We use the following assumptions for the cutting model shown in Figure 7:

- The target object is sand and the criterion of failure is expressed as $\tau = \sigma \tan\phi$.
  (ϕ: internal friction angle)
- The state of cutting is approximated as two-dimensional.
- Cutting is done horizontally with constant velocity.
- Cutting angle α is constant.

Before considering the reactive force of cutting, the initial state and steady state should be distinguished. Once the cut soil grows into a certain shape, it cannot become larger because it cannot keep its shape any more. This state is defined as steady state, and the state before it comes to steady state is defined as initial state.

Hatamura et al.[2] studied the cutting mechanism of soil in steady state. According to them, the reactive force of cutting **F** is almost proportional to the depth of cutting. They also say **F** does not have any relation to the cutting velocity. We take it as a steady state model.

At the beginning of cutting, on the other hand, cut soil continues to grow until it comes to steady state. Hatamura's model cannot apply to this state.

We assumed two models for initial state and observed which expresses the behavior best. Figs. 8(a) and (b) illustrate these two models.

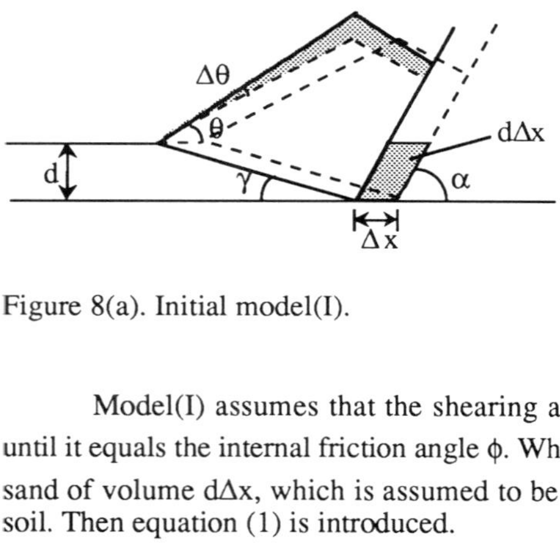

Figure 8(a). Initial model(I).

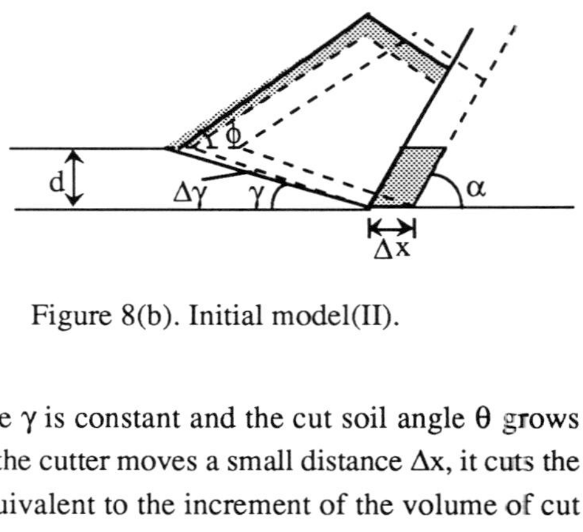

Figure 8(b). Initial model(II).

Model(I) assumes that the shearing angle γ is constant and the cut soil angle θ grows until it equals the internal friction angle ϕ. When the cutter moves a small distance Δx, it cuts the sand of volume $d\Delta x$, which is assumed to be equivalent to the increment of the volume of cut soil. Then equation (1) is introduced.

$$\Delta x = d\{1/2\tan^2 \gamma + 1/\tan \gamma(\tan \alpha + \tan \phi)\}\Delta\theta \tag{1}$$

Figs. 9(a) and (b) show estimated θ and **F** from the formula above. Simulation is done with the following conditions.

- Angle of the cutter from horizon: $\alpha = 60$ degrees
- Depth of cutting: $d = 10$ cm
- Internal friction angle of sand: $\phi = 38$ degrees
- Length of cutter: $L = 20$ cm
- Density of sand: $\rho = 0.00146$ kg/cm^3

Model(I) predicts that the cutting process reaches the steady state at about 40cm from the beginning of cutting, and cutting force grows to about 2.5kg.

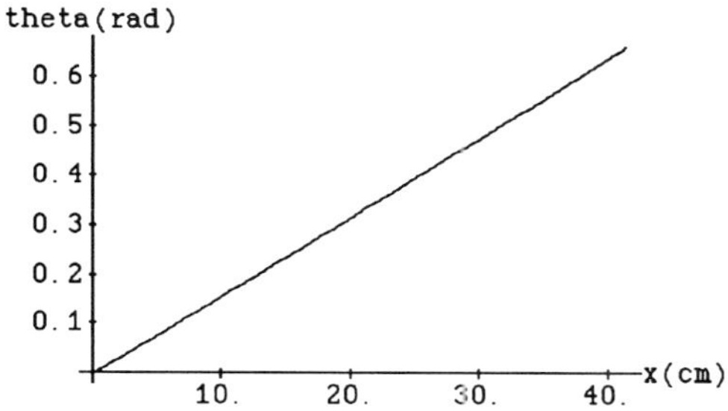

Figure 9(a). Estimated angle of cut soil θ.

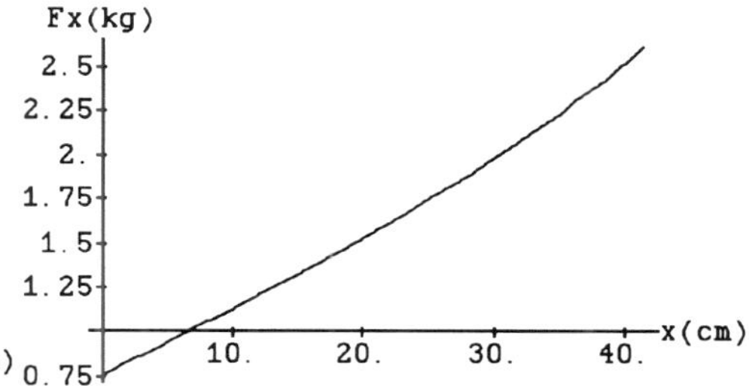

Figure 9(b). Estimated reactive force of cutting F_x.

Model(II) assumes that the cut soil angle is constant and equal to the internal friction angle ϕ, and that the shearing angle γ becomes smaller as the cutter moves. As for model(I), the cut volume is assumed to be equal to the increment of cut soil. This introduces the equation (2).

$$\Delta x = -[\{d\tan\phi + d^2\tan\phi(2\tan\alpha+\tan\phi)/(\tan\alpha+\tan\phi)^2\}1/\tan\gamma + d\tan\phi(2\tan\alpha+\tan\phi)/(\tan\alpha+\tan\phi)^2]\Delta\gamma \qquad (2)$$

Figs. 10(a) and (b) show estimated γ and **F** from formula (2). Simulation is done under the same conditions as model(I). This model shows that steady state is reached at a cutting distance of about 15cm, and that the reactive force is about 1.2kg.

According to Hatamura's model, the reactive force of cutting at a depth of 10cm is about 1.3kg. Observation of the cutting distance and behavior of the cut soil also indicates that model(II) is more similar to the real process of cutting soil.

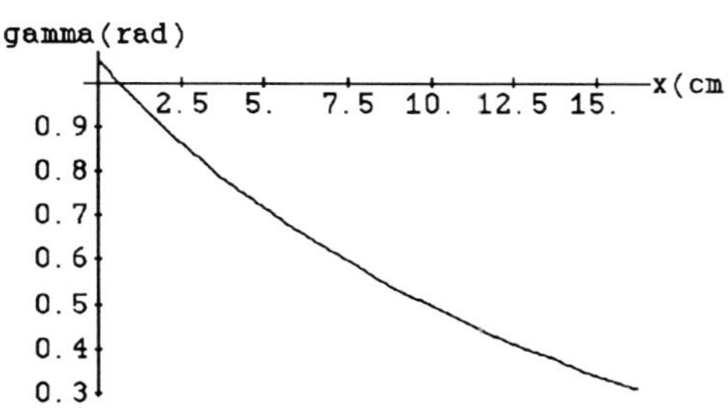

Figure 10(a). Estimated shearing angle γ.

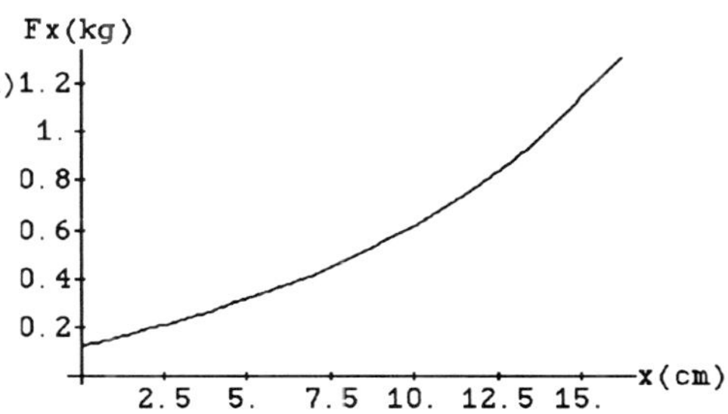

Figure 10(b). Estimated reactive force of cutting F_x.

4.4. Task planning with cutting model

What we ultimately want to do is to determine optimal parameters of cutting. In the case of horizontal cutting of sand, the main parameters are depth and speed of cutting.

Three cost functions L(i), V(i), W(i) are introduced. L(i) is the length of i-th cutting. V(i) is volume of cut soil of i-th cutting. And W(i) is amount of work of i-th cutting. Total amount of each function is expressed with L, V, and W. L and W should be minimized under the condition of constant V.

If we assume that the cutting process mainly consists of steady state, V(i) is expressed with the following formula (3).

$$W(i) = \int_{L(i)} \mathbf{F} \cdot dx = k \cdot V(i) \qquad (k : \text{coefficient}) \tag{3}$$

Since V is defined as constant, W is also constant and independent of the cutting path. Therefore the depth is determined to minimize L, that is, the maximum depth that the cutter can afford.

Optimization can also be done relative to actuator power. Optimal torque and velocity are determined to maximize the power.

Thus the depth and speed of cutting can be determined and excavation can be carried out.

5. SUMMARY

This paper describes research on three components of an underground excavation robot: the manipulation system, the locomotion system, and the task planning system. For the manipulation system, a teleoperation/autonomous control system with parallel link manipulator has been developed. For the locomotion, an experimental vehicle with a hierarchically controlled hybrid legged/crawler system has been developed. For task planning, a method using contact data and a cutting model of soil has been proposed and simulation has been carried out.

REFERENCES

1 Gocho, T., Yamabe, N., Hamaguchi, T., Hitomi, S., Ono, T., Oshima, H., Mori, M., and Fukuda, M. "Automatic Wheel-loader in Asphalt Plant", Proceedings of the 9th International Symposium on Automation and Robotics in Construction, pp.803-812, 1992.

2 Hatamura, Y. and Chijiiwa, K. "Explication of cutting mechanism of soil (5)", Transaction of the Japan Society of Mechanical Engineers vol.42 no.359, pp.2258-2268, 1976(written in Japanese).

Automation and Robotics in Construction X
G.H. Watson, R.L. Tucker and J.K. Walters (Editors)

Autonomous Dump Trucks System for Transporting and Positioning Heavy-duty Materials in Heavy Construction Sites

H. Sugiura[a], S. Yuta[b], K. Nishide[c], O. Hatakeyama[a], S. Nishigaki[a]

[a]Technical Research Institute, Hazama Corporation, 515-1 Nishimukai Karima, Tsukuba-City, Ibaraki 305, Japan

[b]University of Tsukuba, 1-1 Tennoudai, Tsukuba-City, Ibaraki 305, Japan

[c]Tokyo Aircraft Instrument Co., Ltd., 35-1, Izumi-Honcho 1-Chome, Komae-City, Tokyo 201, Japan

Abstract

This paper describes an autonomous dump trucks system being developed to overcome worker shortage problems and to prevent accidents in heavy construction sites (i.e., dam constructions, road constructions, etc.). The major distinctive functions of this system are:

(1) the autonomous driving function:
- this system enables unmanned dump trucks to autonomously drive at 40 km/h as planned. The difference between the planned path and the trajectory is within ±90 cm;
- the vehicle command system can recognize obstacles and prevent their being crashed or destroyed by, utilizing a color image processor;

(2) the advanced measurement function:
- the laser navigation system provides measurement function (i.e., the measurement errors regarding the position and the direction are within ±50 cm and ±1 degree, respectively).

This system enables two operators to manage dump trucks without drivers. It is anticipated that this system could reduce approximate 30% of the workers required at heavy construction sites in the future.

1. INTRODUCTION

The Japanese construction industry has made research efforts to develop on-site automation in order to overcome declining productivity, increasing labor costs, hazards in job sites, and scarcity of skilled labor. Generally speaking, developing the modules of manufacturing, assembling and inspecting rather than transporting has received much attention and has been forced so far in the Japanese construction industry. Transportation of earth, sand and rocks in earthwork has frequently reconfigured operations and often severe environmental conditions. Dump trucks in earthwork do not depend on a fixed path such as railroad. Dump trucks operations are repetitive, tedious, boring, physically dangerous and critical to productivity in earth work.

Faced by a need to improve productivity and safety in earthwork, an autonomous dump trucks system in heavy construction sites is developed as shown in Figure 1. The dump trucks system should have adaptability for changing surroundings, such as long distance driving, high-speed driving, driving within road width, change of route, etc.. To secure the safety of driving, the dump trucks system must recognize specific workers or vehicles and take actions to avoid obstacles. The dump trucks system must at least have 3 driving modules, that is, remote control for loading and unloading, unmanned driving for transportation, and manned driving for movement to parking lot.

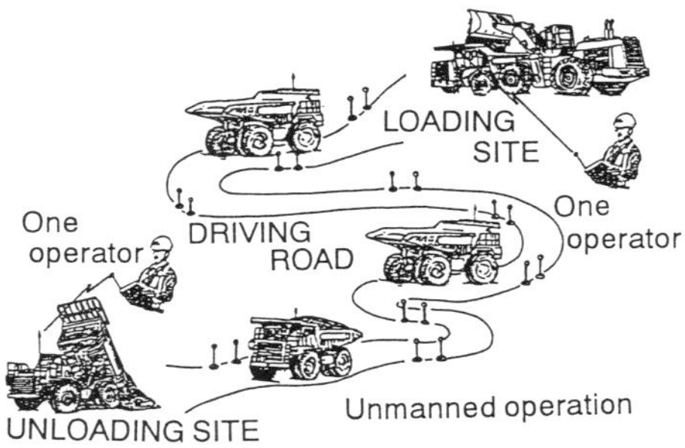

Figure 1. Image of an autonomous dump trucks system for transportation

The autonomous dump truck is equipped with a main unit and 3 sub units as show in Figure 2. These units work for the above 3 driving modules. The technologies required for the above 3 driving modules are developed by using the test vehicle shown in Photo 1.

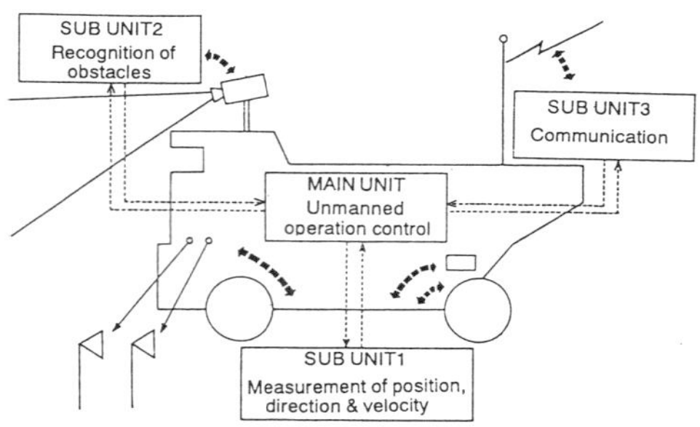

Figure 2. 4 units for an autonomous driving of dump trucks

Photo 1. Test vehicle for development of element technologies

First, this paper discusses the basic principles of the unmanned operation control of the test vehicle. Second, this paper shows the findings derived from the driving experiments of the test vehicle. Finally, we conclude with a summary of this study and discuss anticipated results and further research.

2. TECHNOLOGIES OF THE AUTONOMOUS DUMP TRUCK

It is indispensable for unmanned driving system of vehicle to have a real-time dead-reckoning function and obstacle avoidance function. This section discusses the dead-reckoning and obstacle avoidance functions of the test vehicle.

2.1 UNMANNED OPERATION CONTROL TECHNOLOGY (MAIN UNIT)

The unmanned operation control of the test vehicle is designed for 4-wheeled vehicles. The basic principles of the unmanned operation control consists of line and circle mode which were developed for "Yamabiko" robot. The "Yamabiko" robot is an independent 2-driving wheel steering type moving vehicle which is currently under study at University of Tsukuba.

As shown in Figure 3, a planned route on global coordinate system for unmanned driving is derived from combinations of local coordinates based on line mode and circle mode.

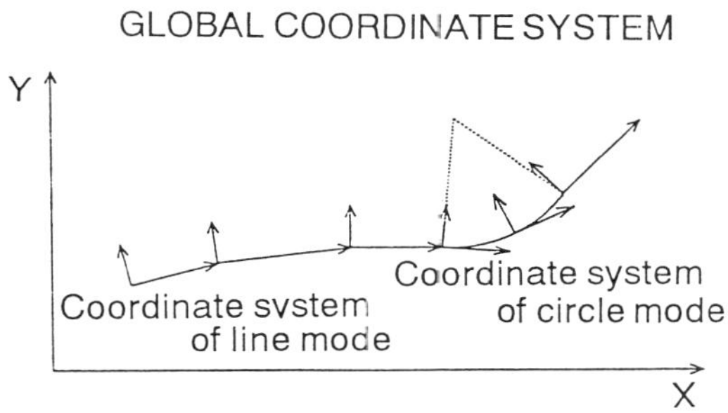

Figure 3. A planned route for unmanned driving

The unmanned control by line mode in Figure 4 requires target velocity $V(t+\Delta t)$ and target angular velocity $\omega(t+\Delta t)$. $V(t+\Delta t)$ and $\omega(t+\Delta t)$ are given by the following equations:

$$V(t+\Delta t)=V_d-K_4|\omega(t)|, \quad (1)$$

$$\omega(t+\Delta t)=\Delta t\{-K_1\eta(t)-K_2\theta(t)-K_3\omega(t)\}+\omega(t), \quad (2)$$

where V_d denotes set velocity; K_i for $i=1, 2, \cdots, 4$ denotes feedback coefficients; Δt denotes interval time between one sampling and the other. Feedback coefficients K_1 to K_4 in (1) and (2) were estimated by experiments so that the test vehicle can follow the planned route in a natural way.

The unmanned operation control by circle mode works for a curved driving along a small radius of curvature. Suppose that a 4-wheel vehicle is driving on a circle of a radius r. The relation among target velocity $V(t+\Delta t)$, target angular velocity $\omega(t+\Delta t)$ and steering angle $\phi(t+\Delta t)$ of the vehicle is shown in Figure 5.

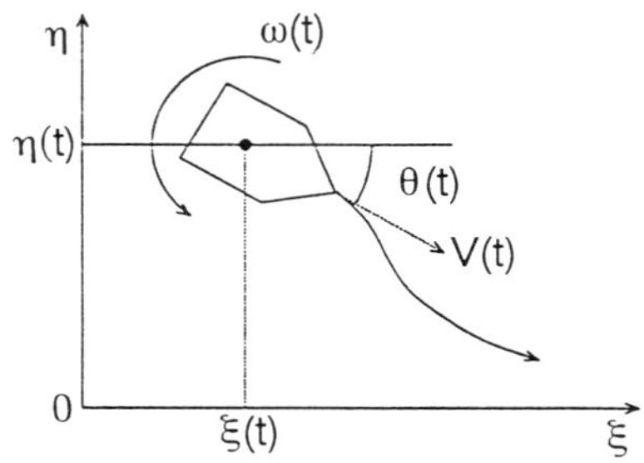

Figure 4. Local coordinate system of line mode

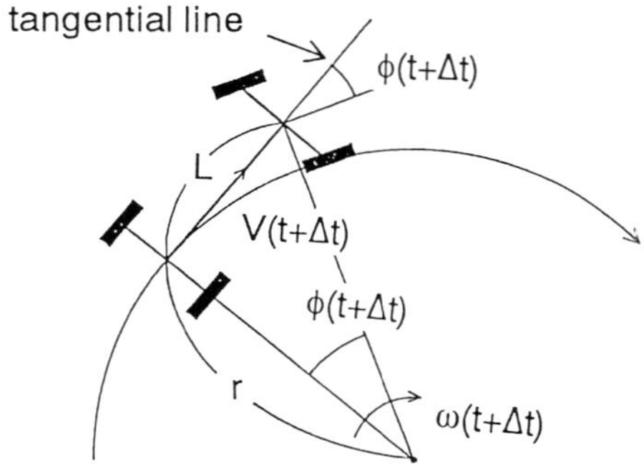

Figure 5. The relation among target velocity, target angular velocity and target steering angle

The target steering angle $\phi(t+\Delta t)$ is given by the following equation:

$$\phi(t+\Delta t)=\tan^{-1}\{\omega(t+\Delta t)L/V(t+\Delta t)\}. \quad (3)$$

Suppose that a unmanned driving is started. First, the main unit receives the driving information $\{\eta(t), \theta(t), \omega(t), V(t)\}$ from the sub unit 1. Second, the target velocity $V(t+\Delta t)$ and the target steering angle $\phi(t+\Delta t)$ are calculated based on the equations (1), (2) and (3). Third, the main unit performs a speed control (brake, accelerator) and a steering control according to those target values. This operation is repeated until the target point is reached.

2.2 POSITION, DIRECTION AND VELOCITY MEASURING TECHNOLOGY (SUB UNIT 1)

See Figure 6. Driving distance and velocity of the vehicle are detected by encoder sensors attached to rear tires. Direction of the vehicle is detected by a fiber optic gyroscope. Positions of the vehicle are calculated successively based on the driving distance and the direction.

A laser transmitter/receiver is equipped at the left side of the test vehicle as shown in Photo 2. Photo 3 shows that poles (2 poles/set) with laser reflectors (corner cubes) are installed along the driving route at an interval of approximately 50 m. Any accumulated positional errors in a long-distance driving are corrected based on the feedback data from the laser transmitter/receiver and the poles.

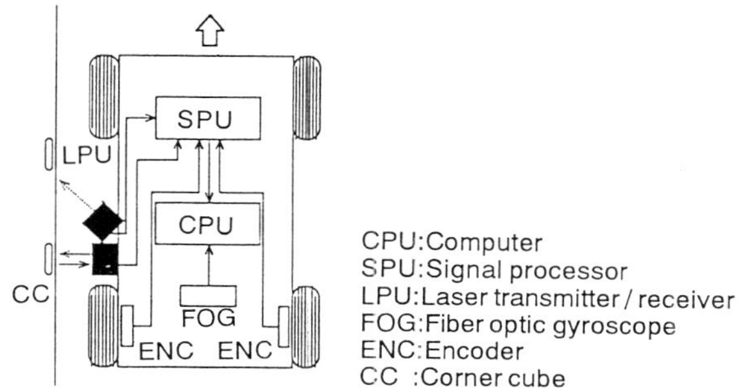

Figure 6. Position, direction and velocity measuring system

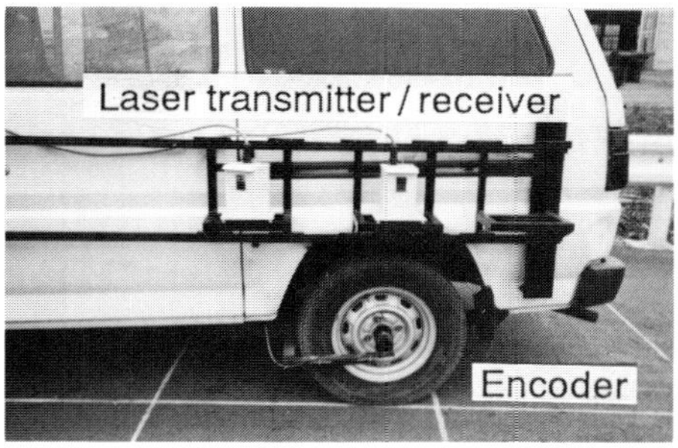

Photo 2. Encoder and laser transmitter/ receiver

Photo 3. Laser reflector (corner cube)

2.3 TECHNOLOGY OF OBSTACLE AVOIDANCE (SUB UNIT 2)

To make an autonomous driving of a vehicle safer in outdoor environments, it is necessary to detect any obstacles in the front direction on the driving road and to take some safety measures for autonomous evasive action or stop of the vehicle.

This test vehicle has a front obstacle detecting system. This system uses a method of comparatively easy macro graphic detection. This system can fetch an image of the front view into a personal computer from a CCD camera equipped at the assistant's seat side. The image of the front view can be processed by a transputer loaded on this personal computer.

At the current stage, this system recognizes workers by extracting the helmet worn commonly by all workers. The color of the helmets is usually yellow in many construction sites. First, the processor searches portions of the same yellow color as that of helmets based on the brightness and the hue of the fetched image. Second, it discriminates between the presence and non-presence of helmets by searching their surface areas and shapes among the extracted portions. Thus, it is possible to detect workers within the range of 20 m ~ 60 m in the front direction (Fig. 7).

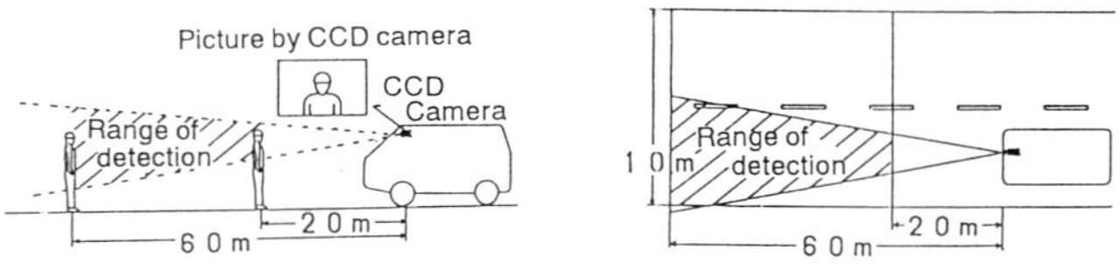

Figure 7. Range of detection of workers by CCD camera

2.4 COMMUNICATION TECHNOLOGY (SUB UNIT 3)

A simple remote-controlled operation by radio controller is available for communication system. By this communication system, it is expected that loading and unloading can be fully operable from remote locations as shown in Figure 1. Furthermore, it could perform to easily switch from unmanned driving module to manned driving module by this communication system.

3. FINDINGS DERIVED FROM THE DRIVING EXPERIMENTS OF THE TEST VEHICLE

The field driving experiment of the test vehicle was conducted in a dam construction site in 1991. Further refinements of those technologies of the test vehicle have been and are being evolved at the Technical Research Laboratory of Hazama. Table 1 shows the current performance of the test vehicle.

Table 1
Performance of the test vehicle

Function	Item	Target performance	Current performance
Driving mode	Manned driving Remote controlled driving Unmanned driving	Free switching Free switching Free switching	Free switching Free switching Free switching
Unmanned operation	Distance of continuous driving Maximum speed Width of meandering Radius of gyration	2km 40km/h ±90cm 15m	1.5km 20km/h ±130cm 40m
Measurement of position & direction	Position accuracy Directional accuracy	±50cm ±0.5degree	±50cm ±1.0degree
Recognition of obstacle	Worker(fine) Worker(Cloudy,Slightly rainy) Other vehicle	100% 100% 100%	80% 100% Under development
Evasion of collision	Preceding vehicle at stop Driving preceding vehicle	100% 100%	Under development Under development
Remote-controlled operation	Radius of operation	Within 50m	Within 50m

The continuous driving distance that the unmanned driving control could achieve is 1.5 km at the maximum speed of 20 km/h. The difference between the planned path and the trajectory was within ±1.3 m.

The unmanned driving experiments of the test vehicle of long-distance (2 km) and high-speed (20 to 40 km/h) are being conducted at the test course of the Public Works Research Institute of the Ministry of Construction.

4. CONCLUSION

This paper presents the major distinctive functions of the autonomous dump trucks system as follows:

(1) The autonomous driving function:
- This system enables unmanned dump trucks to autonomously drive at 20 km/h as planned. The difference between the planned path and the trajectory is within ±130 cm;
- The vehicle command system can recognize workers that worn a helmet by, utilizing a color image processor;

(2) The advanced measurement function:
- The laser navigation system provides measurement function (i.e., the measurement errors regarding the position and the direction are within ±50 cm and ±1.0 degree, respectively).

Fig. 8 indicates an example of composition of workers at a certain dam construction site. Since this system enables two operators to manage dump trucks without drivers in heavy construction sites. It could reduce approximately 30 % of workers in the future. Likewise, the technologies in this study could, with some enhancements and modifications, perform useful functions for remote work vehicles in heavy construction sites.

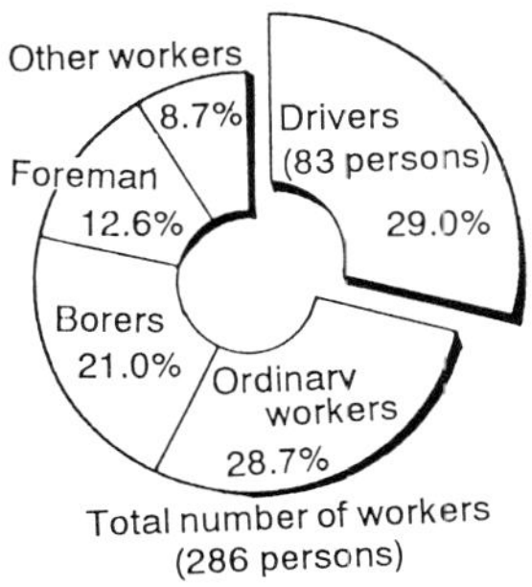

Figure 8. Example of composition of workers at a dam construction site

Further research problems are to develop technologies in respect of:

(1) making the directional accuracy being within ±0.5 degree by utilizing a fiber optic gyroscope;
(2) improving the ability to recognize workers by a laser radar;
(3) reciprocal communication technology among a plural number of vehicles;
(4) recognizing obstacles just before, behind, side of vehicle;
(5) recognizing other vehicles, rocks, holes, etc..

5. REFERENCES

1 O. Hatakeyama, H. Sugiura "An Engineering Approach to Carriage Automation System at Construction Site (the First Report)" proc. The 9th International Symposium on Automation and Robotics in Construction, pp849-857, 1992
2 S. Iida, S. Yuta "Vehicle Command System and Trajectory Control Method for Wheeled Mobile Robots" proc. The First Robot Symposium in Japan, pp85-90, 1991
3 M. Hanawa, N. Hori "A Position and Heading Measurement System Using Laser Beams for a High Speed Vehicle (2nd report)" proc. The 5th ADVANTY Symposium, pp29-32, 1992

Automation and Robotics in Construction X
G.H. Watson, R.L. Tucker and J.K. Walters (Editors)

DEVELOPMENT OF A ROBOT ASPHALT PAVER

R. Umeda * :ADVANCED CONSTRUCTION TECHNOLOGY CENTER
H. Ikeda: TOHOKU DISTRICT BUREAU, THE MINISTRY OF CONSTRUCTION
H.yasu: OBAYASHI ROAD CO. ; M. Fukukawa: KAJIMA ROAD CO.
Y. Masuyama: SEIKI TOKYU KOGYO CO. ; S. Kinosita: TAISEI ROTEC CO.
Y. Takagi: THE NIPPON ROAD CO. ; T. Gocho: NIPPON HODO CO.
F. Goto: NIIGATA ENGINEERING CO.
* OTOWA NS BLDG. 2-10-2 Otowa, Bunkyo-ku, Tokyo 112, JAPAN

Abstract

Frequent changes in the paving conditions of general roads in Japan require considerable labor. This demands a heavy work load and a high level of skill to the asphalt paving operators and screedmen, and also affects the work load of subsequent manual operations, such as shoveling and raking.

In this project, five automation items to be developed were determined, after analyzing the present paving work by Industrial Engineer (IE) analysis and considering the paving work as a total system. The items are asphalt mix feeding, control of paving thickness, screed extension, steering operation, and operation of receiving the asphalt mix. The automation was achieved by using various sensors, an image processing system, and element technology using fuzzy control. The initial aim of labor savings and reduction in the level of skill could be achieved by developing this robot asphalt paver.

1. INTRODUCTION

Lack and aging of skilled workers, and dropping off of young workers from the construction industry have been continuing in Japan today. However, road paving work still depends on manual labor and adequate skill to a large extent. The reasons for this are:

(1) Land space is small and land acquisition for roads is difficult. Therefore, frequent changes in the paving widths are general in Japan.
(2) Ensuring the correct paving thickness, trimming the paving edges and finishing the surface, need manual labor.

To solve these problems, we have jointly developed a one-man robot asphalt paver with the aim of labor savings and reduction in the level of skill. The robot asphalt paver is a total system consisting of preprocesses (material transport), self processes (asphalt paving) and post processes (manual operations such as shoveling and raking).

2. ANALYSIS OF EXISTING WORK PROCESSES

We analyzed the following items to extract functions to be improved and developed:

(1) Analysis of current paving work processes
The functions of the conventional paver, operators, screedmen, and workers were analysed using IE techniques and video recording the paving conditions.

(2) Survey of the conventional asphalt paver and finished shape
Paving work was carried out only by operators and screedmen, and influences on the operations of the operators and screedmen and the finished shape were surveyed.

(3) Analysis of operational functions of the conventional paver
A questionnaire survey was performed on the operations for each work condition for the operators and screedmen, and the level of difficulty of machine operations was analyzed.

The problems confirmed quantitatively from the analysis are as given below.

(1) The steering and asphalt mix feeding operations, which are manual operations done by the operators, account for a large percentage of the work.

(2) The transverse movements of the screedman for screed control, checking and controlling the paving thickness is large.

(3) Raking work load is heavy while shoveling work load is light.

(4) While the conventional paver is docking with the dump truck, the overall work efficiency decreases.

(5) Irregular work (manual work) occurs at the beginning and the end of the paving work (Ex: the tapering to the existing road), resulting in peak work load of the workers.

As mentioned above, (5) cannot be avoided at the design stage. However, (1) to (4) are problems related to the conventional paver functions. These points were taken up as targets for development.

3. DEVELOPMENT POINTS

This development project was planned as a total system centering around the conventional paver from the transportation of materials to shoveling and raking works based on the results of analysis of the present operations. The five points which complied with development targets are given below.

(1) Asphalt mix feeding
Asphalt mix feeding to the front of the screed is an operation that needs maximum concentration of the operators. A uniform feeding over the entire width of the front of the screed lightens the heavy work load in the shoveling, raking and cleaning operations of the paving edges.
Also, it reduces variations in laying thickness and influences the next item.

(2) Controlling the paving thickness
Decreasing the level of skill in screed adjustment operations, which largely depends on experience, skill and intuition, helps one-man operations.
Also, if the laying thickness is uniform, quality of the finished pavement naturally improves and the work load in post processes also decreases.

(3) Extension of the screed
The adjusting work of the screed to match the paving width is an extremely heavy work for the screedmen, and this operation has a considerable influence on post processes.

(4) Steering operation
The steering operation bears considerable importance for the operators. Automation of steering operation plays an important role to develop one-man operations.
Also, the steering operation is closely related to screed width adjustments, which also influences post processes.

(5) Receiving the asphalt mix
The overall working efficiency decreases during the docking and undocking with the dump truck when receiving the materials. This was observed from the results of analysis of the present working processes. If the docking with the dump truck is smooth, there will be an improvement in the post processes. By semi automating docking indications (position, distance) of the conventional paver, and instructions for truck hitching and dump-up angles, uniform feeding of materials is possible.

If the above mentioned items are automated, the operator needs only to monitor the automatic operations and to do a part of the auxiliary operations by himself. Also, automation reduces the number of workers required in post processes and lightens the work load of the workers, surely resulting in overall labor savings and reduction in the level of skill in the total system.

4. DEVELOPMENT FUNCTIONS

Automation is necessary for achieving the development targets. The development items are described in detail below.

(1) Automatic feeding of asphalt mix
Four ultrasonic sensors were placed at the center and the both ends of the screed. The sensors at the center automatically control the asphalt mix feeding quantity of the bar feeder. The sensors at the both ends automatically control the asphalt mix feeding quantity of the screw conveyer.

(2) Automatic control of laying thickness
The paving thickness is measured and indicated with four ultrasonic sensors consisting of two sensors on each end, an inclination sensor and a travel distance sensor. The paving thickness is automatically controlled, based on the data measured by the sensors.

The fundamental technology used in (1) and (2) above, is the same as reported in "Development of Advanced Asphalt Finisher" in the 9th ISARC, and is applied to this project with several application developments. For details, refer to the paper of the 9th ISARC.

(3) Image processing system used for automatic extension of the screed and steering
The automatic extension of the screed and steering is mutually interrelated. The control systems are so made as to be interrelated and the points given below were taken into account.

· Existing structures are used as reference lines to eliminate new reference settings for control, thereby achieving labor savings
· The same sensor is used for sensing the whole objects usable for reference lines
· Common sensors are used for automatic screed extension control and automation of steering to reduce the number of sensors.

Based on the above prerequisites, we have developed a sensor (tentatively named 'road eye') using laser beams and a CCD camera. This sensor detects the edge if there is a level difference in the reference line. For flat reference lines such as white lines, only the CCD camera is used, which detects the reference line by binary values of bright and dark of the image.

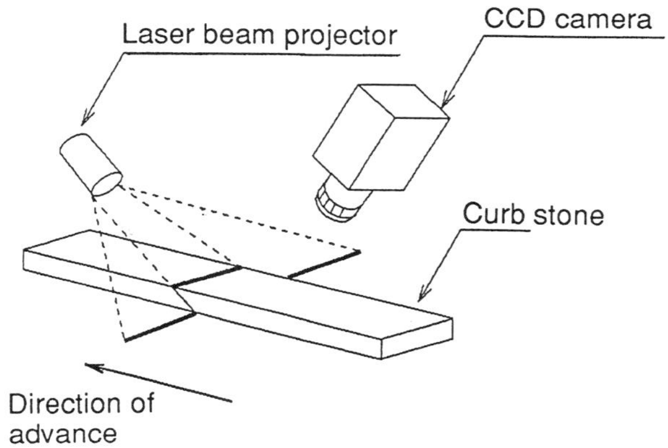

Figure 1.

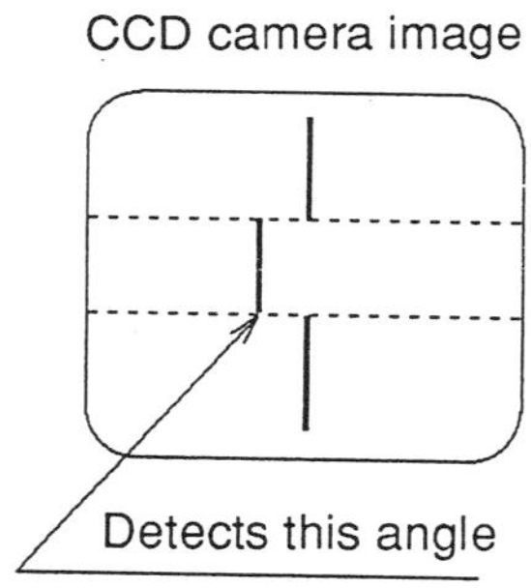

Figure 2.

The example of using a curb stone as the reference line is explained here. As shown in figure 1, a laser slit beam is projected from the front side on the curb stone and the CCD camera captures the image. In this image, the edge of the curb stone can be detected because the laser slit beam is at a different level. Figure 2 shows the monitor image (after converting to binary values) of the edged curb stone.

(4) Automatic screed extension control
Figure 3 shows the screed part of the robot asphalt paver and its movement with time. Sensors are installed at two locations - the front end (sensor A) and the rear end (sensor B) of the screed endplate. Both sensors, A and B are controlled so that they are always on the inside (paving side) the reference line. This is to keep the screed always inside the reference line to prevent collision when the reference line is a structure and to prevent overflowing the materials from the reference line when other lines are used as reference lines. Figure 4 shows the control system block diagram.

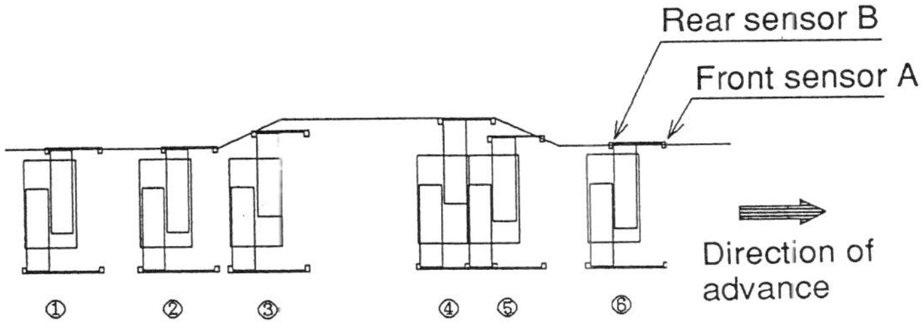

Figure 3. Indication of the screed movements only

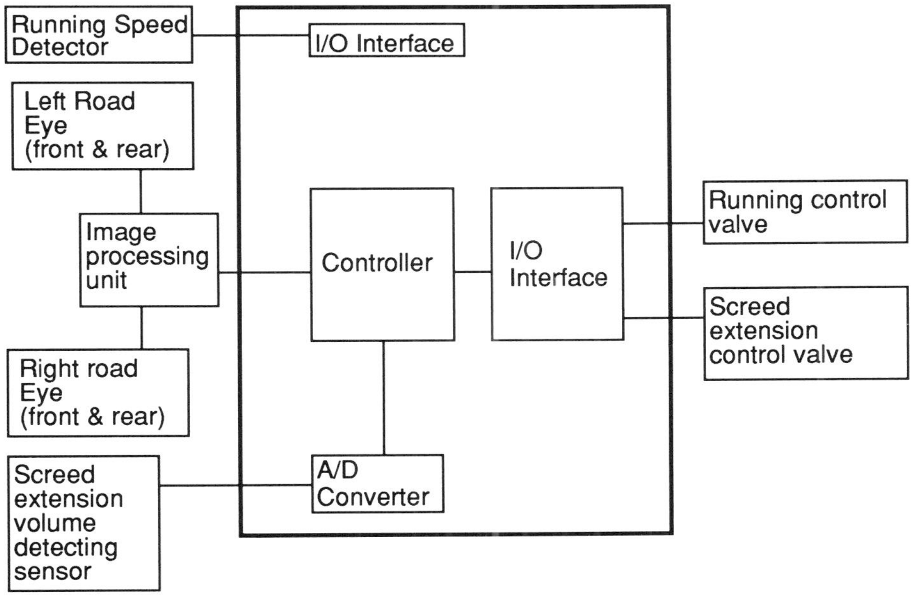

Figure 4. Control system block diagram

(5) Automatic steering device
The sensor used for measuring the azimuth angle deviation of the robot asphalt paver is the road eye, which is used also for the automatic screed extension control. Therefore, the sensing position is behind the swiveling shaft center of the vehicle body, and the road eye moves opposite to the steering direction at the start of the steering operation. Also, the steering angle must be changed according to the displacement of the line of travel from the reference line. Under these conditions, fuzzy control was used because we predicted that framing of control rules by numerical functions would be difficult.

The two fundamental rules for control are given below.

1) Vehicle travel control
In figure 5, if the distance between the front sensor A of the screed endplate and the reference line is taken as X1, and the distance between the rear sensor B and the reference line is taken as X2, then the control is such that (X1 + X2) / 2 = 0.

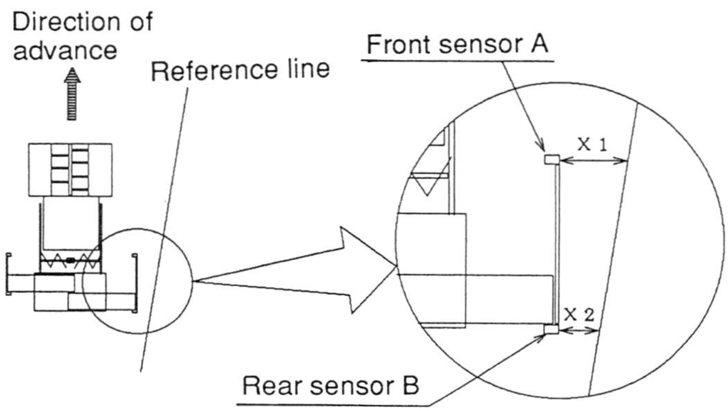

Figure 5.

2) Parallel travel control
The vehicle is always controlled to move parallel to the reference line such that X1 - X2 = 0. This control has the effect of restraining the overshoot control that is likely to occur in vehicle travel control.

To determine these control rules, computer simulation was carried out and the controls rules were applied to the actual machine, after verifying effective working of these rules.

The control block diagram is shown in figure 4.

(6) Semi-automation of asphalt mix receiving process
This development is divided into two parts.

1) Devices for guidance and operation instructions for the dump truck driver
A large indicating panel was installed by the side of the cabin of the robot asphalt paver. The indication of the panel has two modes, the docking mode with the dump truck and the material feeding mode, which are automatically switched each other. During the docking mode, the distance remaining to the stop position, is displayed on the indicating panel by the distance sensor. When the stop position is reached, STOP is indicated on the panel. The robot paver operator can make the lateral displacement displayed on the indicating panel by pressing the button. If a person or an object comes between the dump truck and the robot paver at this time, an alarm is generated. Also, during the material feeding mode, instructions for dump-up, dump-down, and release can be displayed by pressing the button.

2) Automatic truck hitching
The dimensions of the rear wheels of the dump truck and the rear end of the vessel vary depending on the type of the dump truck. The truck hitching device developed in this project starts the hitching operation automatically when the rear wheel touches the push roller. Then, the push roller retracts in the rear direction, and is locked automatically when the position of the vessel rear end becomes the same as the position of the hopper.

5. SPECIFICATIONS OF THE DEVELOPED ROBOT PAVER

Table 1 shows the main specifications of the robot paver developed in this project. Figure 6 shows the outline drawing of the paver.

Table 1. Specification

Total weight:	13000 Kg
Overall length:	6700 mm (Max)
Overall width:	2490 mm (in transportation)
Engine power:	66 Kw/2000 r.p.m.
Paving width:	2.5 - 4.5 m
Automatic system:	
Mix feeding:	Ultrasonic type automatic feeder
Laying thickness control:	Ultrasonic type automatic adjuster
Laying thickness measurement:	Ultrasonic type automatic measuring device
Screed extension:	Image-processing type automatic extension device
Steering:	Image-processing type steering device
Dump truck guiding:	Semi-automatic dump truck guiding device

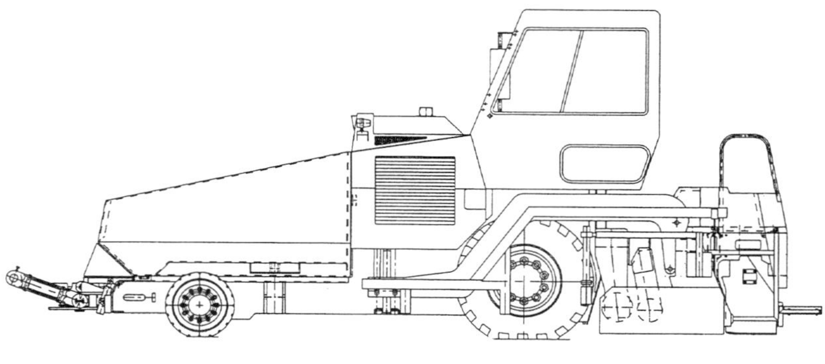

Figure 6.

6. CONCLUSION

Considerable labor savings and reduction in skill were achieved by the robot paver developed in this project. We will make further efforts to improve performance of the robot paver and popularize it to general paving projects in Japan. Also we are planning in future to apply this automation technology to other kinds of construction.

Automation and Robotics in Construction X
G.H. Watson, R.L. Tucker and J.K. Walters (Editors)

TOWARDS THE DEVELOPMENT OF A BRIDGE INSPECTING AUTOMATED DEVICE

Anna McCrea[a] and Denis A.Chamberlain[b]

[a]Senior Lecturer, School of Construction, Economics and Management, South Bank University, London.
[b]Head of the Construction Robotics Unit, Department of Civil Engineering, City University, London.

Abstract

There are, as yet, no commercially viable robotic or automation solutions in the field of reinforced concrete bridge inspection, these structures suffering from deterioration of both their concrete and steel reinforcement elements.
In the manual inspection activity for defect diagnosis and repair specification, visual assessment is supplemented by a range of non-destructive tests. There are a range of such tests, each directed at a particular aspect of the decay problem. For automation to be effectively developed, the operational requirements of each test probe needs to be defined in relation to many combinations of access and location geometry. A library of these 'benchmarks' is in fact needed for use in the research and development of useful automation devices. They can be used in computer based simulation studies to provide work-pieces which test the capability of concept devices and aid their evolution. This is important for the development of both the access vehicle (ROV) and the robot mounted on it. The need for this rational approach is clear in that existing ROV's would fair badly when attempting even the simplest 'benchmark' task. The application and value of the 'benchmark' approach to the automation problem has been demonstrated in the consideration of two device concepts for bridge inspection.

KEY WORDS: Concrete decay, inspection methods, benchmarks, dextrous envelope, inspection facility.

1. INTRODUCTION

The application of automation and robotics in the fields of monitoring and inspection, as a means of achieving greater reliability, productivity and safety, is gathering interest [1]. Such a role has been clearly established in respect to bridge inspection.

However, the emergence of a commercially successful device depends on many technological and human factors. This paper seeks to address some of these, particularly those which will influence research and development strategy.
Experience related from various sources indicates that the inadequacy of the current manual approach to bridge survey is most acute where access is particularly difficult, extensive areas are to be tested or non-standard inspection techniques are required. It relation to access, there are in the UK some 300 important bridges for which this poses extreme difficulties. These are typically tall viaducts, multi-span structures and large bridges over water. The requirements for the remaining 100,000 or so bridges would without substantial progress in cost-effectiveness, be best met by current methods. In the research, a range of bridge and component geometries have been gathered for the development of a 'benchmark' library. The requirements and activity for inspection have been built into these so that prospective automation devices can be evaluated on a correct basis.
After the content and role of 'benchmarks' are introduced, their use is discussed in the context of the simulation and evaluation of inspection automation concepts using the GRASP computer software. In conclusion we see that the 'benchmark' approach to device development is realistic, technically sound and essential.

2. DETERIORATION OF BRIDGES

There are two main and closely associated, forms of decay in bridges [2,3], deterioration of concrete and corrosion of reinforcing steel. Several factors contribute to the occurrence of these phenomena. Where no specials requirements are set in the testing for such decay, information is required on the (i) visual appearance of the surface including the and location of cracks and rust stains, (ii) likelihood of reinforcement corrosion, (iii) delamination , (iv) cover to the reinforcing bars, (v) depth of the carbonation front and (vi) level of chloride contamination.

3. INSPECTION METHODS AND AUTOMATION FACTORS

Of all the available non-destructive tests, only a few are clearly amenable to automation. The factors influencing this are the weight, size and robustness of the probe, method of handling, vibration aspects, reliability, accuracy, data size and interpretative methods [4]. Some useful analysis of these attributes is given elsewhere [5].
Supportive sensing would be required for each of the above. This is necessary for collision avoidance, contact force limitation, positioning and orientation of probes relative to the inspected surface. A distributed processing system combined with A/D conversion capability would be needed to handle the support sensing.
The probe operational characteristics are vital considerations in the correct definition of a robot or automation device. Size and weight of the probe together with the presentation to surface requirements will inevitably influence the actuator and dextrous envelop specification. Here the 'dextrous working envelop' is quite distinct from the 'positional working envelop' commonly used in suppliers descriptions of robots.It accounts for the orientation and positional requirements and is invariably less extensive.

Determination of the dextrous envelop can be achieved explicitly using inverse kinematics or somewhat easier by experimentation with a robot simulation facility. A combination of these approaches have been employed with the aid of simulations built using the GRASP computer software [6].

4. INSPECTION TASK BENCHMARKS

In order to establish a sound basis for task led research and development of inspection automation devices, it is necessary to focus on clearly defined task requirements or 'benchmarks'.

Here a 'benchmark' is taken to comprise geometric descriptions of the zone requiring inspection and permissible access to it, specification of the probes to be handled and the point, line and area coverage requirements. An example is illustrated in Fig. 1, where a robot located in region A is required to detect the steel reinforcement over area B. Impulse hammer readings are also required at positions a-f on B.

Clearly an extensive library of 'benchmarks' is necessary, each translating into the dextrous requirements for the inspection robot. In this work, both 'benchmarks' and robots has been modelled using the robot simulation software.

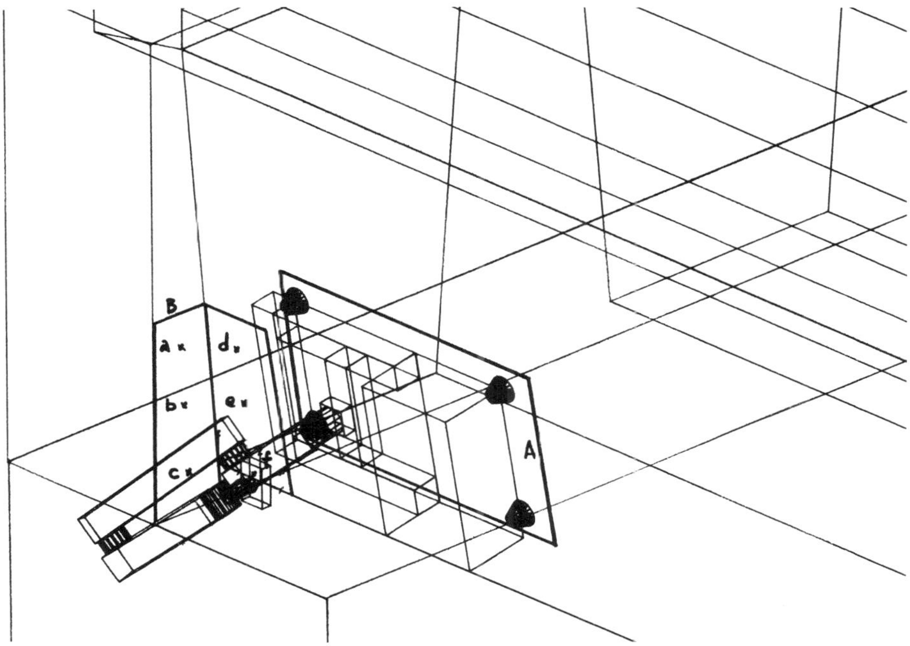

Figure 1. A Benchmark Example

5. SIMULATION STUDIES

5.1 Computer Modelling

The simulation software GRASP comprises a solid modeller with full boolian operations and a process modeller supporting both standard and user defined robot joint configurations. It can be used interactively or by running programmes prepared in a descriptive high level language. It supports 'tracks' which can be used to define a task in terms of motion paths between set points. The object orientated basis of the facility makes these a useful attribute for benchmark development. On executing a track based programme the cycle time and collision risks are determined, the former clearly dependent on the maximum speed and accelerations prescribed for the robot motors.

In simulation studies robots divide into two classes, those matching available industrial robots configurations and non-standard robots for which the user must supply the inverse kinematic relationships.

Robots with more than 6 DOF's can be simulated by setting configuration rules. Once the combination of revolute and prismatic joints are set, the architecture or flesh is added. This is not a cosmetic exercise as collision may depend on it.

The robot tools, its inspection probes, are readily interchangeable in the simulation, these mounted on the robot's TAP (Tool Attachment Point). Each tool, however, has its own TCP (Tool Centre Point) which is required to follow track paths. Depending on the type of probe, this corresponds to its contact point or centre of contacting area with the inspected surface. Probes are also assigned contact axes which force them to align with track owned targets. This is necessary to cover the requirements imposed in benchmarks.

5.2 Dextrous Limits

Figures 2-4 show a concept robot which has been studied in simulated bridge inspection. Its arrangement of revolute joints conforms to a 'Puma' configuration model, though its link lengths and motor travel limits are somewhat different. These variations are necessary to overcome the limited dextrous envelop of this robot when working on planar surfaces. The extent of the dextrous envelop, which was determined by trial and error, is apparent by the boundaries of the accessible areas indicated. As the simulation facility advises the user when motors have reached the limit of their travel for the required TCP orientation, it is a relatively easy matter to determine the lateral extent of the dextrous envelope.

As shown in Fig. 4, the investigated configuration provides limited access to the edge of the member, indicating that an extra degree of freedom or an alternative configuration should be considered. Improved concepts are currently being developed and evaluated to this end.

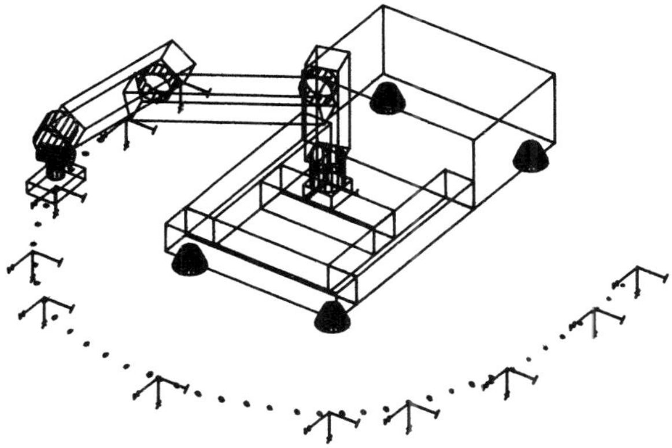

Fig.2 Dextrous Limits To Plan

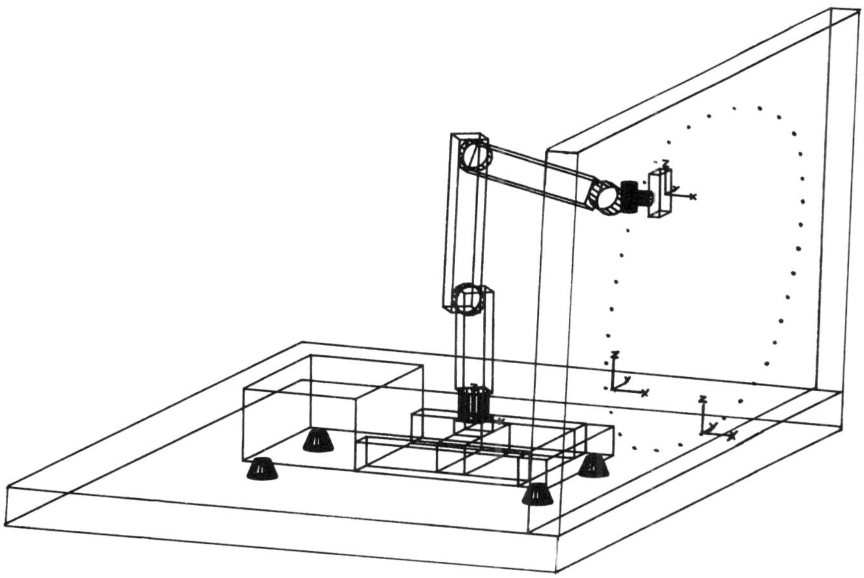

Fig.3 Dextrous Limits To Upstand

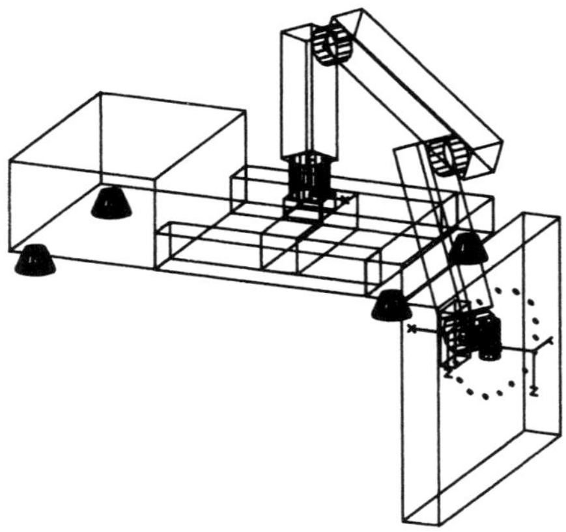

Fig.4 Dextrous Limits To Overhang

5.3 Task Tracks

Determination of the intersection of the dextrous envelope for particular probe orientation requirements and working surfaces is the starting point for assessment [7]. To improve this, the robot can be relocated, have motor limits extended or, more drastically, its configuration modified. Having achieved working boundaries to the extent of the benchmark area limits, the performance of the robot can be evaluated by running the programme tracks. Whilst similar, these tracks are specific to the operational requirements of each probe type.

6. ACCESS SYSTEMS

In establishing realistic benchmarks the means of access and mobility are clearly of prime importance. Here, two classes of solution are apparent these based on the use of (i) a bridge access system and (ii) a surface traversing vehicle or ROV. Considering the former, large bridges and viaducts are generally provided with a permanent access system for maintenance purposes. There are many examples of such systems used for both temporary and permanent access [8,9]. Analysis of these indicates that they could generally support some form of manipulator mounted on a local gross motion axis. However, these supporting structures will require vibration suppression using a combination of energy absorbing materials and local bracing strategies. In some cases active damping provisions may also prove necessary.
Where routine inspection and maintenance is envisaged and no permanent access provided, demountable tracking could be employed, this attached by permanent fixing points and brackets.Such provisions should automatically be considered in the design of new structures. Two concepts based on these ideas are now considered.

6.1 System 1

Figure 5 shows an inspection robot suspended from two trolleys which run on tracks to each side of the deck. This arrangement enables the device to inspect the sides and under-side. A separate arrangement of clamps or tracks along the piers with vertical trolleys would enable pier inspection to be undertaken. The motion of the trolleys and the robot would be provided by an arrangement of vacuum grippers and long stroking pneumatic actuators for which a supply line could be supported on the trolleys.

6.2 System 2

With this system the robot platform is suspended form clamps attached at junctions between the under-side of the deck and the tops of piers. The powered platform would ascend and descend along the piers with the aid of two powered cable winches located at the ends. Where the under-side of the deck requires inspection, the same platform would move by means of the cable winches, these clamped to the structure at the ends of each inter-pier span. This system is similar to one proposed for building inspection [10].

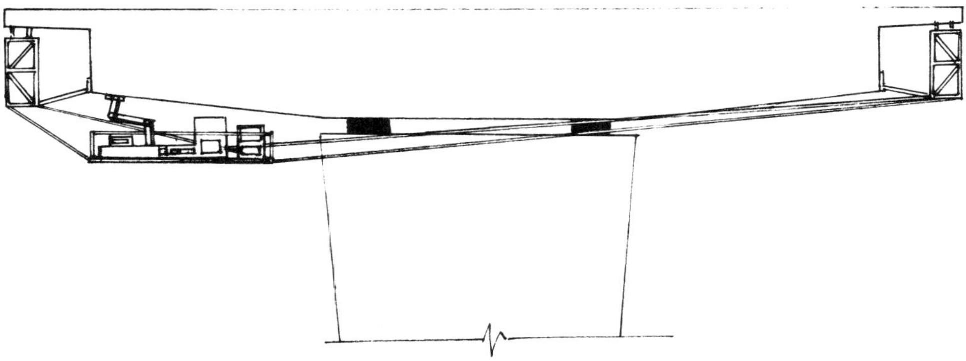

Figure 5. System 1

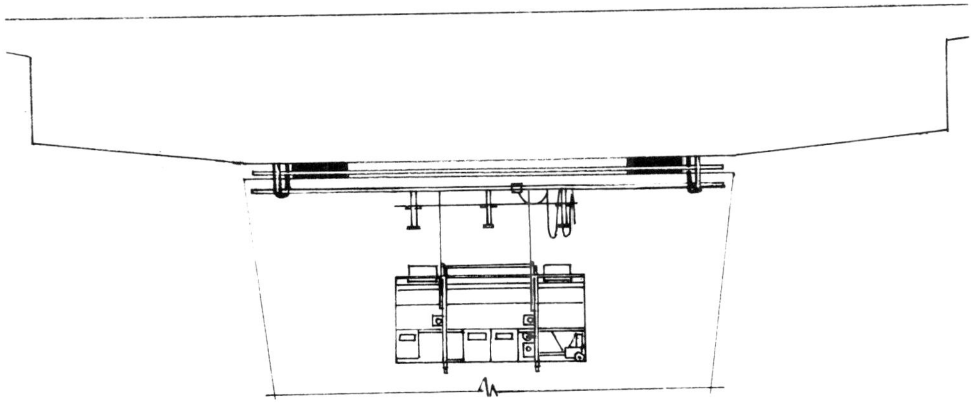

Figure 6. System 2

7. CONCLUSIONS

The background to the robotic inspection of bridges has been presented. In the research and development of effective devices it is important that some sound basis for uniform assessment be determined. To this end the 'benchmark' concept has been introduced, this providing a precise means for evaluating and comparing robot concepts. These 'benchmarks' must cover the use of the major inspection probes, the purpose of which has been set out in summary form. The development of benchmarks and their use in robot evaluation have proved worthwhile. However, this work now needs extending to cover the full range of inspection methods with realistic target geometry. Costing information is currently being analyzed with a view to verifying a method for economic evaluation.

REFERENCES

1 S.M. Levy, Japanese Construction: an American Perspective, Van Nostrand Reinhold, 1990.
2 P. Vassie, Reinforcement Corrosion and Durability of Concrete Bridges, Proceedings of the Institution of Civil Engineers, Part 1, August 1984, pp 713 - 723.
3 DoT, Bridge Inspection Guide, HMSO, London, 1984.
4 Kabedohda, Wall Inspection Robot, Ohbayashi Technical Research Institute, Tokyo, 1990.
5 D.A. Chamberlain, A Remotely Operated Building Inspection Cell, Automation in Construction, Vol 1, Part 1, May 1992, pp 71 - 83.
6 F. Nazy, Report 1: Kajima Corporation, Advanced Robotics, Report of the U.K. Study Mission to Japan, Pub. Advanced Robotics Administrative Support Agy., Manchester, 1986.
7 M. Donner, Computer Simulation to Aid Robot Selection, Robotics Technology, edited by A. Pugh, Control Engineering Series 23, IEE, London, 1983.
8 Tractel Group, Access Equipment for Construction, Inspection and Maintenance, 1991.
9 Beeche Systems Corporation, Modular Space Frame Platforms and Mecano Systems, Scotia, N.Y., USA, 1992.
10 T. Wong, Application of Advanced Automation and Robotisation to Inspection and Maintenance of Large Panel Systems to Local Authority High Rise Residential Buildings, MSc in Construction Management, South Bank University, 1991.

Automation and Robotics in Construction X
G.H. Watson, R.L. Tucker and J.K. Walters (Editors)
1993 Elsevier Science Publishers B.V.

Design of a machine which rotates hoist elements around its vertical axis

F.J.M. van Gassel

Eindhoven University of Technology, Faculty of Architecture & Building, Postbus 513, 5600 MB Eindhoven, The Netherlands.

Abstract

On construction site hoist elements are rotated by the hands of assembly workers. This work is dangerous. To get better working circumstances, the elements can be rotated by a machine. On the principle based on Newton's third law "action and reaction" there is designed and built a machine. The result of the design process is that the principle of action and reaction works for light and small elements. For a short period of time the machine can withstand the wind forces by heavy and sizeable elements. How long this period has to be must be researched. Should this period be too long then other principles might to be considered.

1. INTRODUCTION

The Steering Committee Small-scaled Mechanisation on the construction site in The Netherlands (PKMB) has given the Building Research Institute TNO and the Eindhoven University of Technology TUE the task to develop a machine with which heavy and sizeable hoist elements can be rotated around its vertical axis.
PKMB will improve the working conditions (health, well-being and safety) of the workers on the construction site and without loss of productivity. To reach these goals research centers develop, optimize and promote small-scaled mechanisation tools within the industry. The result is usually a prototype that shows that the idea for an improvement works.

2. PROBLEM

More and more buildings consist of heavy and sizeable products: wall, roof and floor panels. These products are assembled with a crane and workers. The assembly workers now have to carry out physically demanding and dangerous work when they rotate loads with their hands or with a rope.
The health and safety inspectorate demand that workers be tied to the building by a safety line or insist that they work behind safety-rails. These two factors inhibit efficient working practice. Figure 1 shows the need to rotate elements.

3. POSSIBLE SOLUTIONS

The elements to be hoisted have to be rotated by a machine and not by the hands of the assembly workers. Such a machine can work on various physical principles. Some of these principles are:

- jet propulsion [1];
- gyroscop;
- reaction forces in the hoisting cables [2];
- action and reaction.

4. CHOSEN SOLUTION

For the design of the machine there is taken into account the goals of PKMB chosen for the principle of action and reaction[3]. This principle is based on Newton's third law:
"If a body exercises a force on another body, then the force on this body is as large, reversed and in the same line as the force of the first body".
The body is the hoisting element and the other body a flywheel. To rotate these bodies torque is required. According to Newton's second law is the size of the torque:

Torque	=	Moment of inertia	*	angular acceleration
M [Nm]	=	J [kgm^2]	*	"phi [$rads^{-2}$]

To show the working of the principle between a hoist element and a wheel, a simple prototype has been built. See figure 2.

5. EXISTING IDEAS AND MACHINES

In 1965 Lawrence Lawton received a patent of the United States Patent Office for an apparatus for orienting a suspending load [4]. He used a flywheel to get a torque to rotate a load. The Swede Paul Anderson got in 1978 a Dutch patent for an apparatus with two flywheels [5]. Also Smit Slikkerveer in The Netherlands deposited in 1978 and 1979 ideas based on action and reaction for public by the Dutch patent office [6] [7].
End seventies brought HOLEC in The Netherlands a machine on the market to handle sea containers. This machine was based upon the principle of action and reaction. The weights of the machines varied between 530 and 3500 kg.
At this moment the machines are not to be sold because other principles to handle sea containers are being developed.
An idea to win back energy by braking the flywheel is deposit by Nagron in The Netherlands [8]. Peiner in Germany got in 1989 a Dutch patent for a machine to rotate big loads on a crane [9].
There exist some ideas, but hardly machines on the market which can solve the assembly problems.

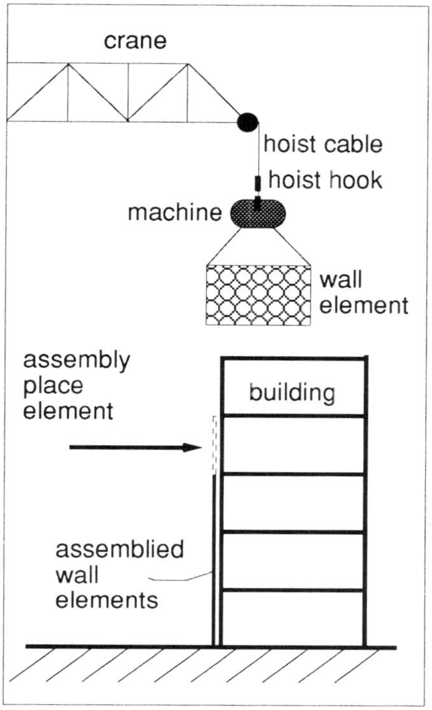

Figure 1. The machine had to rotate the wall element before assembly.

Figure 2. Photo of a simple prototype which shows the working of action and reaction between an element and flywheel.

6. PERFORMANCES OF THE MACHINE

The rotation performances depend on the parameters of the machine and hoist element.

The rotation performances are:
- the average angular velocity of the hoist element without wind force;
- the time to bring a certain angular velocity back to zero without wind force;
- the time to hold the angular velocity on zero with wind force.

The parameters of the machine are:
- the moment of inertia of the wheel (Jw);
- the maximum of the angular velocity of the wheel ('phi w max);
- the shaft power of the wheel (Pw).

The parameters of the hoist element are:
- the weight (Ge);
- dimensions: length, width and height.

7. FORMULAS

There is chosen for braking the wheel to rotate the hoist element. Then the power of the driving can be held small. This means that we need two wheels for rotating in both directions. The diagram of this machine is shown in figure 3. To design the machine for certain performances and hoist elements there are derived some formulas. Results of these derivations are shown in table 1. Two factors need some explanation:

- Available time Tb:
  this is the time that you can get out a certain torque Ma.
- Using factor Vf:
  this is the amount of brakes during Tb.

The performances of the machine are then:

- the maximum angular velocity of the hoist element after Tr1 seconds;
- the angle that the elements rotates.

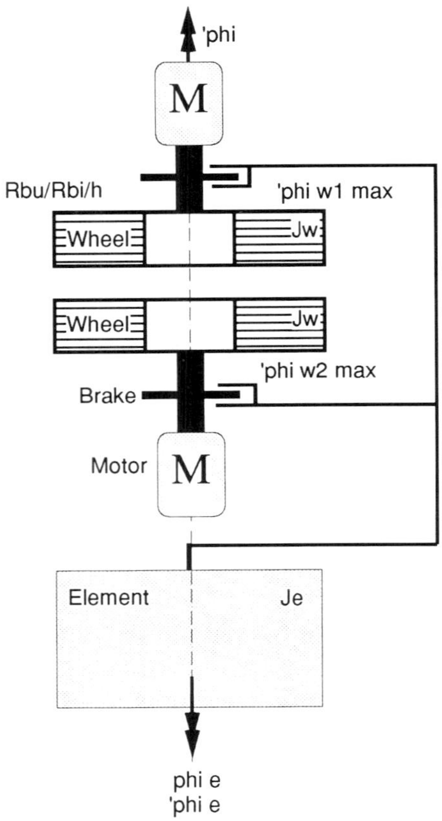

Figure 3 . The diagram of the machine with two flywheels

Table 1.
Formulas action and reaction

Description	Symbol	Dimension	Formulas
Wheel:			
Outer radius	Rbu	m	
Inner radius	Rbi	m	
Height	h	m	
Weigh	Gw	kg	=7900*3,14*(Rbu*Rbu -Rbi*Rbi)*h
Moment of inertia	Jw	kgm^2	=Gw*0,5*(Rbu*Rbu+ Rbi*Rbi)
Max transverse velocity	Vw max	ms^{-1}	
Max angular velocity	'phi w max	$rads^{-1}$	=Vw max/Rbu
Max revolutions per min	Nw max	$revmin^{-1}$	=60*'phi w max/2/3,14
Time to reach Nw max	To	s	
Shaft power	Pw	watt	=Jw*('phi w max)^2/To
Driving torque	MW	Nm	=Jw*('phi w max)/To
Element:			
Weight	Ge	kg	
Lenght	b	m	
Moment of inertia	Je	kgm^2	=Ge*b^2*/12
Ratio Je/Jw	R	-	=Je/Jw
Available time	Tb	s	
Brake torque	Ma	Nm	=Jw*'phi w max/Tb
Using factor	Vf	-	
Using time	Tr1	s	=Tb/Vf
Max angle velocity element after Tr1	'phi e max	$rads^{-1}$	=Ma*Tr1/Je
Max rev per min element after Tr1	Ne max	$revmin^{-1}$	=9,5*'phi e max
Angle element after Tr1	phi e Tr1	rad	=Ma*Tr1^2/2/Je
Angle element after Tr1	alpha e Tr1	degrees	=phi e Tr1*360/2/3,14

8. HEAVY AND LARGE HOIST ELEMENTS

Calculations have been made of an element with the following parameters:
- length: 7200 mm
- height: 3600 mm
- weight: 10000 kg

There is chosen for a steel flywheel with an outer radius of 500 mm, an inner radius of 350 mm and a height of 100 mm.

According to Genta [10] there are some restrictions as to the maximum transverse velocity. In this case we chose 40 ms^{-1}.

The results of the calculations are:
- To give the wheel the maximum angular velocity in 60 seconds you need an shaft capacity of 6286 watt.
- The ratio between the inertia moment of the element and the wheel is 733.
- The available time to brake is one second and brakes one time.
- The maximum revolutions to a minute is one.
- The angle that the element rotates is 3·313 degrees.

In these calculations the influence of wind forces has not been considered.

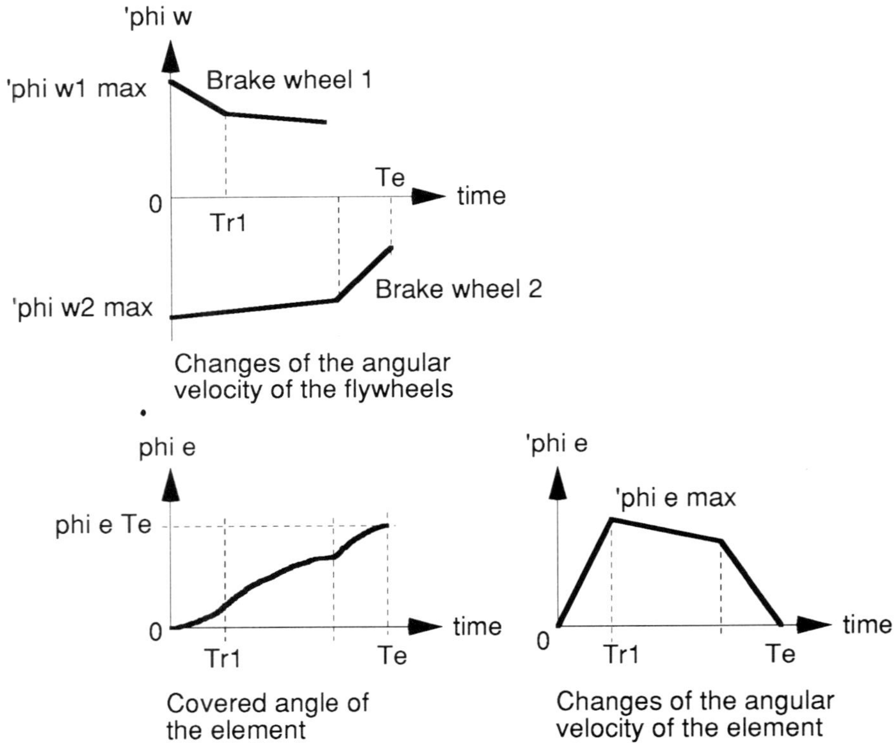

Figure 3. Behavior of the flywheels and element in time.

phi e : covered angle element
'phi e/w : angular velocity element or wheel
Tr1 : brake time wheel 1
Te : end of rotating

9. WIND FORCES

To hold the angular velocity of an element on zero by wind forces a worker can do this by his hands with a force of 300 N during some minutes [11]. By the element described in paragraph eight means this, that the worker can exert a torque of 1080 Nm. The machine can bring up this torque about four seconds. Probably this is too short. Empirical data on the construction site will give more information what time is necessary.
The Dutch norm NEN 2018 says that the loading on the hoist is 0·03 of the weight. By our element means that a force of 3000 N on the hoist cable. What force the assembly worker had to bring up is unknown.

10. PROTOTYPE FOR LIGHT AND SMALL ELEMENTS

To get more experience with the physical principle of action and reaction in practice there a prototype has been built for a lighter and smaller elements with the parameters:
- length: 1200 mm
- width: 400 mm
- height: 1000 mm
- weight: 800 kg

There is chosen for two flywheels of steel with an outer radius of 250 mm, an inner radius of 150 mm and a height of 75 mm. The maximum transverse velocity is 36 ms^{-1}. By experiments with the prototype the machine needs about 10 seconds to rotate 90 degrees. For assembly work is this sufficient. The influence of the wind forces by small elements are a multiple smaller. For a photo of this prototype see figure 5.

Figure 5. Prototype for light and small elements.

11. RESULTS

The result of this design process is that the principle of action and reaction for light and small elements works. Depending on the time that the machine can withstand the wind forces also heavy and sizeable elements are possible but how long that period is, must be researched. Should this period be too long then other principles might have to be considered.

REFERENCES

1. The company Fukushima Seisakusho in Japan produces a machine called "Non-Kuru". The machine can rotate a load of maximal 6000 kg by a turning blower.

2. Pompe, W., *Veilig hijsen met de "Antipsin"*, Bouwmachines 13(1978) nr. 2.

3. Kroon, J. de, and others, 1992, *Samenvattende rapportage oriëntator, PKMB*, TNO, Delft.

4. Lawton, L.,1965, *Apparatus for orienting a suspended load,* Patent number 3,210,114, United States Patent Office.

5. Anderson, P.,1978, *Inrichting voor het draaien van een vrij draaibaar opgehangen last*, Octrooi nummer 156109, Octrooiraad Nederland.

6. Smit Slikkerveer B.V., 1978, *Richtwerk voor het richten van lasten*, Terinzagelegging nummer 7706928, Octrooiraad Nederland.

7. Smit Slikkerveer B.V., 1979, *Bevestigingsinrichting voor het bevestigen van lasten aan een hijswerktuig*, Terinzagelegging nummer 7800205, Octrooiraad Nederland.

8. Nagron Aerolift, 1981, *Inrichting voor het richten van een aan een hijskabel hangende last,* Terinzagelegging nummer 7905858, Nederlandse Octrooiraad.

9. Peiner, 1989, *Inrichting voor het draaien van een aan een kraan opgehangen last*. Octrooi nummer 184610, Nederlandse Octrooiraad.

10. Genta, G., 1985, *Kinetic Energie Storage*, Butterworths, London.

11. (-), *Hijsen bij wind*, Bouwmachines 27 nr. 6/7 (17 juni 1992).

Automation and Robotics in Construction X
G.H. Watson, R.L. Tucker and J.K. Walters (Editors)

Development of an Exterior Wall Painting Robot Capable of Painting Walls with Indentations and Protrusions

Shin Terauchi (a), Toshikazu Miyajima (b), Takezo Miyamoto (c), Kazuhiko Arai (d) and Seiichiro Takizawa (e)

(a) Kajima Corporation, No. 4 Research Division 2-19-1 Tobitakyu, Chofu Tokyo, Japan
(b) Kajima Corporation, No. 7 Research Division 2-19-1 Tobitakyu, Chofu Tokyo, Japan
(c) Kajima Corporation, Construction Business Headquarters, Machines Division 1-1-5 Fujikage Bldg., Motoakasaka, Minato-ku, Tokyo, Japan
(d) Kajima Corporation, Construction Business Headquarters, Construction Engineering Headquarters, Production Engineering Division 1-2-7, Motoakasaka, Minato-ku, Tokyo, Japan
(e) Kajima Mechatronics Engineering Co., Ltd. 1-6-1 Minamidai, Kawagoe, Saitama, Japan

Abstract
We developed an exterior wall painting robot for the purpose of automating this painting operation. The robot is mounted on equipment which permits it to move up and down, left and right along the exterior walls of a building. It is computer controlled and is activated simply by the operator pressing a switch on the control panel located on the ground. The robot is capable of painting a four square meter wall surface (4 m long × 1 m high) at one time. It is also equipped with sensors which measure indentations and protrusions in the wall surface, making it possible for it to paint exterior walls with windows, pillars or other indentations or protrusions.

1. Introduction

In recent years, the construction industry has experienced the effects of an aging work force, including a chronic shortage of skilled construction workers. This has resulted in a tendency for work efficiency and quality to deteriorate. One method of overcoming this problem that has been urged forcefully upon us has been automation and robotization of construction operations. One of the areas where this has been attempted is in painting work using the painting robot.

2. Robot Development Concepts

This robot was developed to meet the following requirements:

(1) It must have a capacity equivalent to or greater than a skilled worker (both in work efficiency and painting quality).
(2) It should save energy.
(3) It should be able to paint walls with indentations and protrusions.
(4) Operation of the robot body and its moving equipment must be possible by one man control from a control room on the ground.
(5) The robot main body must use moving equipment that enables it to move precisely to its proper work position and maintain that position.
(6) It must have safe operation and must not have an adverse effect on the surrounding environment.

3. Description of the Painting Robot System

The automatic painting system using this painting robot consists of the robot main body which sprays paint, robot moving equipment to carry the robot main body to the proper work position, paint supply equipment and a controller to control the system. A view of the entire system is shown in Photo 1 and the painting system configuration is shown in Fig. 1.

Photo 1 View of Automatic Painting System

3.1 Robot Main Body

The robot main body consists of a main frame, a painting gun, a gun driver and a control unit. The robot main body is shown in Fig. 2 and the painting gun is shown in Photo 2.

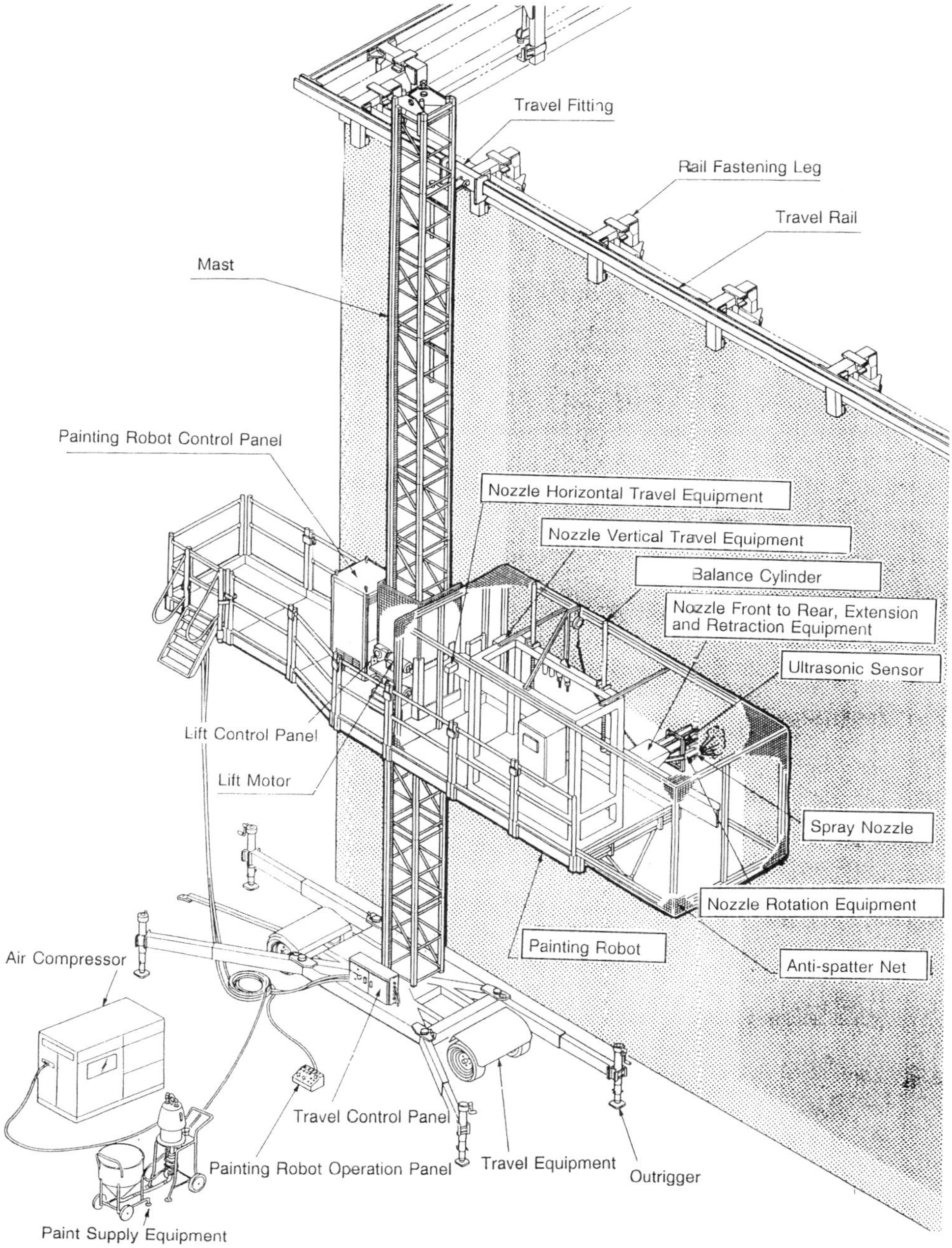

Fig. 1 Exterior Wall Painting Robot System Configuration

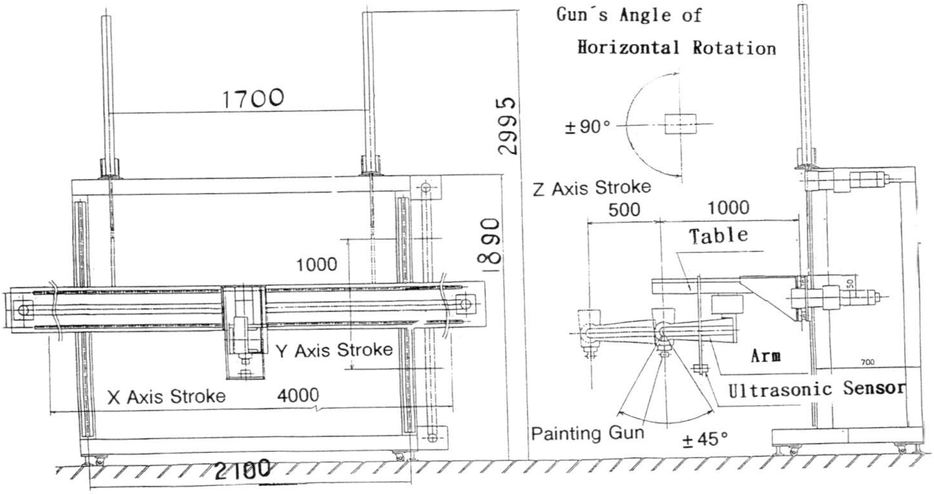

Fig. 2 Robot Main Body

The painting gun is driven in three directions, left and right (X direction), up and down (Y direction) and back and forth (Z direction). It is also capable of rotating left to right and back and forth.

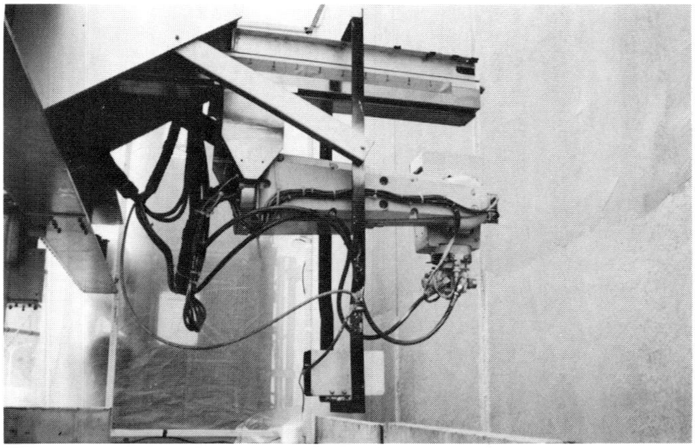

Photo 2 Painting Gun

Movement of the painting gun in the X direction is done by the X axis servomotor along a horizontal axis plate that acts as a positioning table. Movement of the painting gun along the Y axis is accomplished by Y axis servomotors which move the entire frame holding the horizontal axis plate along guide rails mounted in posts set on the left and right ends. A horizontal axis plate balancer cylinder is connected to the frame so that driving force will be applied uniformly when the horizontal axis plate is driven up and down.

Movement of the painting gun in the Z axis direction is accomplished by the Z axis servomotor driving the spray nozzle arm, which the gun is mounted to, along the positioning table. An ultrasonic sensor is mounted to measure the distance between the painting gun and the painted surface. Based on the measuring information from this sensor, the Z axis servomotor is controlled so that the distance from the painting gun to the painted surface is constant. In this way, the painting robot can paint walls which have indentations and projections and by moving around the indentations and projections, the gun nozzle can avoid colliding with the painted surface. Fig. 3 shows the nozzle path according to the painted surface. Rotating of the painting gun left to right and back and forth is done by a dual axis rotation servomotor in the rotating mechanism. Control of the speed of each servomotor is done by means of rotary encoders.

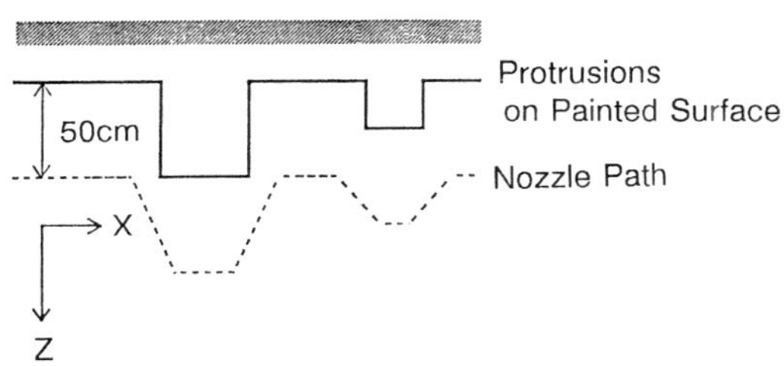

Fig. 3 Painted Surface and Path of Painting Gun

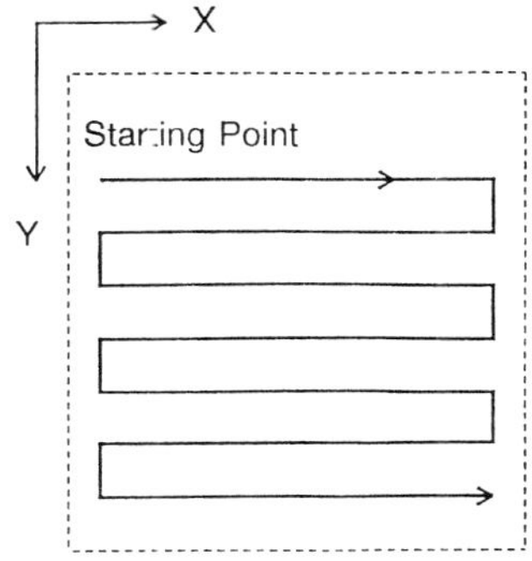

Fig. 4 Painting Pattern

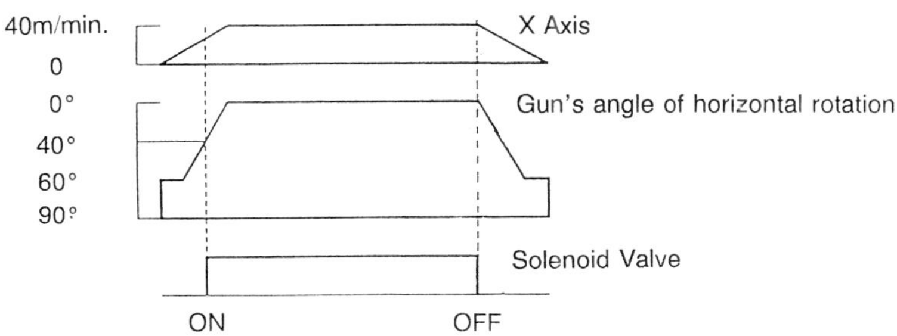

Fig. 5 Operation Timing

On-Off timing and speed of each servomotor as well as timing of the solenoid valves which turn the paint spray On and Off, etc., are controlled automatically by a personal computer unit, control panel and operation panel via control cables. Changes in the painting gun's speed and range of movement are accomplished automatically by the personal computer unit.

However, if a basic work pattern like that shown Fig. 4 is used in a painting job, the moving distance in the Y axis direction can have an effect on the painting pattern, causing nonuniformity. Also, the timing of the start of painting at the starting point and end point in the × axis direction as well as the angle of the painting gun with respect to the painted surface in the horizontal direction will have a great influence on the occurrence of run and sag of the paint. For these reasons, this robot is designed to control the nozzle's Y axis direction moving pitch and the On-Off timing of the solenoid valve at the paint spray starting and end points as well as the painting gun's angle of rotation, at the appropriate value for each type of paint so as to assure good painting quality.

Fig. 5 shows the relationship between the painting gun's angle of rotation in the horizontal direction and the On-Off operation timing of the solenoid valve with respect to the speed of the gun in the X axis direction.

3.2 Moving Equipment

For moving equipment, a movable work base which uses a mast was adopted. It is composed of a transporter which propels the moving equipment along the outside of the building being painted, a work stage on which the robot main body is mounted and which moves up and down automatically, and a mast which serves as a guide for raising and lowering the work stage. The top of the mast is attached to a travel fitting via a support. This fitting moves along a guide rail mounted on the top of the building. (See Fig. 1.)

3.3 Painting Robot Specifications

Main Specifications of the robot are as follows :

1. Painting Area	Max. 4 m^2 (4 m horizontal x 1 m vertical)
2. Painting Capacity (1 gun)	Base, Top Coat 200 m^2/hr
	Intermediate, Texture Spraying: 100 m^2/hr
3. Gun Horizontal Moving Speed	40 m/min.
4. Gun Vertical Moving Speed	8 m/min.
5. Gun Rotation Angle	Left/Right ±90°, Up/Down ±45°

4. On-Site Test Operation Results

The performance of this painting robot was tested in painting of the exterior wall of a building that was being refurbished. The results of that test operation are described below.

4-1 Painted Surface on Exterior Wall of Building
Length 72 m x Height 23m (1,656 m^2)

4-2 Painting Robot System
The configuration of the painting robot system was as shown in Fig. 1.

4-3 Painting Procedure

The flow of the automatic painting robot's painting procedure is shown in Fig.6.

4-4 Painting Results

It was confirmed that the robot demonstrated sufficient work efficiency and painting quality.

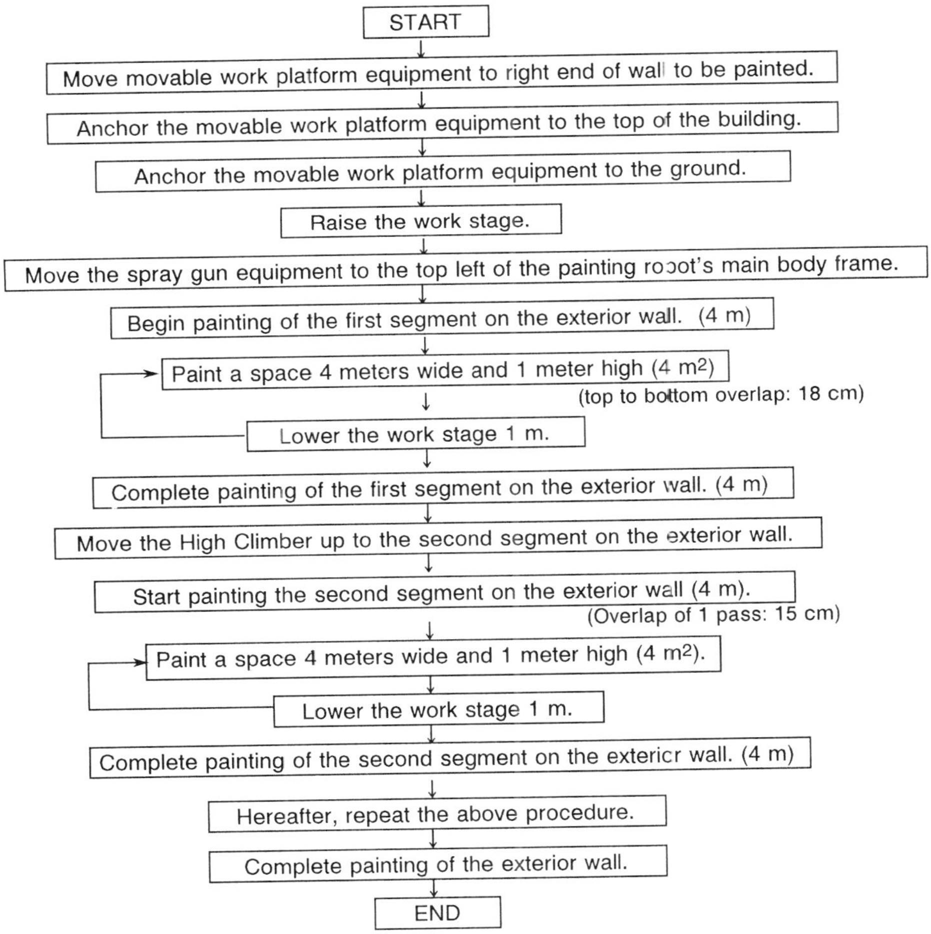

Fig. 6 Flow of Automatic Painting Robot Painting Procedure

5. Features of Painting Robot

The principle features of this painting robot are as follows.

1 It maintains painting quality which cannot be distinguished from the quality of skilled painting personnel.

2 Painting by manual labor on high scaffolding, as done previously, is eliminated, the work efficiency is high and there is little danger.

3 Uneven walls with indentations and protrusions can also be painted automatically.

4 Through automatic painting, the dirty work required with human workers is reduced and the work environment is improved.

6. Conclusion

As a result of on-site trial operation, the effectiveness of introduction of this painting robot to actual construction projects was confirmed.

In the future, we are planning to make dramatic improvements in the painting robot and will be developing a new type which can be suspended from a building roof. We hope to introduce these improved painting robots on actual work sites as quickly as possible.

Automation and Robotics in Construction X
G.H. Watson, R.L. Tucker and J.K. Walters (Editors)

An Evaluation of the Pipe Manipulator Performance in a Material Handling Yard

Hsieh, T.Y., Research Assistant
Fulton, C., Research Assistant
Dr. Gibson, G. E. Assistant Professor
Dr. Haas, C.T., Assistant Professor

Department of Civil Engineering
University of Texas at Austin
Austin, Texas 78712, USA

ABSTRACT

The purpose of this paper is to compare the productivity between a prototype pipe manipulator and a telescopic rough-terrain crane in handling pipes in a piping material laydown yard. Productivity data were collected by video-taping identical operations performed by the two machines in the field. The productivity data are evaluated in terms of the total cycle time and total work-hours. The results show that the pipe manipulator has a slightly shorter cycle time but is outperformed by a cherry picker by a ratio of 5.16:1 in terms of total work-hours, primarily due to the gripper's restriction on lifting one pipe per lift. A discussion of the possible ways to improve the manipulator's productivity in handling pipes reveals two major directions for enhancements: (1) enabling multi-pipe lift and (2) designing a better control system. These enhancements not only augment the manipulator's productivity to be comparative to that of a cherry picker but also reduce the cycle time by 67%.

1. INTRODUCTION

This paper describes the research results of one facet of a larger ongoing research program being conducted at the University of Texas at Austin [7,8]. This research program centers on the issues of developing a large-scale manipulator for construction use. A prototype manipulator, the Grove Pipe Manipulator, is being used as a test bed for various types of experiments. The physical characteristics of the pipe manipulator have been extensively discussed in previous literature [1,2,4,5]. The focus of this paper is in evaluating the productivity of the Pipe Manipulator in handling pipes in a piping material laydown yard versus that of a conventional mobile telescopic crane (cherry picker). It was also recognized that this prototype manipulator can benefit workers' safety to a great extent. However, this aspect will not be discussed in more depth within this context.

This research involves the use of a video camera to record and analyze the productivity data of both machines. Approximately sixty minutes of pipe handling operations were filmed. The productivity data were then recorded manually by replaying the tape. To verify the collected data and gain more insight, formal and informal interviews were conducted. All the productivity data comparisons are based on identical operations with the same size and length of pipes.

The paper first introduces the current practice of pipe handling in a laydown yard. Then, the acquired productivity data are displayed. In data analysis, thorough discussions are provided surrounding the issue of productivity comparison with varying factors. An extension of effort is made in the following section in predicting the potential productivity improvement on the Pipe Manipulator should some enhancements be made. Finally, conclusions are given regarding the significance of the research findings.

2. CURRENT OPERATIONS OF PIPE HANDLING IN A LAYDOWN YARD

The standard pipe length delivered to material laydown yard is typically 40 feet. This extended length, as opposed to a shorter length of 20 feet, reduces the number of welds required of the construction contractors during assembly operations. With this 40 foot length, the current pipe unloading operation requires a crew of five: one cherry picker operator, two riggers on the flatbed, and two workers on the ground for pipe control and placement.

The sequence of operation is as follows. Suppliers deliver one size and type of pipe on standard flatbed delivery trucks in several stacked configurations including:

(1) Pipe stacked into a pyramid formation with no spacers between the layers, and with pipe resting on pipe.
(2) Pipe stacked in layers with wood spacers separating each layer.
(3) Large pipe placed in a single layer, resting on the flatbed.
(4) Smaller pipe in bundles, with metal bands holding the pipe together.

In most situations, standard hold-down chains or nylon straps are used to secure the load while in transit. Also, when pipe is delivered in a stacked configuration, flatbed side stakes are often employed to arrest any lateral motion of the pipe. The configuration of pipe delivery, and the weight of the pipe will influence the type of rigging to be employed during offloading.

From field observations and interviews, several factors which will influence the positioning of the crane relative to the delivery truck during offloading operations were identified and include:

(1) Maneuvering room for the delivery truck.
(2) Space available to position the crane relative to the truck.
(3) Orientation of the pipe on the ground.
(4) Space available to traverse the pipe from truck to ground.
(5) Weight of the pipe.

Consequently, the goals of handling pipes by a cherry picker are to: (1) reduce the movement of the boom while unloading; (2) reduce the horizontal extension of the boom for stability reasons; and (3) reduce the distance required to move the pipe. The reduction in movement of equipment and material can result in higher productivity. This aspect, however, will not be explored further in this paper.

3. DATA ACQUISITION

Two techniques were used to gather data on the equipment's performance: (1) Field tests were videotaped using a standard portable video cassette recorder; and (2) Formal and informal interviews of management and field personnel were conducted. The goal of filming was to record the cherry picker and the pipe manipulator performing identical operations so that an accurate comparison of their performance and productivity could be obtained.

The drawback of a standard VCR is that analysis of the tape is much more tedious and time consuming compared to other methods such as time-lapse film. Time-lapse film can be viewed at a multitude of speeds such that an 8 hour work day can be viewed in as little as one hour, while VCR analysis requires the full 8 hours to review. Nevertheless, because the total recording time is less than one hour in this study, this drawback does not impose any particular difficulty.

A typical pipe handling operation of a cherry picker involves three major steps, namely, attaching sling, lifting and placing, and detaching sling and returning to the original position. In this analysis, six cycles of pipe handling operation were recorded, as shown in Table 1.

Table 1 Productivity Data for Cherry Picker (in seconds)

	Rig Load	Lift and Place	Unrig and Return	Total Time
Cycle #1	101	68	65	234
Cycle #2	75	50	40	165
Cycle #3	30	45	60	135
Cycle #4	125	55	45	225
Cycle #5	140	110	45	295
Cycle #6	30	55	50	135
Average Time	84	64	51	198

A typical manipulator operation can also be divided into three major steps. The "Grasping" involves the motions of approaching the pipe and the actual gripping. The "Lift and Move" consists of lifting the pipe off the flatbed, swinging and placing the pipe to the designated yard location. Finally, the "Release & Return" represents the motions of releasing the pipe from the control of the gripper and moving the gripper back to the arrays of pipe. This fashion of motion classification results in the convenience of recording productivity data from VCR reviews. The manipulator productivity data are shown in Table 2.

Table 2 Productivity Data for Pipe Manipulator (in seconds)

	Grasping	Lift & Move	Release & Return	Total Time
Cycle #1	25	70	49	144
Cycle #2	41	90	50	181
Cycle #3	22	80	42	166
Cycle #4	30	81	40	151
Cycle #5	40	81	61	185
Cycle #6	40	139	55	155
Cycle #7	40	60	64	243
Cycle #8	23	69	73	165
Cycle #9	10	49	51	110
Cycle #10	21	96	58	175
Cycle #11	63	56	55	174
Average Time	32	79	54	165

4. DATA ANALYSIS

From Tables 1 and 2, it can be seen that, in terms of the completion time per lift cycle, the Pipe Manipulator outperforms the Cherry Picker by a ratio of 1.2:1. It is felt that this difference of performance is marginal and may well be balanced out by the experimental errors introduced by the research method.

In addition to total completion time, two other factors need to be used in comparing their performance. Because of the difference in physical capabilities, these two machines work with different sizes of crews and can handle different number of pipes per lift. In the field experiment, the crew size for the Cherry Picker was five, whereas the manipulator has only one, the operator. However, the Cherry Picker can handle 31 pieces of pipe in one lift when they are bundled together, while only one pipe can be handled by the manipulator per lift. With this information, the productivity of both machines is presented in Table 3.

Table 3 Productivity Comparisons

	Crew Size	# of Pipe/Lift	Time/Lift	Man-Sec/Lift	Man-Sec/Pipe
Manipulator	1	1	165	165	165
Cherry Picker	5	31	198	990	32
Ratio (M/C)	0.2	0.03	0.83	0.17	5.16

"Man-Second per Pipe" is a productivity criterion that is closely related to labor costs. Since the cycle completion time of both machines is comparative, the productivity based on this criterion is determined by the number of pipes per lift and the crew size. From Table 3, it is clear that the productivity of the Pipe Manipulator is critically undermined by its limitation of handling one pipe at a time, although the Cherry Picker requires a crew of five.

Understandably, the observations from the field experiment may not reflect all possible configuration of a pipe handling operation performed by a cherry picker and a manipulator. Under other circumstances such as longer or larger pipes to be handled and the congestion of the laydown yard, both the number of pipes to be handled per lift and the crew size may vary. Consequently, the study results derived thus far may not provide enough insight. To grasp the dynamics of many "what-if" conditions, it is useful to establish the relationship of productivity on both machines as follows:

$$\text{Productivity} = \text{Cycle Time} * \frac{\text{No. Pipes Handled per Lift}}{\text{Crew Size}} \tag{1}$$

Assume $\text{Productivity}_{\text{cherry picker}} = \text{Productivity}_{\text{pipe manipulator}}$

$$\text{Thus,}\quad 165 * \frac{P_m}{C_m} = 198 * \frac{P_c}{C_c} \quad \text{or} \quad \frac{P_c}{P_m} = \frac{165}{198} * \frac{C_c}{C_m} \tag{2}$$

Where,

P_m : Number of Pipes Handled per Lift by Manipulator
C_m : Crew Size of Manipulator
P_c : Number of Pipes Handled per Lift by Cherry Picker
C_c : Crew Size of Cherry Picker

With this relationship, the sensitivity analysis regarding the variations of the two variables is performed and shown in Figure 3. From Figure 3, the vertical axis, the capacity ratio, represents the ratio between the number of pipes that can be handled per lift by a cherry picker and that by a manipulator. The horizontal axis, the crew size ratio, represents the ratio between the sizes of a cherry picker crew and a manipulator crew. The "Cherry Picker Region" indicates that a cherry picker will be more productive in handling pipes in terms of total work-hours with any given combination of the two ratios in this region.

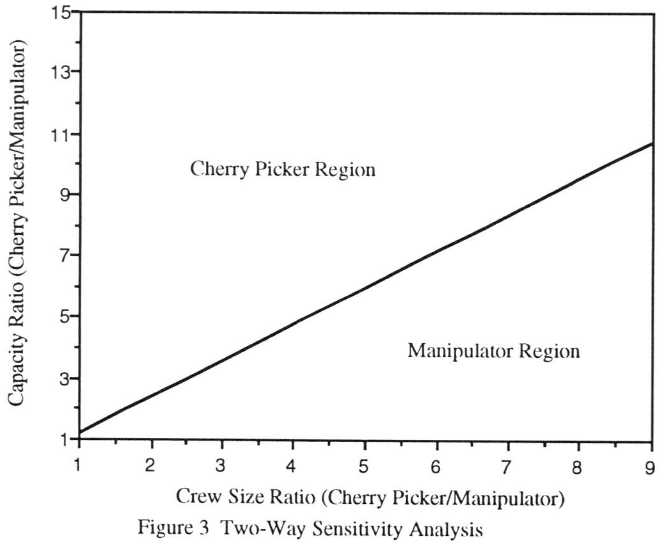

Figure 3 Two-Way Sensitivity Analysis

According to the field data, the crew size ratio between the two machines is 5 and the capacity ratio is 31. It is not difficult to see that the productivity gap between them are significant. It is also apparent that, given that the cherry picker's crew size is fixed at 5 and the the crew size of the manipulator at 1, one possible way for the manipulator to compete with the cherry picker is by decreasing the capacity ratio. In this case, the minimal attractive capacity ratio is 6. In other words, the manipulator would have to handle five pipes per lift to become competitive. Another possible way is to change the linear relationship between the completion times per lift. By so doing, the area of the "Manipulator Region" will increase. The next section discusses these aspects.

5. DISCUSSION

From the previous analyses, it is quite clear that constructors will most likely be satisfied with using conventional cherry pickers. In terms of handling pipes in a laydown yard, the major drawbacks or weaknesses of the pipe manipulator can be summarized as follows:

(1) Its lifting capacity is limited to 1600 pounds, whereas a cherry picker has a capacity of 20,000 pounds (12.5 times);
(2) The gripper can only handle one pipe per lift resulting in a high capacity ratio; and
(3) With the current control system, the operator can only move one joint at a time, resulting in a longer operation time.

As mentioned earlier, to improve the productivity of the manipulator, two major aspects need to be considered. The first is to find ways to reduce the capacity ratio in order to increase the number of pipes handled per lift by the manipulator. Two issues are related to this aspect: (1) increasing the lifting capacity and (2) enabling multi-pipe lift. A second aspect is to reduce the cycle time per lift. which involves the design of a better control system than the existing 8-lever control panel.

A. Reducing the Capacity Ratio

The first obstacle to reducing the capacity ratio is the limited lifting capacity. With the current configuration, the manipulator's lifting capacity is only 1600 pounds. Since this limitation is mainly imposed by the geometric stability of the system, it may be possible to obtain a higher capacity by enforcing certain arm motions with which the moment resulted from the payload at the manipulator base is always below a certain amount. This possibility is, however, considered impractical in the field because of the strict safety requirements for construction job sites. Another possibility is to increase the distance between outriggers. This concept is feasible only if the physical strength of every components of the manipulator is examined.

If a higher lifting capacity is obtained, two means are possible to achieve the goal of lifting multiple pipes per lift. One can create an attachment at the end of the manipulator assembly on which to attach a sling. This attachment could be a hook device such as that used on a cherry picker. This device must have the ability to be removed easily prior to actual manipulator use. It is also possible to modify the jaws such that bundled pipes can be grasped and lifted. It is perceivable that with the modified means of handling pipes the crew size may have to increase due to the same set of reasons for a conventional cherry picker. However, it is quite possible that the productivity of both machines becomes comparative in this configuration.

B. Shortening the Cycle Time

According to the field productivity data (Table 2), the average cycle time for a manipulator operation is 165 seconds. With close examination, it can be seen that this duration does not reflect the actual motion time of the manipulator. Various types of delay exist in the cycle time such as the decision process and the trial-and-error process experienced by an operator. It may be possible to reduce these delays with the aid of a more friendly control system. To comprehend the amount of delay that can be potentially reduced by a better control system, it is essential to know the actual motion time.

Table 4 is an exhibit of a detailed pipe-handling operation by the manipulator. The manipulator motion data in the first four columns in the table were retrieved by numerous sessions of viewing VCR tape in slow motion. A total of fourteen joint motions are involved in the operation. The maximum joint motion rates and force buildup delay time are obtained from field experiments by one of the authors. Based on these data, the actual manipulator motion time is calculated and totaled at 90.14 seconds. Accordingly, 74.86 seconds are attributed to various delays and can possibly be saved by an enhanced control system.

Another opportunity presented by an enhanced control system is the ability to enable the operator to move several joints simultaneously. In the case of handling pipe in a laydown yard, many of the fourteen manipulator joint motions identified in Table 4 can be moved at the same time. For example, the following six segments of motions can be considered:

(1) Segment 1: Swing, Main Telescope, Main Lift, and Main Telescope
(2) Segment 2: Roll, Rotate, Main Lift and Secondary Telescope
(3) Segment 3: Grasp
(4) Segment 4: Main Lift
(5) Segment 5: Swing, Main Lift and Secondary Telescope
(6) Segment 6: Release

Table 4 A Detailed Analysis of a Pipe Handling Operation by the Manipulator (in seconds)

Manipulator Motions	Type	Unit	Motion Range	Maximum Joint Motion Rate	Rate Reduction	Force Buildup Delay	Motion Time
Swing Base	C. Clockwise	deg	90	19.8 deg/sec		0	4.54
Main Telescope	Out	Inch	24	10.7 inch/sec		0	2.24
Main Lift	Up	deg	30	4.6 deg/sec		0	6.52
Main Telescope	Out	Inch	36	10.7 inch/sec		0	3.35
Gripper Roll	Down	deg	15	32 deg/sec	50%	1	1.94
Gripper Rotate	Clockwise	deg	5	13 deg/sec	50%	3	3.77
Main Lift	Down	deg	10	4.6 deg/sec	50%	0	4.35
Secondary Telescope	Out	Inch	5	1.4 - 1.5 inch/sec*	50%	1.5	8.64
Gripper Grasp	Close	Cycle		8 seconds/cycle		1	8
Main Lift	Up	deg	45	4.6 deg/sec		0	9.78
Swing Base	Clock-wise	deg	90	19.8 deg/sec		0	4.54
Main Lift	Down	deg	55	4.6 deg/sec		0	11.96
Secondary Telescope	Out	Inch	7	1.4 inch/sec	50%	1.5	11.5
Gripper Release	Open	Cycle		7 seconds/cycle		1	9
* Telescope-in: 1.5 inch/sec, Telescope-out: 1.4 inch/sec						**Total**	**90.14**

Since several joint motions can move simultaneously, the motion time for each segment is the longest joint motion time within that segment. Consequently, the total cycle time is further reduced from 90.14 seconds to 53.9 seconds (Table 5). With the shortened cycle time, the relationship between the capacity ratio and the crew size ratio has been recalculated and is depicted in Figure 4. Comparing the figure with Figure 3, it can be seen that the "Manipulator Region" is greatly increased, indicating that an enhanced manipulator will be more productive than a cherry picker in most cases.

Table 5 Reduced Manipulator Cycle Time

Motion Segment	Motion Time (seconds)
Segment 1	6.52
Segment 2	8.64
Segment 3	8
Segment 4	9.78
Segment 5	11.96
Segment 6	9
Total	53.9

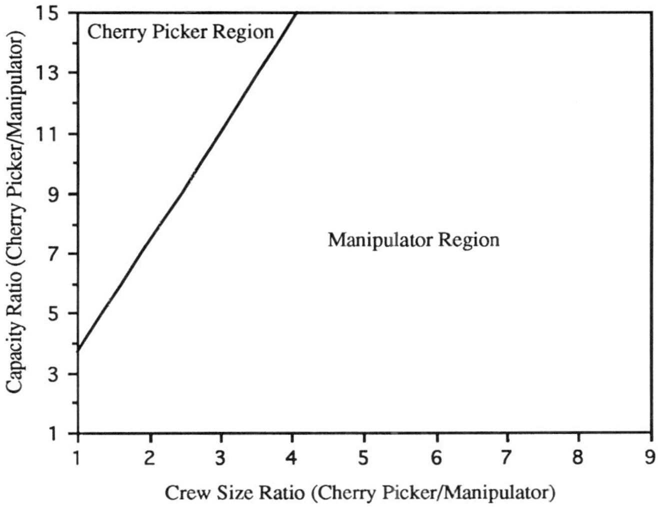

Figure 4 Sensitivity Analysis with A Better Manipulator Control System

6. CONCLUSIONS

This paper has presented the results of a field study that involves the productivity comparison of a telescopic rough terrain crane and a prototype pipe manipulator. With the manipulator's current capabilities, the cherry picker outperforms the manipulator by a ratio of 5.16:1 in terms of total work-hours, whereas the manipulator has a slightly shorter cycle time of 165 seconds as opposed to the cherry picker's 198 seconds.

Nevertheless, an underestimated benefit of the pipe manipulator is its promising safety improvement. Improved safety is gained by reducing the labor employment and removing workers from physically interacting with pipes. The major problem areas of the prototype manipulator found in this study are the limited lifting capacity, single-pipe lift restriction and the difficult 8-lever control system with which an operator generally has problems moving more than two joints at the same time. With these problems, it is highly unlikely that constructors will be interested in using the manipulator for handling pipes.

The failure of the pipe manipulator, in its current configuration, to compete with a cherry picker signifies the need for more research. Two major areas for improvement discussed in this context are the feasibility of multi-pipe lift and the design of an enhanced control system. Both areas have shown very high potential, in that the multi-pipe lift will enable the pipe manipulator to be readily competitive to a cherry picker, and a better control system has the potential to shorten the cycle time from 165 seconds to 53.9 seconds, a reduction of 67%. With these enhancements, the pipe manipulator may become a much better piece of equipment than a cherry picker in handling pipes.

ACKNOWLEDGEMENT

We would like to acknowledge Head & Guild Equipment Co. and Dow Chemical Co. for their support of this research project.

REFERENCES

1. Fisher, D. J. (1989). "Piping Erection Constructability Issues in a Semi-Automated Environment." *Dissertation*, University of Texas at Austin, May.

2. Glass, C. C. (1984). "The Pipe Manipulator: A Complete Assessment of a New Idea in Construction Equipment Technology." *Thesis*, University of Texas at Austin, December.

3. Grove Manufacturing (1981). Operator's Handbook for the Pipe Manipulator, Shady Grove, MD.

4. Hughes, P. J. (1987). "Automated Piping Erection in Process Plants Using a Heavy Manipulator." *Thesis*, University of Texas at Austin, December.

5. Hughes, P. J . (1990). "Construction Manipulator Tele-operation with Ergosticks." *Dissertation*, University of Texas at Austin, May.

6. Parker, H. W. and Oglesby, C. H. (1989). *Productivity Improvement in Construction*, McGraw-Hill Series in Construction Engineering & Management, McGraw-Hill Book Company, New York.

7. Hsieh, T. Y. and Haas, C. T. (1993). "Performance Evaluation Model for Construction Manipulators." *Proceedings of the 10th International Symposium on Automation and Robotics in Construction*, May.

8. Hsieh, T. Y. and Haas, C. T. (1993) "Applications of Large-Scale Manipulators in the Construction Environment." *Proceedings of the 10th International Symposium on Automation and Robotics in Construction*, May.

Automation and Robotics in Construction X
G.H. Watson, R.L. Tucker and J.K. Walters (Editors)

Performance Evaluation Model for Construction Manipulators

Hsieh, Ting-Ya, Research Assistant
Dr. Haas, Carl T., Assistant Professor

Department of Civil Engineering
University of Texas at Austin
Austin, Texas 78712, USA

ABSTRACT

This paper describes a model for evaluating the performance of construction manipulators. It involves four steps: (1) constructing the operation plan, (2) evaluating task and machine related parameters, (3) evaluating time components and (4) total time study. Each of these steps is described in detail, and a hypothetical example is used to demonstrate the procedure. The results indicate that based on the physical capabilities of a prototype pipe manipulator, a well designed control system may reduce the task completion time by a factor of 4.25 to 6.33. The example also reveals that the manipulator's existing 8-lever control system will result in lengthy sensing-and-decision delays. Although this evaluation model is developed for construction manipulators, it is perceived that it will also be useful for a larger group of automated construction equipment with some minor modifications. It can be used to examine performance of an operating system, or to predict the performance of a system design.

1. INTRODUCTION

Construction manipulators are a new class of construction equipment which has just started to be utilized in a number of construction applications. Currently, their applications include fireproof material spraying [1], interior partition installation [2], concrete masonry walls construction [2], piping erection [3], concrete placement [4,5,6], facade/surface finishing [7], reinforcement steel placement [8], palletized load handling [9], and interior finishing [10]. A more in-depth discussion of the characteristics of construction manipulators can be found in [11].

One important issue related to the application of construction manipulators is their performance evaluation. Since this research domain is rather new, a systematic approach to evaluate their performance is still non-existent. Similar performance evaluation studies have been found in the realms of industrial engineering and industrial robotics [12,13,14]. However, significant deficiencies exist for construction applications in these methods, primarily due to the complexity and the unstructured nature of construction.

This paper describes an evaluation model that incorporates the critical factors and variables undefined in previous methods. This model is developed for evaluating the performance of a construction manipulator but can also be useful for similar types of construction equipment. The paper begins by discussing the task environment of construction manipulators in order to establish the basis for understanding the difficulties and uniqueness of evaluating their performance. Then, the performance evaluation model is introduced and discussed. To demonstrate the application of this model, a hypothetical example is used. From this example, sensitivity analyses of the evaluation results are presented. Finally, a summary of the potential of this model and concluding remarks are given.

2. TASK ENVIRONMENT OF CONSTRUCTION MANIPULATORS

The task environment of a construction manipulator consists of the task, the manipulator and human workers. Under ideal conditions, the manipulator operator will employ the strength of a manipulator to perform a specific task such as picking and placing a length of pipe. The completion time is determined solely by joint motion time. In other words, if he or she operates one joint at a time, the total operation time is the sum of all joint motion time. If all joints can be move simultaneously, the completion time is the maximum of all joint motion times.

In reality, errors, machine characteristics and uncertainties complicate the calculation of total completion time. A task may encounter geometric errors of the building envelope and dimension tolerance of the material. More often, the manipulator will have platform positioning/orientation errors, joint actuating errors, arm deflection, attachment changeover time and some inherent lags of joint motion. As well, human performance depends on the experience, the information provided prior to and during the operation, and many other factors. Consequently, numerous forms of delays are inevitable during operation and can be substantial in terms of the percentage of the total completion time. The evaluation results could be highly inaccurate if these variables are not modeled.

3. PERFORMANCE EVALUATION MODEL

The proposed model assumes that the total completion time is the most important criterion for performance evaluation. Therefore, the whole focus of the performance evaluation is on systematically identifying time components and objectively approximating their values. Other likely criteria for performance evaluation such as total cost, quality and workers' satisfaction are not within the scope of this model. Four steps are involved in this model, explained as follows.

I. Constructing the Operation Plan

In nature, the operation of a construction manipulator is similar to that of a human hand. Therefore, there exists a set of pre-defined elemental motions to describe all operations of a construction manipulator. A set of such elemental motions is proposed in Table 1. With this set, an operation plan can be constructed.

II. Evaluating Task and Machine Related Parameters

The parameters related to the task and the machine's physical capabilities are identified in Table 2. In evaluating the task related parameters, the ground conditions dictate the time requirement for platform leveling. The motion rate reduction in the fine arm motion category is a direct result of safety considerations for the case where a collision between the manipulator and other objects may be encountered. The tolerance level[1] determines the number of segments to which a continuous fine motion needs to be disintegrated as high motion precision is required. The machine related parameters are obtainable from either field experiments or manufacturer's specifications.

III. Evaluating Time Components

This step involves the calculations of motion time, including the platform motion, the gross arm motion, the fine arm motion and the end effector motion, and the evaluation of three types of delay as described in Table 3. The calculation of motion times is fairly straightforward. It can be done simply by dividing a motion range such as a distance of travel or degrees of rotation by a given motion rate. Special attention needs to be paid to the motion sequence of arm motions. Sequential and parallel motions have to be distinguished and evaluated separately. If several motions start at the same time and act in parallel, the total motion time is the longest of these joint motion times. If they take place sequentially (only one joint moves at a time), the total motion time is the sum of all joint motion times.

In evaluating the sensing-and-decision delay, two variables need to be estimated. First, all occurrences of sensing-and-decision delay should be identified for the entire operation. This process is done by thoroughly examining each joint motion according to two criteria: (1) the tolerance level and (2) whether it demands external sensing. For example, when an end effector is approaching to the "Pick" position to grasp an object, this fine motion may require frequent external sensing to recognize the end effector's actual distance and relative orientation to the object and to adjust for ensuing motion. This continuous motion is therefore divided into three segments and three sensing and decision processes are involved, each preceding one segment of motion.

Table 1 Description of Element Motions

Motion Type	Elemental Motions	Description
Platform Motion	Re-spot Platform	Re-positioning the machine platform from one location to another on the job site, measured by driving speed and travel distance
	Level Platform	Leveling the machine platform close to horizontal level, involving adjusting the outriggers and/or providing a sound base to support the machine's weight
Gross Arm Motion	Reach	Elemental gross arm motion employed to move the arm and the unoccupied end effector to a destination without payload
	Move	Elemental gross arm motion employed to move the arm and the payload which is secured by the end effector to a destination
	Orient	Elemental gross wrist motion employed to rotate the unoccupied end effector to a certain orientation without payload
	Turn	Elemental gross wrist motion employed to rotate the payload which is secured by the end effector to a certain orientation
Fine Arm Motion	Disengage	Combined fine arm/wrist motions employed to separate one piece of material from a cluster of others involving micro scale of slow translation and rotation of the handled material
	Pre-position	Combined fine arm/wrist motions employed to position and align the end effector to an intangible point with very high precision involving micro scale of slow translation and rotation of the end effector
	Approach	Elemental fine arm motion employed to locate the end effector to a physical point with very high precision involving micro scale of slow translation of the end effector
	Slide	Elemental fine arm motion employed to move the end effector and/or material along a fixed trajectory on a surface with very high precision
End Effector Motion	Grasp	Elemental type I end effector motion employed to secure the control of a piece of material
	Release	Elemental type I end effector motion employed to abandon the control of a piece of material

Table 2 Task and Machine Related Parameters

Motion Type	Task Related Parameter	Machine Related Parameter
Platform Motion	• Travel Distance • Ground Conditions	• Desirable Travel Speed • Outrigger Motion Rates
Gross Arm Motion	• Joint Motion Ranges	• Maximum Joint Motion Rates
Fine Arm Motion	• Joint Motion Ranges • Motion Rate Reduction • Tolerance level	• Maximum Joint Motion Rates
End Effector Motion	N/A	• Grasp Time per Cycle • Release Time per Cycle

Table 3 Delay Factors

Delay Factor	Description
Sensing & Decision (S&D)	The time which is required in one cycle of sensing information of task and environment, waiting for instructions, processing information, and making decisions regarding ensuing motions and is evaluated according to the technology used: 1. Automated sensing-and-decision and 2. Manual trial and error
Force Buildup	The time which is required between the operator sending a motion command and the actual motion response from the machine, usually varied from actuator to actuator and subject to kinematic and dynamic parameters
Attachment Changeover	The time which is required by occasional supportive operations, consisted of end effector changeover time and other attachment assembly/disassembly time

A second variable is the duration of sensing-and-decision. If a sensing-and-decision process is done by manual trial-and-error, the duration depends on the experience of the operator and the means of communication to acquire external information between the operator and other workers involved. If it is achieved by sensors and a computer with possible input from the operator, the duration is determined by the efficiency of this control loop. To construct the evaluation model, one must first establish the base values of delay duration for either case. Then, the likely upper and lower bounds of the duration variation may be considered indicating the most difficult and the simplest cases respectively. Since uncertainties are explicitly dealt with here, more insight is gained by sensitivity analyses.

The force buildup delay can then be incorporated into the calculated joint motion time. The basic rule is that, whenever a joint moves from a static state, i.e. a complete stop, there is a force buildup delay accompanying that joint motion. Finally, the attachment changeover delay can be obtained either from the experience or the suggested data from the manufacturer. Usually, if the operator's skill remains at the same level, the attachment changeover time is close to a constant.

IV. Total Time Analysis

The total time analysis consists of calculating the total completion time and a series of sensitivity analyses. Since the total motion time constitutes a number of estimated values, it is necessary to analyze the sensitivity of these estimates. These values include the maximum joint motion speeds, the reduced joint motion speeds due to safety, the total number of sensing-and-decision delays, the duration of sensing-and-decision cycles, and the duration of force buildup delays. Three techniques can be used to perform sensitivity analyses: (1) the Tornado Diagram, (2) the Two-Way Sensitivity Graph and (3) the Spider Diagram. In this paper, only the Spider Diagram will be used in the example. The description of these techniques can be found in various sources [15].

4. CASE STUDY

This example assumes a hypothetical piping erection task which involves a stationary truck delivering pipes and a 30-foot high T-shape pipe rack. Pipe erection will be performed by a construction manipulator. The physical capabilities of the construction manipulator are obtained from those of the Grove Pipe Manipulator. Detailed information on this machine can be found in [3]. Two levels of control capabilities are assumed available for this machine. The full control capabilities include internal and external sensors for the joints and the end effector, a computer-aided task planner to interface with the operator and provide motion commands, and a servo-controller. The primitive control capabilities have only joint encoders and an 8-lever controller. Two identical manipulators are each equipped with one of the control capabilities. In essence, this evaluation example compares the performance of the same manipulator with different control capabilities.

The first step of the evaluation is to construct an operation plan. By using the elemental motions in Table 1, this plan is described as follows:

(1) Re-spot Platform and Level Platform: Drive the manipulator platform and station to the proper position for handling pipes.
(2) Reach and Orient: Move the manipulator arm to reach the proximity of one length of pipe on the truck.
(3) Pre-position and Approach: Align and move the gripper to the "Pick" position of the pipe.
(4) Grasp: Gain control over the pipe.
(5) Disengage, Move and Turn: Move the end effector away from the pipe cluster and move the arm to a stable configuration.
(6) Move and Turn: Move the end effector to reach the proximity of the designated "Place" location.
(7) Pre-position and Approach: Align the pipe and move the end effector to the "Place" location.
(8) Release: Release the pipe.
(9) Pre-position: Move the end effector back to the previous position, away from the pipe.
(10) Move and Orient: Move the arm back to a stable configuration and complete this cycle.

The next step is to evaluate machine and task related parameters. The values of machine related parameters were obtained from manufacturer's specifications and verified by field experiment, as shown in Table

4. Regarding the task related parameters, the travel distance of the platform is estimated to be 600 feet. The location at which the platform will be stationed is flat and roughly leveled. Also, the tolerance level of this task is considered medium and therefore the reduced fine motion speeds can be 50% of the gross motion speeds. The ranges of joint motion are listed in Table 5.

Table 4 Physical Characteristics of the Grove Pipe Manipulator

Manipulator Function	Left/Up/In/Close	Right/Down/Out/Open	Build Pressure
Outrigger	6 seconds	4 seconds	0
Platform Travel	3.67 ft/sec	N/A	N/A
Swing on Base	19.8 deg/sec	19.8 deg/sec	0
Main Lift	4.6 deg/sec	4.6 deg/sec	0
Main Telescope	10.7 inch/sec	10.7 inch/sec	0
Secondary Lift	2.2 deg/sec	3.2 deg/sec	0
Secondary Telescope	1.5 inch/sec	1.4 inch/sec	1.5 seconds
Roll	32 deg/sec	32 deg/sec	1 seconds
Rotate	13 deg/sec	13 deg/sec	3 seconds
Pivot	2.3 deg/sec	2.3 deg/sec	10 seconds
Jaw	7 seconds	8 seconds	1 seconds

Table 5 Ranges of Joint Motions in the Hypothetical Example

Motion Elements	Ranges of Joint Motion
Reach & Orient	Swing(+90°), Main Lift(-41°), Secondary Lift(+20°)
Pre-position	Roll(+105°), Secondary Telescope(+7")
Approach	Roll(-5°), Secondary Telescope(+2")
Grasp	Jaw Close
Disengage	Roll(-5°)
Move & Turn	Roll(-50°), Main Lift(+41°), Secondary Lift(-20°), Secondary Telescope(+9")
Move & Turn	Swing(-90°), Main Telescope(+115"), Main Lift(-16°), Secondary Lift(+30°)
Pre-position	Roll(-30°), Secondary Telescope(+15")
Approach	Roll(5°), Secondary Telescope(+2")
Release	Jaw Open
Pre-position	Roll(-5°), Secondary Telescope(-2")
Reach & Orient	Main Lift(-25°), Main Telescope(-115"), Secondary Lift(-10°), Secondary Telescope(-17")

As described in the model, to proceed to the evaluation of time components, one has to identify (1) the motion sequence, i.e. whether a set of joint motions is in parallel or sequential, and (2) the occurrence of sensing-and-decision delays. Such findings for both manipulators are presented in Table 6. The duration of the sensing-and-decision delays is estimated according to the observations in field experiments, shown in Table 7.

Based on all the data derived above, the completion time for both manipulators is presented in Table 8. It can be seen that, depending on the complexity of the sensing-and-decision delays, the total completion time varies from 325.57 seconds (full control) or 1384.81 seconds (primitive control) to 460.57 seconds or 2914.81 seconds. In other words, the improvement on performance resulted from a better control system is in a ratio of 4.25:1 to 6.33:1.

Sensitivity analyses on three variables were performed: (1) the maximum joint motion rates, (2) the reduced joint motion rates due to safety and (3) the duration of sensing-and-decision delays. The results are shown in Figure 1. The spider diagram in the left represents the sensitivity of the manipulator with primitive control capabilities and the right the full control capabilities. It should be noted that both diagrams were drawn in the same scale (vertical axis representing the total completion time in seconds and horizontal axis representing the percentage of change); and therefore, the range of changes is comparative. Clearly, sensing-and-decision delays have the highest impact on both cases and the maximum joint rate the second. *In other words, the most significant benefit of investing in a better control system is to reduce the amount of sensing-and-decision delays.*

Table 6 Motion Sequence and Occurrence of Sensing-and-Decision Delay

		Primitive Control Capabilities		Full Control Capabilities	
Operation Plan	Motion Aggregates	Motion Sequence	Sensing-and-Decision	Motion Sequence	Sensing-and-Decision
Re-spot Platform	• Driving Platform • Leveling	• Sequential, each in one segment	• 1 manual trial and error	• Sequential, each in one segment	• One automated S&D
Reach & Orient	• Swing • Main Lift • Secondary Lift	• Sequential, each in one segment	• N/A	• Parallel in one segment	• N/A
Pre-position	• Roll • Secondary Telescope	• Sequential, each in one segment	• 2 manual trial and errors	• Parallel in one segment	• One automated S&D
Approach	• Roll • Secondary Telescope	• Sequential, each in two segments	• 4 manual trial and errors	• Parallel in two segments	• Two automated S&D
Grasp	• Close Jaws	• N/A	• N/A	• N/A	• N/A
Disengage	• Roll	• Sequential in two segments	• 2 manual trial and errors	• Parallel in one segment	• One automated S&D
Move & Turn	• Roll • Main lift • Secondary Lift • Secondary Telescope	• Sequential, each in one segment	• N/A	• Parallel in one segment	• N/A
Move & Turn	• Swing • Main Telescope • Main Lift • Second Lift	• Sequential, each in one segment	• N/A	• Parallel in one segment	• N/A
Pre-position	• Roll • Secondary Telescope	• Sequential, each in one segment	• 2 manual trial and errors	• Parallel in one segment	• One automated S&D
Approach	• Roll • Secondary Telescope	• Sequential, each in two segments	• 4 manual trial and errors	• Parallel in two segments	• Two automated S&D
Release	• Open Jaws	• N/A	• N/A	N/A	• N/A
Pre-position	• Roll • Secondary Telescope	• Sequential, each in one segment	• 2 manual trial and errors	• Parallel in one segment	• One automated S&D
Reach & Orient	• Secondary Telescope • Main Telescope • Secondary Lift • Main Lift	• Sequential, each in one segment	• N/A	• Parallel in one segment	• N/A

Table 7 Duration Estimates of Sensing-and-Decision Process

Control Capabilities	Normal Case	Most Difficult Case	Simplest Case
Automated S&D	10 seconds	20 seconds	5 seconds
Manual T&E	80 seconds	150 seconds	60 seconds

Table 8 Total Completion Time of Both Levels of Control Capabilities

(in seconds)	Motion Time	Force Buildup	Sensing & Decision	Total
Manipulator with			1020	1384.81
Primitive Control	340.31	24.5	1360	1724.81
Capabilities			2550	2914.81
Manipulator with			45	325.57
Full Control	257.57	23	90	370.57
Capabilities			180	460.57

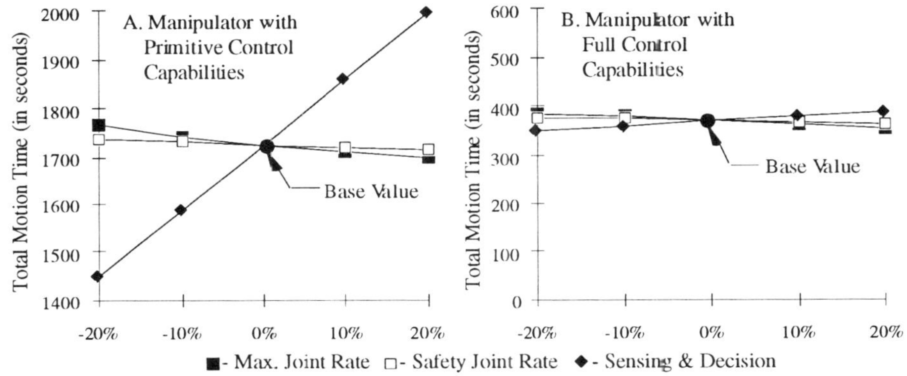

Figure 1 Sensitivity of the Performance of Both Manipulators

5. DISCUSSION AND CONCLUSIONS

This paper has described the proposed performance evaluation model for construction manipulators in a step-by-step fashion and demonstrated one important application of this model which is to predict the potential improvement on task completion time if a better control system is provided for a prototype pipe manipulator. Some insightful findings are revealed in the final portion of the evaluation.

Another potential application of this model is to evaluate the usefulness of a single manipulator for different types of construction tasks. By varying the targeted task, it is possible to foresee what type of construction task is particularly suitable for a given manipulator. By the same token, by varying the physical capabilities and control capabilities, it is possible to examine the type of manipulator that is particularly useful or productive for a given construction task. The immediate benefit generated from this process is the substantial savings of time and money made possible by reducing the necessity for field experiments during a prototyping stage.

As more and more automation technologies now are being introduced to construction, the proposed model can also be used to evaluate the performance of other types of automated equipment such as automatic backhoes, computer-controlled cranes and automatic forklifts. Although this model is developed for construction manipulators, it does provide flexibility for more general use in construction, with some minor modifications such as renewing the terminology of motion elements and re-defining task and machine related parameters. It also provides a key element for a machine performance simulation system for design. Both aspects, however, remain to be explored later.

NOTE

1. For a more in-depth analysis, the tolerance level of a construction task needs to be further defined in terms of accuracy and repeatability, in that accuracy deals with the closeness to the task requirement and repeatability with closeness of replicate manipulator motions.

REFERENCES

1. Yoshida, T. and Ueno, T. (1985). "Development of a Spray Robot for Fireproof Treatment." *Shimizu Technical Research Bulletin 4*, Tokyo, Japan, 48-63.

2. Slocum, A. H. et al. (1987). "Construction Automation Research at the Massachusetts Institute of Technology." *Proceedings of the 4th International Symposium on Robotics & Artificial Intelligence in Building Construction*, Building Reseach Station, Haifa, Israel.

3. Fisher, D. J. (1989). "Piping Erection Constructability Issues in a Semi-Automated Environment." *Dissertation*, The University of Texas at Austin, May.

4. Takenaka Komuten Co., Ltd. (1988). Video Tape, Robots: Automation of Construction.

5. Wanner, M. C. (1987). "Design of a Manipulator with Very Large Reach for Applications in Civil Engineering." *Proceedings of the 4th International Symposium on Robotics in Construction*, 259-267.

6. Benckert, H. (1991). "Computer Controlled Concrete Distribution." *Proceedings of the 8th International Symposium on Automation and Robotics in Construction*, 111-118.

7. Tondu, B. et al. (1987). "Presentation of EUREKA/GEO Project: A Facade Working Robot." *Proceedings of the 4th International Symposium on Robotics in Construction*, 94-102.

8. Sherman, P.J. 1988. "Japanese Construction R&D: Entree into U.S. Market," *Journal of Construction Engineering and Management*, ASCE, 114(1), 138-142.

9. Johnson, D. W. et al. (1991). "Towards an Autonomous Heavy Lift Robot for Field Applications." *Proceedings of the 8th International Symposium on Automation and Robotics in Construction*, 455-464.

10. Warszawski, A. and Navon, R. (1991). "Robot for Interior-Finishing Works." *Journal of Construction Engineering and Management*, ASCE, 117(3), 402-421.

11. Hsieh, T. Y. and Haas, C. T. (1993). "Applications of Large-Scale Manipulators in the Construction Environment." *Proceedings of the 10th International Symposium on Automation and Robotics in Construction*, May.

12. McGovern, D. E. (1974). "Factors Affecting Control Allocation for Augmented Remote Manipulation." *Ph.D. Thesis*, Stanford Univeristy, Novermber.

13. Book, W. J. and Hannema, D. (1980). "Master-Slave Manipulator Performance for Various Dynamic Characteristics and Positioning Task Parameters." *IEEE Transactions on Systems Man and Cybernetics*, Vol. SMC-10, No. 11, November, 764-771.

14. Hill, J. W. (1979). "Study of Modeling and Evaluation of Remote Manipulation Tasks with Force Feedback." *Final Report, SRI International Project 7696*, March.

15. Clemen, R. T. (1991). *Making Hard Decisions: An Introduction to Decision Analysis*. PWS-Kent Publishing Co., Boston.

Automation and Robotics in Construction X
G.H. Watson, R.L. Tucker and J.K. Walters (Editors)

Robot Oriented Modular Construction System - Part II: Design and Logistics

H. Kurita , T. Tezuka , Dr. H. Takada

Institute of Technology, Shimizu Corporation, 3-4-17 Etchujima, Koto-ku
Tokyo 135, JAPAN

Abstract

The 3-Dimensional Modular Unit Construction System involves constructing a) in fixed cycle time using b) 3-D Modular Unit Building Blocks which are assembled at a c) "factory" automated staging area on the jobsite prior to being lifted into place on the building. This system has great potential to facilitate the use of robots on the jobsite and dramatically reduce the construction period. In order for this system to succeed, Logistics and Design become critical issues. We proposed a) a time scheduling model which incorporates the jobsite logistic requirements and b) a concurrent design flow model combining functional design and production design. Production design considerations and the Unit Building Block concept must be introduced in the early stages of design. CAD/CAP/CAM systems must be improved to the point where a) concurrent design is feasible and b) information used during the design process can be integrated into management tools during construction.

1. Introduction

In recent years, many construction robots have been developed with the goal of reducing the construction period. Unfortunately, as of yet there have been few cases where these robots have found application on the jobsite. Even in the cases where robots have been used, the resulting reduction in the construction period has been small. There are two main reasons for this phenomena:

(a) Conventional construction is very labor intensive. In order to use robots on the construction site, we must develop highly sophisticated robots, e.g. self judging, able to perform complex motions, highly mobile. This type of robot is very expensive to produce and difficult to operate and maintain.

(b) The possible reduction in the construction period, even in the case of fully automated construction, is limited by the difference in the work rate of the robot in comparison to the human worker.

Therefore, in order to facilitate the use of robots on the jobsite and reduce the construction period, we must reevaluate the construction process. This means that we must develop systematized construction methods that will 1. facilitate the use of robots and 2. dramatically reduce the construction period.

Based on the issues discussed above, the authors previously presented the concept of a 3-Dimensional Modular Unit Construction System (3-D MUCS) at the 9th ISARC. In this paper, we will review the concept of 3-D MUCS and present two issues which are critical to the success of this construction system. Finally, we will propose a design flow model which integrates function design and production design for the 3-D MUCS.

2. The Concept: 3-Dimensional Modular Unit Construction System (3-D MUCS)

In conventional construction there are many constraints and problems which affect construct ability :

-Many activities are complicated, making the construction complex.
-Materials used on the jobsite represent various levels a sophistication in terms of prefabrication and technology. (raw materials, prefabricated panels, nails and bolts)
-The working area moves about the jobsite, following the working schedule. This makes it difficult to streamline material delivery to the working area.
-There are often many activities proceeding simultaneously in the working area. This creates 1. inefficiencies which reduce productivity, as well as 2. unsafe, unstable working environments high above the ground.

These constraints inhibit the development of automated construction methods. The 3-D MUCS was developed to overcome these constraints.

In 3-D MUCS we construct a building a) in a fixed cycle time using b) 3-D Modular Unit Building Block which are assembled at a c) "factory" automated staging area on the jobsite prior to being lifted into place on the building. (Fig. 1)

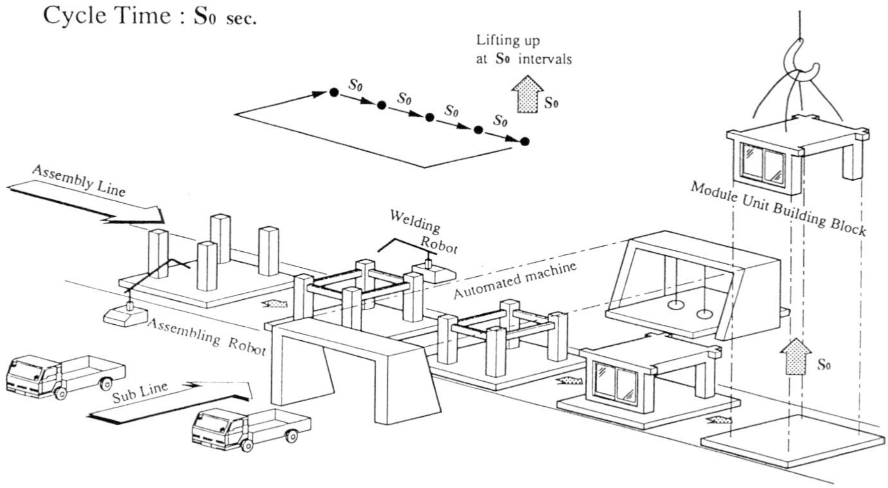

Fig. 1 The Concept Image of the 3-D Modular Construction System

The 3-D MUCS has three main characteristics which facilitate the use of robots on the jobsite:

(1) Work is divided into discrete work packages which are performed repeatedly at one set location.
(2) This method reduces the number of activities on the working floor, simplifying the construction process , streamlining material delivery to the working area.
(3) The activities in the different work packages are synchronized to a fixed cycle time, allowing the construction schedule to proceed at a steady rate.

In order to realize the successful application of the 3-D MUCS, there are two issues that we must consider; a) Logistics and b) the Design Process.

a) The key to supplying the 3-D Modular Unit Building Block in a fixed cycle time is in synchronizing the following activities:

1) the material delivery
2) the jobsite factory automation cycle
3) lifting the building block up to the working floor
4) work area activities on the working floor

Furthermore, each type of building element (raw materials, parts, building block unit, etc.) should be delivered using Just In Time (JIT) scheduling in order to optimize the efficiency of each activity. A time scheduling model has been developed, incorporating the logistic requirements of the 3-D MUCS.

b) Generally speaking, the construction method used on a given project depends on the design of the building system. However, to successfully apply the 3-D MUCS, the opposite must be true. The design process must be changed . In 3-D MUCS, the building system and the construction process must be chosen during the design process

3. Logistics

In this chapter we will present a model of the time schedule and logistics of the jobsite. Our discussion will focus on the activities performed on the jobsite itself. Our scheduling model will not encompass the offsite factory production or the delivery of Building Block elements and materials to the jobsite. Our model is based on the assumption that all parts and materials are delivered to the jobsite using JIT scheduling (Fig. 2).

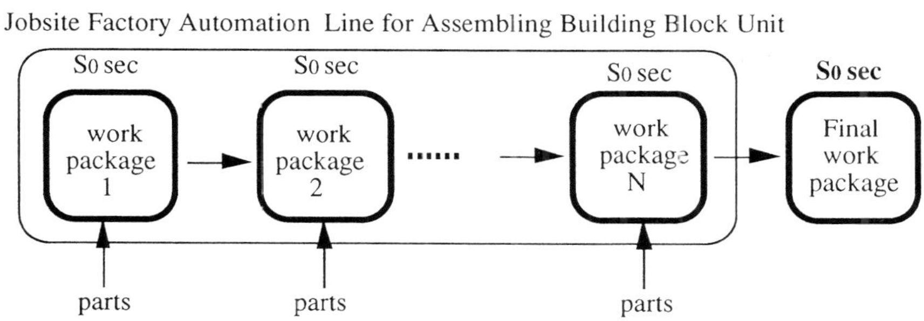

Fig. 2 Logistics of the Jobsite

In 3-D MUCS, activities are divided into discrete work packages with each activity progressing in a fixed cycle time. The "best fit" cycle time is defined by the critical path activities. Other activities are then adjusted to synchronize with the activities on the critical path.

For the purposes of this paper, we will assume that the final work package, which includes the activities of a) lifting and b) connecting the building block units, defines both the critical path and the cycle time. All other work packages are adjusted to synchronize with the cycle of this final work package. Basically, the cycle time is a function of the a) number of robots and equipment used and b) the efficiency of these machines. In order to simplify our model we will assume that all of our machines are equally efficient and assign a ratio of one machine per work package.

We assume that the cycle time of the work packages and the handling time (e.g. time required for a robot to manipulate an object) are as follows:

1) The Cycle Time for Work Packages: $\mathbf{S_0}$ seconds.

2) The Handling Time: $\mathbf{H}$ seconds.

The cycle time, $\mathbf{S_0}$, is defined as follows:

$$S_0 \geq T_{bj} + T_{bl} + T_{bs} + T_{br} \tag{1}$$

T_{bj} : activity time for connecting Building Block Unit to the lifting machine
T_{bl} : activity time for lifting up Building Block Unit
T_{bs} : activity time for setting the Building Block Unit into place
T_{br} : activity time for returning to the starting position

Similarly, the handling time, $\mathbf{H}$, is defined as follows (Fig. 3) :

$$H = T_j + T_l + T_s + T_r \tag{2}$$

T_j : activity time for connecting element to lifting machine
T_l : activity time for lifting element
T_s : activity time for setting element into place
T_r : activity time for returning to the starting position

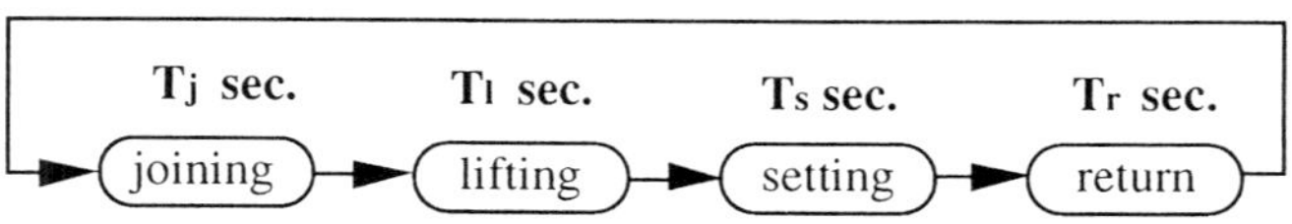

Fig. 3 Handling Activities

If we were to add another machine with the activity of adjusting the elements, that activity would have to be completed before the next element could be set into place (Fig. 4).

$$T_{ad} \leq T_b + T_j + T_r \tag{3}$$

T_{ad} : activity time for adjusting elements or parts

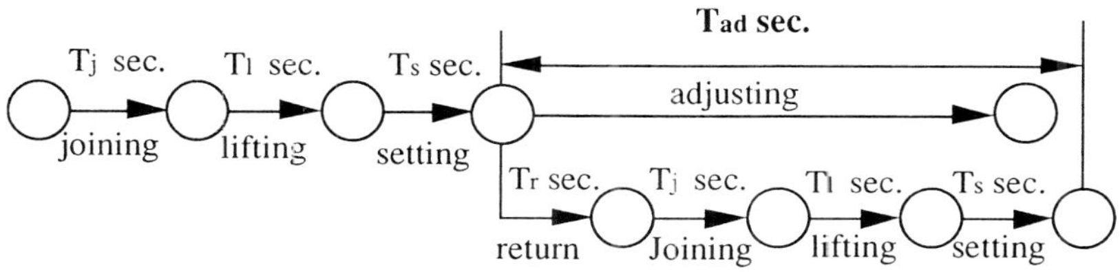

Fig. 4 Adjusting Time

Next, we assume that one handling machine can process **n** elements during the cycle of one work package (Fig. 5). The cycle time, S_0 must also be defined as follows:

$$S_0 = nH + T_{ad} \tag{4}$$

Trucks will deliver elements to the jobsite. The cycle time between the last element being picked from a truck and the next truck moving into place for the next pick is defined by T_t:

$$T_t \leq T_l + T_s + T_r \tag{5}$$

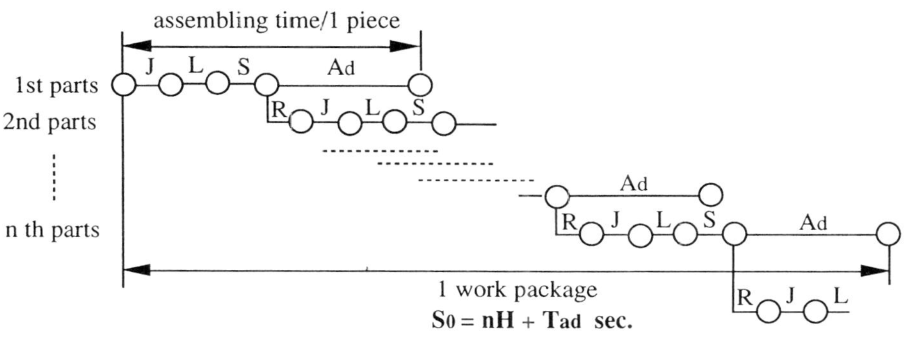

Fig. 5 Schedule Based on Work Package

Each of the above equations must be satisfied in order for all of our activities to be balanced. The number of elements that make up the Building Block Unit affects the efficiency of our 'assembly line'. To achieve 100% efficiency in the use of the handling machines, the number of elements per Unit should be an integral multiple of **n**, the number of elements that one handling machine can process in one work cycle. By satisfying the following equation, machine stand-by time is eliminated:

$$A_p = n \times M \times W_p \tag{6}$$

A_p: number of elements in one Building Block Unit
n : number of elements that one handling machine can process in one work cycle
M : number of machines
W_p: number of work packages

If each work package is designed to meet the above time schedule criteria, the 3-D MUCS will be completed in a fixed cycle time.

4. Design Process

In this chapter we will discuss the new design process required for the realization of 3-D MUCS.

The design of a building is made up of three parts: 1) architectural design, 2) functional design, and 3) production design.
Architectural design encompasses the exterior design (shape, color, texture, etc.) and the design elements which define the atmosphere both inside and outside of the building. Functional design is based on the building purpose or function; the basic structural requirements and safety considerations. Production design, on the other hand, involves designing the shape of the elements and joints within the parameters of construct ability and cost.

In most cases, production design and ease of construction is not generally a consideration during design, because architects and structural engineers design buildings based on the assumption that they will be constructed using conventional construction methods. In designing a standard building to be built within a normal construction period, such as an office building, the designers can easily choose one of several standard building systems. Because the designers are dealing with standard systems and expect the contractor to use conventional construction methods, little or no attention is given to the production design of most buildings. Building element sizes and joint locations are decided based on architectural and functional requirements.

Production design is a key issue in 3-D MUCS. In Chapter 3, we illustrated that the number of elements per Unit impacts the cycle time and the synchronization of activities. In the same way 1) the shape of the unit, 2) the structural system, and 3) the number of units per floor influences the construction cycle. These factors are decided during the design stage, prior to construction. Therefore, to efficiently utilize the concept of 3-D MUCS the building process and the Building Block design must be designed concurrently with the architectural and functional design.
A 3-D Modular Unit Building Block must be a core element of the design process from the schematic design through working drawings. The design must reflect the production requirements of a) the factory assembly line, b) the construction process and c) robot and equipment technology to be used (Fig. 6).

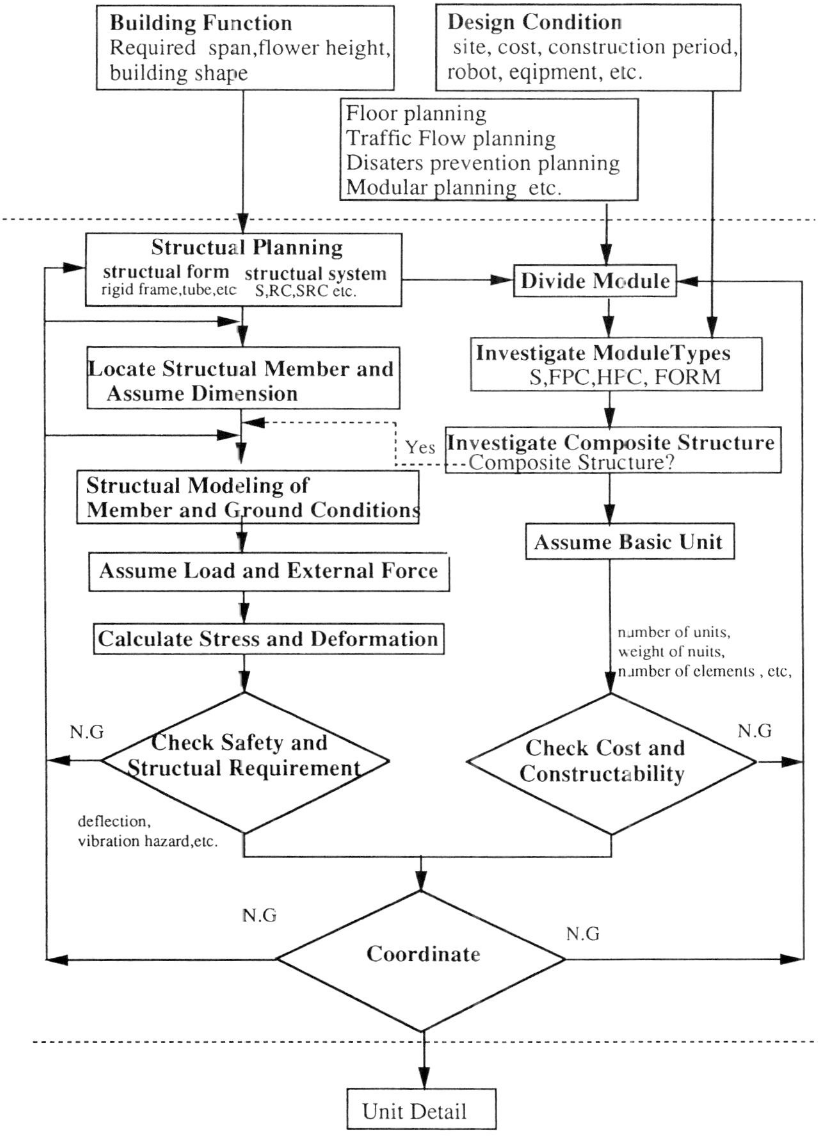

Fig.6 Concurent Design Flow Model (Structual design and Production Design) in Design Development Stage

5. CAD/CAP/CAM

In order to use 3-D MUCS we must be able to estimate cost and evaluate construct ability during the building design. This leads to an iterative, trial and error design process. To optimize the design in terms of function and production and to reduce the design period, computers(CAD) will become a necessary tool for designers.

Computers will also be needed on the jobsite to control robots, equipment and the movement of huge amounts of materials and semi-finished products around the jobsite. Computers will be used to program process activities for robots and allow construction managers strict control of the jobsite "production" scheduling.

Presently, there are two main limitations to the state of the art in construction computer technology:
1) Present CAD systems are merely drafting systems. They are not capable of fulfilling the data retrieval requirements of construction process application software.
2) Most computer systems today are stand alone systems. They are not capable of linking Computer Aided Design with Computer Aided Production.

An advanced CAD system must be developed that is capable of a) building modeling during design b) implement construction process application software and c) integrating CAD/CAP/CAM into a streamline information system.

6. Conclusion

In order for the 3-Dimensional Modular Unit Construction System to succeed, architects and engineers must accept the concept of the concurrent design process. Production design considerations and the Unit Building Block concept must be introduced in the early stages of design. CAD/CAP/CAM systems must be improved to the point where 1. concurrent design is feasible and 2. information used during design can be integrated into management tools during construction.

The 3-D MUCS is just one of many examples of automation and robot oriented construction methods which have been discussed in literature. Among these methods, there seems be wide agreement that the development of concurrent design technology is a critical step toward the implementation of future building processes.

Automation and Robotics in Construction X
G.H. Watson, R.L. Tucker and J.K. Walters (Editors)

Description Method of Required Motion Specification for Construction Modularized Robot

YUKIO HASEGAWA*
TAKUMI ONODA*
KINYA TAMAKI**
ZHU YUWN*

* System Science Institute, Waseda University
3-4-1 Ohkubo, Shinjuku-ku, Tokyo 169, JAPAN

**School of Business Administration, Aoyama Gakuin University
4-4-25 Shibuya, Shibuya-ku, Tokyo 150, JAPAN

ABSTRACT

The research and development of modularized robot are indispensable to successful introduction of robot into building construction, which includes a great many job types. As the first start of the modularized robot research, we previously presented a conceptual design framework, data collection and analysis techniques, and key technologies to be developed at the 9th ISARC. The purpose of this paper is to report a method of specifying the required motion specification for the design of the joint configuration of the modularized robot. This method is applied to the development of modularized robot associated with board application for ceiling, wall, and floor in construction finishing work. Additionally, a simulation system using computer graphics is developed in order to dynamically illustrate the motion trajectories of individual boards.

1. INTRODUCTION

For the robotization in building construction sites, we must solve new problems which include the unique characteristics of construction production systems. Therefore a concept of robot modularization is proposed as a reliable tool for solving these problems. Especially when a new modularized robot is developed in building construction sites, we have to specify joint configuration of modularized robot at the first step of the design. The purpose of this paper is to report a method of specifying the required motion specification for designing joint configuration of modularized robot. The method are to analyze the motion trajectories, which include points from the start positions of the work pieces to the final positions and the positions of obstacle. The positioning motion of modularized robot to place work pieces to the nearest point of the final position is analyzed. These motion trajectories and positioning motions are classified into the several motion elements. The required motion specifications are specified by these motion elements. And in this paper the method of specifying the required motion specifications is applied into the board application for ceiling, wall, and floor, since application task for backer and finish board is one of principal work in construction finish work.

2. DESCRIPTION METHOD OF REQUIRED MOTION SPECIFICATION FOR MODULARIZED ROBOT

Fig. 1 illustrates the process of the description method of the required motion specification for designing modularized robot. Firstly, a standard motion trajectory is defined based on four types of data as follows :

1) Overall design of robotic task system; fundamental system functions to complete a given task, basic size of modularized robot and peripheral machinery, and robotic task sequence.
2) Environmental conditions; overall layout, fundamental and robot coordinate system, working space for robot motions, and existence of obstacles.
3) Start position of the work piece located in the robot coordinate system.
4) Characteristics of the work piece such as weight, size and so on.

The standard motion trajectory means the simplest motion which is to move the work piece from a given start position to a previously determined goal without any collision between the work piece and obstacles.

Secondly, in order to prevent collision between individual work pieces and obstacles in the working space for robot movement and peripheral machinery, collision conditions are classified with the various factors related the final position of work piece, location of obstacles, modularized robot and peripheral machinery. Thirdly, the standard motion trajectory is judged if it can be applied without collision. If it can not, the standard motion trajectory is modified to various trajectory patterns in response to each classified condition. Fourthly, all motion trajectories of each work piece, which is moved and positioned as different working locations, are intensively enumerated. Fifthly the individual motion trajectories are divided into motion elements. Finally, a variety of the motion elements are classified into same types of the motion elements. The robotic range according to these individual types are determined base on the maximum motion quantity and area involved in each motion element. The required motion specification for designing robotic joint configuration is specified from the result of the data concerning the each motion range.

3. TASK SEQUENCE OF ROBOTIC BOARD APPLICATION SYSTEM

The presented description method is applied to boards applying for ceiling, wall, and floor in a construction finish work. Fig. 2 shows the standard motion trajectory for ceiling. The task sequence of the robotic board application system is described as follow:

1) The modularized robot is cooperatively operated with the automated vehicle for providing the required boards.
2) The vehicle follows behind the robot.
3) The robot lifts up a board one by one from the vehicle, and applies it to the final position.
4) When there is no board on the vehicle, it is automatically replaced with another vehicle which is already replenish with the boards.

4. ANALYSIS OF MOTION TRAJECTORIES AND POSITIONING MOTIONS FOR ROBOTIC BOARD APPLICATION

4.1 DEFINITION OF STANDARD MOTION TRAJECTORY FOR BOARD APPLICATION

The standard motion trajectory is defined under the follow conditions:

1) The fundamental measurements of the modularized robot and the board provision vehicle illustrate in Fig. 2.

2) The boards are sucked by the vacuum pads attached to the end effector module of the robot.
3) No obstacle exists during the boards transferred from the start position to the final position.
4) The start position of the board in the robotic coordinate system is shown in Fig.2.
5) The size of the board used in robotic board application are 1820*910*9 mm and 12 mm as the backer board of ceiling and wall, 600*300*9 mm as the finish board of ceiling, and 900*600*2 mm and 3 mm as the finish board of floor. The maximum weight of these boards is less than 15 kg.

Fig.2 shows the standard motion trajectory which satisfies 1)~5) conditions for ceiling board application.

4.2 MODIFICATIONS OF STANDARD MOTION TRAJECTORY IN RESPONSE TO COLLISION CONDITIONS

The standard motion trajectory has to be modified in response to the conditions of existing the obstacles within the robotic working space. In order to prevent collision between individual work pieces and obstacles in working space for robot movement and peripheral machinery, collision conditions are classified with the various factors related the final position of work piece, location of obstacles (column, wall, etc.), the modularized robot and the prevision vehicle. The standard motion trajectory is judged if it can be applied without collision. If it can not, the standard motion trajectory is modified to various trajectory patterns in response to the collision condition. For example Fig. 3 shows it is impossible to apply the board to the final point P_6 by the standard motion trajectory. So the standard motion trajectory is modified that the whole modularized robot moves along Y-axis in the robot coordinate system after the board is moved to the point P_3 in Fig. 3. Fig. 4 shows the all motion trajectories for the ceiling board application. The quantities of motion in Fig. 4 are found during the checks of collision between the obstacles and the board using computer graphics. Fig. 5 shows the example of the motion trajectories using computer graphics. As the result of these analysis the motion trajectories for board application are specified 13 patterns for ceiling, 6 for wall , 8 for floor.

4.3 ANALYSIS OF POSITIONING MOTION FOR BOARD APPLICATION

The positioning motion of the ceiling board is analyzed in Fig. 5. The qualitative motion pattern, with regard to the positioning motion for ceiling board, is common to the wall board and floor.

5. REQUIRED MOTION SPECIFICATION FOR JOINT CONFIGURATION OF MODULED ROBOT

All motion trajectories with regard to every work piece, which a modularized robot must handle in given task, are divided and classified into the same types of the motion elements. The typical types of the motion elements are described as follows:
1) The motion of one dimension along X-axis in the robot coordinate system.
2) The motion of one dimension along Y-axis in the robot coordinate system.
3) The motion of one dimension along Z-axis in the robot coordinate system.
4) The motion of two dimensions on XY-plane in the robot coordinate system.
5) The motion of two dimensions on YZ-plane in the robot coordinate system.
6) The motion of two dimensions on XZ-plane in the robot coordinate system.
7) The motion of three dimensions in the robot coordinate system.
8) The rotation motion of X-axis in the work piece coordinate system.
9) The rotation motion of Y-axis in the work piece coordinate system.

10) The rotation motion of Z-axis in the work piece coordinate system.

Regarding the robotic board application, the individual motion trajectories according to all types of bucker and finish boards for ceiling, wall, and floor, are classified into each kind of the motion element. Fig. 7 groups all motion elements together by the type of the classified motion. Fig. 7 includes data of motion quantity and area concerning each motion element. The maximum quantities and areas within the same classified motion are associated with the required motion specifications. Table 1 and Table 2 show the required motion specification for arm-module and wrist-module, respectively. Both the tables include the individual required motion specifications for ceiling, wall, and floor board applying. These specifications are useful to design joint configuration for arm- and wrist-modules consisting of a modularized robot.

6. CONCLUSION

The proposed method of specifying the required motion specification is essential to design the joint configuration with regard to the arm- and wrist-module involved in a modularized robot. These specifications are defined from the result of analyzing all motion trajectories and positioning motions for completing the given tasks. This method was applied to determining the specifications associated with a modularized robot which applied several kinds of the bucker and finish board for ceiling, wall, and floor by performing a variety of motion trajectories and positioning motions. All motion trajectories and positioning motions for the board application were classified into the same types of the motion elements. The required motion specifications regarding the arm- and wrist-module of the modularized robot were defined base on the maximum quantities and areas within each type of the motion element.

ACKNOWLEDGMENT

This research is carrying out as a part of the WASCOR (WASeda COnstruction Robot) Research Project. For rationalizing building construction systems with robots, this project had been organized by Waseda University and eleven participating companies. The authors wish to express their hearty gratitude to the group members of this project their helpful advice and collaboration.

REFERENCE

1. Japan Industrial Robot Association, "Research Report of Fundamental Specifications for Robot Modularization," 1979. (Japanese)
2. Y. Hasegawa and T. Tamaki, "Robot Module Application for Complicated Construction systems," Proc. 15th International Symposium Industrial Robot, 1985.
3. Y. Hasegawa and T. Tamaki, "Development of Construction Robot Design Methodology with Robot Modularization," Proc. 16th International Symposium Industrial Robot, 1986.
4. Y. Hasegawa, N. Okumura, H. Hoshino, T. Shinozaki, R. Yoshitake, S. Yamashita, H. Asano, T. Kobayashi, and T. Tamaki, "The Development of a New Reinforce Concrete Work System through the Application of the Robot Module Concept," Proc. 6th International Symposium on Automation and Robotics in Construction, 1989.
5. Y. Hasegawa, K. Tamaki, and T. Onoda, "Module Robot Application for Flexible Building Construction," Proc. 9th International Symposium on Automation and Robotics in Construction, 1992.

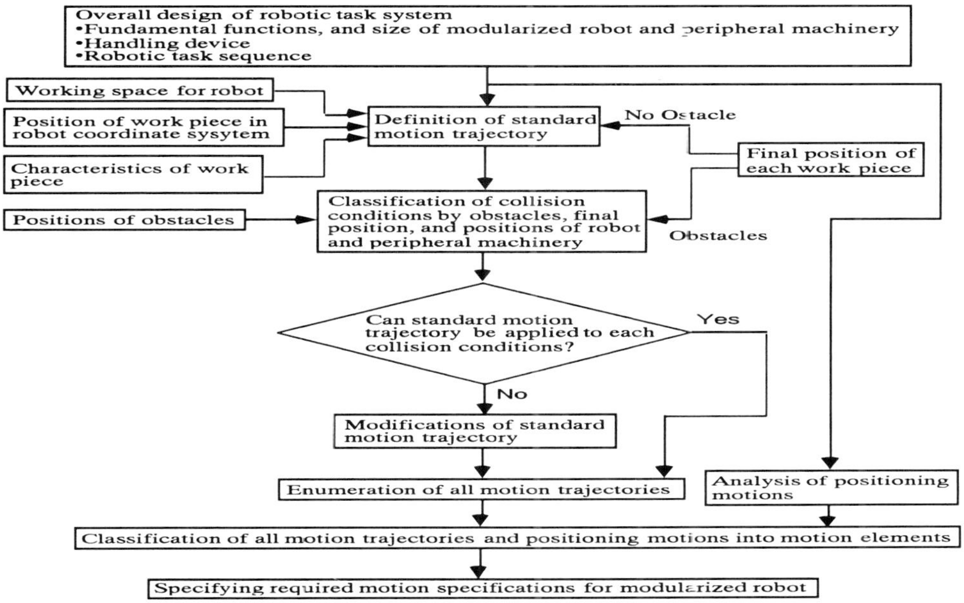

Fig. 1 Process of Description Method of Required Motion Specification

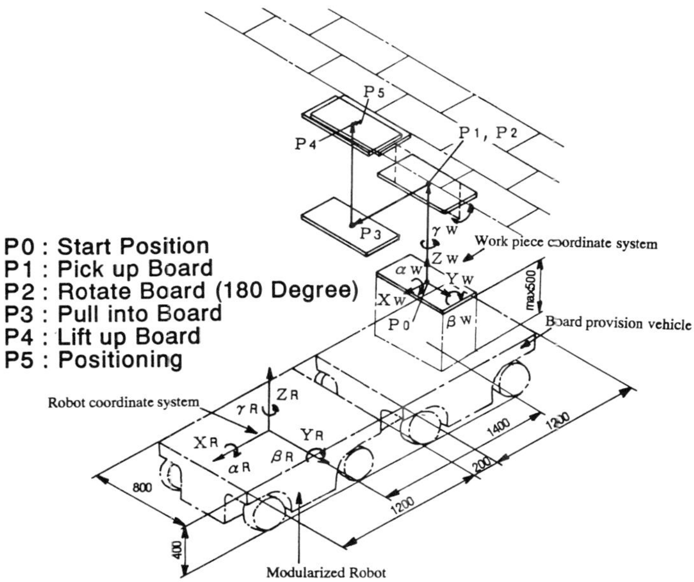

Fig. 2 Standard Motion Trajectory for Ceiling Board Application

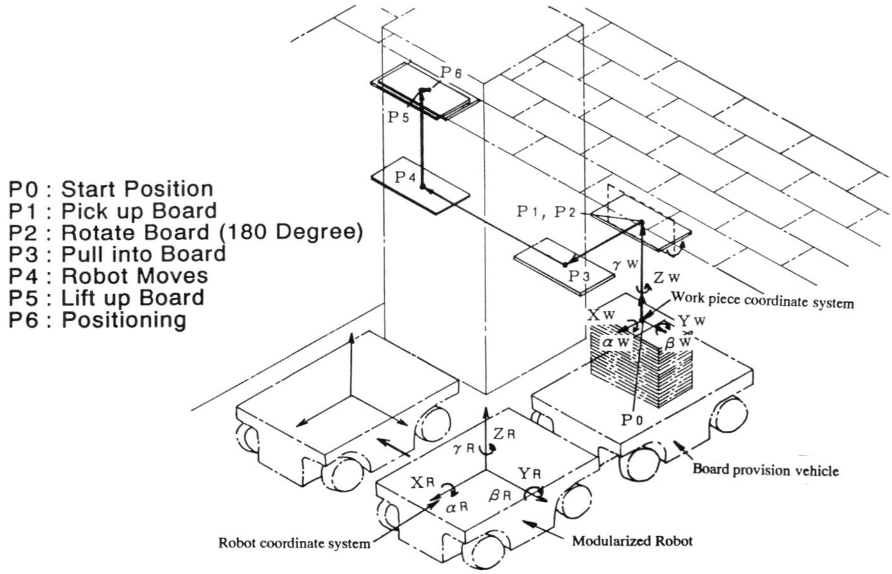

Fig. 3 Example of Motion Trajectory for Ceiling Board Application

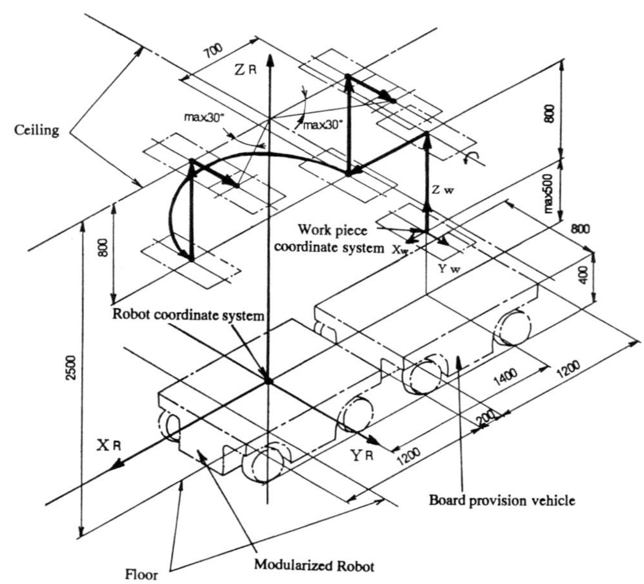

Fig. 4 Variety of Motion Trajectories for Applying All Ceiling Boards

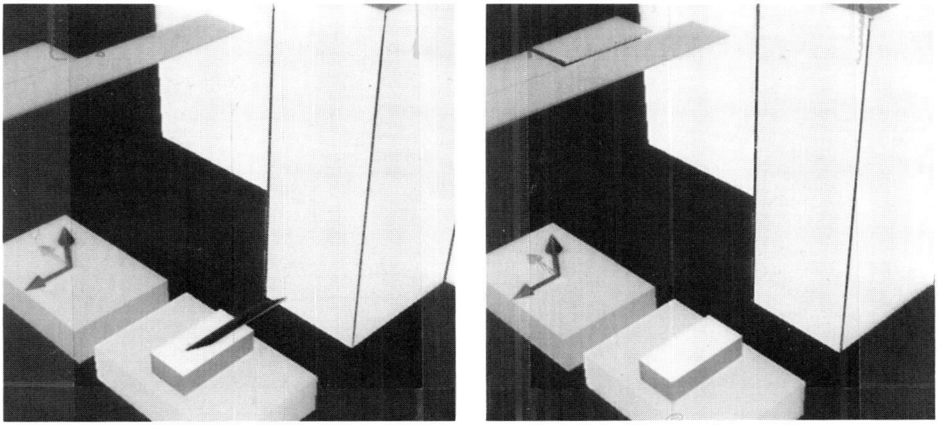

Fig. 5 Example of Motion Trajectory Using Computer Graphics

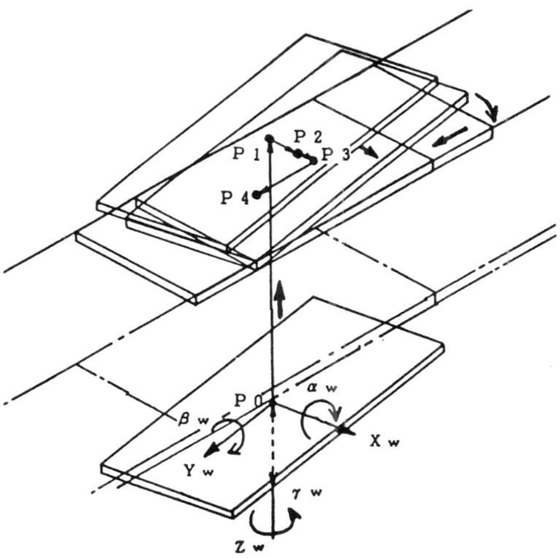

Fig. 6 Positioning Motion for Ceiling Board Application

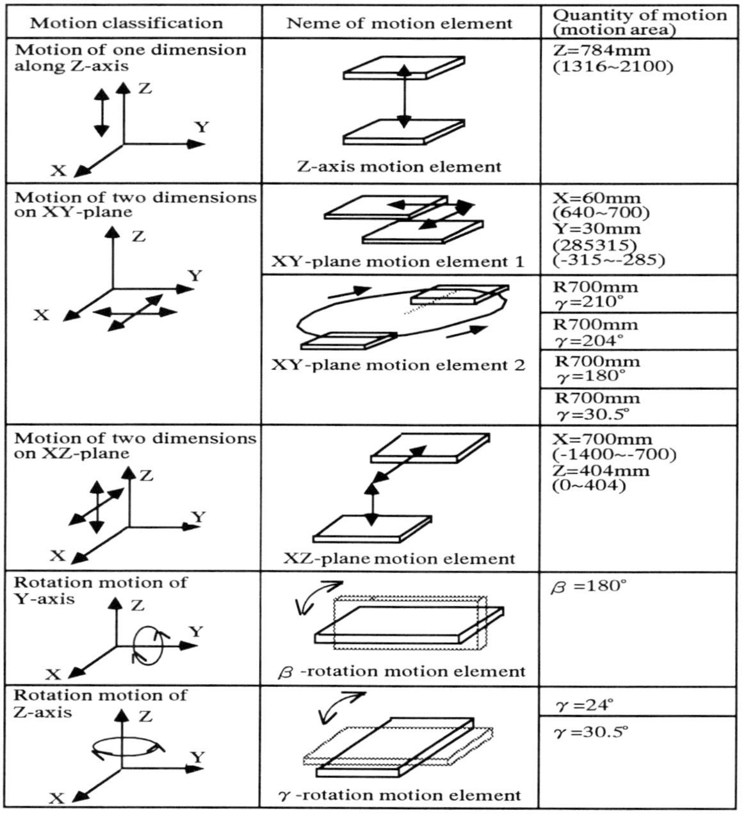

Fig. 7 Group of All motion Elements for Ceiling Board Application

Table 1 Required Motion Specification for Arm-Module

	Ceiling	Wall	Floor
Straight line motion along X-axis(mm)	700(700~1400)	950(700~1650)	700(700~1400)
Rotation motion of Z-axis(°)	±210	±180	±205
Straight line motion along Z-axis(mm)	2100(0~2100)	1600(0~1600)	850(-400~450)

*X-axis, Z-axis:Robot coordinate system

Table 2 Required Motion Specification for Wrist-Module

	Ceiling	Wall	Floor
Rotation motion of Y-axis(°)	±90	±90	±5
Rotation motion of Z-axis(°)	±16	±90	±25

*Y-axis, Z-axis:Work piece coordinate system

Automation and Robotics in Construction X
G.H. Watson, R.L. Tucker and J.K. Walters (Editors)

Automated Access Planning On Construction Sites: An Expert GIS Approach

Koshy Varghese[a] and James T. O'Connor[b]

[a] Department of Civil Engineering, The University of Texas at Austin, Austin, Texas 78712.

[b] Department of Civil Engineering, The University of Texas at Austin, Austin, Texas 78712.

Abstract

The use of prefabrication to enhance constructability necessitates careful planning for access of the prefabricated modules within a construction site. Current planning tools are inadequate for representing the integrated information required for planning access. Hence, developing a reliable plan using these tools is time consuming. This paper presents a computerized approach for planning vehicle access routes on an industrial site. It proposes that a Geographic Information System (GIS) be used to represent integrated information and an Expert System be used to represent the knowledge required to select the optimal access route for prefabricated modules on a construction site. The paper also presents the design of such a system using the GIS Arc/Info and the Expert System Nexpert Object.

1. INTRODUCTION

Prefabrication is an important constructability concept which when properly planned can enhance chances of project success [1]. The primary advantages of prefabrication are that it permits the manufacture of important components in a controlled environment and it permits the acceleration of construction schedule. But, to derive the benefits of prefabrication careful planning is necessary to ensure that the module can be transported to the site, transported to its final location within the site and installed without any hindrances. This study focuses on developing a computerized method for planning the transport of large prefabricated or preassembled modules within the site.

Large prefabricated or preassembled modules form an integral part of the process equipment at industrial facilities. These modules are usually very bulky and can weigh up to 1000T. Figure 1. shows a prefabricated process column which was installed at a plant in Texas. This column weighs 600T and has dimensions of 210'x30'x28'.

Planning the access of such modules within an industrial site is a challenging task for both new and retrofit projects [2] [3]. For new projects, the planners and designers have to synchronize decisions on site layout, facility design, construction sequence and vehicle characteristics to ensure that an accessible route to transport the module to its destination is available when the module is delivered to the site. For retrofit projects, the layout and the design of the existing facility is fixed, but the size of the modules can be varied. As the facility already exists and can be in operation, the planner has to determine the most suitable access route available for the transport of the module.

Figure 1. Transport of a prefabricated process column [Mammoet Transport]

To develop the best overall access plan, it is essential to investigate a number of alternate project scenarios. But the repetitive, tedious and time consuming nature of the analysis using current tools restricts the planners from investigating alternate scenarios. This can lead to the development of non-optimal or even erroneous plans.

The objective of this work is to investigate the feasibility of automating the repetitive planning tasks, thus enabling the planner to focus on the more creative components of the planning process.

The subsequent sections of the paper are presented in five parts. The next section discusses the variables and decisions influencing the planning process. The third section presents the basic concepts of Expert GIS and describes the tools used to implement it. The fourth section presents a conceptual design of the route planning system and the final section presents the conclusions of this paper.

2. ACCESS PLANNING ISSUES

The objective of planning access routes is to determine the optimal route for a vehicle traveling between two points on a construction site. The layout of the site for a new project will be dynamic as it will change as the construction progresses. The optimal route in such a case, can be defined as the route which satisfies both the access objective and overall project objectives. The access objective is to ensure that the transport activity has a minimum potential of being obstructed from the various site features. The overall project objectives are to minimize cost and duration while ensuring a safe work environment.

The project management decisions influencing access are: site layout, facility design, construction sequence and transport vehicle characteristics. These decisions can be varied and combined to generate alternate project scenarios. Each scenario is then be analyzed for access requirements and the scenario which best satisfies the overall project objectives as well as the access objectives is selected for execution.

The human planner develops alternate scenarios based on his/her experience, imagination and creative thinking. To analyze a scenario the planner has to be aware of the project features which influence access and the methods to determine the potential of obstruction from these features. The common causes of obstruction on an industrial site are shown in Figure 2.

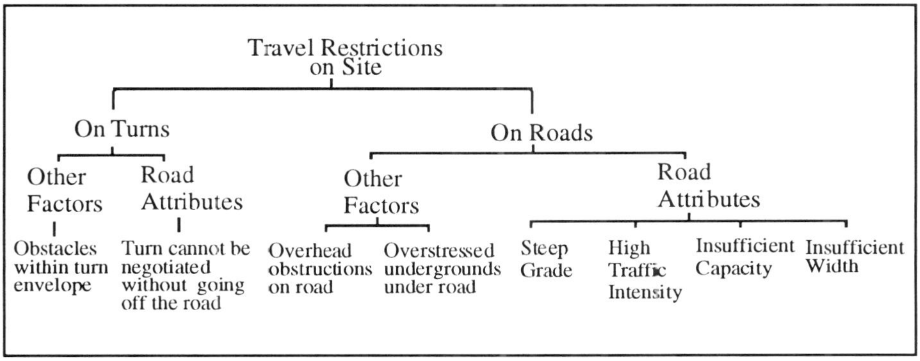

Figure 2. Causes of obstruction on an industrial site

The conventional tools used for analysis are: plot plans or solid models for representing layout of the site, design in the form of drawings, bar charts for representing the schedule, and templates to represent the vehicle/module. To analyze a given scenario, the planner has to refer to the schedule documents and design drawings to determine the layout of the site at the time of delivery and transform this information into a graphic representation. Next, the planner identifies potential access routes and using the template traces the vehicle paths along the routes to determine if they are obstructed. In addition calculations have to be performed to determine obstructions from features such as underground elements. If no accessible routes are found, elements of the scenario have to be modified to create new scenario which is then reanalyzed. If multiple access routes are found the planner has to compare the routes to select the best route. The comparison is based on the potential of obstruction from the features along the route and the route with the least potential of being obstructed is selected as the best route. Thus it can be seen that as the information is not integrated, keeping track of the various pieces of data is a tedious task. Further, the iterative nature of the process requires the planner to repeat many of the mundane tasks, thus reducing the planning productivity.

The successful development of such a system which automates the tedious and repetitive planning tasks will increase the planners productivity. Although the generation of a scenario is a challenging task it is based primarily on the human creative capabilities and no computer methods to date can accurately model these capabilities. Further, there are numerous other factors which influence the acceptability of a scenario. Thus, this work focuses on providing the planner tools using which he/she can use to: easily model alternate scenarios, represent the integrated information for each scenario and automatically analyze the scenarios for the selection of the best access routes.

3. EXPERT GIS CONCEPTS AND TOOLS

An expert geographic information system combines the reasoning capability of an expert system and the spatial data representation and analysis capability of a geographic information system to form a powerful tool for spatial reasoning. The concept of expert GIS has been an active area of research, but only recent advances in hardware and software has permitted its application of to real world problems.

Expert systems is a branch of artificial intelligence, an area in computer science dealing with simulation of human intelligence. Expert systems attempt to simulate human reasoning by representing and processing knowledge. Knowledge can be represented using various methods such as predicate calculus, lists, frames, semantic nets and rules. Of all these methods, the use of rules is the most popular because its format is extremely flexible.

Nexpert Object (Nexpert) is an environment for developing rule-based expert systems. It was developed by Neuron Data [4]. Nexpert is a hybrid system since it has the capability to represent the two basic components needed to make a decision: data and knowledge. The data representation capabilities of Nexpert are based on the object-oriented paradigm. Knowledge is represented in the system using rules. These two capabilities have been called the representation dimension and the reasoning dimension. Figure 3 shows a schematic representation of these dimensions. The intersection of these two dimensions represents the focus of the system at any given time, i.e. which rules are being executed based on which objects. The execution of the rules is controlled by an inference engine called Agenda. In addition to these basic capabilities, Nexpert permits dynamic communication between its object representation and external relational representations.

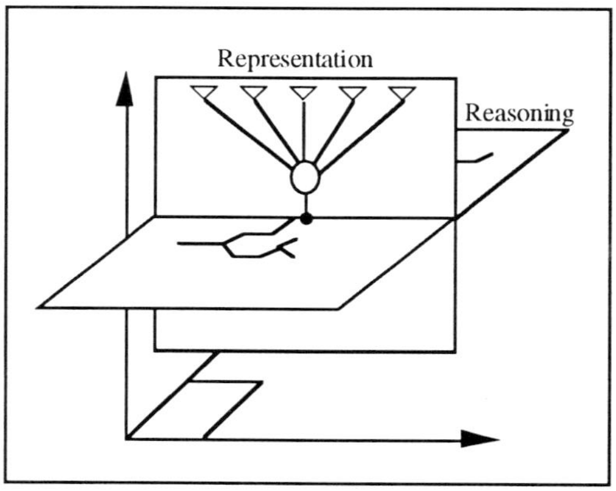

Figure 3. Nexpert Dimensions [4]

A geographic system is defined by Morehouse as a spatial database with a set of spatial operators [5]. In its basic form, it integrates spatial modeling, database management, spatial analysis and computer graphics into a software environment for representing and managing geographic features. Geographic information systems as a technology, had its origins in automated mapping systems about two decades ago.

The primary function of a geographic information system is to integrate and store locational (geometric) information and thematic (attribute) information. Locational information consists of data on the geographic location of points and lines, while thematic information consists of the non geometric attributes of the points and lines. For example, the locational information of a building can be the end coordinates of the lines which form its border, while the thematic information can pertain to its street address. Various sophisticated tools have been built to manipulate, query and analyze the integrated information. These tools along with the tools required to represent the information constitute a geographic information system.

Arc/Info is a commercial vector based geographic information system developed by Environmental Systems Research Institute [6]. Arc/Info's name is derived from the capability of the software to represent locational data (Arc), and thematic data (Info). In Arc/Info locational data is represented using a topological data model while the thematic data is represented using a relational model. This dual data model is called a geo-relational model, as it combines a specialized geographic view of the data with a conventional relational model. The numerous functions of Arc/Info are organized into various components, Table 1 presents the components of Arc/Info and their functions.

Table 1.
Arc/Info component and functions

Component	**Function**
Arccad	Geometry creation and editing using CAD
Arcedit	Geomerty creation and editing using digitizing
Data Convertion	Coversion of foreign geometry formats
Starter Kit	Basic capabilities for representing information
Tables	Editing and adding non-geometric information
Arcplot	Graphical querying of represented information
Overlay	Spatial analysis of represented geometries
Network	Representation and analysis of networks
Macro Language	Creation of macro functions

There are two methods by which Nexpert and Arc/Info can be combined to form an expert geographic information system. The first method involves establishing a data link between the two software packages. Each software would share common data files but function independent of each other. As Arc/Info represents the attributes of a feature in a Dbase file format and Nexpert's databridge is capable of reading/writing information from/to this format, this link can be established by programming the rulebase to read the data, reason based on the data and write the conclusions back into the Arc/Info files.

The second method utilizes the "open" nature of both the systems to build an active command interface. This enables one software package to control the functioning of the other as well as the transfer of data between the two. Such an interface would be capable of dual control: Arc/Info can control Nexpert if the problem consists of a simple reasoning process based on intensive spatial analysis and Nexpert can control Arc/Info if the problem consists of a complex reasoning process using little spatial data [7].

The human planner, when planning for access, uses many capabilities which can be emulated by an expert GIS. Therefore, a computerized system developed using an expert GIS approach can assist the planner by automating many of the route planning tasks.

4. SYSTEM DESIGN

The overall design of the system is based on representing the roads on the site as a network, assigning travel impedances to the roads and turns and using a network search algorithm to determine the path of least impedance between any two points on the network. The overall flow chart required to utilize this approach is illustrated in Figure 4.

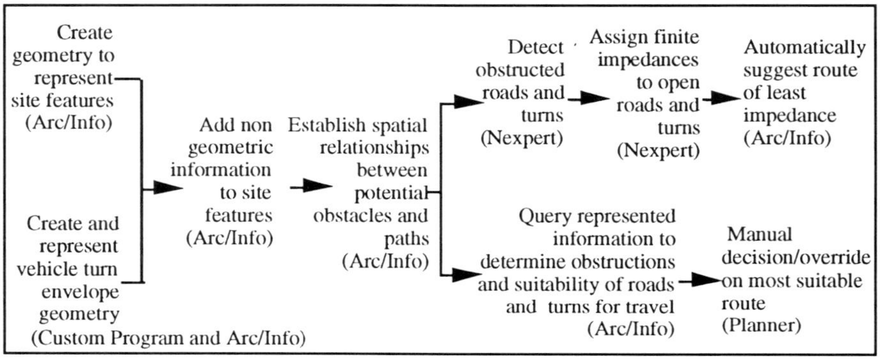

Figure 4. Overall system flow chart

The first step is to develop an integrated representation of the information needed for analysis. Information for a given scenario consists of the layout of the site, construction schedule, facility design and vehicle characteristics. The site layout requires a graphic representation of all the roads, overhead elements, underground elements, permanent facilities and temporary facilities. In addition, the non graphic attributes (design, schedule) of each of these features have to be integrated along with the graphic attributes. The creation, representation and integration of this information is carried out using the basic functions of Arc/Info.

Vehicle characteristics required for analysis consists of its dimensions, weight, kinematics and load distribution scheme. This information is used to test the accessibility of the vehicle on the roads and turns. To test access at a turn, envelopes representing the space required to make the turn have to be generated. The geometry of the turn envelopes is generated by a custom program and this geometry is then represented in an Arc/Info format.

The second step is to establish spatial relationships between the potential obstacles and access paths. The following four spatial relationships have to be identified to determine the potential obstacles on the roads and turns: (i) roads crossed by overhead elements, (ii) roads crossed by underground elements, (iii) turns blocked by permanent facilities, and (iv) turns blocked by temporary facilities. These relationships are established using functions within Arc/Info's Overlay module. It should be noted that the result of these operations are based only on the geometric relationship between the features.

The next step is implemented using a rulebase in the Nexpert environment. In this step the schedule and design information associated with the site features are used to determine the roads and turns which are obstructed when the module is delivered to the site. These roads and turns are assigned an infinite impedance to indicate that the route cannot pass through them. Each of the open roads and turns are then assigned a finite impedance based on its characteristics and the importance assigned to each characteristic. The importance assigned to each characteristic depends on the planners preferences and has to be established prior to the analysis and stored as relative weights in the rule base.

Functions to search the road network to select the route of least impedance can be implemented using Arc/Info's Network component. In addition to selecting the route of least impedance, routines within Network can be used to test user defined routes within the site, generate alternate routes and display the routes.

In addition to the functions described above, provision of a querying capability will allow the system to act in a decision support mode. The querying capabilities of the system are implemented using Arcplot. In the decision support mode the planner can query the represented information, view the results graphically to reach a manual decision if necessary. For example, in response to a query to display the permanent facilities which will obstruct the vehicle at a turn, the system initially selects the set of permanent facilities installed before the specified travel day, and from this set those facilities which are within the vehicle's turn envelopes will be displayed in red.

The final component of the system is its user interface. The interface links both software environments to the user and is structured as a hierarchical menu. This interface is developed using Arc/Info's macro language. Figure 5. shows the overall design of the system. Only an overview of the system and it's basic functions are presented in this paper. A detailed description of the system and test problems used to assess the system are given in [8].

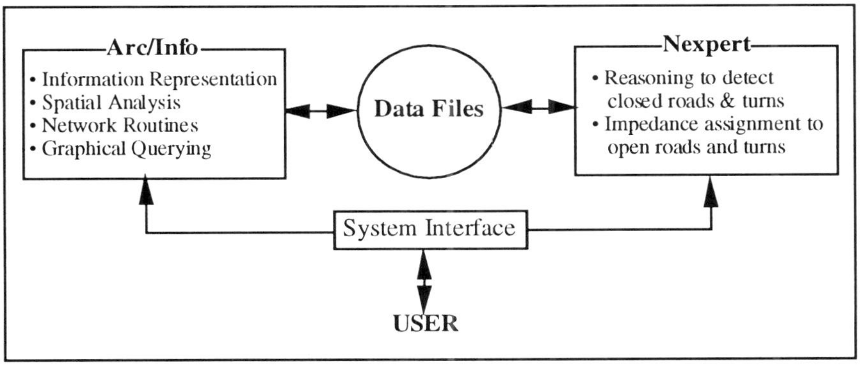

Figure 5. System Design

5. CONCLUSIONS

Planning access routes for large vehicles within a construction site is an important consideration in the development of an effective project execution plan. The constructability and the maintainability of a project can be impaired if access routes are not planned properly. But a majority of the planning effort is spent on ensuring no errors are made in carrying out the tedious and repetitive components of the planning process. The automation of these components will allow the planner to focus on the creative planning tasks thus enhancing the reliability of the plans and improving planning productivity.

This paper presented an approach to automate the tedious and repetitive tasks required to plan for access on a dynamic construction site. The expert GIS approach described in this work allows the creation of an integrated database which can represent both the graphical and thematic information required for access planning. Using this representation as a base, custom programs, functions within Arc/Info and a rule base were developed to automate the necessary planning tasks.

As constructability analysis generally requires integrated representation of information, spatial analysis and reasoning, this technology can be applied to automate a variety of planning tasks. Further, the tools used for this work provide a suitable environment for the development of custom applications with a minimum investment in time and programming effort.

Future work in this area will focus on expanding the system to: (i) Model uncertainty and identify areas of risk, (ii) Develop an example based pre-processor for quantifying planners preferences, and (iii) Represent strategies and knowledge required to modify a scenario to facilitate access.

6. REFERENCES

[1] Construction Industry Institute. "Constructability concepts file". Publication 3-1, Construction Industry Institute, December 1987, The University of Texas at Austin, Austin TX.

[2] Gosch, Donnie Sr. Rigging Consultant, Brown & Root Braun, Personal Interview By Author, August 1990, Houston TX

[3] American Society of Civil Engineers. "Constructability and Constructability Programs: White Paper." The Construction Management Committee of the ASCE Construction Division. Journal of Construction Engineering and Management Vol. 17 (March 1991): p.67-89.

[4] Neuron Data "Nexpert Functional Description v 2.0." Publication, Neuron Data, October 1990, Palo-Alto, CA.

[5] Morehouse, S. "The Arc/Info GIS" A One Week Course Presented as a Part of ESRI Summer Institute Ed. Maidment, D.R. (June 8 1991): Redlands, CA.

[6] Environmental Systems Research Institute. "Understanding GIS". Publication Environmental Systems Research Institute, January 1991 Redlands, CA.

[7] Maidment, D. R. and Djokic, D.[1991] "Creating an Expert Geographic Information System: The Arc-Nexpert Interface." A One Week Course Presented as a Part of ESRI Summer Institute Ed. Maidment, D.R. (June 8 1991): Redlands, CA.

[8] Varghese, K. "Automated Route Planning For Large Vehicles on Industrial Construction Sites". Dissertation, University of Texas at Austin, Department of Civil Engineering., August 1992.

Automation and Robotics in Construction X
G.H. Watson, R.L. Tucker and J.K. Walters (Editors)

A Remote Measuring System for the Mapping of Hazardous Environments

N. S. McPhater

CADCentre Ltd, High Cross, Madingley Road, Cambridge CB3 0HB, UK

Abstract

Close range photogrammetric measurement is a proven method for the mapping of industrial structures and has been widely employed within the process plant and nuclear industries. Unfortunately the radiological conditions encountered within active parts of many nuclear installations may severely constrain any measurement system which depends on photo–chemical imaging.

This paper describes the development of the HAZMAP photogrammetric system, developed under the European Commission TELEMAN project initiative, which is intended to operate within hazardous nuclear environments. The system utilises a remote measurement/inspection unit which is coupled via a photogrammetric module to a process plant design and management system (PDMS). Accordingly 3D CAD plant models can be constructed from digital images acquired photogrammetrically. A number of trials of the HAZMAP system are also described. These have demonstrated the feasibility of the system for constructing accurate 3D CAD plant models which can be used both for conventional engineering purposes and for planning and simulating robotic entry.

1. BACKGROUND

The Hazmap system has been developed under the TELEMAN project initiative which is sponsored by the European Commission in Brussels. The technical objective has been to strengthen the scientific and engineering base for the design of remote equipment for use throughout the nuclear industries in Europe. This is to be done by providing new solutions for manipulation, material transport and mobile surveillance in nuclear environments and by demonstrating their feasibility.

The HAZMAP project was started by University College London (UCL), Department of Photogrammetry and Surveying. They wanted to exploit the proven techniques of close range photogrammetry onto a realistic model of the nuclear environment. UCL appreciated that photogrammetric techniques might match well with 3–dimensional (3D) Computer Aided Design (CAD) modelling technology. Accordingly UCL approached CADCentre with a view to working together on this project.

2. HAZMAP DESCRIPTION

HAZMAP is based on a merging of 2 technologies, close range photogrammetry and 3D CAD modelling, supported on a common UNIX CAD workstation. This workstation acts as a software development environment and also possesses a powerful 3D graphics capability. The philosophy was to closely integrate photogrammetrically–acquired digital images with 3D CAD models of nuclear plant. The technique of close range photogrammetry has already been widely employed within the process plant and nuclear industries for the mapping of industrial structures. It also offers a more flexible approach to 3-dimensional measurement than conventional photogrammetry. The other important reason for choosing digital images is that conventional photo–chemical imaging processes may not be stable in an active hazardous nuclear environment.

The original HAZMAP system is fully described elsewhere (1). However in overview, the system comprises three modules.

The first module is for remote acquisition of digital images. The prototype unit utilises off–the–shelf components. It consists of a pair of standard CCD video cameras mounted on a motorised theodolite pan/tilt unit. This hybrid camera configuration can also be operated remotely via an RS 232 data link to the CAD workstation. Its resolution is 768 x 575 pixels. The first camera carries a wide field of view and a wide angle lens allowing the operator to visually inspect a number of plant components at once. The second camera uses a narrow field of view to enable the high resolution image acquisition of close range details of individual plant components. The operator is thus able to build up and store a number of discrete digital image tiles on the screen of the CAD workstation. Overlapping stored image tiles can then be assembled to build up a complete panoramic photo–mosaic of the wider field of view of the plant. Since the prototype system is built up of off–the–shelf components it is not radiation hardened.

The second module is the photgrammetric module which is resident on the CAD workstation, a Silicon Graphics SG 4D/20. Sophisticated bundle adjustment software is used for localisation of the separate camera viewing points. This determines the exact 3-dimensional positions of the camera at each image acquisition location.

Manual identification of homologous points by the operator interactively on a number of image tiles on the CAD workstation screen enables real–time intersection of key plant components to be established and the corresponding 3-dimensional coordinates to be determined and made available for 3D plant modelling.

The third module is the 3D CAD plant modelling module and the interface into it from the photogrammetric module. The plant modelling system used is the Plant Design and Management System (PDMS) from CADCentre Ltd. This system uses an intelligent engineering database and has the potential to construct a detailed layout of a process plant model. This will now be described.

PDMS DESCRIPTION

PDMS is a sophisticated, configurable 3D design engineering system for the process plant, oil, gas and power related industries. PDMS creates an accurate full colour 3D computer engineering model which can be accessed by designers, engineers and project managers. The heart of the system is a powerful multi–user project database directly coupled to a full scale 3D solid modelling system. This modelling system is based on the construction of 3D solid model primitives e.g. cylinders, cones, pyramids etc.

Piping isometrics, general arrangement detail drawings, bills of materials and other engineering data project reports can be produced quickly from PDMS. A particular feature is that piping components employed in the design may only be selected from engineering piping specifications stored in the database. These specifications consist of predefined lists of all permissible piping components for that engineering project. This ensures design integrity and enables powerful automatic checking for piping data consistency.

PDMS is particularly appropriate for piping–intensive design applications in hazardous environments. In particular these include North Sea oil platform topsides, conventional and nuclear power stations, and nuclear fuel reprocessing plant.

4. 'PROOF OF CONCEPT' TRIALS AND FURTHER DEVELOPMENTS

Initial trials were carried out on a mock–up (i.e. non–hazardous environment) of the core of a British nuclear reactor of the Magnox type. The item selected for modelling was a restraint clamp. Firstly, the exact positions of the HAZMAP camera viewing points were established in the process of localisation mentioned above. Then image tiles were acquired and stored on the CAD workstation from these viewing points. Thereafter, homologous points on the clamp were identified in separate image tiles and 3-dimensional coordinates of these points established. Then a 3D model of the restraint clamp was made in PDMS by generating the constituent 3D solid model primitives.

This trial proved that a single component could be modelled to a precision of 1:5,000 over a range of 1 to 50 metre stand–off distances. However the actual construction of the 3D CAD model itself was not optimal and it highlighted the need for an improved interface between the photogrammetric and 3D CAD modules. Also the trial had not included any pipework which is a particular strength of PDMS.

Although it had been quite a simple trial, 'proof of concept' had been established. It was also clear that this technology had further potential. So it was decided to extend the HAZMAP contract and carry out further trials under more realistic conditions in nuclear installations.

However, before these trials were carried out the HAZMAP system was further enhanced. One development was an improvement in the sophistication of the layout of the photo–mosaic of the digital image tiles on the screen of the CAD workstation. A second development was a closer interface to PDMS to enable the intelligent creation of engineering components within the 3D plant modelling database rather than just building up a simple solids model with little or no engineering design intelligence. Continuous developments were also carried out on the bundle adjustment software to improve the accuracy of the 3D coordinates measured.

5. BENEFITS FROM AN ACCURATE INTELLIGENT ENGINEERING MODEL

Until recently the benefits of such 3D CAD models for process plant have been restricted to the design phase of engineering. The use of HAZMAP, as well as an increasing awareness of using the 3D CAD model as an 'as–built' database record for the benefit of plant operations and maintenance (3), suggest that this intelligent model is about to be used much more widely.

Within the engineering design phase of process plant, the PDMS 3D CAD model with its clash detection capability can be used to manage the space volume of the plant and accordingly help coordinate the different engineering disciplines working on plant design. The data consistency of the piping design can also be checked quickly and engineering drawings, based directly on the 3D model, can be produced automatically with the rule–based draughting package. The 3D 'walk–through' capability can also be used to carry out design reviews and project progressing.

On completion of detailed design, piping isometrics can be produced automatically from the database to assist in the construction of the plant. These isometrics can be separately manipulated within the PDMS environment to produce fabrication, erection and spool isometric drawings with annotation as required. 3D coordinate information can be passed to pipe–bending machines if appropriate. Also erection sequences can be easily prepared to assist in construction. These can be produced as a number of colour shaded images and retained as hard copy or made into animation sequences on video.

During the operational phase of the plant, as described above, the accurate 'as–built' 3D CAD model can be developed from the 'as–viewed' HAZMAP digital images. The 3D plant model can also be used to assist in planning plant maintenance. For example, a removal and installation sequence of a turbine or pump can be simulated. With the benefit of on–line clash detection optimal routes can be established within the 3D CAD model preventing costly mistakes on site. The 3D model can also be used to plan, simulate and execute robotic entry.

To add to this, the 3D CAD model is now being interfaced into existing plant maintenance databases. In this way, the 3D CAD model becomes a pair of eyes into maintenance databases that up until now were not accessible through a graphical interface.

6. TRIALS IN A CONVENTIONAL PLANT ROOM

Before going out to real nuclear plant to validate the HAZMAP system on site, a trial run was carried out to test the HAZMAP enhancements and produce a more intelligent engineering plant model. The site chosen was a hydraulics laboratory at UCL, the environment of which was non–hazardous. The laboratory consisted of a single room 7 metres by 4 metres with associated pumping, equipment, pipework and other plant items. This is described more fully elsewhere (2).

The HAZMAP camera was deployed at a number of fixed locations on a stable tripod. Remote operation was not used on this trial. A photo–mosaic of the plant room was generated interactively on the CAD workstation. Then a 3D plant model was generated off–line in PDMS as now described.

The plant model was made up from three types of engineering items. Firstly, standard pieces of engineering equipment like tanks, pumps and vessels were modelled. They were then correctly positioned and orientated within the 3D space of the plant model. Equipment items were created by building up appropriate 3D solid model primitives. Once created these standard pieces of equipment can be stored in the PDMS catalogue and can be reused again for repeated instancing when necessary. It is also essential to model 'nozzles' attached to the equipment. The particular significance is that nozzles are the attachment points on the equipment, to which pipework is connected within the PDMS system.

Secondly, pipework runs were modelled by establishing centre lines, by locating the centre points of flanges at each end of the straight pipe. PDMS then managed the creation of pipework using a pipework specification to select the appropriate nominal bore. Bends with the same bore and from the same piping specification were then created as appropriate between straight length pipework. Thus quite complex pipe runs could be constructed by the combination of accurate determination of the orientation and single point measurements of flange pairs.

Also valves were placed within the pipework at the correctly measured positions. These valves were selected from the PDMS catalogue. These predefined valve definitions contain adequate geometry to represent the valve graphically for engineering purposes rather than replicating exact photographic detail. One flange measurement is sufficient to position the valve but an additional measurement of valve handle location and extent is necessary to set the orientation relative to the pipe centre line.

Thirdly, other non–standard plant items were built in the model by creating solid 3D primitives, and combinations thereof, that could be manipulated to fit the viewed position. This was also done as a sensible engineering approximation.

The building of this small but representative plant model to an accuracy of $\pm$ 3mm in a static non–hazardous, non–remote control environment with on–line acquisition of digital images, and off–line 3D CAD model generation proved that the technology was now ready for testing in a realistic nuclear environment.

7. TRIALS AT A NUCLEAR POWER STATION

Two separate trials were carried out at a French nuclear power station which was still under construction. This gave the opportunity to carry out real data collection under realistic conditions but in a non–hazardous environment.

The first trial took place in an auxiliary building inside a room of size about 15 metres by 4 metres. It contained much regularly spaced medium bore stainless steel pipework with connected pumps. The camera was operated remotely from the engineering workstation in an adjacent room. Good image acquisition was achieved.

The second trial involved image acquisition and 3D modelling of the heat exchanger bowl at the base of the steam generator building. The bowl acts as a heat exchanger between the primary and secondary water systems and contains a high concentration of small bore piping. The immediate surroundings of the bowl in the base of the building were also modelled. This included floor, walls and other adjacent pipework. In this case the 3-dimensional precision of the HAZMAP measurement system was checked against the proven techniques of industrial photogrammetry and conventional theodolite measurement. This was done with the assistance of theodolite targets which were placed in the bowl environment. These checks confirmed that HAZMAP was achieving accuracies around $\pm$ 2–3mm.

8. TRIALS AT A NUCLEAR REPROCESSING PLANT

The trial was conducted at a BNFL plc facility. The facility was nearing completion of construction prior to becoming active. The HAZMAP equipment was both tripod and gantry crane mounted. On–line acquisition was obtained remotely observing (theoretical) radiation restrictive practices (i.e. all cables were fed through overhead wall plug and shielding with no human intervention). The gantry crane was operated remotely to reposition the camera during the data acquisition stage.

Once on–line digital image acquisition was completed with the construction of the relevant photo–mosaics, the camera was retracted. Then the 3D CAD model consisting of vessels, pumps, building structure and pipework was built up off–line.

The 3D CAD model thus assembled was transferred via an existing program interface to the IGRIP robotics simulation package. The characteristics of the robotic device to be used had already been programmed. Thereupon simulation trials were carried out for robotic intervention.

9. FUTURE DEVELOPMENTS

The HAZMAP system with its 3D CAD model is now set to be at the heart of a continuation phase of the original European initiative. This phase is intended to integrate a number of new emerging technologies into systems that can be transported by a mobile platform into a hazardous nuclear environment. As part of this it is intended to incorporate radiation hardened camera equipment into the system when it becomes available.

The overall objective is to develop and test a light mobile robot in a number of configurations with different sensors/devices/navigation systems for a variety of inspection and surveillance tasks. These are intended to cover a range of maintenance, repair, replacement, post–accident and decommissioning situations.

One such configuration is to mount the HAZMAP camera equipment on–board the mobile platform and drive it remotely from the engineering workstation. A second configuration will mount a separate sensor systen. to HAZMAP on the mobile platform for rapid inspection. Information collected from this separate multiple sensor system is expected to include temperature and radiation levels as well as range data. It is intended thereafter that information acquired from both the HAZMAP and the rapid inspection configurations will be merged together such that temperature, radiation levels and other sensor results obtained can be mapped onto the HAZMAP 3D CAD model.

Practical project trials on these and other configurations are expected to be carried out separately for maintenance tasks and decommissioning work in hazardous areas. They are planned to take place under realistic operating conditions in nuclear plants over the next three years.

10. CONCLUSION

These trials have demonstrated both the feasibility and practical engineering application of HAZMAP as a new form of photogrammetry which is used in close association with PDMS 3D CAD plant modelling. The HAZMAP photogrammetry system can be used for remote acquisition of digital images on–line in a hazardous nuclear environment with satisfactory precision to generate a 3D CAD plant model for remote operation and engineering purposes. The plant model thus constructed is a highly intelligent engineering model which can also be made available to plan and simulate robotic entry.

From a wider context for conventional process plant, this technology will also enable the practical capture of 'as–built' engineering models which hitherto has not been possible cost–effectively. A CAD engineer can be quickly trained to operate this technology and use it as a 3D 'digitiser'. He can also make sensible engineering judgements as to the extent of modelling needed to capture the 'as–built' engineering model. Accordingly not all plant has to be 100% faithfully modelled.

Also the separation of on–line digital image acquisition and storing from off–line 3D plant modelling has two important benefits. Firstly, in a hazardous environment, the 3D plant model can later be constructed away from the hazardous area minimising exposure of the equipment to the hazardous environment. Secondly, digital imagery of a complete plant can be acquired without the immediate need to construct the whole 3D plant model. The 3D plant model, or part of it, need only be constructed as and when required e.g. for the refurbishment and redesign of part of the plant.

11. REFERENCES

1 D P Chapman, A T D Deacon, M A Jones, 1991. A remote measuring system for the mapping of hazardous environments (HAZMAP).

2 David Chapman, Andy Deacon, Asad Hamid, and Rudiger Kotowski, 1992. CAD Modelling of Radioactive plant: the role of digital Photogrammetry in Hazardous Nuclear environments. XVII International Congress for Photogrammetry and Remote Sensing, Washington DC, USA.

3 N McPhater, November 1990. Using Design Information for the Benefit of Maintenance. Institution of Mechanical Engineers Seminar: Computers in Plant Maintenance for the 1990's.

Automation and Robotics in Construction X
G.H. Watson, R.L. Tucker and J.K. Walters (Editors)
1993 Elsevier Science Publishers B.V.

Air Force Construction Automation/Robotics

A.D. Nease and E.F. Alexander

Air Force Materiel Command/Wright Laboratories
Air Base Operability and Repair Section
139 Barnes Drive, Suite 2, Tyndall AFB FL 32403-5319

Abstract

The Air Force has several unique requirements that are being met through the development of construction robotic technology. The missions associated with these requirements place construction/repair equipment operators in potentially harmful situations. Additionally, force reductions require that human resources be leveraged to the maximum extent possible and more stringent construction repair requirements push for increased automation. To solve these problems, the United States Air Force is undertaking a research and development effort at Tyndall AFB, Florida, to develop robotic construction/repair equipment. This development effort involves the following technologies: teleoperation, telerobotics, robotic vehicle communications, automated damage assessment, vehicle navigation, mission/vehicle task control architecture and associated computing environment. The ultimate goal is the fielding of a robotic vehicle capable of operating at the level of supervised autonomy. This paper will discuss current and planned efforts in range clearance/explosive ordnance disposal, hazardous waste cleanup, fire fighting, and automation of routine civil engineering operations.

1. INTRODUCTION

There currently exists in the United States many previously closed military installations, installations currently scheduled for closure, and operational installations encompassing areas that may contain buried/unexploded ordnance (UXO). These installations must be rendered safe by detecting and clearing the UXO. The hazardous conditions surrounding these operations not only inhibit manned operations, but also require additional manpower and equipment for backup support. Current manpower-intensive methods cannot safely complete the detection/cleanup/repair in a cost-effective manner.

The Air Force construction automation/robotics program is involved in a tri-service effort to provide the necessary automation/robotic technology required to solve the range clearance problem. This technology provides a telerobotic means of executing peacetime range clearance as well as post-attack Explosive Ordnance Disposal (EOD) and operating surface repair and recovery during wartime. As manpower becomes more critical to the Air Force of the future, the benefits of unmanned ground vehicles (UGV's) to the peacetime range clearance and wartime base recovery missions become more obvious and important. The UGV can perform these missions in a dangerous environment while human operators and/or combat airbase personnel remain in a safe shelter.

It should be noted that there are other missions, both wartime and peacetime, that lend themselves to application of the technology being developed under the UGV program. Airbase fire fighting/fire detection, hazardous waste handling/cleanup, and the automation of routine Air Force civil engineering operations are other areas that will benefit from this technology **(Fig 1)**.

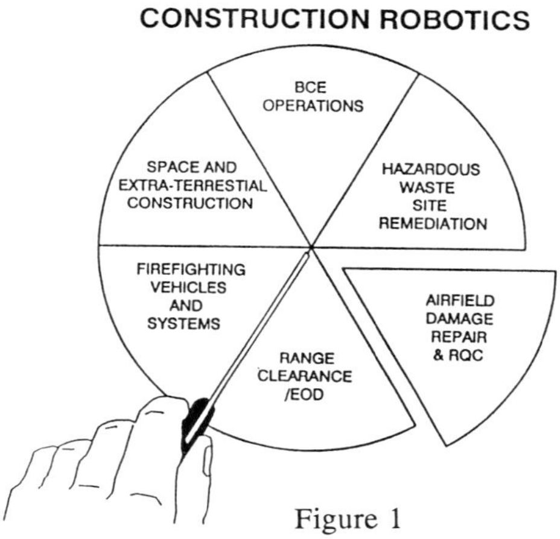

Figure 1

2. CRITICAL TECHNOLOGIES

The Air Force Wright Laboratories has developed a telerobotic excavator, a John Deere 690C, that has a teleoperational capability as well as on-board smarts in the form of preprogrammed functions. Drawing on technological and operational concepts generated, in part, from testing the John Deere telerobotic excavator system, efforts toward producing a mature construction automation/robotic UGV concentrate on several critical technology areas: a) communications, b) navigation/guidance, c) mapping/sensors, d) vehicle/platform, and e) task control architecture/computing environment.

2-1. Communications

The construction automation communications system has been developed in two phases with a demonstration provided in each phase. In phase 1, the fixed operator control station and the existing robotics excavator communication links were developed and tested to evaluate their performance in a PC environment with a single vehicle. Phase 2 involves multiple vehicle communications in a VME-based environment. The command and control link provides a two-way digital data link between the control center and the vehicle(s). This link operates at 34,800 baud at ranges up to 5 miles in the 1433 MHz band. Responses to the control center over the return link occurs only from the specific vehicle addressed by a preceding control center transmission over the C/C link (**Fig 2**). A one-way video link transmits a standard 525 line color or B/W signal at 1795.5 MHz. In addition, the comm system formats and encodes data for transmissions over the identified system links, decodes received transmissions, manages link operation and interfaces the links to the end using equipment.

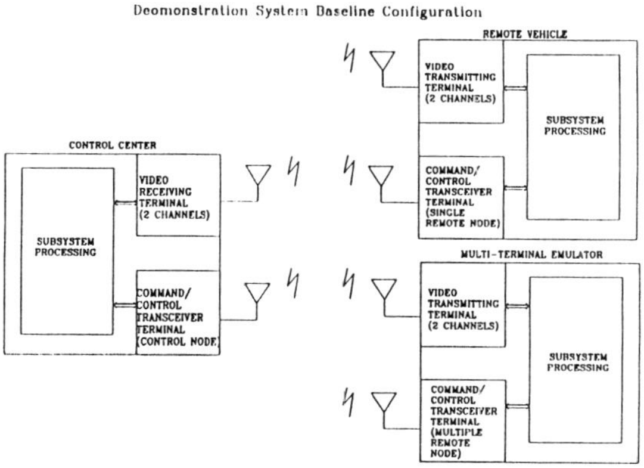

Figure 2

2-2. Navigation

A robotic construction vehicle navigation system is being developed to provide the capability for the UGV to move autonomously from its storage area/shelter to the clearance site, position itself within working range of a previously detected UXO with positional accuracy of +/-6in, to move autonomously among several sites, and then return to the storage area upon completion of the mission. The heart of the system is a Modular Azimuth Positioning System (MAPS) ring laser gyro inertial navigation system. The MAPS will be updated at approximately 3 minute intervals by a kinematic differential Global Positioning System (GPS). The kinematic differential GPS builds on accuracy of standard differential GPS by comparison of phase differences between reference station and remote receiver. The nav system will be given initial vehicle position/orientation, UXO location (digital map), and first site position and required orientation. A plan for vehicle motion will be generated off-line and executed with the aid of obstacle detection sensors (**Fig 3**). The nav system is being developed on a "mule", a small 4-wheel vehicle used to provide an "anywhere" test capability. The INS portion of the system has been successfully demonstrated. The complete INS/GPS system will be demonstrated in July, 1993.

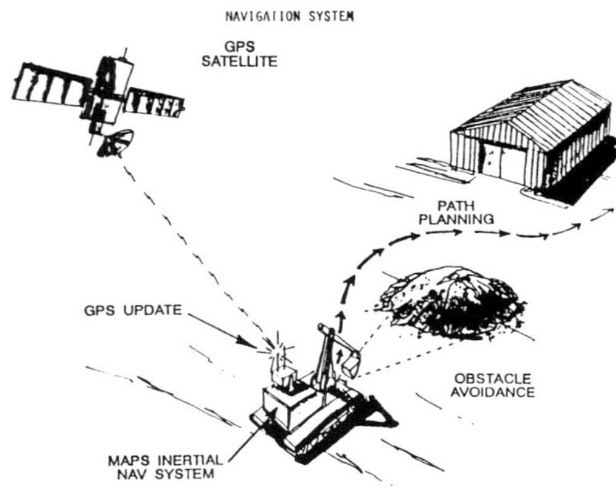

Figure 3

2-3. Mapping/Sensors

The objective of the mapping/sensor development effort is to provide an autonomous capability to characterize the UXO removal/repair site. It is given that the vehicle reference point remains constant and that displacement is known from the reference. First, a 2-dimensional camera and an electronic distance measurement (EDM) device will be used to size the site from a stationary vehicle. Multiple images will be spliced together to form the composite 2-D image. Image analysis of the collected data of the site will be used to estimate the location of the UXO as it is exposed and determine the ordnance/fuse type. This information will be verified using EDM to locate and confirm the actual site perimeter versus false edges created by lighting problems. A 3-D wire frame, perspective view, model will be laid over the 2-D image with vertices at all confirmed points. The volume of the crater/repair site will be estimated using first and second order approximations of the wire frame model. The multiple 3-D images will be spliced together to form the work space environment containing as much of the repair site area as possible. The information displayed will be the classification and centroids of objects of interest in the area as well as the edge of the repair site. The data will be provided to the vehicle control system. Analysis and tests will be made to incorporate information from the second ambiguity region of the scanner. Rules for use of the laser scanner will be developed where appropriate as well as scene requirements and conditions. The initial demo will be done from a fixed base and will show the capability of this technology and the associated software to solve the problems involved with mapping/characterizing a repair site for a remote operator and/or an autonomous vehicle.

2-4. Testbed Vehicle

The new robotic testbed vehicle is being designed and built from the ground up as a next-generation construction robot (**Fig 4**). This vehicle will be an all-terrain, low ground pressure, rubber tracked machine powered by a 275 hp Cummins diesel engine.

A 5kw generator that can provide AC and DC power for the on-board electronics as well as for field testing of systems will be driven from by the main diesel engine. The vehicle hydraulic system will be a closed-center, load-compensating system that will feature in-cylinder position/force sensors and servo valves mounted directly on the cylinders/actuators. The VME-based, on-board computer will be housed in a shock-proof, climate controlled enclosure designed for easy access. The key to this vehicle is that it will build on the lessons learned with the existing robotic excavator.

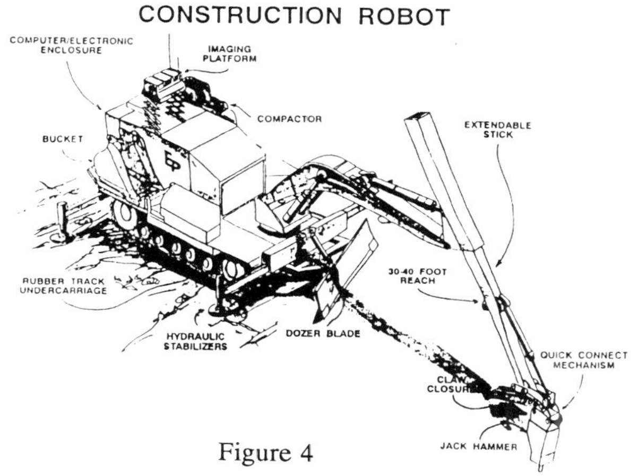

Figure 4

2-5. Architecture and Computing Environment

The system development platform consists of Sun Sparc/2 workstations which provide a UNIX platform for compilation and development. Graphics/simulation work is being done on a Silicon Graphics IRIS 4D/310VGX workstation. All units are interconnected via Ethernet communications links (**Fig 5**). The VXWorks Operating System provides a UNIX-like realtime multitasking platform for VME targets. The OS allows for a small, reconfigurable kernel and provides the necessary speed and multitasking capabilities. VXWorks enjoys wide R&D community support as do the Sun and Silicon Graphics platforms.

The control station and remote vehicle platforms act as VME targets that allow for multiple CPU processing and exchange of information across a common bus. The multiple CPU's process concurrently while individual CPU's process multiple tasks concurrently. The individual CPU's are dedicated to functional areas or tasks and exchange information with the common memory area.

The control station functional structure provides a communications module that communicates between remote vehicles and the control station. This link also provides for communications between CPU's across the bus. This system provides separate channels for emergency messages, remote control functions, and vehicle status information. The operator interface module provides functions for remote control, displays status information, provides the interface to autonomous functions and routines for emergency or error handling. The task coordination module provides coordination between the remote vehicles. The navigation module incorporates UXO site information into a global repair area map.

The remote vehicle functional structure consists of a system planning and arbitration module that performs overall task planning, task control, error handling, and arbitration of system resources for contending CPU's and tasks. The navigation & guidance module integrates INS, GPS, and embedded vehicle sensor information and performs path planning. This module also integrates IR/acoustic sensor information and performs rudimentary obstacle avoidance. The vehicle control module integrates sensor information to initiate and control vehicle movement. The imaging module integrates laser scanning and video images to produce a local repair map for the system planning module. The sensor management module samples all sensors and updates common memory.

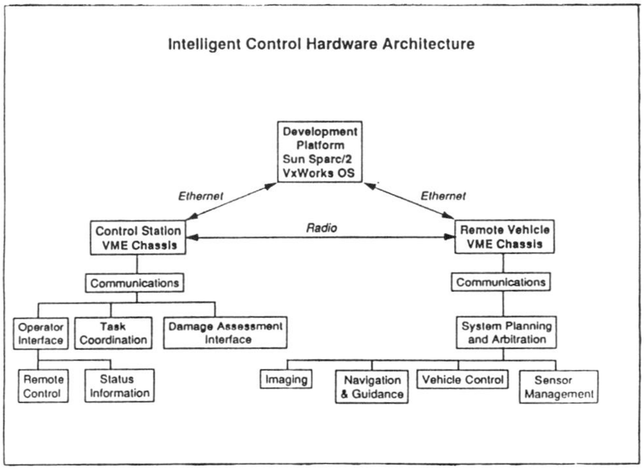

Figure 5

3. CONCLUSION

System integration will be conducted as an in-house effort. The AFMC/WL/FIVCO robotics program currently has 8 engineers and support personnel and a laboratory dedicated to the successful completion of the program. The Air Force realizes the critical nature of robotics development for missions that are hazardous, repetitive or both. The array of missions goes beyond range clearance and EOD to include robotic fire fighting/surveillance, hazardous material handling, and space construction activities as well as the automation of routine Air Force Civil Engineering tasks. The bottom line requirement for robotic construction equipment can be summed up in three words: FASTER, SAFER, CHEAPER.

Automation and Robotics in Construction X
G.H. Watson, R.L. Tucker and J.K. Walters (Editors)

The development of a drilling robot for the installation of railway tracks

Ronald Krom[a,b] , Ruud Kloek[a] and Prof. Charles Vos[b,c]

[a]TNO-Building and Construction Research, PO. Box 49, 2600 AA, Delft, The Netherlands

[b]Delft University of Technology, Faculty of Civil Engineering, PO. Box 5048, 2600 GA, Delft, The Netherlands

[c]Hollandsche Beton en WaterBouw (HBW) B.V., PO. Box 268, 2800 AG, Gouda, The Netherlands

Abstract

TNO Building and Construction Research is developing an autonomous drilling robot. This robot is to be used for drilling holes for anchor bolts which fasten rails to concrete slabs. Special features of this drilling robot are: it's high precision and the relative large diameter of the holes it can drill. This paper presents the background of the project, the design of the drilling unit and robot vehicle and the experiments carried out so far.

1. INTRODUCTION

Currently a joint venture of Dutch contractors (called the "Schiphol Railway Tunnel Combination": KSS) is working on the expansion of the railway station at Amsterdam airport. This railway station is an underground station which will be doubled in capacity over the next years. One of the activities to be carried out is the drilling of approximately 20 000 holes for the fastening of the rails to the concrete slabs on the floor of the railway tunnel. Besides the drilling of the 20 000 holes for the rails, another 100 000 holes have to be drilled in the tunnel floor to insert starter bars for the concrete footpath edges at both sides of the tunnel outside the station area.

An initiative was taken to develop an autonomous drilling robot for this task. This will be discussed in more detail in section two. A special foundation called 'Robouw' was formed. The participants in this foundation are:

- HBW B.V.: a general contractor part of the HBG group, member of the KSS joint venture,
- Strukton Groep N.V.: a general contractor, specialised among other activities in rail road building; member of the KSS joint venture,
- Hilti Nederland B.V.: the Dutch sales representative of the drilling machine and mounting devices manufacturer Hilti in Liechtenstein.

The actual development of the drilling robot is carried out by TNO Building and Construction Research in co-operation with the Robouw Foundation.

1.1 The application

Most railways are built using a ballast bed system. In this system, the rails are fastened to (concrete) sleepers which are laid in a gravel bed. However, in stations this system has a drawback. The metal particles produced by the wear of the rails and wheels collect in the gravel bed and are blown around and cause rust stains in the station. Therefore, a ballast-less system, called slab track system is used. In this system the rails are fastened to concrete slabs on the tunnel floor every 60 cm using base-plates. The concrete slabs are 0.85 m wide and 3.45 m long. Figure 1a shows a cross section of the tunnel. Figure 1b shows the layout of the slabs and the base-plates in a section of the tunnel. Between the slabs there is a gap of 0.15 m in lateral direction and 0.60 m in transversal direction.

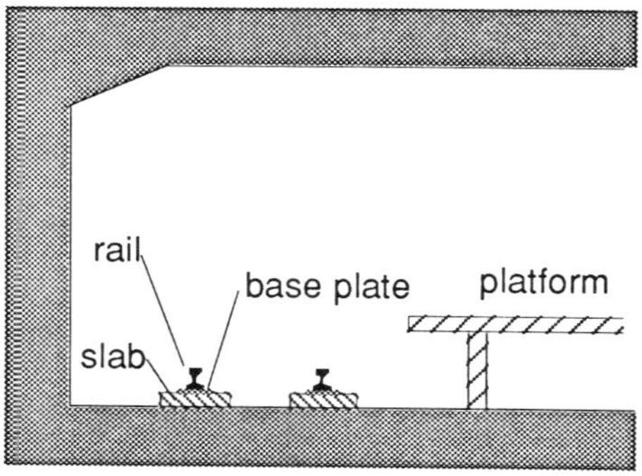

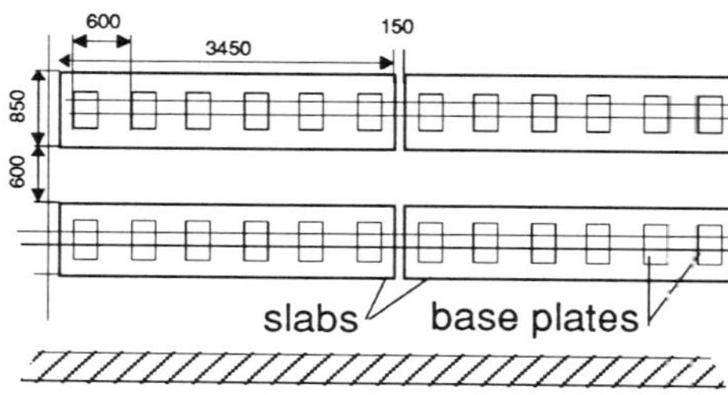

Fig.1a: Cross section of tunnel Fig.1b: Top view: Layout of slabs and base-plates

The base-plates are fastened to the slabs with two resin anchor bolts. These anchor bolts are glued into holes drilled in the concrete slabs. A pre loaded spring and nut hold the base-plate to the floor. The holes needed for anchor bolts have a diameter of 37 mm and are 130 mm deep. In total three railway tracks of each approximately 1000 m length are to be built using this system.

Although the drilling robot will de designed primarily as a tool for the installation of rails, reconfiguration of the robot for other drilling jobs must be possible.

2. HOW IT STARTED

Being aware of numerous experiments all over the world, the interest rose within HBW to gain experience and knowledge on the benefits and costs for construction robots. Simultaneously TNO Building and Construction Research had shown interest to develop robots for the building industry.

2.1 Suitable construction robotics projects

After the initial contacts between TNO Building and Construction Research and HBW it was decided to start a joint exercise. In a series of brainstorm sessions with designers, construction managers, site superintendents and equipment engineers of HBW, a list was made of potential applications for successful construction robots. The list had to be in-line with the interests of both parties. These interests were:

- commercially interesting
- development in co-operation with the eventual user
- in line with business activities
- preferably using the TNO developed positioning sensor called CAPSY[1] (Computer Aided Positioning SYstem)
- perspective for further development
- able to be used on other projects

Based on these requirements the interviews and brainstorms took place, gradually eliminating ideas that were not in line with the above mentioned points.

2.2 The selection of a project

The possible projects were further examined in a multi criteria analysis against a set of relevant characteristics. These characteristics were:

- the amount of repetition
- the increase in productivity when being used
- the increase in construction quality
- the possible benefits for the environment
- the improvement of health and safety for workers
- the simplicity of the project, assuring a short time realisation of a prototype

These criteria, together with the interests of the participants led to the selection of the drilling robot, for the following reasons:

- TNO's positioning system CAPSY could be used
- the automation of the drilling process is of limited complexity
- the number of holes in a railway tunnel is quite large (>100 000)
- future railway tunnels are likely to be on the market
- surveying, marking and drilling is one activity
- unmanned operation is feasible
- human handling of heavy drilling equipment is avoided
- in case of success, further developments are possible, such as:
  * horizontal and vertical upwards drilling
  * positioning and fixing ceiling support systems

In this stage Strukton and Hilti were asked to join HBW and form the Robouw Foundation.

3. THE PROJECT; DEVELOPMENT OF THE DRILLING ROBOT

After analysing the task of a drilling robot it was concluded that two items of the required functionality of the robot played a key role:

- the ability of the robot to position itself accurately enough to drill the holes at the required location with a tolerance of ± 2.0 mm.
- the ability of the drill unit to drill 37 mm holes automatically.

To make sure that these requirements were met, it was decided to carry out two experiments which should prove the feasibility of the required functionality be-

---

[1] After the development of the concept of CAPSY by TNO, the further development was taken over by Spectra Physics Laserplane Inc. who will produce and sell CAPSY. See also [deVos93]

fore the development of the complete drilling robot. In this section the motivations behind these two experiments are described.

3.1 The accuracy of CAPSY in a railway tunnel

The positioning sensor CAPSY was expected to fulfil the required positioning accuracy. However, the special shape of the work area (long and narrow) in relation to the accuracy demands, was a concern. An experiment was planned to investigate how accurate positioning with CAPSY was in practice.

The non-linear behaviour of CAPSY's triangulation position calculation algorithm is complex. It is impossible to formulate simple accuracy behaviour rules. To investigate CAPSY's accuracy behaviour a computer simulation program was developed . This program calculates for each location within a selected area what effect angle measurement errors have on the calculated X and Y position. Figure 2 shows the accuracy behaviour in the tunnel area.

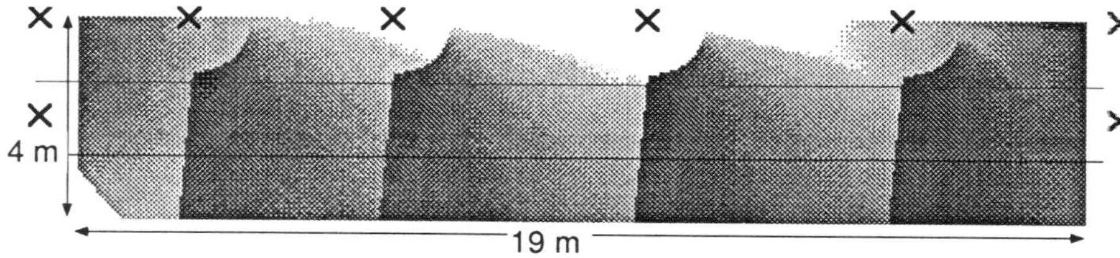

Fig. 2: Accuracy behaviour of CAPSY in tunnel (white = low accuracy)

After the verification of a number of possible reflector locations an experiment was carried out to investigate the accuracy in practice. Section 5.2 describes this experiment and its results.

3.2 The development of an automatic drilling unit

For drilling holes in concrete two systems can be used. A hammer drill is normally used for small diameter holes. For holes with a larger diameter a core diamond drill is used. This drill is cooled with water. For the drilling of the 37 mm holes both systems can be used. A major disadvantage of the core drilling process is that after the drilling the concrete core needs to be broken off and removed. This is a very difficult process to automate. Another disadvantage is that water needs to be supplied and (if possible) removed. A major drawback of using a hammer drill is that it generates vibrations. Eventually it was decided to use a hammer drill due to it's low price, reliability and simplicity. A test drilling unit has been built to get a better insight into the automatic (computer controlled) drilling process using a hammer drill.

4. THE DESIGN OF THE DRILLING ROBOT

This chapter presents the construction of the test drilling unit and the conceptual design of the robot vehicle. After the completion of the experiments with the test drilling unit, a new drilling unit will be designed which will accommodate two drills. This design will be optimised using the experience gained with the test unit.

The main components of the drilling robot are:
- the drilling unit
- the robot vehicle
- the position measuring system (CAPSY)
- a control and monitoring system
- a reinforcement detector
- auxiliary support equipment such as: vacuum cleaner, air compressor etc.

The next two paragraphs explain the design of the test drilling unit and the robot vehicle.

4.1 The test drilling unit

The picture in figure 3 shows the test version of the drilling unit. It consist of the following components:
- a hammer drill, Hilti Model TE 74
- a vertical guidance mechanism
- a pneumatic cylinder to control the vertical drive of the drill
- a pneumatic safety break
- a dust suction nozzle

The drilling process is controlled by sensors:
- a (vertical) drill position sensor
- a pressure difference sensor
- a power consumption sensor connected to the hammer drill.

The hammer drill is mounted on the drilling frame and is guided in the vertical direction along two cylindrical columns. The vertical motion of the drilling machine is activated

Fig. 3: Test drilling unit

and controlled by a pneumatic cylinder. One of the main reasons for using a pneumatic cylinder is that it absorbs the vibrations generated by the percussion mechanism in the hammer drill. During the drilling process the pressure difference (i.e. the drilling force) in the upper and lower half of the cylinder is controlled. A drilling jig is used to guide the drill during the initial stage of drilling the hole. The bore dust is removed by a suction nozzle connected to a vacuum cleaner. The drilling force (i.e. air pressure), the consumed (electric) power and the vertical position are measured in order to control and to monitor the drilling process.

A sensor measures the vertical position of the drill. In case of any calamity, the air pressure in the pneumatic safety break is released, activating the brake and the drill is instantly held at its current position.

A personal computer is used to control the drilling unit. A special interface card controls the air pressure valves and activates the drill and vacuum cleaner and reads the sensor values for air pressure, power consumption and vertical position and takes the appropriate action.

4.2 The robot vehicle

The robot vehicle will integrate all components into one drilling robot. There are a number of requirements for the vehicle. These are:

- it must accommodate the drilling unit, CAPSY, reinforcement bar detector, control unit, control computer, vacuum cleaner and air compressor.
- it must fit on the slabs
- it must be able to cross the gaps between the slabs
- it must be as light and small as possible to make transportation simple

Figure 4 shows a drawing of the conceptual design of the vehicle. The vehicle will be 2.3 m long, 0.9 m wide and 1.4 m high. The total weight will be approximately 750 kg.

In the middle of the vehicle the drilling unit is placed. The unit contains two drills which can drill two holes simultaneously. The relative positions of the drills are set to the corresponding positions of the holes in the base-plates. Each

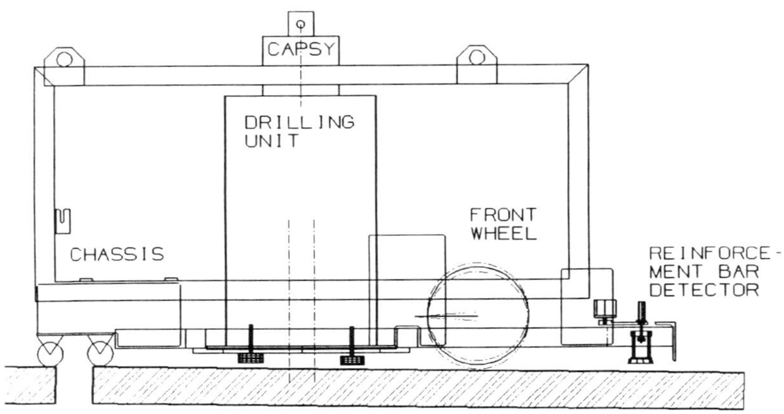

Fig. 4: Side view on vehicle

drill has its own guidance and control system. The cladding of the drilling unit reduces the noise level and provides stability for the drill guidance construction.

At the front of the vehicle a reinforcement bar detector is placed. This detector is placed 1.20 m in front of the centre of the drill unit and is protected by a metal bumper. The detector verifies that no reinforcement bars will be hit during the drilling. Although the concrete slabs are designed so that no reinforcement is present at the locations of the base-plates, presence of any reinforcement bars must be verified because electrical contact between the rails and the reinforcement bars is strictly forbidden.

At the front of the vehicle, behind the reinforcement bar detector, a single front wheel is placed. This front wheel is driven by an electric motor and is mounted in a special suspension which has two degrees of freedom. A special characteristic of the applications for which the drilling robot is being designed is that the holes to be drilled are in (nearly) straight lines. Therefore, the front wheel is fixed in a 'straight ahead' direction. Instead, the wheel can be raised and lowered and be moved to the left or to the right. During drilling the drive wheel is raised and the vehicle will lower itself and rest on the back wheels and the two pads near the drilling unit (see fig 4). When the front wheel is lowered and the pads are off the ground, the front of the frame can move 40 mm to the left or 40 mm to the right. This way the vehicle can turn itself ±1.3° to the left or to the right. When the required direction has been reached, the wheel is retracted, is set in the centre position and lowered again. This steering mechanism can also be used to adjust the position of drill unit in transverse direction.

At the back of the chassis a double wheel system is mounted which enables the vehicle to cross the gaps between the slabs.

5. EXPERIMENTAL RESULTS

At this stage two experiments have been carried out. Paragraph one describes the positioning verification tests. Paragraph two describes the experiences with the test drilling unit.

5.1 Experiments with CAPSY

The most important criteria for the placement of the base-plates are the gauge and alignment of the rails. Therefore the allowed accuracy of the positions in transverse direction must be within ±2.0 mm. In longitudinal direction a lower accuracy of the base-plate positions is acceptable.

As described in section 3.1, an experiment was carried out to verify the accuracy of CAPSY in practice. In a section of the railway tunnel a surveying team marked 36 locations of base-plates using traditional techniques. They also marked eight locations for the reflectors. In the experiment the pre marked locations of base-plates on six slabs were measured using CAPSY. At the time the experiment was carried out with a July(1991)-prototype version of CAPSY.

The measurements of the pre marked locations with CAPSY showed only small differences. The maximum measured differences in transverse direction of the base-plate locations were within the allowed limit of ±2.0 mm. The average difference was 1.1 mm. The expected performance enhancements of the production version of CAPSY and potential increase in accuracy by changing the reflector set-up promise further improvements in CAPSY's positioning accuracy.

5.2 Experiments with the drilling unit

The aim of the development of the test drilling unit is to obtain experience with automated hammer drilling in concrete and verify the design of the test drilling unit. Currently test drillings are carried out to obtain measurements of the pressure difference sensor, electrical power sensor and position sensor are registered and evaluated to obtain the control parameters for the drilling process.

The vibrations caused by the percussion mechanism are absorbed by the pneumatic cylinder. Nearly no vibrations are noticeable in the drilling unit. The first experiments with the test drilling unit showed that automated drilling in concrete with a hammer drill is feasible.

6. CONCLUDING REMARKS

The experiments carried out until now show the feasibility of the development of an autonomous drilling robot. The position measurements of CAPSY are accurate enough for the application. The test drilling unit drills holes smoothly without producing much dust or vibrations. The development of the complete drilling robot has started.

7. REFERENCES

[deVos93] de Vos, L.B.C., Hasara, B. "Field applications with CAPSY", Proceedings 10th ISARC, Houston, May 1993.

Automation and Robotics in Construction X
G.H. Watson, R.L. Tucker and J.K. Walters (Editors)

Robotic rock handling during backhoe excavation

X. D. Huang[a] and L. E. Bernold[b]

[a]Department of Mechanical and Aerospace Engineering, Box 7910, North Carolina State University, Raleigh, NC 27695-7910, USA

[b]Department of Civil Engineering, Box 7908, North Carolina State University, Raleigh, NC 27695-7908, USA

Abstract

Excavation of soil and rock is a high volume and repetitive construction operation. Research in robotic excavation has been focusing mainly on path planning and path control. Very little is known about how to detect and handle underground obstacles such as rocks or utility lines robotically. This paper introduces a research facility to study robotic backhoe excavation using a heavy-duty hydraulically powered and computer controlled manipulator. Conventional excavation methods and an existing telerobotic excavator are introduced. The control system for this robotic backhoe excavator is based on multiple sensors for force, acceleration and position measurements. The same set of sensors are also used for obstacle detection, recognition and mapping. An approach for the detection and recognition of removable obstacle is discussed. Contour mapping procedures have been developed to provide a partial surface map of a buried obstacle.

1. BACKGROUND

1.1. Introduction

Almost all production oriented robots today are utilized within the manufacturing industry. One main reason why true robots are hardly seen in the construction environment is because of the existed unstructured and complex conditions. In addition, construction mostly takes place in an uncontrolled environment, exposed to the weather, dust and noise. The attributes of the materials to be handled range from large, heavy, bulky, and non-homogeneous to light, fragile, and homogeneous. In addition, although one of the traditional materials handled in construction is soil, the mechanics of excavating soil and rocks is poorly understood.

Despite the many difficulties, opportunities for applying high technology in construction are abundant. For certain applications and situations, such as construction in hazardous areas (i.e., nuclear waste disposal, space construction), robotic technology is unavoidable [1]. However, applying high technology in these areas requires empirical as well as theoretical research.

Excavation of soil and rock is a high volume and repetitive construction operation. It is estimated that a fully automated excavating machine performing at 25% the efficiency of an expert human operator would be commercially feasible [2]. The exploration of space offers still another challenge to excavation. NASA studies conclude that excavation will be one of the first activities necessary to establish planetary habitats [2]. Robots are expected to do much of the work.

Studies on the applications of robotic excavation have been undertaken by several researchers. Vaha in 1990 studied and established a kinematic and dynamic control model for a robotic excavator [3]. Bullock and Oppenheim in 1992 developed an approach for force-cognitive robotic excavation [4]. Tochizawa et al. in 1991 reported about an automated excavator for excavating a trench for drainage using laser guidance [5]. The authors showed that the laser guidance helped to decrease the labor hours and increase the digging accuracy.

1.2. Conventional Excavation Method

Some 30% of the earth's crust is shale or mudstone, much of which can be excavated by using a backhoe, front-end loader, or other heavy mechanical excavators. Because of their versatility, backhoe excavators are especially popular on construction sites. During excavation, the equipment operator adjusts the digging angle, depth of penetration and length of digging in order to fill the bucket most efficiently under the given constraints. His/her hands and feet, necessary to operate the equipment, are coordinated intuitively and integrated with his/her physical senses such as seeing, touching, hearing, etc. The accurate and efficient maneuvering of the boom, stick, and bucket of a backhoe excavator require many years of practice. The handling of obstacles is yet another problem. Especially when interfering with an unexpected object, the operator's experience and craftsmanship are crucial in avoiding damages. Water mains, gas lines, or telephone cables are just a few of the obstacles which have to be avoided. In the case of a removable obstacle such as a rock, the operator has several options for dealing with it. He/she can a) remove the rock by loading it into the bucket, b) dig around the rock if it is too big, and c) push, slide/roll the rock from its original place. The operator analyzes the situation, decides what to do based on the gathered information (e.g., visual inspection) and executes the decisions on the basis of his/her previously established rules.

1.3. A Teleoperated Excavator

Langreth reported about an advanced teleoperated hydraulic excavator — Haz-Trak by Kraft TeleRobotics [6]. This excavator operates with a master-slave control and incorporates the force-feedback control so that the human operator can not only control the mechanical arm very easily, but also "feel" the obstacles that the bucket hits. Provided with video images from three cameras, the operator sits remotely inside a building or a vehicle, manipulating and observing the operations. The Haz-Trak incorporates many features useful for cleaning up hazardous chemical and radioactive wastes at weapons labs or removing unexploded munitions buried at military installations. Because of the force sensors inside the hydraulic valve manifold, resistance at each joint are measured and transmitted back into the controller. This enables the operator to probe for buried obstacle and waste without accidentally puncturing a barrel and releasing radiation or toxic chemicals.

2. AN EXPERIMENTAL FACILITY FOR ROBOTIC EXCAVATION

A Multipurpose Robotic Manipulator Platform (MRMP) has been built within the Construction Automation and Robotics Laboratory of North Carolina State University. This robotic manipulator has a reach of 6 meters. It is driven by one hydraulic motor for base rotation and three hydraulic actuators (cylinders) which provide a total of four-degree-of-freedom (DOF). Figure 1 depicts an overview of the experimental device. Many different end effectors (i.e., bucket, gripper) can be mounted at the end of the arm to reassemble this device into different experimental configurations. Presently, the MRMP is equipped with a bucket to operate as a computer controlled backhoe excavator.

Two kinds of sensors are utilized in the data collection system. One force sensor — load cell, is mounted at the rod of the third hydraulic actuator. Three accelerometers are mounted on the boom, the arm and the end-plate (connection between the bucket and arm) respectively to detect accelerations both in the digging plane and perpendicular to the digging plane (it is assumed that the base does not rotate during one digging path so that movements of the boom, arm and bucket stay in one plane). A metal detector search coil will be adopted to mount on the bottom of the arm to detect some utility lines and electric wires. By using an analog/digital board with a sampling rate over 4K Hz, the 386 computer reads at real-time the changes from the force sensor and the accelerometers. Control commands from the 386 PC computer are sent to the actuators through a digital/analog board. The commands act on four electrohydraulic proportional flow control valves.

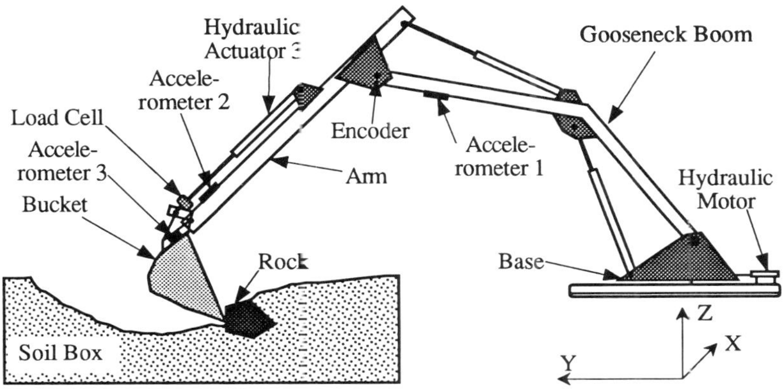

Figure 1. Experimental facility for robotic backhoe excavation

3. IN PROCESS OBSTACLE DETECTION AND RECOGNITION

3.1. Background

Obstacles encountered during excavation can be divided into two basic categories: 1) removable rigid obstacles like rocks or lumber pieces, and 2) non-removable objects like utility lines. In 1989, the United Kingdom reported about 70,000 instances of damage to buried services during excavation [7]. It becomes more and more critical that during excavation, underground buried obstacles such as utility lines can be detected and distinguished from removable obstacles such as rocks to avoid accidents.

However, tasks which humans find simple and trivial are sometimes complex and difficult for a robot to perform. A human can find an arbitrary object visually in a cluttered environment and proceed to grasp the object and move it at will, avoiding obstacles along the way. This task is beyond the capability of most robots in use today.

One approach to building robots for tasks such as object recognition, manipulation, and collision avoidance is to study human performance, hoping to gain insight into the interaction of human sensors, cognition, and motion control. This is still an active research area, but results directly applicable to actual present day robotic tasks are few [8]. A second approach is to exploit existing technology. Most of the researchers are focusing on the applications of 3-dimensional vision systems and tactile sensing devices, which depend on expensive and sensitive equipment. However, construction usually takes place in a harsh environment where rugged and dependable tools and devices are required.

Fortunately, a variety of other sensory devices have been developed and tested for heavy-duty application. The following discussions present an idea of integrating several introduced standard industrial sensors for removable obstacle detection and recognition.

3.2. Detection of the Removable Obstacle

Based on the above discussion, a robotic backhoe excavator has to be able to discriminate between utility lines and removable obstacles. One possible solution to this problem is the use of electromagnetic detectors capable of detecting metal pipes and electrical wires. By attaching such a sensor to the arm close to the actual location where the bucket will interfere with a metallic obstacle, high accuracy and dependability could be achieved. However, such a sensor must be able to discriminate between the steel of the manipulator itself and a buried metallic object. If this could be accomplished, then the sensor could send out a warning signal and stop the excavation before the bucket cuts the line.

The action of digging in a uniform soil with a bucket can be compared to cutting cheese with a knife. The force caused during bucket-soil interactions increases gradually (Figure 2, before 75 samples). If a large obstacle buried in the soil is interfered, the motion discontinues when the measured resistance force is larger than normally expected. Thus, the force pattern during digging is an excellent indication of the soil conditions and the presence of a large obstacle. This phenomenon was observed during the experimental work with the robotic backhoe excavator at NCSU. Figure 2 shows the output of the load cell during a bucket-rock collision. In this figure a negative force represents compression while a positive value represents tension mode in the hydraulic actuator 3.

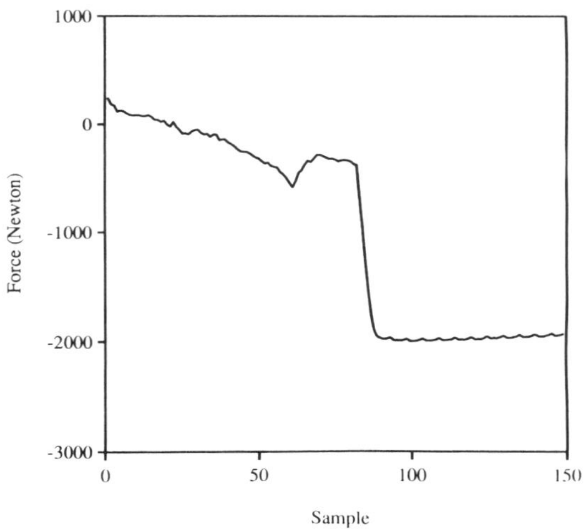

Figure 2. Load cell force pattern during bucket-rock collision

Figure 2 shows that after the compression force undergoes a normal increase, an abrupt and drastic increase can be observed. Actually at exactly this point, the bucket hit a buried rock in the soil box. The slope of the curve at this point is almost vertical. As the compression force reaches -2000.0 N (-450 lbs), the control system would then stop the excavation motion and switch to the obstacle handling mode. The impact position will also be recorded in the computer data base. The horizontal line in the force profile then indicates a stable force.

4. PARTIAL SURFACE CONTOUR MAPPING

4.1. Surface Representation

While it is important to know the presence of an obstacle in the digging path, more information is needed before a decision can be made about how to handle the buried object. If the obstacle is known not to be part of a utility line, it is still not clear if it is removable at all. In other words, an encountered rock surface could actually represent bedrock. Thus, it is necessary to acquire more data to describe the obstacle, particularly its dimension. As mentioned earlier, vision sensors would have great difficulties in providing this information since the obstacle is mostly buried and covered with the soil. The tactile sensing concept has been chosen as a first approach to develop a contour surface map of the obstacle.

The obstacle's surface can be represented by X, Y, Z coordinates of all contact points. Figure 3 shows a parametric representation of a surface patch $\mathbf{P}(u,v)$, where the surface parameters u,v range from 0 to 1. The patch has four knot points, $\mathbf{P}(0,0)$, $\mathbf{P}(0,1)$,$\mathbf{P}(1,0)$, $\mathbf{P}(1,1)$ which determine the boundaries of the patch.

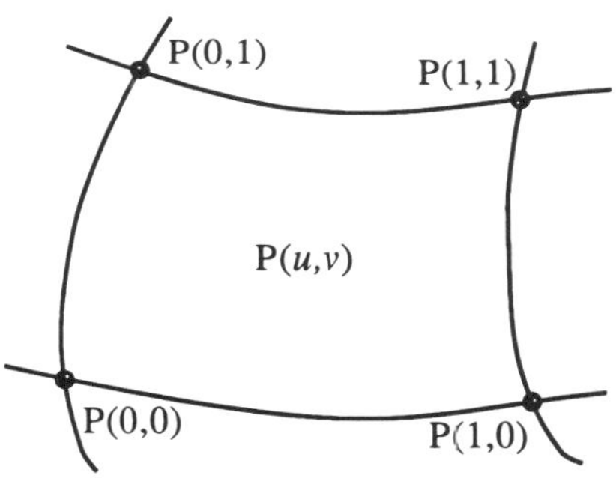

Figure 3. Parametric surface patch

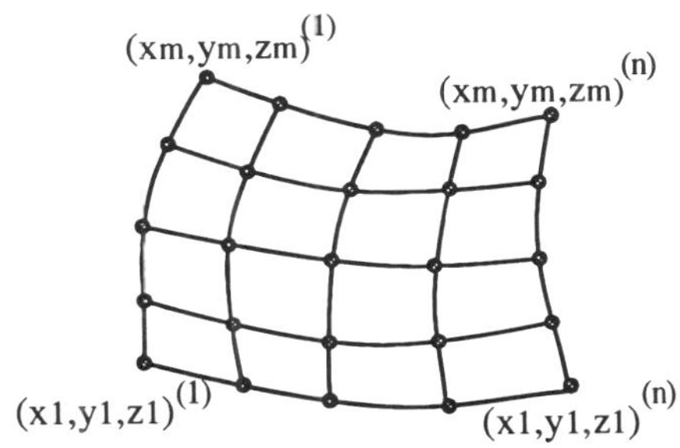

Figure 4. Composite surface with x-y-z coordinates

The knot points are obtained from the position sensors (e.g., joint encoder) while the bucket makes tactile contact with the obstacle. The notion of a single patch can be extended, as more and more data will be available, to a composite surface containing a series of curvature-continuous patches defined on an arbitrary knot set (Figure 4). These data or knot points first need to be organized before further surface mapping can be built up. In the obstacle surface contour mapping experiments, the collected data are put in a matrix format, which later can be used to generate key dimensions of the obstacle. For example, the following is a set of matrices which can be used to represent the surface patch(es) shown in Figure 4.

$$\begin{bmatrix} x1, y1, z1 \\ x2, y2, z2 \\ \vdots \\ xm, ym, zm \end{bmatrix}^{(1)}, \quad \begin{bmatrix} x1, y1, z1 \\ x2, y2, z2 \\ \vdots \\ xm, ym, zm \end{bmatrix}^{(2)}, \quad \cdots\cdots, \quad \begin{bmatrix} x1, y1, z1 \\ x2, y2, z2 \\ \vdots \\ xm, ym, zm \end{bmatrix}^{(n)}$$

In the matrix representation, an X-Y-Z coordinate is used to define an impact point on the contact surface. One matrix records all the contact points within one digging plane (path).

Therefore, several different matrices have more data points and may produce a partial surface description. The total set of n-matrices establishes a grid pattern of points (and patches.) Once the entire surface has been built up, the dimensions of the obstacle, such as its width and height can be easily derived.

4.2. Partial Surface Map Derivation --- Analysis and Results

Position and acceleration sensors are presently utilized in the derivation of the surface map of a rock buried in the soil box. A partial obstacle contour mapping algorithm has been developed and tested. It utilizes the force and position control principles to command the bucket tip around the obstacle while probing and storing the coordinates of the impact points. After a drastic and unexpected increase in force output, a mapping procedure is invoked that comprises five steps: 1) Identification which point of the bucket made contact with the obstacle, 2) storage of the impact position into the matrix of the path, 3) execution of a small back track motion, 4) movement of the bucket tip up vertically by 0.02 m, and 5) continuation of forward motion parallel to the original trajectory. If the rock is hit again after the first adjustments, the above five-step mapping procedure is repeated until the obstacle will be bypassed at its top and the current digging path will be continued at a new level.

During the execution of one digging path, only one set of points is stored into a matrix. It is apparent that further data are needed in order to create the complete grid map either by digging deeper in the same digging plane or by moving to a new plane next to the first one. This is accomplished by rotating the base for a certain degree. As a consequence, the bucket may come in contact with the same rock again and it will adjust its path and hence start a new matrix or append to a previously initiated path matrix.

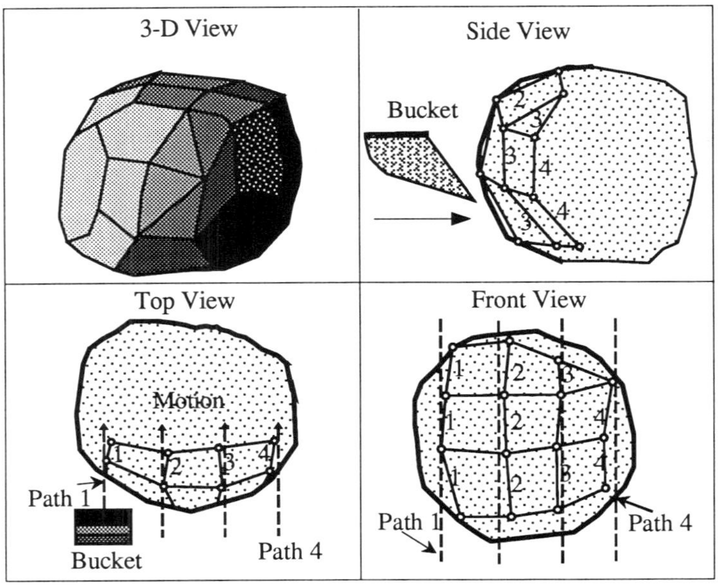

Figure 5. View-based surface grid maps of a hypothetical obstacle

Figure 5 depicts a buried rock from different view points. The circles show the contact points while the grid lines indicate the interpolation web structure between the contact points. A total of four digging paths are necessary to map the surface. Each path is represented by a dashed line (see front view.) Since the bucket has a certain width, it may hit the rock at different places on its tip. One accelerometer has been installed to detect different deflections of the bucket caused by an eccentric force. By gathering data from that accelerometer, the actual point of impact on the bucket can be derived. Therefore, the actual impact position may vary from the path, as shown in Figure 5.

Figure 6 presents graphically the actual obstacle surface map derived through the above processes. Each line in the Y-Z plane represents a partial cross section and the actual trajectory followed by the tip of the bucket. Plotted on the X-axis are the rotational advancements between the four paths. The Z-axis represents the vertical path adjustments. This partial surface map is built from the elements in the data matrices which contain those coordinates where the bucket collided with the obstacle. The resulting matrices can be arranged as follows:

$$\longleftarrow \text{Width direction} \longrightarrow$$

$$\text{Height Diret.}\downarrow \quad \begin{bmatrix} x1, y1, z1 \\ x2, y2, z2 \\ : \\ xm, ym, zm \end{bmatrix}^{(1)}, \quad \begin{bmatrix} x1, y1, z1 \\ x2, y2, z2 \\ : \\ xm, ym, zm \end{bmatrix}^{(2)}, \quad \cdots\cdots, \quad \begin{bmatrix} x1, y1, z1 \\ x2, y2, z2 \\ : \\ xm, ym, zm \end{bmatrix}^{(n)}$$

Each of these matrices represents a vertical plane (or path). It is very seldom the case that all the points in a matrix will be generated during one path alone, due to the fact that excavation progresses from top to bottom. Thus, one matrix may have to be appended with impact points which lie deeper then the original set. Sorting algorithm can be used to rearrange the matrices after the completion of the mapping procedure, which will enhance the recognition efficiency of border points in one plane.

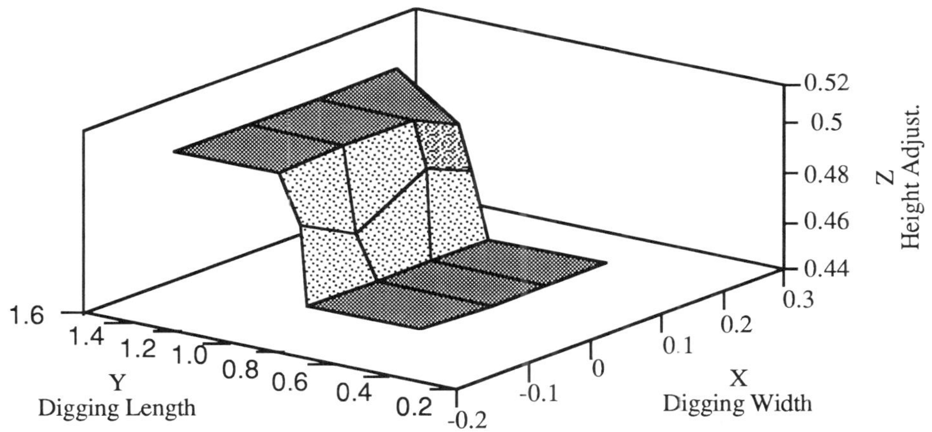

Figure 6. Partial surface map of a buried rock

A removable obstacle can be handled in a variety of ways, depending on the objective of the excavation and the characteristics of the obstacle itself (such as size, shape, etc.). Strategies have to be developed for this purpose. Finding the "best" strategy is a decision making process which may take advantage of a rule base using some input conditions. These conditions include: 1) the data about the dimensions of the obstacle, 2) the excavation requirements, 3) the mechanical system configurations, and 4) relative position of the bucket/excavator to the obstacle. Ongoing research in robotic excavation investigates different obstacle handling strategies that are using the information from the implemented contour mapping procedure.

5. CONCLUSION

This paper presents concepts and experimental results of research on robotic excavation, obstacle detection, and recognition. Multiple sensors such as a load cell, accelerometers, and angle encoders have been installed on a large-size computer controlled backhoe excavator for empirical work on problems related to robotic excavation. The control system is designed to be able to detect obstacles during digging and to map their surface contours. This information is critical for deciding how to deal with a removable obstacle robotically. The ongoing research effort utilizes the presented obstacle recognition approach for the establishment of a rule base for the selection of an appropriate manipulation strategy. The initial research results support the effectiveness of using multiple sensors for robotic excavation in unstructured environments.

6. ACKNOWLEDGMENT

This research has been supported by the College of Engineering at North Carolina State University and the Space Studies Institute at Princeton, New Jersey.

7. REFERENCES

1 R. *COLBAUGH* and M. JAMSHIDI, "Robot Manipulator Control for Hazardous Waste-Handling Applications," *Journal of Robotic Systems, Vol. 9, No. 2, pp. 215-250,* (1992).

2 S. SINGH and R. SIMMONS, "Task Planning for Robotic Excavation," *Proceedings of the 1992 IEEE/RSJ International Conference on Intelligent Robots and Systems, Vol. 2, pp. 1340-1349,* Raleigh, NC (1992).

3 P. VAHA, "Modelling and Control of Cognitive Excavation," Technical Report, School of Civil Engineering, Purdue University, West Lafayette, Indiana (1990).

4 D. M. BULLOCK and I. J. OPPENHEIM, "Object-Oriented Programming in Robotics Research for Excavation," *Journal of Computing in Civil Engineering, Vol. 6, No.3, pp. 370-385,* (1992).

5 M. TOCHIZAWA, S. TAKEDA et al, "Automatic Excavator," *8th International Symposium on Automation and Robotics in Construction, Vol. 1, pp. 277-284,* Stuttgart Killesberg, Germany (1991).

6 R. LANGRETH, "Smart Shovel," *Popular Science, Vol. 240, No. 6, pp. 83-85,* (1992).

7 D. A. BRADLEY, D. W. SEWARD, and R. H. BRACEWELL, "Control and Operational Strategies for Automatic Excavation," *6th International Symposium on Automation and Robotics in Construction, Vol. 1, pp. 237-244,* San Francisco, California (1989).

8 P. K. ALLEN, *Robotic Object Recognition Using Vision and Touch,* Kluwer Academic Publishers, Boston (1987).

Automation and Robotics in Construction X
G.H. Watson, R.L. Tucker and J.K. Walters (Editors)

MULTI-JOINTED PILE DRIVING MACHINE WITH A COMPUTER-ASSISTED GUIDING SYSTEM

Toru Uchino , Junichi Narisawa , Yuhei Sato , Koji Kumazawa

Hitachi Construction Machinery Co.,Ltd. Tsuchiura Works
650 Kandatsu-machi, Tsuichiura-shi, Ibarakiken 300, JAPAN

ABSTRACT

Today, in big cities of Japan such as Tokyo, a number of urban renewal and infrastructure development projects are on. Because of restrictive work site conditions in urban areas, such as overhead roads or railways, compact construction machines with high mobility are favored. To meet this need, we developed a new type of pile driving machine: the HITACHI RX 2000.

With this machine, a piling attachment, such as an earth auger or a vibratory hammer, is directly connected to the tip of the multi-jointed pile driver arm. The pile driver arm uses the computer-assisted guiding system that allows the arm tip to move in perpendicular or horizontal directions with ease. We call this new system "arm tip locus control."

As of today, the RX 2000 has been applied to many H-steel-piling and sheet-piling work in the field.

1. INTRODUCTION

Prior to the development of the RX 2000, operators of multi-jointed pile driving machines had to manually control the combined movement of two arms in order to achieve linear movement of the arm tip. This is a very difficult task, requiring a great deal of special skill and dexterity. (See Fig. 1.)

Therefore, development of an easy-to-operatecontrol system, such as our "arm tip locus control", has long been awaited. The requirements of such a system include: highly-accurate piling, including both positioning and piling angle; fast arm tip movement; tough, able to withstand adverse conditions; and a reasonable production cost. Our computer-assisted multi-jointed arm control system meets all of these requirements.

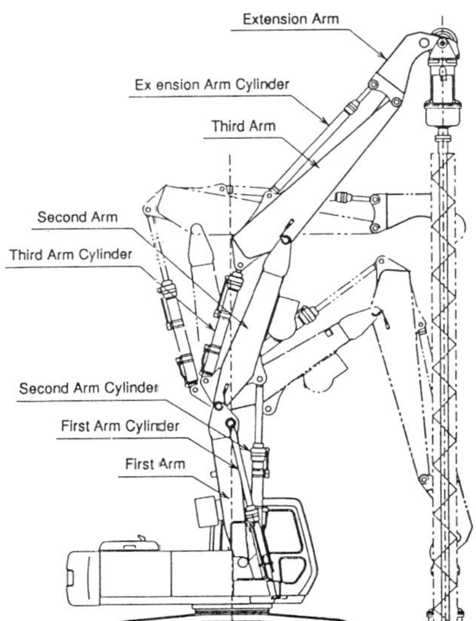

Fig. 1. RX2000

Herein we report on this new system, focusing on the overall system design, the control lever operation system, the proportional solenoid valve operation system, as well as a few examples of RX 2000's field applications.

2. SYSTEM DESIGN

The layout of the control system is shown in Fig. 2. Fig 3 shows a block diagram of the system. The position of each arm is detected by the angle sensors provided at each joint of the arm. Then, the coordinates of the arm positions are calculated. With the operative velocity control signal generated in proportion to the control lever stroke and the coordinates of the arm positions, a target control angular velocity of each arm is calculated. Each arm's control angular velocity can be converted to the arm cylinder operating speed, by which the oil flow rate necessary to operate each arm is calculated.

To extend the service life under adverse job site conditions, and to lower the production cost of the control system, a proportional solenoid valve actuated by an electric signal, given in the form of pulse width modulation[PWM], is employed in the oil flow rate control section.

The arm tip locus control system employs both an open loop and a feedback system. As for the feedback system, this new system employs a potential feedback system well suited to arm tip locus control. As show in Fig. 3, using divergence ΔX in the arm coordinates and velocity control signal(arm lowering and raising) vector $\dot{Y}$, the arm tip locus velocity(arm retracting and extending) correcting vector $\dot{X}$ is obtained. Then, vector $\dot{X}$, perpendicular to $\dot{Y}$, is used as a velocity control signal in the potential feedback system together with $\dot{Y}$. Thereby, steady feedback performance in a wide range of speeds, from slow to high, is possible.

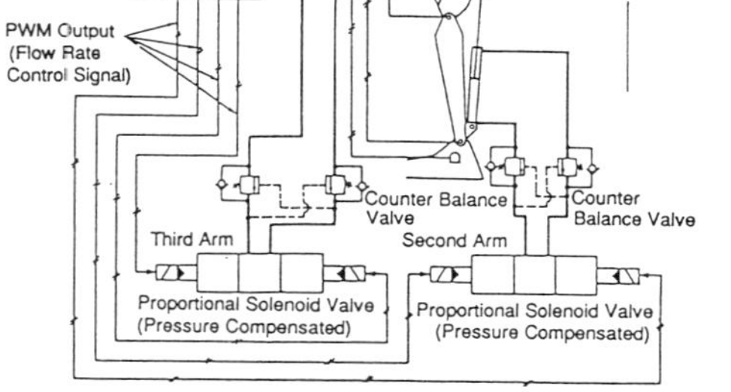

Fig. 2. Layout of The Control System

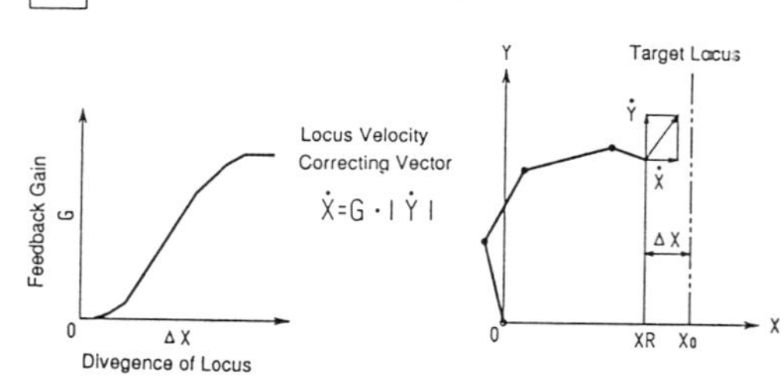

Fig. 3. Arm Tip Locus Control

3. CONTROL LEVER OPERATION SYSTEM

Piling or boring jobs are mostly done on rough, unimproved land. To increase work efficiency under such adverse worksite conditions, ease and speed of positioning the machine and front attachment in the most ready-to-work place is a priority. Next, once the work is started, the machine must offer accurate workability. For these reasons, a vector control lever system was employed (Fig. 4).

This system allows the operator to select either linear movement control,which employs the feedback system, ensuring operative accuracy, or free movement control, which ensures high speed machine control by employing the open loop system. The selection of either system is automatically made. When linear movement control is required, the operator moves the control lever in a single direction either forward or backward (lower/raise) or left or right (extend/retract). However, for combined, free arm movement, the operator can move the control lever at angles between the forward/backward, left/right directions. Therefore, accuracy and speed can both be obtained with this system.

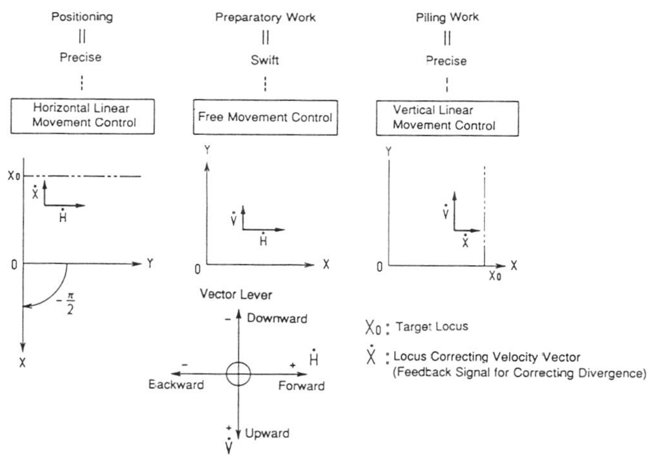

Fig. 4. Operative Pattern of Vector Lever

4. PWM CURRENT FEEDBACK CONTROL

Proportional solenoid valve activation method using PWM electrical signal has come to be widely used because proportional solenoid valves can be directly operated by digital signals from a microcomputer, at a low cost. However, with this control, the coil current is hard to be maintained constantly, as heat generated in the coil causes the coil resistance to vary.

To solve this problem, we have developed a highly accurate PWM current feedback control which is directly controlled by a microcomputer, as shown in Fig. 5.

As shown in Fig. 6, the coil current is stabilized and becomes indifferent to changes of coil temperature. Thus, we were able to greatly enhance the accuracy of the proportional solenoid valve control.

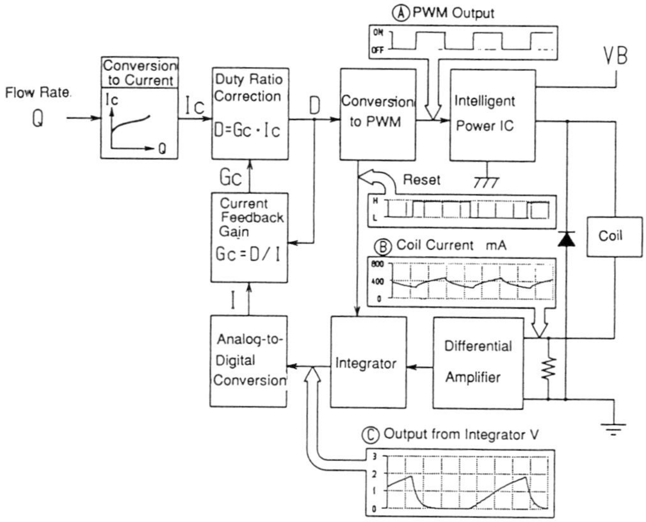

Fig. 5. PWM Current Feedback Control

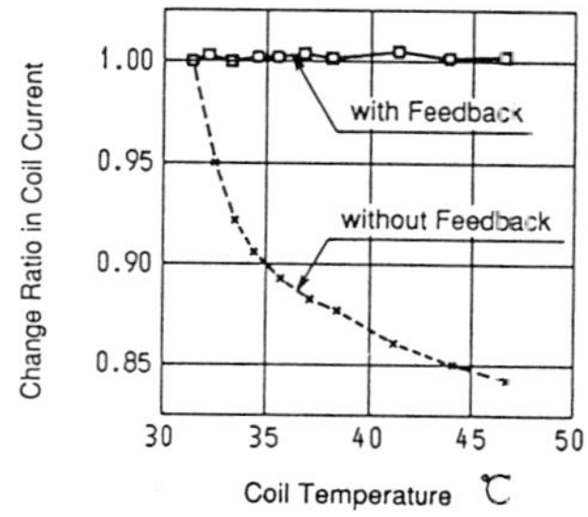

Fig. 6. Effect PWM Current Feedback

5.AUTOMATIC CALIBRATION OF THE PROPORTIONAL SOLENOID VALVE FLOW CHARACTERISTICS

Since the arm tip locus control is achieved by controlling the motion of the two arms relatively to each other, divergence in the flow characteristics of each valve has a great effect on controllability of the arm movement. Especially, difference of the flow start-up point between the two proportional solenoid valves causes a great divergence of the arm movement control, as well as ruining start-up smoothness and steadiness of arm control. Therefore we invented an automatic calibration method which functions to automatically conform the flow start-up point of the valve to that of the standard flow characteristics stored in a microcomputer (Fig.7).

When calibration is conducted, correction current $\Delta I(I_o\text{-}I')$ is acquired and stored in the memory. When the arm tip locus control is operated, the directive current calculated based on the standard flow characteristics is added to the correction current ΔI. This combined current is used as the control signal to activate the proportional valve. Consequently, displacement at flow start-up points in the two proportional solenoid valves is minimized, enhancing operative smoothness.

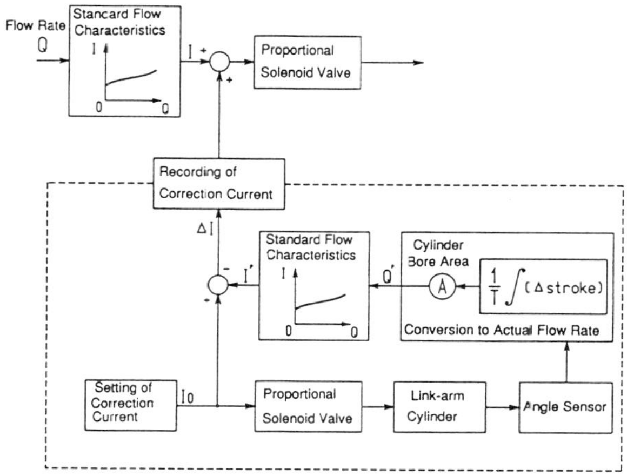

Fig. 7. Self-calibration by Flow Characteristics

6. CONCLUSION

This newly developed system can move the arm tip at the following velocities:

Free movement control: 25 m /minute

Linear movement control: 15 m/minute within ±50 mm errors along the arm tip locus.

Fig. 8 shows one sample of the divergence of the arm tip movement for linear movement control.

Fig. 8. Errors Along The Tip Locus

7. FIELD APPLICATIONS AND PROSPECTS

Photo 1 to 4 show some field applications of the RX 2000.

Photo 1: Railway Shoulder Reinforcement Work

Photo 1 shows H-steel piling work on uneven terrain in a narrow location. The machine is standing by ready to resume the work as soon as the train has cleared the area. In a situation like this, a conventional machine with a leader would take a considerable amount of time to resume the work as its leader assembly would have to be removed from the railway shoulder while waiting for a train to pass by.

The compactness of the machine body and its leaderless front attachment make efficient piling work possible in this narrow, busy-railway traffic location.

Photo. 1. Railway Shoulder Reinforcement Work

Photo 2: River Improvement Work

Photo 2 shows sheet piling work on a riverbed under a railway bridge. Three prominent advantages of the RX 2000 working under these kinds of situations are:

1. The height of the arm tip can readily be changed with this multi-jointed arm.
2. By utilizing the height-limiting auto-stop device, the arm tip does not hit the bridge, in creasing the safety.
3. Because the machine requires no leader, it has high mobility, making getting in and out of the site easy, and not much ground stabilization work is needed.

Photo. 2. River Improvement Work

Photo 3: Foundation Piling Work on Housing Construction in a Mountainous Area

Because the housing construction site was located on a slope in a mountainous area, no foundation work equipment, but the RX 2000, could get into the site. The mobility of the RX 2000 to get into and out of the site almost matches that of an excavator without any preparation work to get into the site necessary.

Photo 4: Foundation Work for Railway Bridge Pier in a Railway Expansion Construction (from Two Tracks to Four Tracks)

As the footing for the machine in this case was a two-lane road, a crane suitable for the work would occupy the whole width of the road, interfering with the traffic. Therefore, the RX 2000 was used because of its compactness, in which the machine fits within one lane of the road. As for accuracy and efficiency of the sheet piling work, RX 2000 matched those of a conventional machine.

In addition, this vibratory hammer has a special center hole chuck that firmly chucks the middle part of a sheet pile or a H-steel pile. Because of this special chuck, pile length is not limited by the base machine's dump height as in the case with a conventional vibratory hammer.

Photo. 3. Foundation Piling Work on Housing Construction in a Mountainous Area

Photo. 4. Foundation Work for Railway Bridge Pier in a Railway Expansion Construction

As shown in the above examples, this machine can perform boring work using a leaderless type earth auger, or highly precise piling work using a vibratory hammer. Because of its ease of applicability, preparatory work, and controllability, the machine is gaining high acclaim from customers.

Lifting loads and working ranges available from multi-jointed arm pile driving machines can by far exceed those offered by general industrial robots.In addition,the construction field in Japan is recently suffering from lower number of workers, aging of the workforce, and an influx of inexperienced workers. Therefore, applications of this type of machine will expand further, as a measure of coping with the need for automation in construction jobsites, a need which robotization of construction machines can help meet.

Automation and Robotics in Construction X
G.H. Watson, R.L. Tucker and J.K. Walters (Editors)

Images of the Future Development of Automated Construction Technology

Kazuhiko Arai

Architectural Institute of Japan, Research Committee on Building Material and Construction Procedures, The Committee on Technology for Robotized Building Production. Kajima Corporation, Architectural Technology Division, Manager,
2-7, 1-chome, Motoakasaka, Minato-ku, Tokyo, Japan 107

Abstract

The purpose of this study is to indicate the future direction of automation technology as applied to the rationalization of building construction. To achieve this purpose, we administered questionnaires to building engineers who had on-site experience with advanced construction technologies. We were thus able to clarify the approach we should take to technological development and other related themes. We also received valuable suggestions on the direction of technological development in the field of construction automation. This research was carried out as part of the study activities of the Architectural Institute of Japan's Research Committee on Building Materials and Construction Procedures, and by the Committee on Technology for Robotized Building Production.

1. INTRODUCTION

In response to the strong interest within Japan concerning rationalization of building production, a variety of automation trials of building-production processes are being carried out at present. Technologies that have already been developed and their applications are discussed in detail in references 1 and 2. The purpose of this study is to explore the future requirements for automation technology as well as to indicate the direction that the development of automation technology should take.

Prior to this study, we administered questionnaires to technicians and engineers who were directly involved with advanced construction technologies. Others who responded to the questionnaires included designers and support personnel in technological control and construction control departments. Based on the results of this research, we clarified the upcoming goals for technological development and automation technology. We also obtained valuable opinions on the future direction of technological development as related to construction automation (hereafter referred to as “CA”).

2. RESEARCH ITEMS AND METHODS

2.1. Research items

Our research focused on the following four items: (a) basic attitudes toward the development of CA; (b) purposes of the development of CA and the automation of construction work; (c) technological themes proposed for future study; and (d) reasons why the proposed technological themes should be studied.

This research was conducted in September and October 1991.

2.2. Research subjects

The main subjects were construction technicians and engineers employed by major contractors in Japan and directly involved with advanced construction technologies. We also included support personnel, as shown in Table 1. Altogether, we received 262 responses, approximately 78% of which were from construction supervisors.

Table 1
Research subjects and number of responses

Correspondent's assignment	Number of replies	Response rate (%)
Field managers	173	66.0
On-site supervisors	31	11.8
Technicians in management departments	32	12.2
Technicians in CA promotion departments	3	1.1
Technicians in R&D departments	19	7.3
Technicians in design and design control departments	3	1.1
Other technicians	1	0.4
Totals	262	100

3. RESEARCH RESULTS AND ANALYSIS

3.1. Attitudes toward the development of CA

The questionnaires addressed the topics, concerning attitudes toward the development of CA and the automation of construction work. About 70% of respondents agreeing with statement (a) thought that the development of CA and the automation of construction should be carried out more actively. When we included those agreeing with statement (b), which says that developments that are clearly advantageous should be the first to be actualized, we found that about 99% of respondents have a positive attitude toward the development of CA and the automation of construction work. These results suggest that active technological development will be necessary to achieve CA and the automation of construction work.

3.2. Purposes of developing CA

Here, we obtained responses to the seven items shown in Figure 1, relating to the goals of the development of CA. We permitted up to three answers for this question, and received a total of 772 answers.

These results show the top-ranking answer was "labor saving and energy saving," which constituted 30% of all responses, followed by "elimination of dangerous jobs" (22%), "shortening of construction terms and work hours" (16%), and "improvement of work environment" (15%). These results are essentially the same as those obtained for technologies that are applicable at present, as described in document 2, except that a slightly higher percentage mention "the elimination of dangerous jobs." On the whole, however, current and future aims do not vary significantly.

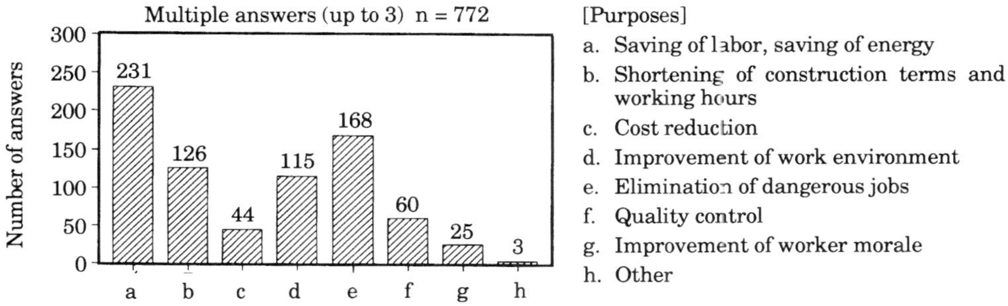

Figure 1. Purpose of development of CA

3.3. Proposals on technology to be developed in the future

Here, we asked those respondents who made proposals on automation technology to describe which themes should be focused on. We classified the proposed technologies according to our automation technologies classification code table (see Table 2). We also asked respondents to explain why the proposed technology will be needed. Respondents were allowed to propose various technologies in up to three different fields.

We received a total of 258 proposals on technological themes to be worked on in the future. Of these, 145 were "single proposals" involving only one of the automation technology classification codes (as shown in Table 2), 53 were "double proposals," covering two classification codes, and 60 were "triple proposals," covering three classification codes. Even if the suggested technological development was to be carried out in a single field, we found its relation to technology in other fields was also considered important.

3.3.1. Total number of technologies proposed according to automation technology classification codes

The proposed technologies were counted according to their automation technology classification codes. A single proposal counted as one case, double proposals counted as two cases, and triple proposals counted as three cases. These results were then totalled as shown in Table 3 . In Table 3, the proposed technologies are indicated in relation to the classification codes shown in Table 2. Altogether, 431 technologies were proposed. A rough classification of the responses shows that proposals regarding construction work account for 46% of the answers, making it the largest group. The next largest group of proposals, 24%, was related to preparation for construction work, followed by proposals regarding production design (15%) and control systems (14%).

In the same way, when the five top-ranked technologies proposed were considered according to their classifications, 24% of all answers referred to proposals for building frameworks in the Construction Work category, followed by proposals for finishing work (15%) and shop drawing (10%) in the same category, construction system planning the the area of Production Design (5%), and scheduling in the Construction Control category.

Based on these results, we found that many respondents considered that the automation of construction work (category D in Table 2) constituted the core of CA. We also found that many respondents had a strong interest in the automation of the "upstream" portion of the production process, which belongs to the Construction Planning (category B in Table 2).

Table 2
Automation technology classification code chart

A. Production design →	B. Construction planning →	C. Construction control →	D. Construction work →	E. Maintenance
a. Estimation	a. Project budget	a. Cost control	a. Building frame construction	a. Maintenance
b. Construction planning	b. Scheduling	b. Scheduling control	b. Finishing work	b. Inspection
c. Construction method planning	c. Construction method planning	c. Safety control	c. Electrical and mechanical installation work	c. Repair work
d. Structural planning	d. Temporary work plannning	d. Quality control	d. Others	d. Demolition work
e. Others	e. Shop drawing	e. Others		e. Others
	f. Others			

Table 3
Relationships between proposed technologies and classification codes (total number of answers)

	A: Production design	B: Construction planning	C: Construction Control	D: Construction work	E: Maintenance	Totals
a	19	7	11	103	2	
b	5	8	21	65	2	
c	15	16	9	19	0	
d	23	20	9	11	2	
e	2	42	11	—	0	
f	—	9	—	—	—	
Subtotal	64	102	61	198	6	431

3.3.2. Analysis of proposed single-coded technologies

Here, we analyzed each theme of the proposed technologies, assigning each a single-technology classification code. Altogether, 145 single technologies were proposed, as shown in Table 4, which includes the subtotals for each classification code. The largest number of proposals (94 cases) concerned Construction Work (D), followed by Construction Planning (B) (23 cases), then Construction Control (C) (18 cases). There were relatively few proposals regarding Production Design (A) or Maintenance (E).

The contents of the most frequently proposed technologies are as follows.

- Shop drawing in Construction Work (B-e): CAD technology for the preparation of shop drawings, and technology for the standardization of work planning and shop drawing

Table 4
Number of single proposals for each classification code (number of cases)

	A: Production design	B: Construction planning	C: Construction Control	D: Construction work	E: Maintenance	Totals
a	2	1	3	52	1	
b	0	2	6	31	1	
c	1	2	0	4	0	
d	3	2	1	7	2	
e	0	11	8	—	0	
f	—	5	—	—	—	
Subtotal	6	23	18	94	4	145

- Building frame construction, in Construction Work (D-a): Development of building systems and work methods suitable for automation, the integration of industrialized methods, the automation of all aspects of the building frame construction, development of all-weather roofs, and the automation of individual jobs.
- Finishing work, in Construction Work (D-b): Automation of the installation of autoclaved, lightweight concrete panels and precast concrete panels, and automation related to various types of finishing work

3.3.3. Analysis of proposed double-coded technologies

Figure 2 shows the distribution of technological proposals extending over two classification codes. This diagram shows that the greatest number of proposals (10) were related to building frame construction (D-a) and finishing work (D-b) in Construction Work. Next were the six proposals related to both temporary work planning (B-d) and shop drawing (B-e) in Construction Planning; and the six related to both estimation (A-a) in Production Design and shop drawing (B-e) in Construction Planning. As mentioned above, we found that Construction Work proposals separated into two types. One consisted of proposals that integrated past work with subsequent work within the same technological classification, and the other proposed the integration of two forms of work belonging to different technological classifications. The contents of the technologies mentioned most frequently are summarized in the following:

- Proposals concerning building frame construction (D-a) and finishing work in Construction Work: Automation of work related to both building construction work and finishing work
- Proposals concerning temporary work planning (B-d) and shop drawing (B-e) in Construction Work: Proposals suggesting comprehensive handling of two types of business functions—planning and shop drawing preparation—in a comprehensive shop drawing preparation system
- Proposals concerning estimation (A-a) in Production Design, and shop drawing (B-e) in Construction Planning: Proposals for the unification of business functions (e.g., a CAD system for the continuous operation of design and construction work) in a comprehensive shop drawing system (including design)

3.3.4. Analysis of proposed triple-coded technologies

There were 60 proposals concerning technologies that encompassed three classification codes. We concentrated mainly on analyzing relationships between technological

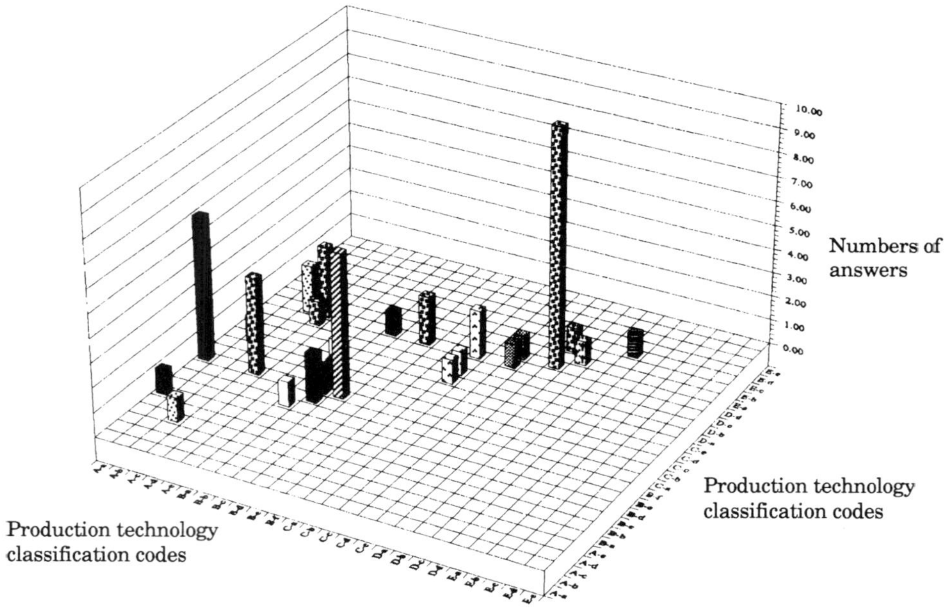

Figure 2. Numbers of proposed technologies covering two production technology classification codes

fields. When we began this analysis, we examined the relationships between different technological areas roughly based on the classification shown in Table 2. Figure 3 summarizes the study results for the relationships between Construction Work and each work subcategory.

When we observed the relationships between the work in each subcategory of D and other work in Figure 3, we found that D-a: building frame construction; D-b: finishing work; and C-c: electrical and mechanical installation work were the most closely related. D-a: building frame construction was closely related to A-d: building system planning, C-a: cost control, and C-c: safety control. In Category D, the actual names of the proposed technologies related to classification items a, b, and c were as follows: fully automatic building-construction systems, robotic technology, measuring systems, and material transportation technology.

As mentioned above, the main trend in technological development has been a concentration on individual technologies. We found, however, that there is currently great demand for systematic technological development both within and across various technological fields, as outlined above.

4. PURPOSES OF DEVELOPING PROPOSED TECHNOLOGICAL THEMES

Here, we asked the respondents to cite one or more of the reasons shown in Table 5 to explain why each of the proposed technologies would require future development. We

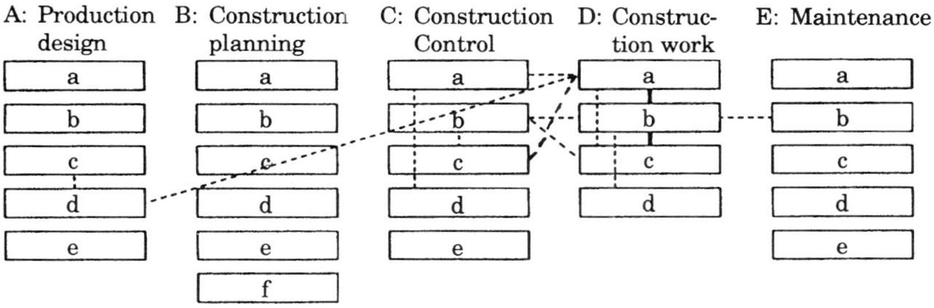

Figure 3. Relationships among triple proposal technologies (classified on the basis of Construction Work-D)

classified the reasons into four categories: "general proposals," "single proposals," "double proposals," and "triple proposals," in the same way that the proposed technologies themselves were classified. We then totalled the answers for each response.

As shown in Table 6, 42.6% of the answers cited "a", saying that the proposed technology was needed as a basic technology; 18.6% cited "b", that the proposed technology was needed in order to integrate existing technologies; and 36.8% chose "c".

The following are the characteristics of the technologies proposed for each reason:

i. The answers citing reason "a", that the proposed technology is needed as a basic technology, mainly concentrated on shop drawing in the category of Construction Planning, as well as on building frame construction, finishing work, and electrical and mechanical installation work in the category of Construction Work.

ii. The answers citing reason "b", that the proposed technology will be needed to integrate existing technologies, centered on estimation and building system planning in the category of Production Design, construction planning and shop drawing in the category of Construction Planning, and cost control in the field of Construction Control.

iii. Answers citing reason "c", that the proposed technology will be needed to expand the applicable range of existing technologies, placed priority on construction system planning and building system planning in Production Design; shop drawing in Construction Planning; and building frame construction, finishing work, and electrical and mechanical installation work in Construction Work.

Table 5
Proposed technologies and reasons why they are required

Reason why technology is required	a	b	c	Unknown	Totals
Single proposals	71	21	50	3	145
Double proposals	17	9	27	0	53
Triple proposals	22	18	18	2	60
Totals	110	48	95	5	258

Notes: a: The proposed technology is needed as a basic technology.
b: The proposed technology is needed to integrate existing technologies (e.g., by providing access to information).
c: The proposed technology is needed to expand the applications of existing technologies.

5. CONCLUSIONS

The following results were obtained from our survey:

a. Consciousness of automation technology is very high among those who work in the field of building construction.
b. Many proposals for technologies related to the construction of building frames in the field of Construction Work were similar to those already in use, and we realize that the development of technology in this area will be in demand both now and in the more distant future.
c. There were many proposals on systematic technologies and wide-ranging technologies that were similar to those already in practice. As mentioned above, a wide range of technological development leading to the use of CA systems has been proposed, and we have obtained valuable suggestions on the future course that development should take. We need to evaluate the automation technology related to CA on the basis of these research results, and we must create an appropriate system in order to develop future CA systems.

6. ACKNOWLEDGEMENTS

We wish to express our sincere gratitude to all persons associated with the eight general contractors that we surveyed as part of our research.

7. REFERENCES

1 The Committee on Technology for Robotized Building Production, AIJ; Research Study of the Building Construction Technologies Aimed at Automated Construction, The Committee on Technology for Robotized Building Production, AIJ, March 1992 (Japanese).
2 Y. Hirabayashi; Analysis of the Development and Application of Automation Technology in Building Production, Proceeding of the 10th ISARC, May 1993.

List of Committee Members

The Committee on Technology for Robotized Building Production, AIJ

[Chairman]
Yasushi Tamura (Waseda University)

[Secretary]
Kazuhiko Arai (Kajima Corporation)
Nobuyasu Miura (Kokushikan University)

[Members]
Norimoto Kashino (Building Research Institute, Ministry of Construction)
Junichiro Maeda (Shimizu Corporation)
Eiji Muro (Takenaka Corporation)
Hiroshi Naito (Japan Association of Representative General Contractors)
Shigeru Sakamoto (Taisei Corporation)
Kunio Shiina (Tokai University)
Toyokazu Shiire (Kanagawa University)
Tooru Shinozaki (Toda Corporation)
Tatsuo Terai (Chiba Inst. of Technology)
Masashi Wada (Kumagai Gumi Co, Ltd.)
Tatsuya Wakisaka (Obayashi Corporation)
Ryoji Yoshitake (Fujita Corporation)
Yasuo Inoue (Obayashi Corporation)
Katumi Kobayashi (Kumagai Gumi Co., Ltd.)
Toshifumi Sato (Kajima Corporation)
Hiroshi Fujita (Nihon Bisoh Co., Ltd.)
Muneyoshi Matuo (Fujita Corporation)
Takashi Miyazaki (Takenaka Corporation)
Takashi Shiokawa (Obayashi Corporation)
Osamu Nagayasu (Fujita Corporation)
Yukio Haga (Taisei Corporation)
Yuji Hirabayashi (Shimizu Corporation)
Masahiro Morita (Takenaka Corporation)

Automation and Robotics in Construction X
G.H. Watson, R.L. Tucker and J.K. Walters (Editors)

The Study on The Development of Building Construction Robots in Japan

Fumio Shintani and Masaki Yamamoto

Senior Consultant, Technology Department, Integrated Research Division,
The Japan Research Institute, Limited

Kioi-cho Building, 3-12, Kioi-cho, Chiyoda-ku, Tokyo, 102, Japan

Abstract

At The Ninth ISARC held in Tokyo, we introduced thirteen construction robots for the building construction being developed by "Construction Robots Consortium in Japan". This is the second report to introduce the progress of the development of each robots and the possible program for the robots to be used at building construction sites.

1. ROBOTS UNDER DEVELOPMENT (Figure.1)

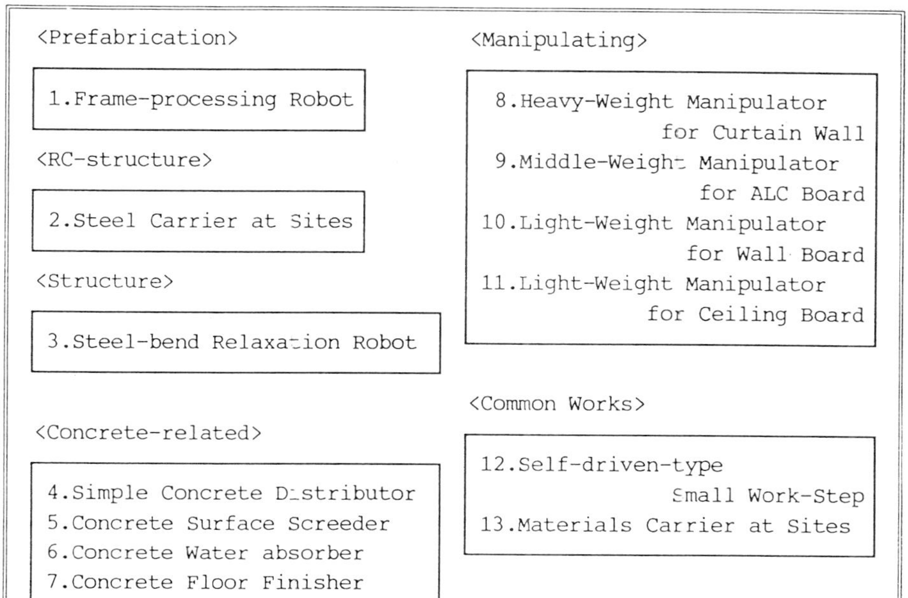

Figure.1 Robots Under Development

2. THE WAY OF DEVELOPMENT

We have taken the unique method for collaborative development. So far, development of construction robots has been conducted by each general contractors independently and this method resulted in the delay of development mainly because, in most cases, they could not create the market of construction robots for the robot manufacturers. This matter is very important because if there is no supplyer there is no possibility for construction robots to exist.

Thus, we organized ten groups for development of construction robots and each group consists of one manufacturer and twenty-nine general contractors. What is good with this method is that we can eliminate the market limitation caused by the independent development of one general contractor. We named group for development as a "Development Unit (DU)", and substantial development has been conducted by each Development Unit shown in Figure.2.

The members of each Development Unit agreed with the basic policy as below;
(1) Manufacturer is responsible for the development cost.
(2) General contractors are responsible for setting up the site for test.
(3) Intellectual property is basically owned by manufacturers.

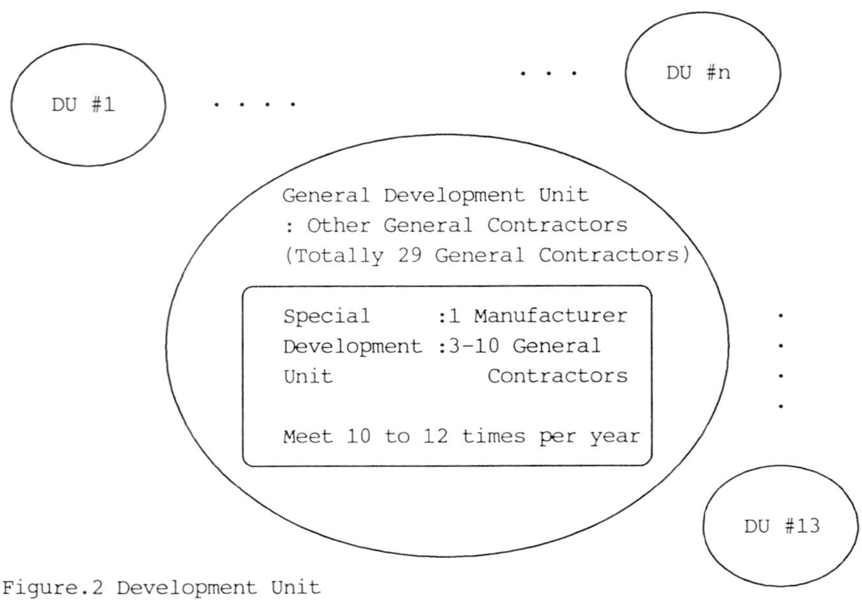

Figure.2 Development Unit

3. PROGRESS OF THE DEVELOPMENT

Consrtium project is now in third phase to introduce the proto-type robots to actual construction site. In first phase, we studied what kinds of robots are needed at construction sites and in second phase, we organized ten Development Units and started designing the specification of thirteen construction robots.

Because of the characteristics of each construction robots, progress of the development of each robots are in various stages as shown in Figure.3. However, most of robots are scheduled to be demonstrated by the end of this year because demonstration in front of users of construction robots is the one big target of Construction Robots Consortium.

One exception of the schedule mentioned above is "Frame-processing Robot". This is the prefabrication system which provides wood frames to mold concrete structure such as column and floor of (S)RC buildings instead of cutting wood plate at site. Actually this system is ready to be produced, however, it may cost two million dollars so that it is necessary to arrange the governmental support or simillar support to introduce this system into sub-contractors of carpenter work. Getting support is now under processing in the collaboration with governmental agency who is responsible for incubation of small companies.

	1993 / 1 2 3 4 5 6 7 8 9 10 11 12
1.Frame-processing Robot	<diffusion phase> ···········
2.Steel Carrier at Sites	····T ···M ···D ··················
3.Steel-bend Relaxation Robot	·······M ···D ····················
4.Simple Concrete Distributor	·P ···T ···M ···D ················
5.Concrete Surface Screeder	··········P ······T ···M ···D ······
6.Concrete Water absorber	·······T
7.Concrete Floor Finisher	·T ···M ···D ·····················
8.Heavy-Weight Manipulator	··········P ···T ···D ··············
9.Middle-Weight Manipulator	·····················P ···T ···M ···D
10.Light-Weight Manipulator(Wall)	(Schedule under planning)
11.Light-Weight Manipulator(Ceiling)	·············P ···T ···M ···D ······
12.Self-driven-type Small Work-Step	·······P ···T ···M ···D ············
13.Materials Carrier at Sites	··········P ···T ···M ···D ·········

P:Finish creating proto-type T:Finish test at site
M:Finish modifying robots D:Demonstration

Figure.3 Progress of Development

4. PROGRAM FOR DIFFUSION OF CONSTRUCTION ROBOTS

After the development it is important to introduce construction robots to the actual construction site. To do this, we have to solve the problems regarding three matters described below;

(1) Cost accumulation of building construction is based on the rate of human workers so that there is no measure for the work by robots.
(2) Most sub-contractors, who are the real users of construction robots, can not afford to buy construction robots because of financial reason.
(3) Recently, few general contractors own machines (includes construction robots) but use the lease/rental system when they need machines.

Thus we have two programs to introduce construction robots into actual building construction sites. First program is to make the standard of cost accumulation when using robots. We have started exploratory meeting with Ministry of Construction for possible system which enhance the automization. Second program is to make good examples of using construction robots at actual sites (Figure.4).

5. CONCLUSION

As I reported, development of constrcution robots by the consotium is now in final effort. Result will be reported next year.

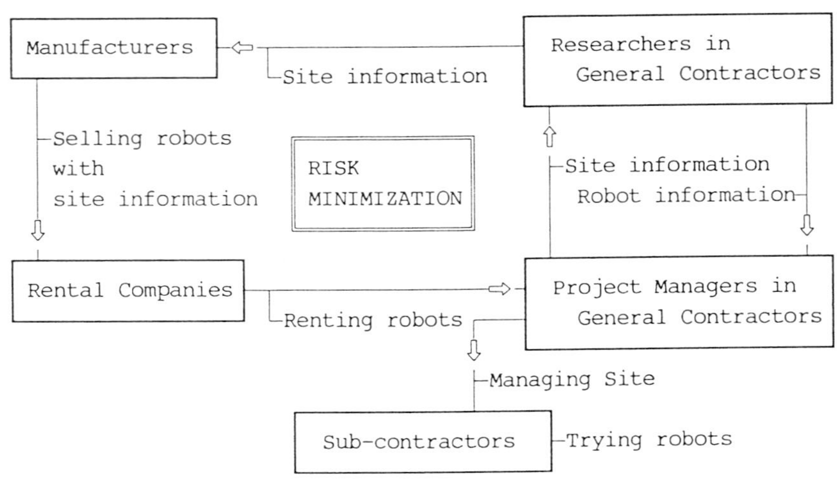

Figure.4 Program for Making Good Examples

Automation and Robotics in Construction X
G.H. Watson, R.L. Tucker and J.K. Walters (Editors)

FEASIBILITY ANALYSIS OF ROBOTIZED VS MANUAL PERFORMANCE OF INTERIOR FINISHING TASKS

Abraham Warszawski and Yechiel Rosenfeld
Technion - Israel Institute of Technology and National Building Research Institute

Abstract
The paper presents results of a feasibility analysis of a robot developed for performance of nterior finishing tasks. Productivity and cost of robotized construction is compared with manual onstruction of the same tasks.

INTRODUCTION

The TAMIR - a construction robot shown in Fig.1, has been developed at the Technion, Israel nstitute of Technology following a multistage process, described in [Warszawski, Navon 1991]. ts pre-prototype version adapted to performance of full scale construction work is based on he S-700 model of an industrial robot made by GMF. The arm of the robot has 6 degrees of reedom, a nominal reach of 1.62 m and a payload of 30kgf.

The robot travels between static work stations and executes at each station the work assigned to . For this purpose it was mounted on a three wheel mobile carriage measuring 0.85 by 0.85 m nd had effectors, sensors , feeding and control systems adapted to performance of various uilding tasks. At present the robot can be guided by a remote control box or follow a preprogrammed ath. At a later stage it will be equipped with sensory devices and a control mechanism which ill enable it to interact with environment, map it, and even plan and monitor its own work Warszawski, Rosenfeld, Shohet 1992].

gure 1 - The interior finishing robot

e robot is designated to perform eventually any type of interior finishing work in residential, mmercial and similar buildings with single or multiple floor levels and an interior height of 0 - 2.70m . It has been adapted so far, and tested, to execute the following tasks:

a. Painting walls and ceilings.
b. Plastering walls and ceilings.
c. Tiling on walls.
d. Building walls and partitions.

The adaptation of the robot to performance of these tasks has been described in [Rosenfeld Warszawski, Zajicek 1991].

Preliminary results of the economic performance of the robot were presented in [Rosenfeld Warszwski 1992]. This paper brings revised and updated results, in a general context of robotized construction .

THE COST OF ROBOTIZED WORK

The feasibility of the robot must be examined by comparing the cost of its work with the cost of performance of similar tasks with conventional manual method. The cost per unit of robotized work includes all pertinent components which are affected by the employment of the robot technology namely:

a. the direct cost of the robotized system - the robot, its operator and other auxiliary labor.

b. the differential cost of materials , whenever materials used in robotized construction are more expensive than those used in conventional work for the same task.

c. the cost of the robot's movement between work stations. This cost depends upon the length of robot movement per work unit , its speed of movement , and the cost of robotized system.

d. the cost of the robot's positioning at work station. This cost depends upon the time of setup per work unit, and the cost of robotized system per time unit.

e. the cost of the robot's transfers between different work areas on site (on the same building or in different buildings). This cost depends on number of transfers per work unit , the time required per transfer , the cost of the robotized system , and the cost of other resources involved in the transfer.

f. the cost of installation of the robot on the building site. This cost depends on the transportation time to site, the setup time , the system cost , and the cost of other resources involved in the transfer.

The major parameters determining the cost of the robotized work can be therefore divided into three main groups:

a. parameters dependent on the robotic system - its cost, its work envelope (and hence the work area it can cover from a work station) , its speed of movement and mode of operation (extent of human control).

b. parameters dependent on the nature of the building site. These include the nature of the tasks to be performed, their quantity, the number of transfers necessary between work sections, and the location/distribution of work in each section.

c. parameters dependent on tasks to be performed by the robot - the output per hour for each task, the materials, and the auxiliary works needed.

PARAMETERS DEPENDING ON THE ROBOTIZED SYSTEM

The most important parameters of the robotic system are its cost and output.

The cost of the robotic system depends on the nature of the robot to be used , and the mode of its operation. It was assumed that at the present stage of development the robot will need a continuous supervision by a human operator . Consequently, the cost of a robotic system includes at present the cost of the robot and the (full or partial) wages of the operator.

The cost of the robot includes depreciation, interest on investment, maintenance, and operating expenses. In the particular case examined here the initial cost of the robot GMF S-700 was $100,000, and together with additional features - carriage, effectors, sensors, and adaptations - $130,000. In order to calculate the annual cost a rate of interest of 7% , and an economic life of the robot - 5 years without salvage value - were assumed. The cost of repairs and high level maintenance was assumed as 10% of the initial investment per year, and its other operating cost (including replacement of some wear affected parts) as $2 per hour. A routine - low level maintenance has been also taken into account (in the context of site activities).

he total cost per hour depends very much on the number of hours the robot will be employed er year. Since the robot will in fact replace skilled human labor which is continuously needed n building , under satisfactory environmental conditions (which can and should be provided on he building site) 1,500 - 2,000 hours of employment per year for a multipurpose robot can be easonably provided.

Assuming $1,500 hours of employment per year, initial investment of $130,000 (based on the ost of the developed robot), 5 years of economic life, 7% interest and maintenance expenses , e cost of robot per hour amounts to $ 32. For 2,000 hours per year it amounts to $24 per hour . ssuming that an operator will continuously supervise the robot's work, his wages , examined t two levels - of $12.50 and $25 per hour, must be added to obtain the total cost of the system er hour.

he output of the robot for each task depends on the nature of the task, and the amount of time hich the robot has to spend on indirect activities, which will be discussed later.

ASK DEPENDENT PARAMETERS

ask dependent parameters determine the direct time input of the robotized system for a articular task execution , the manual input of the human auxiliary activities necessary for the sk performance, the cost of materials, and other resources. The performance of selected tasks y the robot was described in [Rosenfeld, Warszawski, Zajicek 1991].

ctivities auxiliary to robotic system include the support which the system has to receive with uman labor for performance of the designated tasks. These activities are preparations before e work execution - hoisting of materials or adjustment of the work area, and the assistance to e system by human labor, during or after the work execution.

ARAMETERS DEPENDING ON THE BUILDING SITE

rameters dependent on the building site are the quantity of building work on the site and its stribution. They determine the time which the robot has to spend on activities not directly lated to its work, and therefore the efficiency of its utilization.

e indirect time as explained earlier includes the time of installation, movement, transfers, sitioning at the work stations, and routine maintenance. The higher is the ratio between this ne and the actual task execution time, the less efficient the robotic system becomes, when mpared to manual labor.

e installation time includes the transfer of the robot to the site, and its preparation for work. It assumed that this should take somewhere between 4 and 8 hours, depending on the transfer tance and the efficiency of the user and operator organizations. The share of this cost in the al depends on the total amount of work on the site. The other cost - the truck, the loading and unloading, were not considered here, since they are incurred also in transfer of various types o equipment needed for manual work.

The time of positioning at work station depends on the characteristics of the system, and its contribution to total indirect time - depends on the amount of work to be done from the work station.

The positioning time includes self support of the robot, its self levelling, and the search for the task starting point. The measured time for these activities with the particular robot employed was estimated, based on the performance of the experimental robot, as 3 minutes per station. The amount of work per station depends on the wall (or ceiling) area covered by the robot from the work station.

The effective work envelope of the robot's arm sufficed to cover - for painting, plastering tiling, building etc., from a minimal distance - of 0.60 m (considering the width of the carriage) an area of 2.70m height and a length of 1.70m (4.60 sqm), or an area of 2.60m height and length of 1.90m (4.95 sqm) . The effective area covered will be generally less than the maximum considering the layout of the wall and the openings . The full work area from a station will be utilized in a very well structured spaces like for example long, straight walls. In general it ma be assumed that the work covered from a station will be somewhere between 0.75 and 0.90 c the maximum . In special cases , the work area from a station can be more than the maximum stated before. One such case may occur if the robot can perform the work from a station on tw opposite surfaces - as in a corridor. Another, when the robot performs a task on the wall and th ceiling above it, as may be in painting or plastering.

The time of movement between work stations depends on the speed of robot movement and th distance between stations. The speed of the present robot - 4 m/minute, has been determined b the capacity of the carriage motors. The speed can be easily increased to 10 m/minute. Th distance between work stations is determined by the length of area covered from a station. In case of continuous work - this will amount, as explained earlier, to 0.75 - 0.90 of the maximum width (in this case 1.9 m). However, in cases where the work to be done is scattered betwee separate locations , the distance to be traveled between these locations has to be added.

The transfer of the robot between adjacent floors or other locations on the building site shou not exceed 20-30 minutes of the robot time. This includes the travelling of the robot to th transfer point from the last, and to the first, work area on the floor. The time of the crane or ext labor needed for transfer is negligible.

The routine maintenance time is probably somewhat different for each particular task. It h been arbitrarily determined here as 0.5 hour per working day of 8 hours, or approximately as 6 of the total working time.

The "burden" of the indirect activities i.e. the ratio between the duration of the indirect and t direct activities for a particular building task on site depends therefore on the following paramete the total amount of work on site (for the task), the amount of work on each floor, and the amou of work per work station, when related to the direct robot input in the task.

THE ECONOMIC COMPARISON

The feasibility of robot employment for each particular task depends on the building s parameters. The results presented here pertain to a typical office building of eight stories with story area of 1,000 sqm, and an interior height of 2.6 m. Such building may typically inclu (see for exmple Means 1990) approximately 600 sqm of partitions and 1,500 sqm of wall co per floor.

The "burden" of robot movement, with its associated unproductive cost of the robot and operator will depend on the extent of the robot utilization at a work station. It will be assum here that the configuration of the work to be done allows the robot utilization of 75% of maximum reach from an average work station. It is presented for each of the examined tasks in figure 2.

The feasibility of robotic vs manual work in view of these site parameters has been examined for each task. by comparing it to a conventional, manual execution of the same work.
The direct labor of the robotic system for each task has been determined from experiments with the robot . The input of auxiliary labor has been determined from the analysis of each task. The labor necessary for performance of similar task with a conventional, manual, method was determined from accepted productivity rates (Means 1990, Israeli Institute of Productivity 1980).

The indirect labor input - for movement, positioning, transfer, and maintenance has been determined as explained before in the section dealing with parameters of the building site.

The comparisons between total labor input in robotized vs. conventional execution of the various tasks are presented in figure 3 .

The cost of the robotized system - of the robot and the operator was determined considering the direct and the indirect time input and the cost of the system per hour as was explained before. The full cost analysis has been performed in [Warszawski, Rosenfeld, Zajicek 1993] for different wage rates and different annual utilization times (in terms of hours per year). The sensitivity analysis with respect to various cost parameters was also performed.

The result of cost comparison between robotized and manual construction for two wage levels one low of \$12.50 per hour for skilled labor (including operator's wages) and \$9.40 per hour for unskilled labor per year and the other high - of \$25 for skilled labor and \$18.80 for unskilled labor are shown in figure 4. The cost comparison assumes 1,500 work hours of the robot per year, and only those materials which differ between the two construction methods.

FINDINGS AND CONCLUSIONS

The following general conclusions can be drawn from the economic analysis of performance of the multipurpose interior finishing robot:

1. Most of the interior finishing tasks in building can be effectively executed by a mobile robot of jointed configuration with six degrees of freedom and 1.60 - 2.00 m reach. The robot can be adapted to each task by appropriate adjustment of the end effector and of control program.

2. Although it is possible to program the robot to move between its work stations, identify its tasks and execute them almost autonomously it will be more advantageous under the present conditions to have it acting under a continuous supervision of a human operator. The operator will guide it to work stations, identify the work initiation points and deal with various problems which may occur during the operation. Although the operator may be employed during his idle time to perform various auxiliary tasks needed to complement the robot's work this seems undesirable considering his qualifications and the continuous attention required by the robot . The cost of the operator may be reduced by assigning, in the more time consuming tasks such as building or tiling, two robots to one operator.

3. The efficiency of the robot employment will affected to a great extent by the burden of indirect activities associated with its work. These activities include transfer to project site, transfers between various locations on site, traveling to work stations, positioning at the work station, and routine maintenance. The amount of time spent on these activities may exceed in certain tasks the time spent on direct work in station. The robot will be used most efficiently on sites with ample amount of work at each floor level (4-8 hours of employment on floor seem reasonable in this respect), and work distributed in a manner which allows for maximum utilization of the robot capacity at each work station. A utilization of less than 60% - 75% of the capacity , especially in the less time consuming tasks- painting and plastering, may adversely affect the economy of robot employment.

4. The cost of the robotized work will be much affected by the utilization of the robot in terms of its hours of employment per year. Since the robot will in fact replace skilled human labor which is continuously needed in building , under satisfactory environmental conditions (which can and should be provided on the building site) 1,500 - 2,000 hours of employment per year for a multipurpose robot can be reasonably provided.

5. The robot can bring considerable saving in the manual labor on site. The saving in skilled labor ranges between 60% and 90% for typical tasks. Even the remaining skilled labor will be employed in supervisory and technical rather than manual tasks.

6. The costwise feasibility of robotized vs. manual work depends upon the prevailing wages of labor. For wages of $25 per hour 1,500 - 2,000 robot employment hours per year, suitable site and proper organization of materials packaging and handling, savings of 20% - 50% in the cost of work , or roughly 10 - 15% in the total cost including material, can be realized in several examined tasks. The savings will be considerably smaller, and in some tasks nonexistent when considering lower rate - of $12.5 per hour for skilled labor.

7. Some special requirements of robotized performance are difficult to assess in economic term but must nonetheless be taken into account in the comparison. One of them has to do with much stringer precision and quality control than the one which is customary today in traditional construction. Another has to do with precise packaging and positioning of building material which have to be used by the robot. A third has to do with a superior planning and organization of work on site which will allow the robot to move unobstructed to its work stations, and operate there efficiently. Any permissiveness with respect to these requirements will most probably make the robotized system non-operable.

Summing up, it may be concluded from the study that employment of a robot for interior finishing works has considerable potential for productivity improvement on the building site. It appears that economic savings can be also realized from robots employment. Other non quantifiable benefits which can be also obtained include increased safety, reduction of strenuous and unpleasant tasks, and better quality of building. Realization of these benefits depends however on very high precision of the building shell and a very high level of materials packaging and handling, and work organization on site.

BIBLIOGRAPHY

Israeli Institute for Productivity of Labor "Catalog of Labor Input in Building", Tel Aviv, 1980

Mean's Estimating Handbook, R.S. Means Co, Kingston, MA, 1990

Rosenfeld, Y., Warszawski, A., and Zaijcek, U.," Interior Finishing Building Robot",The 8th International Symposium of Automation and Robotics in Construction, Stuttgart 1991

Rosenfeld, Y., Warszawski, A., and Zajicek, U., "Economic Evaluation of Robotized Interior Finishing Works based On Full Scale Experiments" The 9th International Symposium of Automation and Robotics in Construction, Tokyo 1992

Warszawski, A., "Industrialization and Robotics in Building", Harper and Collins, 1990

Warszawski, A., and Navon, R., "Development of A Robot for Interior Finishing Works", Journal of Construction Engineering and Management , Vol 107(2) , 1991

Warszawski, A., Rosenfeld, Y., and Zajicek, U., "Feasibility Analysis of Robotized Performance of Interior Finishing Works", National Building Research Institute, Haifa, 1993

Warszawski, A., Rosenfeld, Y., and Shohet, I., "Autonomous Control System in Interior Finishing Robot" The 9th International Symposium of Automation and Robotics in Construction, Tokyo 1992

APPENDIX (FIGURES)

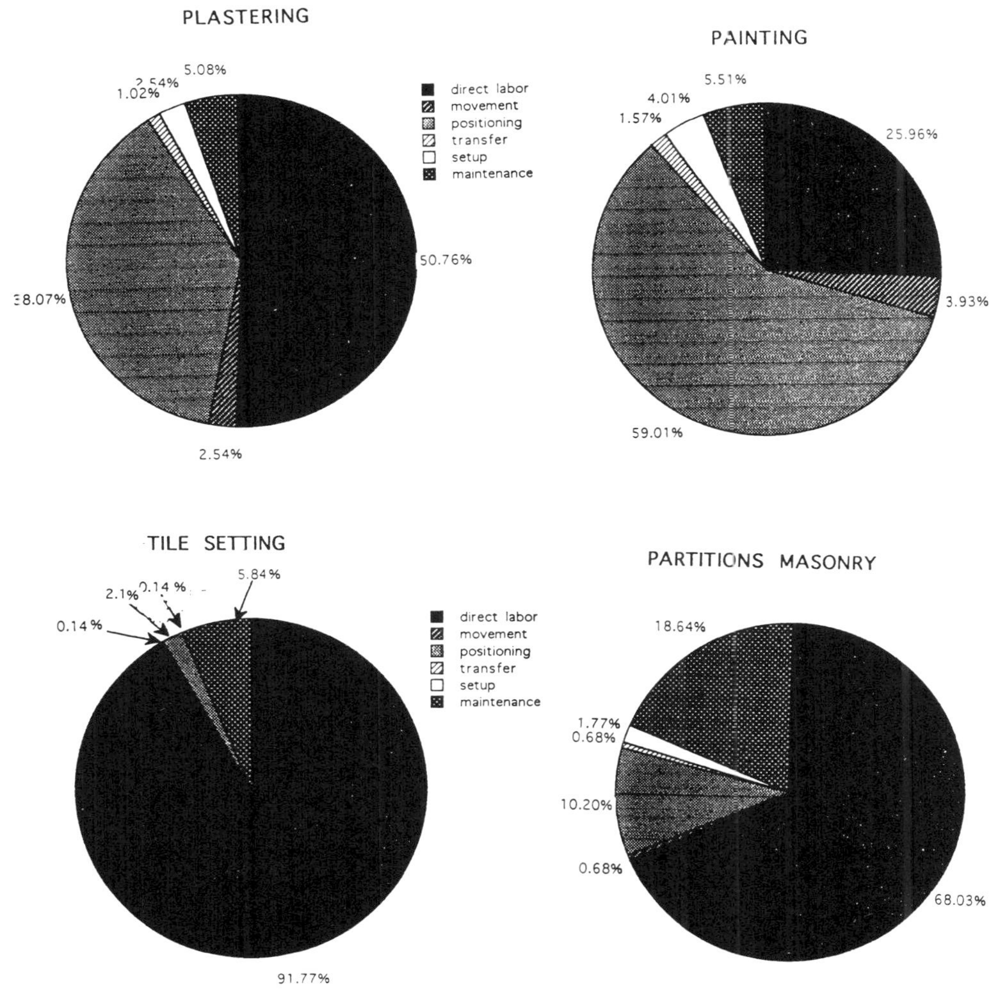

Figure 2 - Breakdown of work time of the interior finishing robot

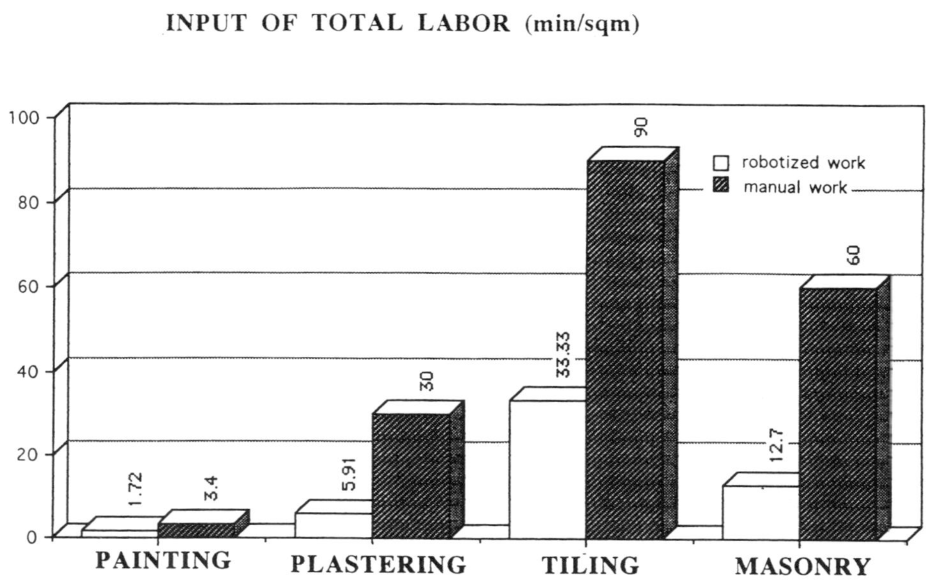

Figure 3 - Comparison of labor input in robotized vs. manual construction

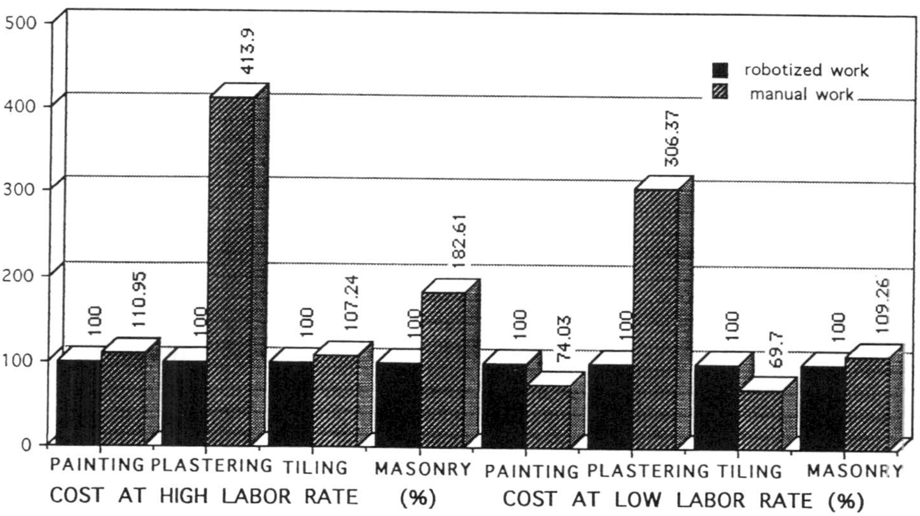

Figure 4 - Comparison of cost in robotized vs manual construction

Automation and Robotics in Construction X
G.H. Watson, R.L. Tucker and J.K. Walters (Editors)

Design-Fabrication Interface: Construction vs. Manufacturing

by John G. Everett

Department of Civil and Environmental Engineering, University of Michigan, 2352 G. G. Brown Laboratory, Ann Arbor, MI 48109-2125 USA

Abstract

Fundamental cultural differences between industries helps explain why automation and robotics have been successful in manufacturing but have largely failed in construction. In construction there has traditionally been a clear separation between a project's design phase and its field operations. In manufacturing, engineering design and shop floor fabrication are highly integrated.

This paper examines the differences between construction and manufacturing in the context of hierarchical structures. Each industry is divided into hierarchical levels including, in order of increasing level of detail: Project, Division, Activity, Basic Task, Elemental Motion, and Orthopedics. In construction, the design-fabrication interface occurs between the Activity and Basic Task levels. Architect/engineers, the product designers prepare plans and specifications at the Activity level of detail and turn them over to field personnel, the process designers who organize and perform the work from the Basic Task level down. In manufacturing, the design-fabrication interface occurs much lower in the hierarchy, between the Elemental Motion and Orthopedic levels. Product and process designers collectively decide not only what the finished product will look like, but how the product will be manufactured, step-by-step. The shop floor worker provides physical input, but little organizational input for the work.

Researchers should formulate their strategies for introducing automation and robotics in construction field operations recognizing the differences between construction and manufacturing. Automation techniques which have succeeded in manufacturing cannot be forced upon the construction industry in its current organizational structure. It will be far easier for construction automation researchers to develop rational strategies than to alter the organization of the construction industry.

INTRODUCTION

Any problem in construction requires examination of the industry at the appropriate level of detail. For example, if the federal government decided to formulate a national industrial policy, it might compare the construction industry to the manufacturing industry, to the aerospace industry, or possibly to the construction industries of other countries. This analysis would be performed at the most general level of detail. A scientist developing hearing protection devices for jackhammer operators might study cells of the inner ear. This investigation would be at a microscopic level of detail. National industrial policies and hearing loss are both significant issues in construction, but they obviously require different kinds of attention.

An earlier paper presented a taxonomy of the construction industry, shown in Table 1 (Everett 1990). Starting from the most general perspective, the industry is broken down into nine levels, each an order of magnitude more detailed than its predecessor.

Table 1.
Taxonomy of the Construction Industry

Level	Examples
1. Industry	Construction, manufacturing
2. Sector	Building, industrial, heavy
3. Project	Office building, power plant
4. Division	Concrete, electrical, mechanical
5. Activity	Drywall partition, HVAC duct
6. Basic Task	Position, Connect, Measure
7. Elemental Motion	Reach, grasp, eye travel
8. Orthopedics	Muscle, bone, joint
9. Cell	Muscle fiber, nerve

Level 1 is the entire construction *Industry*. The construction industry can be compared to other industries such as manufacturing, aerospace, mining, or agriculture. The construction industry can be divided into *Sectors* (Level 2) representing fundamentally different types of constructed facilities such as general building, heavy construction, and special trades.

Construction differs from most other industries because of its *Project* (Level 3) orientation. "Projects are distinguished by their relatively short time frame; definite starting and stopping points; non-routine, often unique, interrelated activities; and a limited time, budget, and resource allocation for the projects performed" (Frankel 1990). Typical projects include petrochemical plants, office buildings, single family homes, and highways.

Level 4 represents a breakdown of on-site construction work into the major *Divisions* of work or trades. The most commonly used classifications in the U.S. are the sixteen Construction Specifications Institute (CSI) divisions, such as concrete, masonry, mechanical, and electrical work (CSI 1988). Level 5 breaks Divisions into specific units of work, or *Activities*. Using the CSI Format, an Activity would correspond to a specification subsection. An Activity represents all the field work which results in a recognizable, completed unit of work. Examples include install drywall partition, install HVAC duct, and erect steel column.

Level 6 is *Basic Tasks*. Basic Tasks are the fundamental building blocks of construction field work, each representing one in a series of steps which comprise an Activity. Any productive work performed in the field can be categorized into one or more Basic Tasks. The set of Basic Tasks is mutually exclusive and collectively exhaustive for all on-site construction work. Basic Tasks can be performed by human craftsmen, machines, or both. The Basic Tasks are: Connect, Cover, Dig, Finish, Inspect, Measure, Place, Plan, Position, Spray, and Spread (Everett 1991).

Construction can be divided into still finer levels of detail. Level 7 examines *Elemental Motions*. Industrial engineers have been studying manufacturing work at the Elemental Motion level of detail for nearly a century. In 1903, Frederick W. Taylor, the Father of Scientific Management, divided shoveling into several simple motions for his time studies (Taylor 1947). In 1920's, Frank and Lillian Gilbreth developed time and motion studies by dividing human movements into seventeen components called *therbligs* (backwards spelling of Gilbreth, almost), such as reach, grasp, and eye travel (Gilbreth and Gilbreth 1924). Another example of Elemental Motions is Methods-Time Measurement (MTM) (Maynard et al 1948). MTM classifies human movements into a set of motions similar to therbligs but also applies standard units of time to each motion to arrive at the total theoretical time required to perform a given task using a prescribed method. Both therbligs and MTM are still widely used in manufacturing.

At the *Orthopedics* level (Level 8), physicians and physiologists analyze human motions by studying muscles, joints, and nerves. Ergonomics and occupational biomechanics examine work at this level of detail to optimize safety and efficiency of the human machine. At the *Cell* level (Level 9), investigators examine individual muscle fibers, nerves, and cellular metabolic activity to understand how the human machine functions and how occupational injuries and illnesses are caused.

The objective of this paper is to describe at which levels in the taxonomy the interface between design and fabrication occurs for construction and for repetitive manufacturing operations. Too often in the past, construction automation and robotics researchers have approached the development of construction technologies from a manufacturing perspective. An understanding of the differences between construction and manufacturing will help future researchers develop more realistic strategies for construction automation.

MANUFACTURING LEVELS VS. CONSTRUCTION LEVELS

Industrial Engineers have been able to analyze manufacturing work down to the Elemental Motion level of detail for two major reasons. First, the repetitive nature of many types of manufacturing work allows consistent detailed analysis. Second, the design and engineering of the manufactured product are intimately connected to the manufacturing process.

Repetitive Motions

Historically, efforts to study manufacturing work have approached the work at the Elemental Motion level of detail. Industrial engineers developed MTM and therbligs to optimize the time and efficiency of repetitive human motions. The motions described by MTM and therbligs are characteristic of mass production assembly lines, where workers typically sit or stand in a fixed position and the work comes to the worker.

Through Elemental Motion analysis, industrial engineers attempt to simplify the work so the human worker can perform with minimal amounts of thinking and/or wasted movement. In highly repetitive manufacturing operations, the "largest common denominator" is the Elemental Motion cycle.

Many parts of construction work are repetitive, but the cycle is much more complex than in mass production manufacturing. While each construction work cycle may be similar to its predecessor and successor, the cycles are not identical. In construction, the craftsperson typically must plan and layout the work, gather appropriate tools and materials, move his/her body to a new location, and then perform the work. Construction craft workers continuously adapt to the changing geography of the workplace, changes in the positions of their bodies relative to the work, and minor differences between successive repetitions of the work.

From an industrial engineers' perspective, construction work varies so much from cycle to cycle that Elemental Motions cannot be considered the largest common denominator. One must move up to the Basic Task level in the Taxonomy of the Construction Industry to reach the fundamental unit of work.

Products and Processes: the Design-Fabrication Interface

Construction and manufacturing exhibit fundamental differences in where the interface or transition occurs between product design and process design or fabrication. Manufacturing product design is closely related to process design. The designers of the final product have a great deal of control over the processes used to create the product and process designers help determine the final product design. Much of Industrial Engineering is dedicated to improving manufacturing processes and the relationship between product and process. Typically, the product designers and the process designers work for the same manufacturing firm.

In repetitive manufacturing operations, the product-process design team controls products and processes all the way down to the Elemental Motion level of detail. Labor controls the fabrication part of the work up to the Orthopedics level. Labor can dictate how hard it will work or possibly how fast it will move, but the shop floor workers follow carefully choreographed routines. The right side of Figure 1 graphically depicts the design-fabrication interface occurring between the Elemental Motion and the Orthopedics levels of detail. The rapidly growing field of occupational ergonomics may represent an effort to push process design down even further, to include the Orthopedics level as well.

Figure 1

Design-Fabrication Interfaces for Construction and Manufacturing

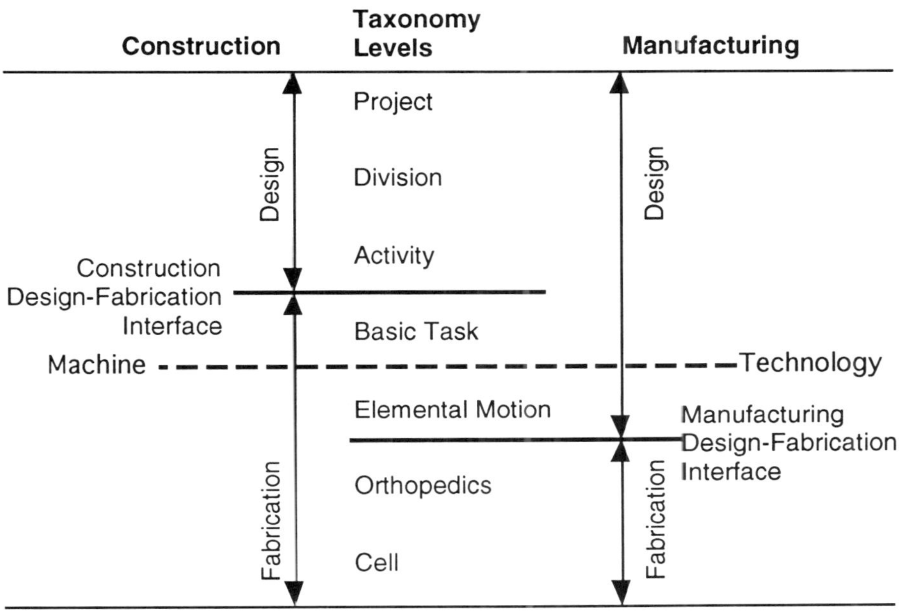

In construction, there is little overlap between product design and process design or fabrication. Architect/engineers control product design down to the Activity level, but do not get involved in the building process other than to inspect the finished work for conformance to design specifications. Constructors control the fabrication process design but generally have little or no input into product design. Constructability reviews incorporate constructor input into product design to improve practicality and cost effectiveness of major product design decisions, but they seldom affect the details of how the work will be performed in the field. A distinct separation exists between the product designers or architect/engineers, and the process designers or craft workers. In construction, the product designers and process designers are almost always separate organizations with different, often conflicting, objectives.

In construction, the product designers control the work only down to the Activity level of detail. The fabricators or craft workers determine how the work will be performed all the way up to the Basic Task level. The left side of Figure 1 shows the design-fabrication interface for construction occurring between the Activity and the Basic Task levels. For example, the product design may specify where an Activity such as an eight inch concrete block wall is to be built, but the mason controls Basic Tasks such as which blocks to lay in which order, how many courses to lay before raising the staging, how often to check for plumbness and alignment, when to tool the joints, etc.

Implications for Automation

Significant improvements in production have been achieved by substituting machines for human workers in many types of manufacturing work. Machines are capable of performing simple Elemental Motions which do not require thinking or adaptation to minor changes. The dashed line in Figure 1 shows the state of the art in machine technology occurring between the Basic Task and Elemental Motion levels. Current machine technology is well above the design-fabrication interface in manufacturing, so application of automation and robotics in manufacturing has been relatively effective. Manufacturing product and process designers control the work down to a level where today's machines can take over.

The culture of the construction industry has evolved differently. Product designers design only down the Activity level of detail. There has been no historical need or incentive for architects and engineers to concern themselves with the intricacies of the work in the field, so finer levels of detail have not been incorporated into design. While this industry characteristic has its advantages, it poses a major problem for developing automation and robotics technology for construction field operations. There is a substantial gap between the lower limit of design detail and the upper limit of machine technology.

The gap occurs at the Basic Task level. Today's machines are capable of performing physically intensive Basic Tasks such as Connect, Finish, Position, and Spray but have trouble with information intensive Basic Tasks such as Inspect, Measure, and Plan. Virtually all construction Activities require performance of information intensive Basic Tasks, so very few practical examples of construction robotics or highly automated machines have been developed. The successful advances in construction automation have almost always been machines that perform one Basic Task and are controlled by a human operator who decides when and where to start, and when and where to stop. Examples abound of machines which connect, cover, dig, finish, place, position, spray, and spread, all single Basic Tasks.

In order to make construction robotics and automation feasible, the gap between design and machine technology must be closed. This can happen in two ways: by increasing the level of design detail or by improving the information processing capabilities of machines. Fortunately, the gap is narrowing from both directions. Advances in computed aided design may one day allow the same computer which stores all the product design details to control a robot or machine that performs the fabrication. Technological improvements in sensors, controls, and actuators may allow smart machines and tools to continuously report their progress and location to the central computer.

Until the product design and process design are integrated, it will be extremely difficult to develop practical fully automated construction machines.

Human operators are still more productive and cost effective than machines and computers for the information intensive Basic Tasks.

CONCLUSION

This paper has shown where the design-fabrication interface occurs in the context of a Taxonomy of the Construction Industry for construction and manufacturing. In manufacturing, product design and process design are closely interrelated. The interface between design and fabrication occurs at a fine level of detail in the taxonomy. Machines can perform many repetitive manufacturing operations because the state of the art of machine technology is above the level of the design-fabrication interface.

In construction, product design and process design are distinct from each other and the interface between them occurs at a high level of detail in the taxonomy. Current machine technology is below the construction design-fabrication interface, and machines cannot perform all of the steps necessary to be fully automated.

Until the gap between construction product design and technology of the machines that perform the fabrication is closed, the promises of practical robotics and highly automated machines will remain unfulfilled. Successful advances in construction automation will continue to be developed at the Basic Task level.

REFERENCES

Construction Specifications Institute (CSI) (1988). *MASTERFORMAT - Master List of Titles and Numbers for the Construction Industry*, CSI, Alexandria, VA.

Everett, J. G. (1990). "Back to Basics in Construction Automation." *Proceedings of 7th International Symposium on Automation and Robotics in Construction*, Bristol, England, June 5-7, 583-90.

Everett, J. G. (1991). "Construction Automation: Basic Task Selection, and Development of the CRANIUM." Ph.D. thesis, Massachusetts Institute of Technology at Cambridge, MA.

Frankel, E.G. (1990). *Project Management in Engineering Services and Development*, Butterworths, London.

Gilbreth, F.B. and Gilbreth, L. (1924). *Classifying the Elements of Work*, in Matteson, M. T. and Inancevich, J. M. (1977) *Management Classics*, Goodyear Publishing Co., Santa Monica, CA.

Maynard, H.B., Stegemerten, G.J., and Schwab, J.L. (1948). *Methods-Time Measurement*, McGraw-Hill, New York.

Taylor, F.W. (1947). *Scientific Management*, Harper and Brothers, New York.

Automation and Robotics in Construction X
G.H. Watson, R.L. Tucker and J.K. Walters (Editors)

Site Layout of Construction Temporary Facilities Using Enhanced-Geographic Information System (GIS)

Min-Yuan Cheng[a] and James T. O'Connor[b]

[a]Department of Construction Engineering & Technology, National Taiwan Institute of Technology, Associate Professor

[b]Department of Civil Engineering, University of Texas at Austin, Associate Professor

Abstract

This paper focuses on developing an automated site layout system for temporary construction facilities. The site layout of temporary facilities (TF) is an important preplanning task that can either enhance or adversely affect construction productivity. An efficient TF layout can significantly reduce construction conflicts and improve project efficiency. In this research, the system, ArcSite, was developed to assist the project manager in identifying suitable areas to locate TFs. ArcSite is comprised of a Geographic Information System (GIS) integrated with a database management system (DBMS). The system includes the knowledge specific to construction site layout, TF databases, Arc/Info databases, and GIS geoprocessing system. In the development of the system, industry rules of thumb and the state of the art in layout modeling were reviewed and compiled in a systematic form. ArcSite uses the concept of Searching by Elimination to generate the potential sites for each TF. A heuristic algorithm was developed to model the process of human decision-making and to identify the spatial relationships between the different data layers which represent the site geographies. Through both qualitative and quantitative modeling of facility relationships, an objective function for calculating the Proximity Index (PI) was used to determine the optimal site of each TF. The primary features of the system are as follows: (1) integrate construction schedule and design information to lay out TFs for both short term and long term usage, (2) allow for the use of the PF as the TF and reallocation of space over time, (3) automatically generate potential sites for each TF, (4) identify optimal site, and (5) locate the TF within the optimal site. ArcSite improves the site planning efficiency by integrating spatial and thematic information into a single environment. Also, compared with current methods, ArcSite significantly improves the computational effort and increases the data accuracy and consistency. Therefore, this study achieves the goal of developing an automated information system to enhance the TF layout design.

1. INTRODUCTION

1.1 Present Environment

Owing to the competitive bidding structure of the construction industry, owners and bidding contractors are reluctant to spend the necessary money to cover the effective planning and management. While it is evident that the cost of site planning and construction of temporary facilities increases project overhead, unfortunately, there is little documented evidence that such an investment will improve overall project efficiency. Particularly, for contractors, the investment of the site planning is not viewed as critical. Therefore, for projects with tight budgets, it is easy to justify neglecting the planning of TF altogether. As a result, project managers generally do not have adequate staff or time to plan TF well. On construction sites, project managers usually accomplish the TF layout based on previous trial-and-error and in an ad-hoc fashion. This was substantiated in a previous survey that found that most project managers layout TF based on experience, common sense, and adaptation of past layouts to present project [Rad 83]. Therefore, it is difficult to detect the layout problems until conflicts occur and relocation of temporary facility is required. Also, the inefficiency of correcting mistakes after-the-fact and dollars lost due to the lack of preplanning go undetected by management.

1.2 Problem Statement

Most of the project managers realize that the layout of a construction site has a considerable influence on the cost performance of the project. But how much influence? How can it be measured? There are no standard methods of measurement. Therefore, the manager can not quantify the productivity loss due to improper layout. Also, the manager can not evaluate the worth of good site layouts. Productivity loss is one of the main concerns in this study. Previous surveys show that productivity is lost in frequent worker travel on site for tools or materials, waiting for tools, material and other members of the crew, job interruptions, and working space conflicts. It should be realized that when conflicts occur and relocation of temporary facility is required, the loss is not only the cost of relocation of facility, but also includes time loss of waiting, job interference, labor morale and work discontinuity. This impact has a tendency to increase when the scope of project increases.

For the layout design, the project manager has to consider the characteristics and spatial constraints of the site and collect information from different resources such as A/E engineering, general contractors, subcontractors, and superintendents. Each party has his own interests and concerns. Therefore, the organization and analysis of the collected information as well as the decisions on tradeoffs between the interested parties have to be made by the project manager. After reviewing the problem area, the problems of TF site layout can be summarized as follows: (1). Design is based on incomplete and ill-structured information. (2). Site layout design depends on labor intensive manual approach. (3). Design relies heavily on individual experience. (4). There are diversity of solutions. (5). There is no exact rules or methods to follow. (6). Analysis lacks rigor: - dominant selection criteria, - driven almost exclusively by visual representations, - very difficult to link spatial features with thematic information, and - appropriate tools to help users to record the previous arrangements. Based on the problems defined, this study focuses on the development of an automated and integrated computer system to solve the TF layout as an ill-structured problem. In addition to replacing the manual methods, the system produces TF layout solutions with the pre-defined levels of analytical rigor.

1.3 Research Purposes and Objectives

The primary purpose of this paper is to investigate the potential for automating the site planning tasks of locating temporary construction facilities. The sub-objectives required to achieve the primary objective are the following: (1) Gain the knowledge and experience of the experts in laying out the temporary facilities (TF); (2)Structure the knowledge and procedures for TF site layout in a systematic form; (3) Define the dominant variables and develop an evaluation method to determine the TF locations; and (4) Develop an automated site layout system using Geographic Information Systems (GIS) to implement the objectives mentioned above.

1.4 Scope Definition

In this study, the system, ArcSite, was developed to design the TF site layout. Using the same methodology and algorithms with appropriate modifications, ArcSite can be applied to most construction sites and be suitable to different site locations, such as remote, suburban, and urban areas. The integration of construction schedules with both PFs and TFs enables ArcSite to lay out TFs for both short term and long term usage.

2. SYSTEM DEVELOPMENT

2.1 Knowledge Acquisition And Representation

This section achieves the task of modeling the expert's knowledge and experience of site planning and expressing it in a systematic form. The procedures for modeling the experts' knowledge and experience for TF site layout were classified into three phases: (1) compiling the experts' knowledge and experience for TF site layout, (2) interpreting the knowledge into the knowledge base, and (3) translating the knowledge base into the ArcSite implementation forms (see Figure 1).

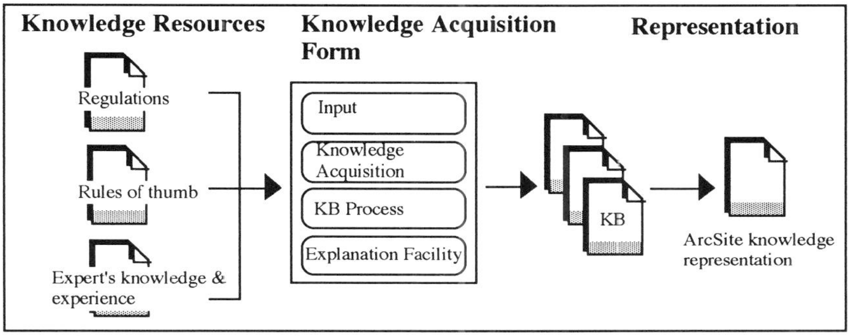

Figure 1 Procedures of Knowledge Acquisition and Representation

To develop the knowledge base, literature such as previous research work, and company handbooks were reviewed and knowledge was compiled according to three categories: (1) rules of thumb or experience of sizing TF, (2) TF design priority, and (3) regulations and safety standards.

After compiling the expert's knowledge and experience in site planning, the next task was to systematically interpret and express it into a knowledge base (KB). The knowledge base consists of four parts, viz., Input, Knowledge Acquisition, KB Process, and Explanation Facility. In the KB, Input is a mechanism designed to obtain input parameters to determine TF size and user inquiry. Knowledge Acquisition is a device to acquire the knowledge of site planning. The KB Process not only interprets the knowledge obtained in the knowledge acquisition phase into a systematic format, but also shows the process of reasoning. The Explanation Facility contains information which is not covered in the other three modules or advanced considerations and explanations required to improve the TF layout design.

Once the site layout knowledge base was established, it is necessary to translate the knowledge base into the ArcSite implementation forms suitable for analysis.

2.2 System Usage

System usage essentially involves identifying the potential users of the system. ArcSite can be used in three construction phases: (1) design, (2) pre-construction planning, and (3) during construction. During the design phase, ArcSite can be used to assist both experienced and inexperienced designers. Furthermore, since the layouts of PF and TF influence each other, designers must repetitively adjust and update the layouts as construction progresses. This iterative process can be accomplished efficiently using ArcSite. Before the construction commences, ArcSite can be used by project managers and superintendents to organize and analyze the information required for site planning. Using ArcSite, the spatial relationships between different data layers can be identified and criteria such as distance, adjacency, position, accessibility, and space can be evaluated to determine the areas in which the TF can be located. During construction, ArcSite can assist the project managers and superintendents in tracking and updating the layout of the site by evaluating and recording the PF and TF construction schedules stored in the database.

2.3 System Functions

The functional modules of the system are developed according to the capabilities and information needed for the user to conduct the site layout design. After assessing the needs of the user, the requirements of the tasks are translated into system functions. A hierarchy of the system functions is shown in Figure 2. Six functional modules, namely File Manager, TF Layout Design, Spatial Analysis, Graphical Query, Database Query, and User and Program Interface are developed. The menus associated with each function are illustrated in the function menu column. This section mainly discusses the function of TF Layout Design.

The TF Layout Design module is a function which allows the user to identify suitable areas for the TF. A hierarchy of the sub-modules of the TF Layout Design function is shown in Figure 3. The **Design** menu activates the design operations.

Temporary facility (TF) library: Functions such as TF checklist, input parameters, equations and rules of thumb for sizing TFs, and user interface functions are included in the TF library. The library which covers most of the TF used in site planning provides project managers a design base in listing TF for any construction project.

Arc/Info databases: The Arc/Info database consists of two parts: (1) the graphic database which contains a number of coverages to represent the geographies of the construction site, and (2) the tabular database which comprises the thematic information integrated with the coverages. The Arc/Info database provides the flexibility for the user to add new fields and attribute values into the tabular database.

Use of Permanent Facility (PF) as Temporary Facility (TF): The system allows for using a PF to temporarily support construction operations. A PF with a finish date earlier than the start date of the candidate TF and with a larger area than the required TF area can be used as a TF once approval from the user is obtained.

Design TF layout that changes over time: As construction progresses, some of the TFs are finished and removed from the site. Thus, the space occupied by the TF is available again. After evaluating the duration of TF on the site, the system deals with the flexibility of reallocation of space over time. Thus, the space occupied by the previous TF can be reutilized to locate successive TFs.

Potential site generation: Using the concept of Searching by Elimination, the system develops an algorithm for generating the potential sites for each TF. Considering the constraints and selection criteria, ArcSite identifies the spatial relationship between the data layers which represent the site geographies. The heuristic approach initiates searching the available space to locate temporary facilities, and then eliminates the areas occupied by PF, working areas adjacent to the PF, and the areas closed for safety consideration. A number of alternatives which satisfy the searching criteria are generated. In addition, ArcSite provides users the flexibility of changing searching criteria.

Identification of the optimal site: Through both qualitative and quantitative modeling of facility relationships, an objective function for calculating the Proximity Index (PI) was developed to rank the identified potential sites. The calculation of PI for each potential site is based on the four relationship matrices: PF/TF travel frequency matrix, TF/TF travel frequency matrix, PF/TF attract/repel matrix, and TF/TF attract/repel matrix. The PI is the sum of the product of the distance and relationship rating between facilities. The potential site with the minimal PI is the optimal site. In addition, ArcSite allows users to select their preferences of the potential sites.

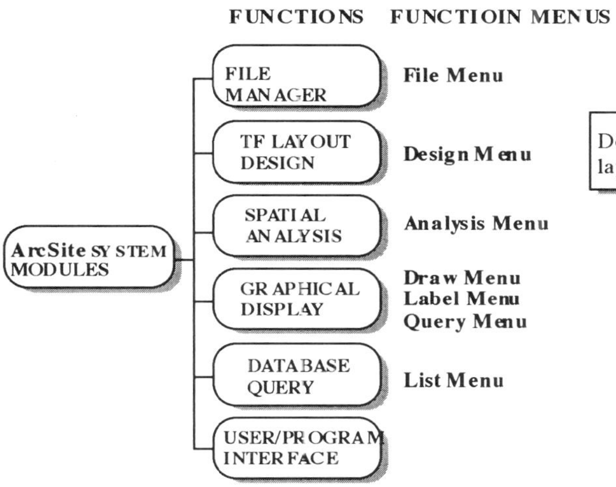

Figure 2 Hierarchy of System Functions

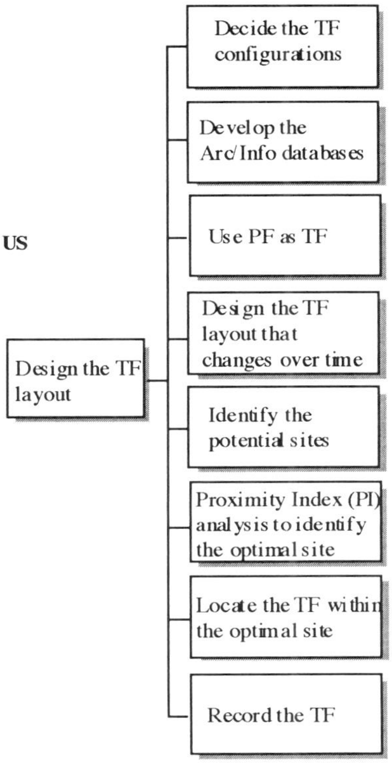

Figure 3 TF Layout Design Functions

Locate the TF within the optimal site: Usually, the area of the optimal site identified by ArcSite is not equal to the size of the candidate TF. If the area of the optimal site is more than 1.5 times of the TF area, ArcSite locates a rectangle to represent the TF within the selected site. The functions included to locate the TF within the optimal site involve: (1) spatial operations to identify the vertices which intersect with the access roads, (2) identification of the optimal and second optimal vertices based on the PI rankings of the vertices, (3) generation of the base line based on the optimal and second optimal vertices, (4) user preference to decide the TF orientation, and (5) layout of the TF along the base line.

Record TF: The area assigned to the TF represents an occupied space which may need to be eliminated from the available space for the subsequent layout design. Thus,ArcSite dynamically generates a coverage to contain the TF and records the name of the created TF. Considering the PI analysis for the successive process, the coordinates of the TF are also recorded.

2.4 System Architecture

The tools used to develop the system functional modules are discussed in this section. Figure 4.4 shows the architecture of the system. Using SML, the user and program interface is integrated at three levels, namely application user interface, command user interface, and program data interface.

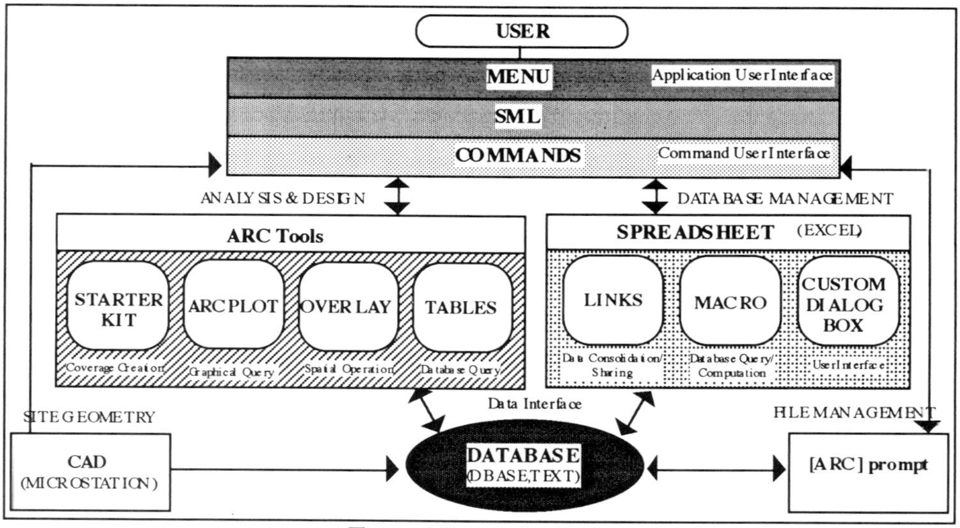

Figure 4 System Architecture

In the system, four prime components, viz., CAD, ARC tools, Excel spreadsheet, and [ARC] prompt, are integrated at the command level and the data level. Site geometry is created using MicroStation, a CAD software package. The data layers drawn to represent the site geographies are saved as a DXF file format. ARC tools such as Starter kit, Arcplot, Overlay, and Tables are used to conduct the TF layout analysis and design. After the site geometries are created, the Starter Kit is employed to model the site coverages. Modeling site coverages is a process to construct topology and integrated feature attribute table (FAT) and make the spatial data usable for analysis. Arcplot is used to perform the graphical query and display the analysis results. Overlay provides the basic operators for spatial analysis. Tables conducts the database query during the TF layout design. Using Excel spreadsheet, the functions of database management system such as adding of the tabular attributes, data consolidation and sharing, database query, and user interface are accomplished. Linking is useful for consolidating and sharing data from several worksheets and thereby designing an effective model. To automate the system commands for the worksheets to do database query and computation are recorded using macros. The custom dialog boxes are created using functions within Excel. Access to the [ARC] environment enhances the file management capabilities of the system. This component allows the user to temporarily suspend the execution of the SML macros and exit to the [ARC] prompt to use the Arc/Info functions.

3. SYSTEM RESULTS AND VALIDATION

This study applied the data from the Bull Run Steam Plant (BRSP) project to ArcSite for testing the system. BRSP is a steam-electric generating plant consisting of four units of total 3.6 megawatts. It is located on the east bank of Melton Hill Lake in Anderson County, Tennessee. The result of the TF layouts generated by ArcSite is shown in Figure 5. Note the system identifies the optimal site of each TF on the site and provides the user with options to select his/her preference site. In addition, the TF layout solution is sensitive to the TF design sequence. Different TF design priorities can result in various layout solutions. Therefore, by changing the TF design sequence, the user can conduct "what-if" analysis to solve both real and hypothetical on-site design layout problems.

Figures 5 shows the layouts designed by ArcSite. When comparing the layouts designed by ArcSite with the conventional approach, the reader will find that the two layouts are not identical. The layout discrepancies can be summarized into five reasons. First, ArcSite included not only the knowledge, information, and procedures stated in the BRSP technical report for TF site layout but also other experts' knowledge and information in the development of the system. Secondly, ArcSite might fail to represent all the factors that the project manager used in generating the layout, or perhaps the technical report did not include all the factors that the project manager considered in the development of the original TF layout. Thirdly, for practicality, this study clustered the TF with the same characteristics so that they would not clutter up large, open spaces on site that might be used for other purposes. The fourth difference relates to the diversity of the solutions. Often, it is difficult to tell which arrangement is best. As discussed in the problem statement, there are no standard guidelines or methods used to evaluate how good a site layout is. Finally, the knowledge and experience that the project manager brings to the TF site layout is nearly impossible to quantify. This represents a communication problem which can be classified in two phases: knowledge acquisition and representation. In the acquisition of knowledge, the training and expertise may lead different engineers to interpret data and terminology differently which can lead to differences of opinion on the job site. Regarding the realm of knowledge representation even on a heuristic level, there are many ways to represent the layout knowledge. It is difficult to ascertain if the knowledge representation is appropriate and complete.

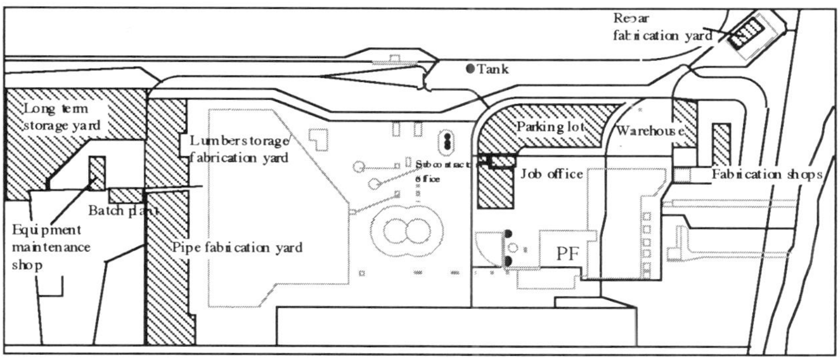

Figure 5 Final Temporary Facility Layout

4. CONCLUSIONS

The hypotheses of this research are validated by experimenting and implementing the system. Four conclusions are made as follows:

(1) <u>TF layout problems can be effectively solved using a computerized system.</u>

In the development of the system, the information required to select the optimal location for the TF was successfully represented and integrated in a computer environment. In addition, an automated site layout system was developed to identify the satisfactory solution for TF layout.

(2) <u>GIS - Arc/Info forms an effective foundation for TF site planning.</u>

The basic functions of Arc/Info including information representation, spatial operations, graphical display and query, and adding functionality form an effective foundation for carrying out TF site layout. Using Arc/Info, the investment of time and programming effort in developing the system was significantly reduced.

(3) A robust system is required to solve the complicated TF layout problem. Over-simplified systems can result in unrealistic plans for actual projects.

ArcSite was developed to replace manual methods and assist planners in assessment the TF layout design. The system identified the rules of thumb and experience for sizing the TFs and used state of the art industry practices to lay down the TFs. Also, a heuristic approach for the facility layout problem was developed to eliminate the site choice of the TF as an ill-structured problem. For practical site planning, ArcSite is a promising decision support system with a high utility value.

(4) The system improves both productivity and quality of site planning.

The advantages of applying ArcSite in solving TF layout problem are identified as follows:

- ArcSite systematically organizes the ill-structured information used for TF site layout.
- Owing to Arc/Info's ability to integrate locational and thematic information, ArcSite improves the site planning efficiency by eliminating the need for extracting data from various resources. Also, using ArcSite, the data integrity and accuracy are ensured.
- The computational information such as facility coordinates and distance between facilities are considered in ArcSite to lay out the TF.
- Compared with current methods, ArcSite significantly improves the computational effort and increases the data accuracy and consistency.
- ArcSite performs a series of complicated spatial operations and database queries to identify the optimal site. These tasks are difficult to perform manually.
- To identify the optimal site, ArcSite considers all of the relationships between the facilities, while in the conventional approach, the project manager may only consider the relationships of the most effective facilities which will determine the TF location.
- ArcSite provides the capability for the project manager to easily track and update the site layout by integrating schedule and design information.
- ArcSite allows the user to change the project attributes and searching criteria in order to conduct "what-if" analysis.
- Due to the significant reduction in paper work and the ease of system accessibility and modification, the quality and efficiency of TF site planning are improved using ArcSite.
- ArcSite can be applied to multiple projects with appropriate modifications.

5. REFERENCES

1. Cheng, M.Y., "Automated Site Layout of Temporary Construction Facilities Using Geographic Information Systems (GIS)", Ph.D. Diss., University of Texas, Dec. 1992.
2. Cheng, M.Y., Varghese, K., and O'Connor, J.T., "Management of Spatial Information for Construction Planning and Design Using Geographical Information Systems (GIS)" The 9th ISARC, Vol. 2, pp. 393-402, June 1992.
3. ESRI, 1989a "PC Arc/Info Reference Manual." Environmental Systems Research Institute, January 1989: Redlands, CA.
4. Handa, V. and Lang, B., "Construction Site Planning.", Construction Canada 88 05, pp. 43-49, 1988.
5. Hamiani, A., "ConSite: A Knowledge-Based Expert System Framework for Construction Site Layout" Ph.D. Diss., University of Texas, December 1987.
6. Neil, J.M., "Teaching Site Layout for Construction.", ASCE Meeting, Portland, OR, April 14-18, pp. 1-11, 1980.
7. Popescu, C., "Large Construction Site Design and Organization in Developing Countries.", SAE paper No. 780533, Society of Automotive Engineers (SAE), 1978.
8. Rad, P.F. and James, B.M., "The Layout of Temporary Construction Facilities.", Cost Engineering, Vol. 25, No. 2, pp. 19-27, April 1983.
9. Tommelein, I.D., "SightPlan: An Expert System that Models and Augments Human Decision-Making for Designing Construction Site Layouts.", Ph.D. Diss., Stanford University, August 1989.

Automation and Robotics in Construction X
G.H. Watson, R.L. Tucker and J.K. Walters (Editors)

MOTION PLANNING FOR AUTOMATED CONSTRUCTION

By: Raghavan Kunigahalli,[a] Jeffrey S. Russell,[b] and Miroslaw J. Skibniewski,[c]

[a] Ph.D. Candidate, Dept. of Civ. & Env. Engrg., University of Wisconsin-Madison, Madison WI 53706.

[b] Asst. Prof., Dept. of Civ. & Env. Engrg., University of Wisconsin-Madison, Madison WI 53706.

[c] Assoc. Prof., School of Civ. Engrg., Purdue University, West Lafayette, IN 47907.

Abstract: Increasing research and development in the area of construction robotics and computer-aided-construction has resulted in a large number of automated construction equipment applications. Full potential of such construction equipment can be realized by providing the collision avoidance motion planning capabilities. Models used in motion planning can be categorized into two types: (1) planning with complete information and (2) planning with incomplete information. State-of-the-art general algorithms appropriate for the motion planning process are highlighted. Motion planning with complete information are typically NP-complete and hence are modeled by approximate approaches or by exploitation of special conditions. Motion planning with incomplete information requires sensory feedback. An algorithm for motion planning in an automated concrete placement operation that exploits the special conditions provided by structural engineering floor design theory is also provided.

1. INTRODUCTION

In the last five years, there has been an increasing trend in the application of automation technology to the construction industry. A large number of automated construction equipment applications have been developed for field operations such as concrete placement, concrete finishing, wall inspection, slurry wall construction, and tunneling. A survey of automation applications in the construction industry can be found in Skibniewski and Russell [1].

A significant challenge in construction automation is the ill-structured constantly changing environment of the construction site. In order to address this challenge, automated machines must be equipped with sophisticated collision avoidance motion planning capabilities. The collision avoidance motion planning is formulated as a static problem. Only the path (i.e., position and orientation information) is being sought, with no consideration for the dynamics along the path. There are two formulations of collision avoidance motion planning problem: (1) path planning with complete information describing the environment and (2) path planning with incomplete information for an uncertain environment. With complete information an optimal path can be obtained. With incomplete information, the path is computed continuously or piece by piece, based on the incoming information from sensors and hence an optimal solution cannot be found.

This paper describes the generic challenges associated with motion planning of automated construction equipment. A review of the classical formulation of the mover's problem is provided along with a discussion of its complexity. State-of-the-art of planning algorithms for complete and incomplete information are also described. A motion planning algorithm with complete information for an automated machine in concrete placement operation is also provided.

2. MOTION PLANNING WITH COMPLETE INFORMATION

In motion planning problems with complete information, the general approach is to compute the configuration space in which the moving object shrinks to a point, and obstacles are grown to compensate for the shrinking object. Any path that lies within the complement of the grown obstacle is feasible. General algorithms for motion planning with complete information are summarized below.

Generalized Mover's Problem. The classical formulation of mover's problem in d-dimensional Euclidean space has been provided by Rief [2]:

Input-- (O, B, P_s, P_t) where O is a set of polyhedral obstacles fixed in Euclidean space, and B is a rigid polyhedron with distinguished positions P_s and P_t.

Problem-- Can B be moved by a sequence of translations and rotations in d-space from positions P_s to P_t without contacting any element in O?

The mover's problem may be generalized to allow B, the object to be moved to consist of multiple polyhedra freely linked together at various distinguished vertices. The generalized mover's problem in 3-space is p-space hard, by a direct log-space reduction from the acceptance problem for polynomial space bounded by Turing machines. The generalized mover's problem is one of few known p-space complete combinatorial problems. Reif [2] has presented a polynomial time algorithm to the classical mover's problem by mapping the problem in lower dimension to a simpler problem in higher dimension.

Piano Mover's Problem. The general mover's problem has been reformulated by Schwarz and Sharir [3] in abstract algebraic terms so as to reduce it to the problem of decomposing certain algebraic varieties into their connected components. This reformulation has been referred to as the "Piano Mover's" problem.

Algebraic formulation of the problem includes one or more hinged bodies B. Each body B consist of a finite number of rigid compact subparts B_1, B_2,........., each bounded by algebraic surfaces. These subparts may be connected to each other by various types of attachments such as revolute and sliding.

Body B is required to be moved from initial to the final position in empty space bounded by a finite collection of walls, each of which is an arbitrary algebraic surface. The rotational component of body B's motion is a smooth 3-dimensional algebraic submanifold of 9-dimensional Euclidean space. If B is a single rigid body, its position is described by a Euclidean motion T that takes B from some standard position to its given position. This transformation $Tx = Rx + X_o$ is defined by a pair of $[X_o, R]$ consisting of a point X_o in 3-dimensional Euclidean space E^3 and of a 3x3 rotation matrix R. Hence T is a point in a smooth 6-dimensional algebraic submanifold G of 12-dimensional Euclidean space E^{12}. Irrespective of the manner in which subparts of the body are connected together, overall position of all its parts can always be defined by a point belonging to smooth algebraic manifold G lying in a Euclidean space of some appropriate dimension.

The theoretical existence of solutions to the "Piano Mover's" problem was shown by Schwartz and Sharir [3]. The techniques to obtain the solution were based on Tarski's algorithm for deciding statements in the theory of real numbers and Collin's algebraic decomposition of Euclidean spaces. An algorithm that solves the "Piano Mover's" problem by computing a connectivity graph with complexity $O((2n)^{3^{r+1}} \cdot m^{2^r})$ was provided. In the complexity equation, r relates to the number of degrees of freedom for all subparts B_i of body B, n is the maximum degree of the polynomial defining a single geometric constraint and m is the number of polynomials defining the sets G and F; where, G is a smooth manifold in the Euclidean space E^r in which transformation T_i are given and F is the manifold describing forbidden configurations of body B due to non-collision requirement.

Visibility Graph Approach. An algorithm presented by Lozano-Perez and Wesley [4], for motion planning with complete information is closely related to an optimization approach. The constraints on the position of an arbitrary reference point on the moving object are computed. Polyhedral obstacles in two or three dimensions give rise to sets of polyhedral forbidden region. This transformation reduces the problem of finding a safe path for the polyhedron to the simpler problem of finding a safe path for a point. The last task is accomplished by finding a path through a graph called *visibility graph* VG(N, L) defined as follows: the node set N is $V \cup \{S, T\}$ where V is the set of all vertices of obstacles and L is the set of all links (n_i, n_j) such that a straight line connecting the ith element of N to the jth does not overlap any obstacle [4].

Configuration Space Approach. Another widely accepted approach presented by Lozano-Perez [5] is based on characterizing the position and orientation of an object as a single point in a configuration space in which each coordinate represents a degree of freedom in the position or orientation of the object. The configuration forbidden to this object, due to the presence of other objects, can then be characterized as regions in the configuration space called *configuration space obstacles*. Efficient algorithms for computing *configuration space obstacles* have also been developed by Lozano-Perez [5].

3. MOTION PLANNING WITH INCOMPLETE INFORMATION

In motion planning with incomplete information, an element of uncertainty is present and hence proof of convergence is essential. The missing data are typically provided by some source of local information through sensory feedback such as ultrasound range finder or a vision module. Hence, path planning becomes a continuous on-line computational operation. A universal lower bound for path-planning in an uncertain environment has been provided by Lumelsky and Stepanov [6]. The bound was formulated in terms of the length of the path (P) generated by the automaton on its way from the starting point (S) to target point (T). The bound is given by $P \geq D + \sum_i p_i - \delta$; where D is the distance between S and T and p_i refers to the perimeters of obstacles intersecting the disc of radius D centered at T.

Considering the motion planning with incomplete information model, when the environment can be modeled as a finite or infinite maze, Maze Searching Algorithms can be used to plan the path. Additional resources of input information are assumed to be the following: (1) automaton has a capability to mark corridors already traversed, (2) automaton knows direction in which traversal took place, and (3) corridors that led to an intersection for the first time.

Maze Searching Problem (MSP). Maze Searching Problem can be formulated as follows: describe a general algorithm that constructs a closed covering walk (W) in a connected graph G, such that in the course of constructing W, the algorithm can handle only local information available at any vertex reached by W [7]. There are many maze searching algorithms. Select algorithms are provided below.

Tarry's Algorithm. This algorithm works with local information only. The algorithm has the following hypothesis: At any vertex $v \in V(G)$ reached in the course of the construction of walk W, the set $E^0_{v,W} \subseteq E_v$ of edges already passed from v is known. Moreover, the edge $e_{in}(v)$ by which v has been reached for the first time is known. For the initial vertex v_o of W, the set $\{e_{in}(v_o)\} = \phi$. The algorithm can be formalized as follows [7, 8]:

Step 1: Set i=0 and choose $v_o \in V(G)$. Set $W = v_o$.

Step 2: Beginning at $v_i \in V(G)$ walk along an arbitrary
$e_i \in \widetilde{E_{v_i}} := (E_{v_i} \cup \{e_{in}(v_i)\})$. Set $W = W, e_i, v_{i+1}$ ($e_i \in E_{v_{i+1}}$).
Set i=i+1.

Step 3: Suppose $W=v_o, e_o, \ldots, e_{i-1}, v_i$ has been constructed. If $\widetilde{E_{v_i}} = \varnothing$, go to Step 2. Otherwise, go to Step 4.

Step 4: If $\{e_{in}(v_i)\} \subseteq E^o_{v_i, W}$ go to Step 5. Otherwise, set $e_i = e_{in}(v_i)$,
$W = W, e_i, v_{i+1}$ ($e_i \in E_{v_{i+1}}$), i=i+1; go to Step 3.

Step 5: W is a bidirectional double tracing in G.

Fraenkel's Algorithm. This algorithm is an improvement over the Tarry's algorithm. It has the following hypothesis in addition to the hypothesis mentioned in Tarry's algorithm: a bijection c: {W}->N $\cup$ {0} is given, where {W} denotes the set of walks produced algorithmically, such that $c(v_o)=1$, and for W and $W'=W, e_i, v_{i+1}$ we have $|c(W) - c(W')| = 1$. The algorithm can be formalized in the following steps [7, 9]:

Step 1: As long as c(W)>0 proceed as in Tarry's algorithm.

Step 2(a): If $v_{i+1} \neq v_j$ $0 \leq j < i$, set c(W')=c(W')+1.

Step 2(b): If $v_{i+1}=v_j$ for some j<i+1 and if $|E_{v_{i+1}} - E(W)| \geq 1$, while
$|E_{W'o}| \leq 1$ for the same v_{i+1}, set c(W') = c(W)-1.

Step 3: Suppose c(W)=0, If $E_{W,o} \neq \phi$ at v_j, proceed as in Tarry's algorithm. Otherwise, set $e_i = e_{in}(v_j)$ if $v_i \neq v_o$.

The outcome of Fraenkel's Algorithm is the final W, a closed covering walk such that $\lambda_W(e) \leq 2$ for every $e \in E(G)$.

Pledge's Algorithm. In addition to the assumptions made in the earlier algorithms for motion planning with incomplete information, Pledge's algorithm assumes that the automaton is equipped with a compass that allows a specific direction, say north, to be maintained. The automaton is equipped with a counter that is activated when it encounters an obstacle and begins integrating the turning angles. The algorithm is given by the following steps [10]:

Step 1: Walk straight toward the north until an obstacle is encountered or target (T) is reached.

Step 2: Turn left leaving obstacle on the right.

Step 3: Follow obstacle boundary until the value of the counter is zero. Go to step 1.

The outcome of the Pledge's algorithm guarantees eventual escape from any finite maze, but its path length performance is not bounded.

Bug1. In addition to the assumptions mentioned in the earlier algorithms, the Bug1 algorithm developed by Lumelsky and Stepanov [6] assumes that the mobile automaton (MA) knows its own coordinates and those of the target at all times. The goal of the algorithm is to generate a path from Start (S) to Target (T). When meeting an ith obstacle, MA defines a hit point H_i for i = 1,2,....... When leaving the ith obstacle, to continue its travel toward the target, MA defines a leave point L_i; initially, i=1 and L_o=S. The procedure uses three registers R_1, R_2, and R_3 to store intermediate information. All three are reset to zero when a new hit point H_i is defined. R_1 is used to store the coordinates of the current point, Q_m, of the minimum distance between the obstacle boundary and T, R_2 integrates the length of the obstacle boundary starting at H_i, and R_3 integrates the length of the obstacle boundary starting at Q_m. The algorithm consists of following steps:

Step 1: From point L_{i-1}, move toward Target (T) along a straight line until one of the following occurs:
- (a) T is reached and the procedure stops.
- (b) An obstacle is encountered and a hit point H_i is defined. In this case, go to Step 2.

Step 2: Turn left and follow obstacle boundary. If T is reached, then stop. After having traversed whole boundary and having returned to H_i, define a new leave point $L_i = Q_m$, and go to Step 3.

Step 3: Based on contents of registers R_2 and R_3, determine shorter way along boundary to L_i. After having defined a point L on an obstacle, if MA discovers that the straight line segment (L, Target) crosses the obstacle at point L, then target is not reachable.

Bug1 tends to be overcautious and never covers less than full perimeter of the obstacle. Another algorithm Bug2, takes advantage of simple situations, but may become inefficient in more difficult cases.

Vision with Motion Planning. Lumelsky and Skewis [11] provided two motion planning algorithms labelled VisBug-21 and VisBug-22 that incorporate vision information. Due to space limitations, a brief overview of only VisBug-21 algorithm is presented. VisBug-21 consists of the *Main Body* that does the proper motion planning along the path, and a procedure called *Compute*T_i*-21*, that performs the test for target reachability and produces the next intermediate target T_i for the given current position C of the automaton. The main body consists of two steps. The first step S_1 consists of moving towards T_i, while executing *Compute*T_i*-21* and performing the following tests: (1) target reached/not reached, (2) target reachable/not reachable, and (3) intermediate target is reached/not reached. The second step S_2 is executed only in those cases when the automaton is moving along a (locally) convex boundary of an obstacle and so it cannot use vision for defining the next intermediate target. Step S_2 consists of moving along the obstacle boundary while executing *Compute*T_i*-21* and performing similar tests to step S_1 [11].

4. PATH PLANNING ALGORITHM FOR AUTOMATED CONCRETE PLACEMENT

An algorithm to plan the path of a concrete placement pipe in a computer-controlled concrete placement system has been developed by Kunigahalli et al. [12]. Apart from providing an efficient placement pipe path, the algorithm integrates design data with the construction operation by obtaining input from a 2-D wire-frame CAD model. Wire-frame data corresponding to a line diagram of the floor plan is processed by employing computational geometry techniques. Processed data are then stored in a hypergraph G=(V, E, H), similar to the *face adjacency graph* structure normally used in Boundary Representation (B_Rep) schemes. Thus, the information stored in the hypergraph is more complete in terms of adjacency relation among beams and rectangular slabs of the given orthogonal polygonal floor.

The algorithm obtains information on beams above, below, to the left, and to the right of any given beam at any given beam to beam intersection. Based on this information, obstructing columns along the beams are identified. The algorithm then performs a rectangular partitioning of the orthogonal polygonal floor excluding obstacles such as elevator shafts. The information regarding obstructing columns combined with partitioned rectangular adjacency information is stored in a hypergraph G = (V, E, H). An efficient traversal of the hypergraph along with direction parallel path within each node of the graph can provide a detailed path plan for the automated concrete placement pipe.

5. CONCLUSION

Motion planning issues with complete and incomplete information have been discussed. The computational complexity of algebraically formulated mover's problem is double exponential in degrees of freedom of the moving body. Hence, approximate algorithms or exploitation of special condition and circumstances are required to solve the motion planning problem with complete information. Proof of convergence and sensory feedback for local information are essential for motion planning with incomplete information. A path planning algorithm with complete information for an automated concrete placement system has been described.

6. ACKNOWLEDGMENT

The second author wishes to thank the National Science Foundation for Grant No. MSM-9058092, Presidential Young Investigator Award, for financial support of this effort.

7. REFERENCES:

1. Skibniewski, M. J. and Russell, J. S., "Robotic Applications to Construction," *Cost Engineering*, Vol. 31, No. 6, (1989), pp. 10-18.
2. Reif, J. H., "Complexity of the Mover's Problem and Generalizations," *Proc., 20th IEEE Symp. on Foundation of Computer Science*, (1979), pp. 421-427.
3. Schwartz, J. T. and Sharir, M., "On the "Piano Movers" Problem II: General Techniques for Computing Topological Properties of Real Algebraic Manifolds," *Advanced Applied Mathematics*, Vol. 4., (1983), pp. 298-351.
4. Lozano-Perez, T. and Wesley, M. A., "An Algorithm for Planning Collision-Free Paths Among Polyhedral Obstacles," *Communications of the Association of Computing Machinery*, Vol. 22, No. 10, (1979), pp. 560-571.
5. Lozano-Perez, T., "Spatial Planning: A Configuration Space Approach," IEEE Transactions on Computers, Vol. c-32, No. 2, (1983), pp. 108-120.
6. Lumelsky V. J. and Stepanov, A. A., "Path-Planning Strategies for a Point Mobile Automaton Moving Amidst Unknown Obstacles of Arbitrary Shape," *Algorithmica*, Vol. 2, (1987), pp. 403-430.
7. Fleischner, H., Eulerian Graphs and Related Topics: Part 1, Vol. 2, North-Holland, Amsterdam, (1991).
8. Ore, T., Theory of Graphs, American Mathematical Society, Providence, Rhode Island, (1962).
9. Fraenkel, A. S., "Economic Traversal of Labyrinths," *Mathematics Magazine*, Vol. 43, 1970, pp. 125-130.
10. Abelson, H., and diSessa, A. A., Turtle Geometry, MIT Press, Cambridge, (1980).
11. Lumelsky, V. and Skewis, T., "A Paradigm for Incorporating Vision in the Robot Navigation Function," *Proc., Int. IEEE Conf. on Robotics and Automation, Philadelphia*, (1988), pp. 734-739.
12. Kunigahalli, R., Russell, J. S., and Veeramani, D., "Path Planning Algorithm for Computer-Controlled Concrete Placement," *ASCE Journal of Computing in Civil Engineering*, (1993), under review.

Automation and Robotics in Construction X
G.H. Watson, R.L. Tucker and J.K. Walters (Editors)

Space Scheduling for Construction Progress Planning and Control

I.D. Tommelein[a] and P. P. Zouein[b]

[a, b] Civil and Environmental Engineering Department,
The University of Michigan, Ann Arbor, MI 48109-2125, USA

Abstract

Space scheduling pertains to allocating work space to resources associated with activities in a schedule as it changes over time. This is an important aspect of materials management in dynamic environments in which many pieces of equipment, crews, and possibly robots move about, as congestion and interference would seriously hamper production. Advance space scheduling is then warranted to reduce the need for real-time sensing and adjusting. Reactive space scheduling may be needed when problems have arisen and must be alleviated. In any case, space-schedule data must be made available in a timely fashion to all parties who need it. The MovePlan model for dynamic layout planning is described to illustrate the overall significance of space scheduling to construction progress planning. MovePlan uses a two-dimensional representation of space to keep the computational costs associated with space-time calculations low. Accordingly, the model does not require a powerful workstation, but runs on a laptop computer—and is expected to run on a palmtop—that could easily be taken out into the field. MovePlan has been loosely integrated with a space-schedule conflict resolver, named ConRes, which delays activities or changes their resource needs to lower the demand for space in so-called problematic time intervals. Issues pertaining to data collection for space scheduling, scheduling progress control, and data dissemination are discussed. A futuristic scenario is presented to illustrate the use of palmtop computing for space scheduling on site.

1. INTRODUCTION

In the last decade, microcomputer development has progressed tremendously and computer hardware costs have plummeted. Personal computing devices are now ubiquitous in construction home offices and on construction sites.

Several site materials management programs, including those described in this paper, are currently being developed to run on networked laptop computers. While such programs exhibit many useful features, they are limited in use due to the hardware on which they run. To make on-site computing become reality, wireless communication is a necessity and robust hardware must be developed to resist the harsh site conditions (e.g., large temperature gradients, dust, impact, and variable weather) under which construction is performed.

The further decrease in size from laptop to palmtop computers, together with the standardization of wireless communication protocols, will change site management techniques. Their combination is likely to have a major impact on field productivity. In anticipation of the advent of such low-cost devices, the data needs of field personnel should be re-engineered to improve data collection, planning and scheduling, control, as well as data dissemination. Specifically, the focus of attention should be on characterizing construction resources in terms of their handling, including the space they require and the timing of their presence on site (Tommelein et al. 92). Research must investigate new methods to effectively model resources tied to site space needs and scheduled time to augment existing materials management systems (Bell et al. 87, Bell and Stukhart 86, 87; Lundberg and Beliveau 89).

2. TERMINOLOGY

When space is limited, tradeoffs between activity sequencing, timing, method selection, resource allocation, and space allocation are necessary to plan executable schedules and feasible layouts. When space is abundant, similar tradeoffs exists as, for example, travel distance and reach of selected hoisting equipment affect materials handling efficiency. The resource *space* as a variable in the scheduling process has not been modeled with much detail, possibly because of the complexity of the interaction between spatial and temporal aspects of resource needs and availability. To define the possible degree of interaction, two terms are used:

Dynamic layout planning (Tommelein and Zouein 93) refers to the unidirectional interaction between the scheduling and space allocation tasks. After a sequence of activities has been defined and resources with their dimensions have been associated with each activity, the schedule is considered to be final. It serves as input to the space allocation task.

Layouts can then be constructed for selected time intervals, each layout using the resources that are scheduled to be present during the interval it spans. It may be that no feasible layout can be constructed to accommodate all resources in each interval. In that case, the dynamic layout planning problem cannot be solved, possibly—but not necessarily—because the schedule was frozen.

Space scheduling (Zouein and Tommelein 93) refers to the bidirectional interaction between the scheduling and space allocation tasks. (Dynamic layout planning is thus a subtask of space scheduling.) A sequence of activities must be defined, resources with their dimensions associated with each activity, and space on site must be identified for each resource. Schedule construction is performed concurrently with layout construction. When layouts are found to be infeasible, schedule changes might alter the problem so that it becomes solvable. For example, activities can be delayed or have their resource use level changed to shorten or lengthen in duration.

Dynamic layout planning and space scheduling both are limited resource allocation problems involving the inherently two- (actually three-) dimensional variable *space* in addition to the scalar variables equipment, labor, materials, time, and possibly money. In general, limited resource allocation problems can be solved using heuristic methods only; no optimization techniques exist. This is one reason for which people who use such models conduct what-if analyses and experiment with alternate problem formulations, before accepting an automatically-generated solution. The development of interactive, heuristic decision support tools to aid them with dynamic layout planning and space scheduling is therefore advocated.

It is generally the case that construction schedules are too abstract (esp. executive-level milestone- and work package-, but even task-level schedules), i.e., they do not represent enough detail to model the need for, use, and availability of space. Only overall site arrangements can be studied early on in the planning process. Not enough decisions have been made at that time to allow for detailed space schedule development. Sufficient detail at the process level must be available (e.g., during daily crew-level planning) to warrant detailed space scheduling. Accordingly, it is at the process level that spatial and temporal data must be collected and used to enable individual workers and crews to organize their work on site and to gauge their progress. The term **progress planning** is used to denote this.

3. SPATIAL AND TEMPORAL DATA

3.1. Data Collection

The collection of large amounts of real-time field data has been made possible by the development of new tools such as laser-controlled and bar-coding systems. Laser-controlled systems assist in positioning the blades of grading equipment. This benefits individual equipment operation. Bar-coding of construction components (e.g., Bell and McCullouch 1988, Rasdorf and Herbert 1990a, 1990b) facilitates component identification and inventory status tracking. This benefits site management (performing tasks such as purchasing, inventory control, and accounting), but there is little mention in the literature as to how these tools and the collected data can be used to improve crew-level operations and construction planning overall.

Data needed at the crew level should provide information to individual workers or superintendents when they are planning or executing work. It should provide answers to questions such as:
Where is the formwork?
Have the windows already been delivered? When will they arrive?
What is the nearest supply of two-by-fours?
Could I use the sheet rock that is over there? What will it be used for?
What is the best way to get to the management trailer?
Could I use that forklift to move the brick pallets closer to my work area?
Will that area be available tomorrow to set up a fabrication shop?

Clearly, much of the data needed to answer these questions changes over time. Materials may be brought to the site by one person, who needs to inform others of their arrival. Supplies get replenished during the course of the day, so workers who are waiting for them must be readily informed of their availability. The nearest supply depends on where a worker is standing and varies as work progresses and that worker moves around. A material may appear to be available to one worker, but may have been designated for sole use by others. Congested and dangerous work areas must be circumvented if possible. Utility equipment gets used opportunistically, it does not get scheduled; what appears to be idle at one time, may be tied up later. Crews need to communicate their plans to avoid future competition for resources or interference. Not only is it necessary to make spatial data available in a timely fashion to all people who need it; logging data as it changes over time must be done by all (or at least many) people on site as it is beyond the scope of what a single individual can do.

3.2. Planning and Scheduling

Few tools, beyond databases and traditional CPM programs, are currently available to aid with materials management as a true construction scheduling task. Construction materials come in great variety and arrive to the site under uncertain conditions. This poses challenges to management, beyond those of materials management in a steady-state manufacturing environment. The space resources require on site, their bulkiness, etc. may be a major hindrance in executing work according to plan. Managing materials accounts for a substantial part of construction cost, however, and research should investigate which models lend themselves to use on site.

3.3. Progress Control

Spatial data is particularly important at the process level. This is the level at which interferences can be prevented, excessive travel times identified, and appropriate material handling means selected. Three-dimensional CAD tools have been developed in recent years for checking interferences during the on-site movement and placement of exceptional loads. Little is formalized about the handling of most other materials on site, which are in the majority. The process level is also the level at which work progress can best be measured and hurdles overcome. It is therefore advocated that spatial data be included in control systems, which all too often limit themselves to modeling only durations and costs. An augmentation of simulation tools with spatial abstraction mechanisms might fill the need for tools to study the space-time-cost trade-off.

3.4. Data Dissemination

Materials data should be made available to all on site who need it. As different people have different data needs, appropriate abstraction mechanisms must be developed. Discussing these is beyond the scope of this paper. One issue worthwhile raising is that everyone on site will need a communication device to enter new and query existing data, but tight controls must be kept to guarantee data consistency and correctness. Wireless palmtop computers equipped with intelligent interfaces are expected to be able to meet those needs.

4. ENVISIONED FIELD IMPLEMENTATION

4.1. Software Needs

Few existing computer-based scheduling programs represent the resource *space* as an explicit variable. New software is therefore being developed to keep track of space schedules and to answer questions such as those mentioned in section 3. Our work includes three models, which address the space scheduling task with increasing degrees of complexity.

4.1.1. MovePlan

Laying out temporary facilities on site all too often means to manually mark up a site arrangement drawing, showing different overlapping laydown areas that will be present for consecutive time frames. This results in a complicated drawing that can be interpreted only by the one who made it, and that by no means lends itself to updating as construction progresses. Its use for task coordination and materials management is thus very limited.

The MovePlan interactive dynamic layout planning tool (Tommelein and Zouein 1993) exactly addresses this problem. MovePlan needs as input a schedule of activities annotated with their resources, the two-dimensional and rectangular space each resource will need on site, data describing the permanent resources on site, and the site's boundaries. It enables users to create layouts for time intervals of any length, related to parts of the early-start project schedule.

The tool aids in solving this dynamic layout planning task by providing templates representing the resources that are on site for the selected interval, and a worksheet in which those templates can be moved around. The worksheet is scaled to the site boundaries and shows where permanent and previously-positioned resources are. When time-overlapping layouts are created, MovePlan maintains consistency among them. A resource that has been assigned to a position in one layout, is shown in the same position in any layout that overlaps in time with the first one and that includes the resource. MovePlan freezes the schedule as soon as the first layout has been saved. A play-back of the dynamic layout sequence shows how the site space use changes over time. The tool thus provides decision support to field engineers who arrange resources on construction sites and coordinate subcontractor or crew activities.

The MovePlan user may find that the layout for a chosen time frame is unsatisfactory (e.g., because it is too congested or no space is available to locate a critical resource). In this case, it may be possible to delay the arrival of some resources in order to free up space for others, or to assign different resources to activities. Pertinent data must then communicated to a conflict resolver that makes the appropriate schedule changes. Schedule changes are subsequently returned to MovePlan for the user to continue with the layout task.

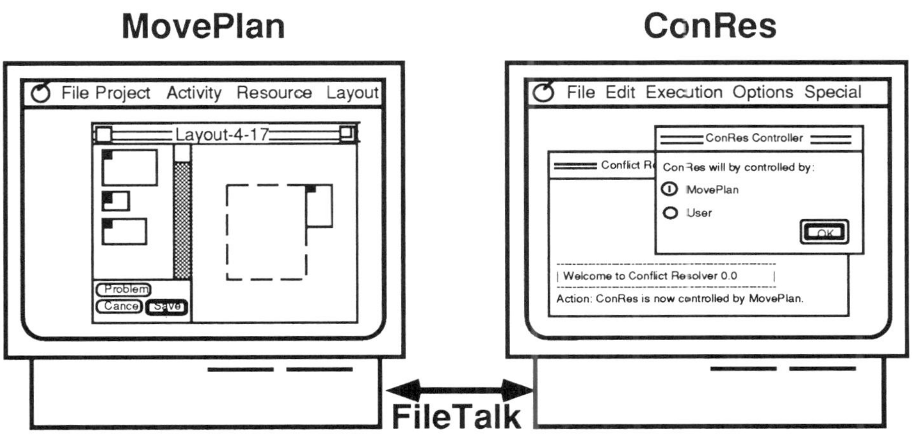

Figure 1. MovePlan–ConRes Interface

4.1.2. MovePlan-ConRes

MovePlan has been loosely integrated with a space-schedule conflict resolver, named ConRes (Figure 1). ConRes opportunistically chooses one of several strategies for modifying a schedule so that space needs can be better accommodated in time intervals that are identified as problematic (Tommelein et al. 1993). Such strategies delay one or several activities to remove a resource from the current time frame while minimizing project delay or minimizing the increase in resource idle time, or, alternatively, change an activity's resource use level while minimizing project delay or maximizing the freed-up space. These strategies are by no means exhaustive, nor do they guarantee that a better schedule will be produced. Additional strategies for space schedule conflict resolution are being investigated.

4.1.3 MoveSchedule

The MoveSchedule program for space schedule construction tightly integrates and automates the space allocation and scheduling tasks. It is being developed by Zouein and Tommelein (1993). One key feature is that resources are modeled with lead times (the time period over which materials are on site before the activity starts) and safety stocks (the minimum amount of material on site that is maintained at all times until the material is no more needed; when a supply falls below the safety stock, it is replenished). The model also reflects resource depletion during the execution of an activity. It can thus represent just-in-time and just-in-case deliveries.

MoveSchedule will lend itself well to integration with a materials inventory control system, possibly tied to a bar coding system for tracking material arrival and placement in the permanent facility.

4.2. Hardware Needs

In 1983, the BRT pointed out that "Too little use is made of new, cheap micro-computers and software." (BRT 1983, p. 56). While this may still be true, one needs to look ahead and assess how the advent of palmtop computers is to change construction materials management. The above three programs all run on laptop computers. We plan to reimplement them in a language that efficiently uses palmtop capabilities. As soon as palmtops will be marketed at a reasonable price, which is expected to be the case in the near future, this software will be available for site use.

4.3. Futuristic Scenario

It is envisioned that individual workers or crews will be equipped with palmtop computers so that they can obtain and log space schedule data. Figure 2 shows a worker who informs management in the trailer about the supply of materials upon their arrival to the site. This information is rebroadcast to others. A graphical interface can be used to sketch the laydown location relative to other items on site, or even better, hand-held surveying equipment can measure the materials' actual position on site (Lundberg and Beliveau 89). This setup provides workers with all data needed to answer their questions and enables them to point out problems to management so that remedial action can be taken.

In addition, schedules and space needs is entered into the palmtop space scheduling tool by crews who plan their next day's work. Pertinent data is abstracted from it and broadcast to others who might detect interference problems or resource conflicts. Mediation software them helps them negotiate priorities.

This setup assumes that data can be entered and retrieved by all palmtops on site. Relatively powerful computers are therefore needed. How such a distributed communication system for on-site materials management will work within the current organizational structures and materials handling practice (e.g., assigning the appropriate responsibility and authority for adding, deleting, and modifying data) is a wide open area of research.

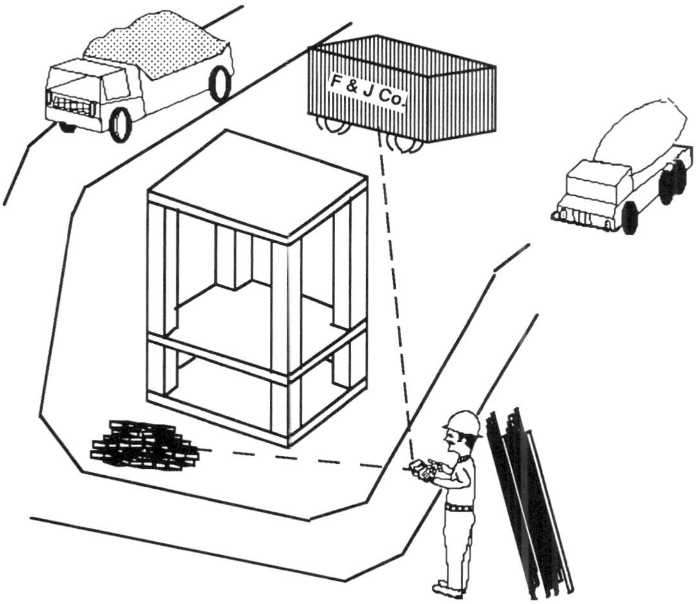

Figure 2. Futuristic Scenario for Use of Space Schedule Data

5. SUMMARY

Ten years ago, the Business Round Table concluded that "The material and equipment built into industrial and commercial facilities and power plants cost an estimated $82 billion in 1979—some 60% of the total cost of those structures. But a CICE survey concludes that a significant part of the labor cost that went into that construction was wasted because materials and equipment were not available at the site when they were needed. That is the kind of foul-up modern management systems can minimize." (BRT 1983, pp. 29-30). At present, microcomputers and management techniques have not been exploited to the maximum of their potential. In the mean time, the technology has leaped forward and even more radical improvements to construction practice can be expected from the laptop and palmtop computer revolution. The desired use of such devices on site must therefore be researched, in anticipation of their wide-spread availability. Accordingly, the MovePlan, ConRes, and MoveSchedule models were briefly introduced, to show how space and schedule data can effectively be used in construction operations planning.

Issues pertaining to data collection for space scheduling, scheduling progress control, and data dissemination were discussed for obtaining an original and adjusting an existing space schedule if plan deviations become intolerable. A futuristic scenario illustrated the use of palmtops for space scheduling on site.

6. ACKNOWLEDGMENTS

The development of MovePlan and ConRes was funded by Grant MSM-9010394 from the National Science Foundation, whose support we gratefully acknowledge. Any opinions, findings, and conclusions or recommendations expressed in this material are those of the authors and do not necessarily reflect the views of the National Science Foundation.

7. REFERENCES

Bell, L.C., Bishop, R., Booth, F.W., Bush, R.J., Carmody, J.M., Carr, R., Cato, J.C., Dodd, G., Early, J.W. (1987). *Project Materials Management Handbook.* Matls. Mgmt. Task Force, CII, Univ. of Texas, Austin, TX.

Bell, L.C., McCullouch, B.G. (1988). "Bar Code Applications in Construction." *J. Constr. Engrg. and Mgmt.*, ASCE, 114 (2), 263-278.

Bell, L.C., Stukhart, G. (1986). "Attributes of Materials Management Systems." *J. Constr. Engrg.and Mgmt.*, ASCE, 112(1), 14-21.

Bell, L.C., Stukhart, G. (1987). "Costs and Benefits of Materials Management Systems." *J. Constr. Engrg.and Mgmt.*, ASCE, 113(2), 222-234.

BRT (1983). *More Constr. for the Money. Summary Report Constr. Industry Cost Effectiveness Project.* Business Roundtable, 96 pp., New York, NY.

Lundberg, E.J., Beliveau, Y.J. (1989). "Automated Lay-down Yard Control System—ALYC." *J. Constr. Engrg. and Mgmt.*, ASCE, 115 (4), 535-544.

Rasdorf, W.J., Herbert, M.J. (1990a). "Bar Coding in Construction Engineering". *J. Constr. Engrg. and Mgmt.*, ASCE, 116 (2), 261-280.

Rasdorf, W.J., Herbert, M.J. (1990b). "Automated Identification Systems—Focus on Bar Coding". *J. Computing in Civil Engrg.,* ASCE, 4 (3), 279-296

Riley, D., Sanvido, V. (1992). "Site Material Management System." *Proc. CIB 92 World Bldg. Congress*, 596-597, Natl. Research Council Ottawa, Canada.

Sanvido, V., Paulson, B.C. (1991). "Site Analysis using Controller-Function Charts." *J. Constr. Engrg. and Mgmt.*, ASCE, 117 (2), 226-241.

Tommelein, I.D., Castillo, J.G., Zouein, P.P. (1992). "Space-Time Characterization for Resource Mgmt. on Constr. Sites." *Proc. 8th Conf. Comp. Civil Engrg.*, ASCE, New York, NY, pp. 1042-1049.

Tommelein, I.D., Dzeng, R.Y., Zouein, P.P. (1993). "Exchanging Layout and Schedule Data in a Real-Time Distributed Environment." *Proc. 5th Intl. Conf. Comp. in Civil and Bldg. Engrg.*, ASCE, New York, NY, June.

Tommelein, I.D., Zouein, P.P. (1993). "Interactive Dynamic Layout Planning." *J. Constr. Engrg. and Mgmt.*, ASCE, New York, NY, June 1993.

Zouein, P.P., Tommelein, I.D. (1993). "Space Schedule Construction." *Proc. 5th Intl. Conf. Comp. in Civil and Bldg. Engrg.*, ASCE, New York, NY, June.

Automation and Robotics in Construction X
G.H. Watson, R.L. Tucker and J.K. Walters (Editors)

INTEGRATION IN DESIGN AND SIMULATION

by

Robert F. Sullivan, Assistant Manager
Advanced Construction Systems
Stone & Webster Engineering Corporation
245 Summer Street
Boston, Massachusetts 02210

ABSTRACT

Today, three dimensional (3-D) solids modeling design technology and the deployment of integrated relational databases on projects in the construction industry have provided the opportunity for advancing the level of achievable integration and automation in the design, construction, and operation of major industrial facilities. The key development has been the linking together of these technologies into one system to allow for the exchange of information between the model and the database (geometric versus nongraphic data), while providing an advanced medium for the accessing, retrieval, and presentation of a project's data. This graphically linked database technique allows the design model to be utilized as a simulation tool to display construction sequences, material availability, installation progress, and other functions, in addition to its primary function of drawing production.

INTRODUCTION

Since 1987, Stone & Webster Engineering Corporation (Stone & Webster) has been utilizing 3-D graphically linked database systems to integrate the design process, and to provide graphical simulation capabilities to a number of projects. Integration is brought about through programs developed by Stone & Webster that link IBM's DB2 relational database with the CATIA 3-D solids modeling application.

Three-deminsional solids modeling, by itself, is a powerful design tool that integrates the various design disciplines to detect interferences before the production of drawings. However, its linkage to a relational database expands the capabilities of the process by providing a repository for geometric data associated with the solid elements of a model. These data can then be imported into application programs for processing. Likewise, data contained in the database can be related to the components in the model to allow the user to obtain information about a facility by prompting the model or to use these data to construct simulations that are displayed using the model. Stone & Webster uses a number of these applications, applying these techniques in the areas of project controls, facility engineering, construction sequence modeling, and material management.

APPLICATIONS

Stone & Webster has recently deployed its 3-D modeling system for facility design, and its graphically linked **CO**nstruction **MAN**agement **D**isplay **S**ystem (COMANDS) for planning, estimating, scheduling, and material management on a 550 MW pulverized coal fossil retrofit project. The scope of the project consisted of replacing the existing scrubber and precipitator units, and replacing more than 800 tons of exhaust duct during a 12-week outage.

The facility was one of two attached units on a congested site. Due to the congestion restraints, tight schedule, and penalties for delays associated with the project, it was decided that the 3-D modeling system would be used during the proposal stage as a planning tool to help determine whether the work scope could be achieved in the allotted time. The existing configuration and the owner's conceptual duct design were modeled by two individuals over a 3-week period based on existing drawings and data gathered during site walkdowns.

The purpose of this approach was to create a model that could be used as a communication tool in the planning process which would allow construction specialists to visualize various construction strategies. The geometric data in the model also provided estimated quantities used in the bid preparation and for determining durations of construction activities. Use of the model allowed construction specialists to position construction equipment, to verify access, and to develop sequences of construction that maximized the content of preoutage work. The result of this planning process was a pictorial display of the optimum construction sequence simulating progress over time. This product was then used during internal management reviews and in presentations to the client to convey the scope and execution plan of the project.

After the contract was awarded, detailed design of the facility was conducted using 3-D for all new structural and mechanical items, starting with the model developed during the proposal. This was inclusive of foundations, structural steel, ductwork, mechanical equipment, and piping. The project's design drawings were then produced from the model by employing the 2-D capabilities of the system.

During this process, construction specialists and project control engineers were able to utilize copies of the model for constructibility reviews, planning, and procurement. As the various engineering disciplines completed design, personnel used the design model to customize products for their specific needs.

The construction specialists first used the proposal model to refine the demolition sequence of the existing precipitators and ductwork. Crane studies were performed to analyze crane reach, capacity, and equipment access. The 3-D system allowed for weight densities and center of gravity to be applied to the solids representing the existing ductwork so that the sections to be removed could be sized and rigging pick points determined in accordance with the position, lift capacity, and reach of the crane. Using the geometric data and the graphic images of the model, a demolition sequence was developed using Primavera's Finest Hour scheduling package.

As design approached completion, the construction group used copies of the model to produce sequence models of the installation process. Construction also sized the ductwork sections to be fabricated using the geometric data features of

3-D model to calculate weight and rigging pick points. When these tasks were completed, isometric piece mark sketches were produced detailing the weight of each duct section, piece number, and priority for fabrication (Figure 1). These data were then forwarded to the duct fabricator to coordinate the manufacturing and delivery schedule.

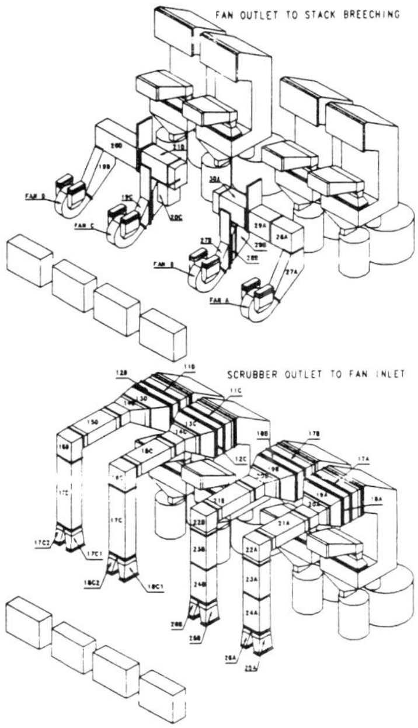

Figure 1. Duct Work Piece Mark Sketch

Isometric sketches were also produced for piping drawings which detailed the location of field welds, spool numbers, hanger locations, and bills of material. These data were incorporated into the prospective pipe fabricator's bid package.

After the Finest Hour construction schedule was completed, the output was electronically passed to the modeling system for integration with the construction sequence model. The resulting document (Figure 2) displayed a construction sequence model linked to a schedule where each week's activity had a corresponding view of the model representing progress arrayed above the scheduled activities.

UTILIZATION OF COMANDS

The material management module of the COMANDS system merges piping specification data contained in the relational database with data extracted from CATIA piping models to produce bills of material that are used to define piping material requirements. What makes COMANDS unique is the graphic data interface that allows the model to be used as a medium for displaying information. In this case, the user is supplied with two graphic data interface functions. The first permits the user to place the curser on any piping component to view an on-screen display of the component's description or a bill of material of all the components on the selected component's piping line number. The second function displays material shortages on the model, and lets the user prompt the model to obtain a component's material status (Figure 3). The material shortages are calculated by identifying all material that will not be on site to meet a user input schedule date; these are displayed in red on the model.

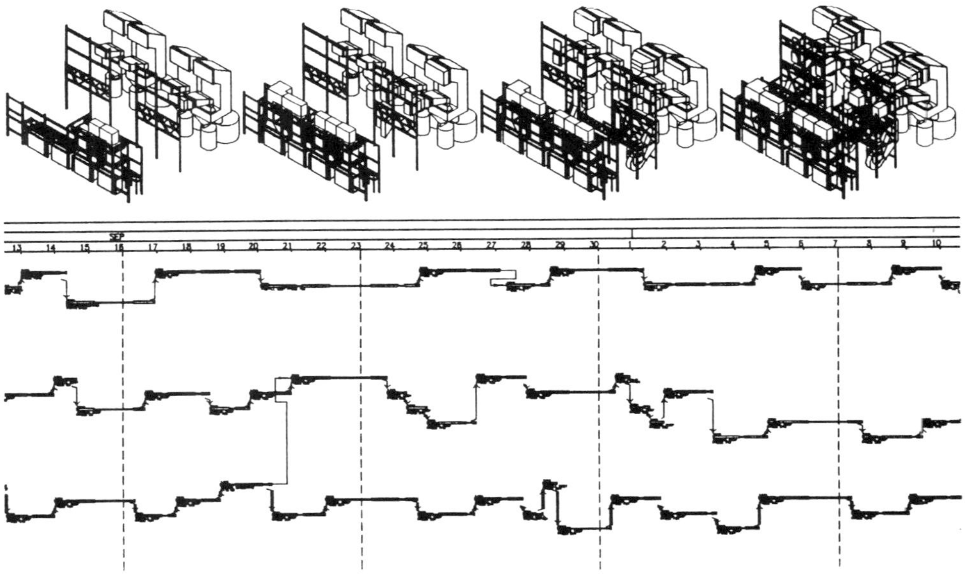

Figure 2. Construction Sequence Model Linked to Schedule

The COMANDS material management module summarizes the material requirements into material summary reports that cross-reference a project's material requirements to their source line numbers and models. These material requirements can be allocated to requisitions within the system by comparing the totals on a commodity basis to materials already on requisition. The requisitions are then processed into purchase orders and an inventory system established to track material delivery, receiving, storage, on-hand availability, and issuance status.

COMANDS was initially deployed to provide piping bill-of-material takeoff data to the pipe fabricator from the engineering models and to provide estimated piping quantities when developing the project's budget and schedule. Later in the project it was used to take off and requisition field-installed piping systems (fire protection deluge and small bore piping) and to track piping material availability and status.

The COMANDS Project Controls module uses the design model for quantity development and for reporting and recording of installed quantities. The application employs traditional work breakdown structure and earned value methods for monitoring a project's progress and is integrated with Primavera network planning

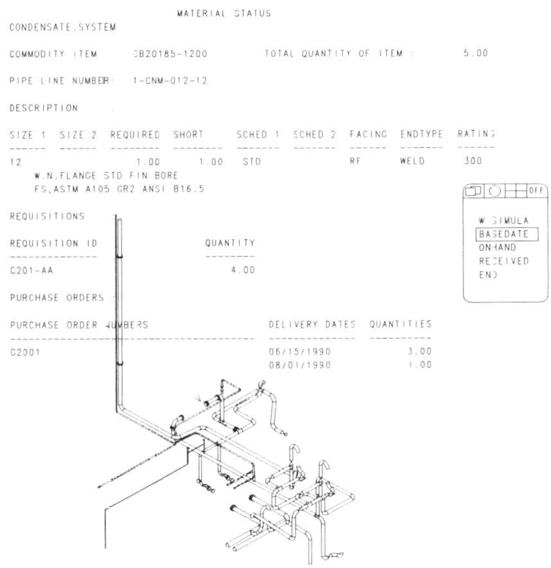

Figure 3. Trial Withdrawal Material Shortage Status

software. The system utilizes a graphic data interface that allows the user to report the installation of a facility's components by positioning the curser on the installed component on the design model. The system then passes the geometric data concerning the item to the database to calculate the design quantity of the installation while highlighting the installed component in the model to provide a visual record of installations on the model. Similarly, the user can also prompt the design model through the graphic data interface to display schedule, budget, and installation status of components in the model (Figure 4).

The Project Controls application of COMANDS is presently being used to monitor steel and duct replacement. The version being deployed was first established on the mainframe where the work breakdown structure was used to cross-reference the solids in the model to the accounting structure and activities in the network. Data extractions were then used to pass the geometric information of the design model to the account structure for establishment of the budget quantities used to track construction progress.

After the database was established, views of the model, along with the database, were accessed, via modem, on a personal computer (PC) located at the site. Stone & Webster's in-house developed software, STONE*view*®, which allows the PC to communicate with the design model and database across regular telephone lines, was utilized. The installed components were then updated on line via the PC and passed by modem back to the mainframe database. The user can also use this method to view the status on line by prompting the model to identify a component's design quantity, installed status, and schedule status. The advantage of this approach is the ability to provide the graphic data interface capability without use of a site-

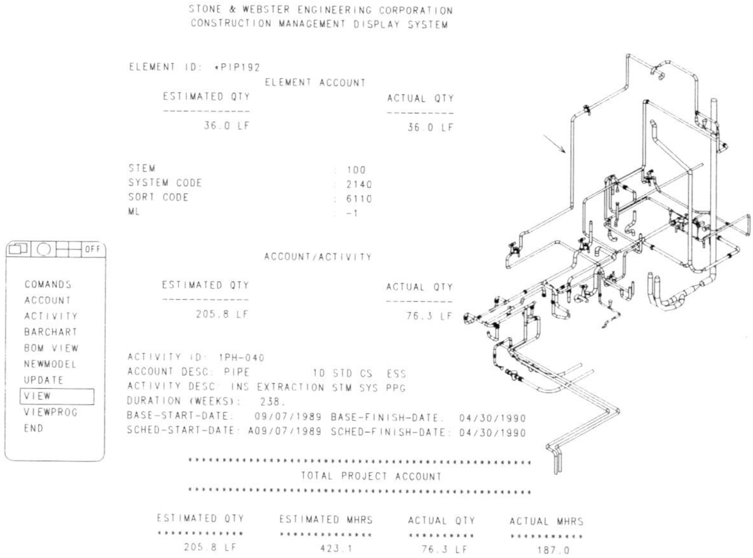

Figure 4. Schedule\Account Status

installed, high-end graphics workstation, which would have necessitated the installation of a dedicated data communication link to the jobsite.

OTHER APPLICATIONS

Recently, Stone & Webster utilized its 3-D solids modeling system, coupled with animation software, to assist the Corporation in obtaining a major contract award. After award, the system was used as an integral part of the process selection work scope. The project award consisted of a $430-million emission control system for a major western coal-fired power plant consisting of three 750 MW units. The work scope includes process selection, engineering for refurbishment of existing stacks, installation of crossover duct work, and installation of a scrubber system for each of the three units.

As part of the proposal effort, the existing site facilities were modeled in CATIA. This modeling effort consisted of envelopes of the main powerhouse structure and yard structures combined with previously modeled generic scrubber configurations. The modeling was performed over a 2-week period, and the resulting product was then animated utilizing Wave Front software to display on video a computer simulation of the construction process. Construction sequence models integrated with Primavera schedules were produced and utilized as part of the proposal presentation to communicate Stone & Webster's project execution plan.

After the award of the engineering contract, the 3-D modeling system was used extensively as a simulation tool to evaluate alternatives during the process selection. As part of this scope, 10 scrubber vendors submitted multiple proposed designs for the project. These configurations were then modeled with the proposal model, resulting in a total of 25 separate model designs. The evaluation process used these models to review the scrubber designs from a constructibility, capital cost, access, schedule, and safety perspective with the client.

The models provided geometric data used to calculate quantities of duct, piping, and steel, as well as a visual display for the various layouts. In turn, these quantities were used to formulate the capital cost estimates. This procedure provided the basis for selection of the process and configuration layout of the project.

In preparation for client review, STONE*view*® was used to transfer isometric views of the various model, configurations to a laptop PC. The presentations were made at the client's offices with the use of overhead projection equipment. STONE*view*® allowed Stone & Webster personnel to visually display the various designs to aid the client in process and layout selection. The model views could be panned, zoomed, and rotated. Comments could be annotated on the views, thus providing a superior medium for communicating the various project configuration options and layouts.

When the final selection was agreed upon and comments had been incorporated, a construction sequence and walk-through animation video were produced. These were incorporated into a informational videotape about the project that was used in presentations to the public.

CONCLUSION

Today's 3-D solids modeling design tools coupled with relational database, network, and animation software products are expanding the integration and portability of the design and construction functions. Computerized design models are replacing the traditional plastic models as a tool for detecting design interferences and for performing constructibility and operability reviews. Examples of this integration are the electronic interface of a models geometric data into engineering structural analysis and stress programs, and the importation of quantity data for procurement and estimate development.

The design model is being incorporated into the construction planning and scheduling process to refine construction sequences through simulation of the construction process. This provides a communication tool for presenting project execution plans to clients, as well as construction forces. Animation software allows the model to be used for the creation of videos that provide a portable, dynamic means for displaying a project's scope for internal reviews and to the general public.

In addition to STONE*view*®, other software products are allowing the importation of model data to PCs for conceptual and detailed design reviews or processing of data locally. The integration of graphics systems to relational databases is providing intelligent design models that can display data in an advanced medium for applications in project controls, material management, and plant operations. These applications can visually simulate material shortages and construction progress on the model. They provide a graphic data interface that acts as a window to a facility's database for obtaining project information.

The results of this technology are automating the engineering/construction process, while improving coordination and communication among engineering, construction, and client organizations. This is increasing the quality and productivity of the construction industry today.

Automation and Robotics in Construction X
G.H. Watson, R.L. Tucker and J.K. Walters (Editors)

FIELD TESTING OF AN AUTOMATED SURFACE FINISHING SYSTEM FOR LARGE DIAMETER STORAGE TANKS

Dailey, C. M., Traver, A. E., Wesley, G. L., Haas, C. T., O'Connor, J. T.

Abstract

Surface finishing of large storage tanks represents a typical area in which construction automation techniques can yield significant improvements over conventional methods. Current techniques are costly and laborious, and they expose workers to significant health and safety risks. Automation offers potential improvements in each of these areas and is well suited to the large, uniform surfaces of the tank wall. The University of Texas has developed a prototype automated surface finishing system for use on large diameter tanks. This system uses a computer-controlled motion module to refinish the vertical exterior walls of a tank. The module is configurable for both blasting and painting and utilizes conventional surface finishing equipment for these processes. The University of Texas' automated paint sprayer was field tested in January, 1992 at an Amoco pipeline facility in Alvin, Texas. These tests, which were used to demonstrate automation techniques to members of the construction community, are described in this paper. The successes of the system are reported along with the problems, and recommendations are given for improving the system. An economic analysis is presented along with future plans for the system.

1.0 INTRODUCTION

Automation of surface finishing techniques for large storage tanks can yield significant improvements over conventional methods. Current techniques are costly and laborious, and they expose workers to significant health and safety risks. A typical refinishing operation involves sandblasting a tank to remove old paint and rust, and then painting it to protect the surface from the environment. This work is performed manually using a painting scaffold or similar device to access elevated work areas. Heavy equipment and protective clothing are required, and process byproducts are released into the environment. The procedures are hazardous, exposing workers to silica dust, harmful paint components, and possible injuries due to falls. Strenuous working conditions and worker fatigue contribute to an inconsistent quality of the applied coating. This hazardous surface finishing procedure is well suited for automation techniques. The large, uniform working areas are accessible with remotely operated equipment. Development in this area of construction automation can be justified by the potential improvements in working conditions, surface finish quality, and reduction in cost. Reduced environmental contamination serves as additional incentive.

The University of Texas has developed a prototype automated surface finishing system for use on large diameter tanks. This system uses a computer-controlled motion module to refinish the vertical exterior walls of a tank. The module is configurable for both blasting and painting and utilizes conventional surface finishing equipment for these processes. An overspray hood is included to reduce contamination of the surrounding environment during painting. The module attaches to a tank's wind girts with steel cables, and its position and velocity are controlled by servo-driven hoists. This automated system also controls the initialization and termination of the blasting and painting operations. This system replaces one person of the three-person crew generally required to refinish a tank and removes the operators from the immediate vicinity of the work area. A description of the automated paint sprayer and the results of preliminary tests were presented at the 8th ISARC.[1]

The University of Texas' automated paint sprayer was field tested in January, 1992 at an Amoco pipeline facility in Alvin, Texas. These tests were used to demonstrate automation techniques to members of the construction community. The field test results were favorable, revealing that automation of surface finishing techniques for large uniform structures is feasible and represents considerable improvements in worker safety. A consistent, high quality surface finish was obtained with minimal risks to the operators. The system has the potential to reduce the costs required to refinish a tank. The field tests of the automated paint sprayer are described in this paper. The successes of the system are reported along with the problems, and recommendations are given for improving the system. An economic feasibility analysis is presented, and future plans for the system are discussed.

2.0 MOTIVATION FOR AUTOMATING THE TANK PAINTING PROCESS

There are five commonly recognized primary motives for automating construction operations: (1) increased productivity/reduced cost, (2) increased quality, (3) safety - minimization or elimination of a hazard (to humans), (4) the need for super-human capabilities (such as lifting heavy objects, or ability to withstand severe work environments, such as arctic conditions), and (5) worker substitution given an inadequate supply of labor. In this development, the primary motives for automating tank spraying are increased productivity, reduced cost, and increased safety.

Sandblasting and painting operations present a multitude of potential health hazards to the workers. Perhaps the most obvious safety hazard is the inherent danger of working above grade on scaffolding. With this approach, workers are exposed to the risk of fall at all times. According to the National Safety Council, 70% of all serious injuries to coating crews are caused by falls. [2]

In addition to the risk of falls, blasting and painting crews are exposed to long-term health hazards posed by process by-products. These include silica dust (from blasting), old surface paint, air-borne atomized paint particles, and the chemical fumes of curing paint. Crew protection methods include auxiliary breathing equipment, protective helmets, heavy clothing, and safety lines and lanyards. The automation of surface finishing can significantly reduce many of these safety risks.

In addition to eliminating or minimizing safety concerns, labor costs and task duration can also be significantly reduced with an automated operation. With current manual methods, human labor contributes significantly to the overall cost of operations. The labor costs alone of blast cleaning and painting a square meter of tank surface in 1990 ranged from $9 to $17. [3] Of course, with an automated approach, much of this required labor can be eliminated. Considering the vast amount of surface area to be coated, the cost savings can be substantial.

Increased productivity can also serve to reduce task duration. With conventional methods, the daily productivity rate of a three-person crew is approximately 100 square meters for blasting or 200 square meters for spray painting. Preliminary estimates of daily productivity for the proposed automated system are 220 square meters for blasting and 2000 square meters for painting, assuming a 90% duty cycle. This assumes a two-person crew which is primarily responsible for equipment set-up and monitoring. [4]

3.0 FIELD TESTS

3.1 Prototype Description

The automated paint sprayer consists of an aluminum motion module which rolls on the surface of a tank (Figure 1). The module attaches to the wind girt of a tank via two 1.0 cm diameter steel cables. The position and velocity of the module are determined by the amount and speed of the cable splayed through two hoists located at the rear of the module. Squirrel cage assemblies mounted beneath the motion module retain the excess cable from the hoists. [5, 6]

A closed loop system is used to control the operation of the hoists. An IBM PC provides a means to create a control program for either a blasting or a painting operation. The program is sent to an axis controller which performs the control operation using feedback from a motor-mounted encoder. An amplifier is included to boost the control signal to the motors. All computer components are mounted in a portable console which remains on the ground during operation of the unit.

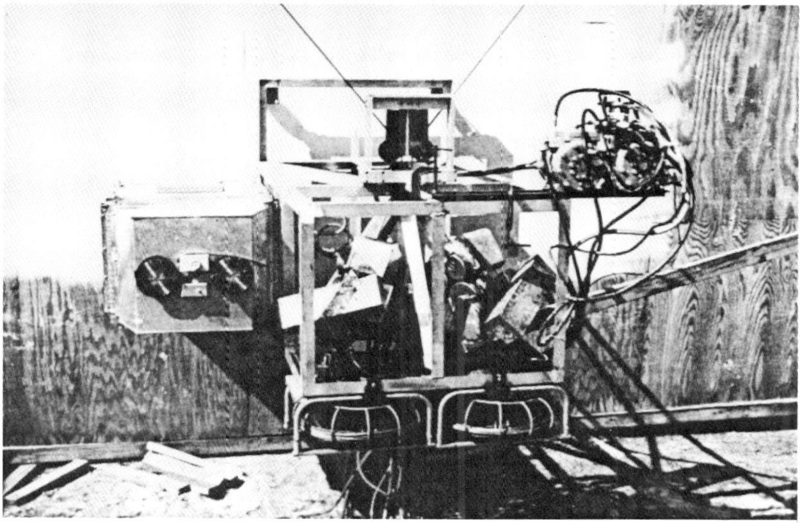

Figure 1. Automated Paint Sprayer

The paint system can be configured for either blasting or painting by modifying the process module attached to the left side of the motion module. This process module contains the linkages and motor used to oscillate the nozzles for the blasting and painting processes. A separate set of linkages is used for each process, and the system must be configured accordingly. An overspray unit, used for the painting process, mounts to the back side of the process module.

The automated paint sprayer is used to refinish the large uniform areas of the tank. The top and bottom of the tank, as well as the areas around the piping and gauge lines, are still refinished using conventional, manual methods.

3.2 Setup

The field tests were conducted in January, 1992. The automated paint sprayer was delivered to the site and positioned in an accessible area inside the tank dike. Special anchors were fabricated to attach the cable ends to the wind girts. Electric power for the controllers and hoists was supplied by Amoco, and a 750 SCFM diesel-powered compressor was rented to supply the air for process equipment operation.

The distance between the cable mounts was approximately nine meters, allowing a work area six meters wide and six meters high. While this was smaller than the area desired for final production, it was a good size to test the functionality of the components. The system was configured for sandblasting and attached to the tank. Counterweights were added to balance and stabilize the motion module.

3.3 Sandblasting

The blasting nozzles were positioned 66 cm from the surface of the tank, providing a vertical path 51 cm wide. A coarse, sharp-edged blasting slag was selected as the blasting medium. The speed of the motion module was adjusted to produce the specified surface finish. A vertical speed of 3.8 cm/sec was used to produce the SSPC's SP-6 Commercial Blast Cleaning Standard as required by Amoco's specifications.

The blasting operation was considered to be satisfactory once the sand was flowing smoothly. Initially, the sandblasting operation was hampered by non-uniform flow of the sand from the blast pot. The pattern was interrupted when the flow of sand stopped. With a consistent flow of sand, it was possible to clean a uniform area to the specified finish. Blasting efficiency at the welds was also considered to be satisfactory.

3.4 Painting

The system was configured for painting by modifying the process module and adding the overspray hood. The paint hood was positioned approximately 2.5 cm from the tank surface to clear surface irregularities in the tank wall and a new control program was generated to facilitate the painting process. Overlap parameters were adjusted to provide sufficient coverage of the tank surface and the module speed was adjusted to 15.2 cm/sec.

The automated paint sprayer applied a uniform, high-quality coating 0.076 mm thick as specified. In general, the system performed satisfactorily. The wheels were able to traverse most of the welds with little difficulty. The system did encounter several obstacles that caused it to momentarily hang up. A warp in the tank wall and a 12 cm thick port cover both caused the system to deviate from its programmed motion. Though the system did get over these obstacles, the painting pattern was interrupted.

Although the quality of the painted surface was acceptable, the boundaries of the work area were not uniform. The overall pattern dipped as painting progressed. This dip, shown in Figure 2, was severe enough that painting operations were halted to prevent the system from hitting the ground. This problem was traced to the design of the drive pulleys on the hoists. The effective radius of these pulleys varies as the applied load varies. Since the system controller monitored the position of the hoists and not the position of the motion module, no compensation was made for the deviations between the desired and actual positions.

Figure 2. Paint Pattern

3.5 Results/Future Plans

The automated paint sprayer worked as expected and proved the design and process concepts. Both sandblasting and painting were accomplished with proper adjustment of the parameters. The problems identified during the tests served as a good indication of the general problems with automation of construction processes.

Future plans for the system include direct monitoring of the position of the motion module to compensate for the dip in the pattern. Plans also exist to monitor the paint pressure to predict clogging of the nozzles. An increase in pressure would immediately halt the painting process and allow the operator to clean the system, thus minimizing the interruptions in paint film quality.

A containment system for the blasting byproducts is also planned. Such a system would prevent silica dust and paint chips from being released into the environment and allow recycling of the blast media.

4.0 ECONOMICS

As mentioned previously, a primary motive for automation of the blast and paint processes is reduced costs. Field tests of the automated paint sprayer indicate productivity rates for blasting and painting to be better than the initial specifications. One automated paint sprayer and two operators can directly replace two standard work crews of three persons each and produce a net savings in production costs of approximately $0.45 per square foot. [4]

An analysis performed by Rowland [4] uses a hypothetical scenario to evaluate the relative worth of automated versus manual surface finishing methods. The increased annual cash flow generated by the automated system was calculated using the equation:

$$\text{Annual Cash Flow} = (\Delta C_{\alpha} - \frac{\Delta C_{OH}}{\alpha_{ave}})\alpha - \Delta C_{year} - \Delta P\left(\frac{i(1+i)^n}{(1+i)^n - 1}\right)$$

where

ΔC_{α}	=	Operating cost reduction ($ 0.45 per square foot)
ΔC_{OH}	=	Additional overhead costs ($ 500)
ΔC_{year}	=	Additional annual costs ($ 1500)
α_{ave}	=	Average job size (30,000 square feet)
α	=	Annual system production (100,000 square feet)
ΔP	=	Additional purchase price ($ 50,000)
i	=	Interest rate (10%)
n	=	Useful system life (2 years)

A baseline case which used the parenthetical values above revealed an increase in annual cash flow of $13,000. A sensitivity analysis was performed by systematic variation of ΔC_{α}, ΔP, α, and n in the economic model. The results showed the most significant factors in the economic viability of the system to be yearly production volume and savings in operating costs. Variations in the purchase price and useful life of the system had moderate impact on system economics. Effective interest rates, which were independently varied, had a marginal effect on the cash flow generated by the automated paint system.

Initial field tests of the automated paint system validated the hypothesis predicting the economic advantages of automation. These advantages, along with improvements in worker safety and product quality, augment the initiative for development in this area of construction automation. [4]

5.0 CONCLUSIONS

The work done so far has been carried out by a small cadre of engineering graduate student operators. The characteristics and performance of these individuals may not be typical of the people who would normally perform these operations. The conclusions that we have been able to draw from our design and test experience to date are summarized below:

• It is quite easy to produce acceptable blast cleaning with automated equipment; the quality of the finished surface is high and the improvement in the operator's work environment is dramatic. No insurmountable obstacles have been encountered thus far.

• A sand pot vibrator improves the flow of the sand and should be included as a component of any blast cleaning system. Careful storage and handling of the sand improves the flow characteristics of the sand and results in higher productivity.

• A larger sand pot would improve the productivity of the blast cleaning operation but would require a heavier vehicle. A heavier vehicle is potentially troublesome around some tank berms and the muddy low lying areas within the berms. A larger sand pot will require a greater capital investment for both the tank and vehicle.

• The high noise level associated with blast cleaning requires ear protection even with a remote operator station. "No hands" radio communication between crew members would be helpful.

• The blast cleaning operation should include a test procedure for lead and other toxins. The remote operator station enabled the crew to stay upwind of the blast effluent; the operators felt that this alone made the automation worthwhile. Lack of a natural breeze might require use of a respirator at times; but with a modest breeze the operators felt no need for a respirator. The system should include a simple test or alarm to insure against operator complacency.

• The development of a blast cleaning effluent system will require a system that can process large air volumes and extract large amounts of sand. The size distribution of the blast effluent is widely distributed.

• It is possible to produce a uniform high quality paint film with automated equipment.

• The paint over-spray containment system worked well. The small amount of air handled and the lack of detectable overspray were encouraging. A roll type filter for continuous filter renewal should be a part of any commercial system.

• The air assisted-airless paint spray system produces a good quality job, is easily controlled and gives efficient paint deposition.

• The high solid epoxy paint favored by the petroleum company-owners is more difficult to work with than some other paints. A search for a more easily managed paint might be worthwhile. Stable single component paints with a long pot life would be easier to handle and reduce the time required for clean up.

• The current control system was able to provide usable accuracy. Greater accuracy, while desirable, is not essential, and should be achievable.

• Graphical pre-processing would be a significant programming aid. The controller needs to be able to switch modes in the middle of a job. The operator would like to be able to interrupt and redirect the machine in the middle of a control sequence.

• The process is insensitive to winch location. Productivity of the crew would be enhanced by developing cable latching and anchor point hardware to minimize the time required when setting up.

• Significant market appeal to owners would result if they could be given the assurance of uniform paint film quality by incorporating on-line monitoring and self shut down feature so that their time and attention could be more easily utilized in support activities while the operation is underway.

• Generous allowance for cooling of the electronic motor drive circuits is needed to deal with operating and environmental conditions encountered.

• While most of the economic potential of the system lies in the huge number of existing tanks world wide, new tanks designed with automated surface finishing in mind will reap the benefits from this new technology.

Development of the automated paint sprayer is ongoing. Research efforts are continuing to refine the system and enhance quality and productivity. The advantages realized by automation of the prototype surface finishing system are encouraging and industry interest in automation is increasing.

ACKNOWLEDGMENTS

The authors would like to thank the Texas Advanced Research Program and the Construction Industry Institute for their support in developing this technology. Thanks also to Amoco Oil Company for allowing the team to use their pipeline facility for the system tests.

REFERENCES

1 Freund, M. A., Traver, A. E., Wesley, G. W., O'Connor, J. T., Rowland, R. S., "Automated Surface Finishing of Large Diameter Storage Tanks", Proceedings of the 8th International Symposium on Robotics in Construction, 1991.

2 Smith, Lloyd M. (ed.), "Identifying and Controlling Job Hazards during Climbing and High Work", Journal of Protective Coatings and Linings, volume 7, number 10, October 10, 1990.

3 Means, Means Building Construction Cost Data, 49th Annual Edition, Kingston: R. J. Means Company, Inc., 1991.

4 Rowland, R. S., Automation of Blast Cleaning and Painting Operations, A Case Study in Construction Automation, Thesis, University of Texas, Department of Civil Engineering, August, 1991.

5 Wesley, G. L., The Design and Construction of an Automated Surface Finishing Machine for Large Petrochemical Storage Tanks, Thesis, University of Texas, Department of Mechanical Engineering, December, 1991.

6 Freund, M. A., The Navigation and Control of an Automated Surface Finishing Machine for Steel Petrochemical Storage Tanks, Thesis, University of Texas, Department of Mechanical Engineering, August, 1991.

Automation and Robotics in Construction X
G.H. Watson, R.L. Tucker and J.K. Walters (Editors)

Automatic Machine Welding of the Primary Circuit of the Sizewell 'B' Nuclear Power Station

M.D. Austick[a] and S. Thompson[b]

[a]General Manager, Construction Division, GEC ALSTHOM Engineering Systems Limited & representative of European Construction Inst.

[b]Vice President, Sales and Marketing, PCI Energy Services

Abstract

GEC Alsthom Engineering Systems, who were required to weld the Main Loop Pipework as part of their Mechanical Erection contract on the Sizewell 'B' PWR, were approached by its Client to make a significant reduction in the programme schedule for this particular task. This paper will describe how, following a detailed assessment of the process, a change in approach from manual to machine welding enabled a successful acceleration of the programme schedule to be achieved.

1. INTRODUCTION

Sizewell 'B' Nuclear Power Station in Suffolk, England, is Britain's first PWR to be constructed. GEC ALSTHOM Engineering Systems (GECA-ESL), a member of the European Construction Institute, was awarded a contract in 1989 for general mechanical installation of the Nuclear Island Plant which included in the scope a requirement to install the Primary Circuit Components.

Sizewell 'B' is a 4 loop PWR with each loop requiring a total of 8 field welded joints to connect the Steam Generator and Reactor Coolant Pump to the Reactor Pressure Vessel, and 7 field welds to complete the Pressuriser Surge Line. The original GECA-ESL approach was to use traditional manual methods to complete the 39 field welds in the original programme schedule of 45 weeks.

2. THE TASK

The joint type used for the loop was a compound angle, single sided 'J' prep (see Fig. 1) with weld configurations in the horizontal position (8 total), vertical position (16 total) and 45 degree position (8 total). The pipe/nozzle sections were nominally 38 inches diameter with wall thicknesses of 3½ to 4" for the main loop welds and 14" diameter, 1¼" wall thickness for the Pressuriser Surge Line. The materials involved were forged pipe sections in 304LN and castings in CF3 with 308L weld filler material.

The welding and inspection had to satisfy the ASME Code Sections III and IX, the requirements of which were supplemented by additional criteria defined in the customer specification. The main impact of this was to require the performance of additional procedure qualification tests and an enhanced NDE programme.

The NDE requirements were particularly demanding. The ASME code specifies a liquid penetrant (LT) and radiographic examination (RT) of the finished joint, this customer also required ultrasonic examination (UT). These needs, coupled with those of progressive examination as the welding progressed, resulted in the following NDE programme being applied to each joint:

a) Cleanliness check and 100% LT of all weld preparations
b) 100% LT of the Root and Hot Pass
c) At 25mm depth of fill a 100% LT and RT
d) At 50mm depth of fill a 100% LT and RT
e) On completion of the joint the surface was machined and ground to give a surface finish of 250 CLA and flatness of 0.5mm in 50mm, followed by 100% LT, 100% RT and 100% UT. An extremely rigorous set of criteria.

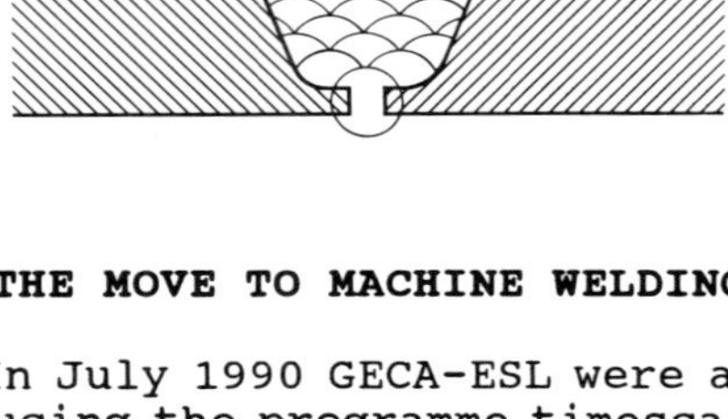

Fig. 1
Joint Configuration and Weld Bead Placement

Approx. 190 passes per Butt

3. THE MOVE TO MACHINE WELDING

In July 1990 GECA-ESL were asked to examine the possibility of reducing the programme timescale in order to accelerate this part of the overall site schedule within the Reactor Containment Building. A revised timescale of 24 weeks was identified as a target. Following a detailed review of the position it was clear that one possibility for securing a reduced programme was the adoption of a mechanised approach to the welding operations. The longer arc times, reduced operator fatigue, repeatability, and cleanliness offered the potential benefits of higher overall deposition rates with reduced repair levels, thus matching the task needs.

The limited timescale available for the review, validation and approval of a revised method and procedure, and the probable requirement for a significant amount of equipment to support the revised schedule, led GECA-ESL to seek out companies with previous experience of this type of work. In December 1990 PCI Energy Services were contracted to demonstrate suitable equipment at the GECA-ESL facility at Leicester, England. This demonstration proved that the mechanised Gas Tungsten Arc Welding (GTAW) process could cope with the maximum joint gaps and mis-matches and the detail inspection of the completed demonstration weld produced acceptable results.

Following this successful demonstration, PCI were further contracted to provide equipment and manpower to conduct a complete Welding Procedure Qualification (WPQ) at the GECA-ESL Leicester facility. The purpose of the WPQ was to prove that the welding process and procedures would achieve the strict quality requirements imposed by Nuclear Electric, the owner/operator of Sizewell 'B'. The WPQ was conducted on two complete assemblies from pipe test pieces cut from the actual contract pipe. One test piece was welded in the horizontal position, the other in the vertical position. The test pieces produced were also used to manufacture the U.T. test blocks, required to prove the UT inspection procedures.

The opportunity was also taken at this time to obtain information on other processes associated with the weld which were to be used later at site. Site cleaning procedures were developed and tested to ensure that the joints would be chloride free prior to fit-up. Punch marks were located on both sides of the weld prep and dimensions were recorded in order to gauge shrinkage as the joint was welded. This information was necessary to ensure that the accurate fit-ups required at site could make appropriate allowance for weld shrinkage.

The WPQ was also used to verify welding NDE and equipment set up times to assist in the eventual preparation of a detailed site schedule.

Following satisfactory completion of the WPQ, GECA-ESL decided to adopt the process for Sizewell 'B' and contracted PCI to provide the welding service at the site.

4. EQUIPMENT DESCRIPTION

The mechanised GTAW pipe welding system provides the capability for joining all types of metal and tubing assemblies to the most critical requirements. The system consists of a power supply with integral programmer, weldhead, track and control pendant.

The power supply used was a standard Dimetric Gold Track II modified by PCI to output 300 amperes welding current. Along with housing the power source for welding, it also houses motor servos, motor power supplies, a water cooler, recirculation pump, output gauges, cooling fan, dual gas solenoid and bulkhead connectors. The approximate physical size is 18" wide x 60" long x 36" tall. The weight is about 415 pounds. There are no operational controls on the power supply. All are located at the remote control pendant which allows the real time control of welding parameters at the location of the weld.

The following functions are programmable to obtain maximum quality and repeatability; carriage travel speed; wire feed speed; automatic voltage control (torch height control); torch oscillation speed and width; and primary and background welding current.

The weld head or carriage is a specially modified Dimetric 'F' head. The weld head is driven by means of a high torque DC motor and drive rollers mounted in the track assembly which is clamped around the pipe circumference. Mounting of the weld head is accomplished by placing it on the track and tightening the latching arm which compresses the head to the track and engages the drive mechanism. The mounting procedure takes less than 5 minutes.

The mounting track assembly is designed to separate into two halves that are joined by two captive socket head screws. Centering and clamping screws are provided on the track to allow for minor variations in pipe concentricity and to firmly clamp the track to the pipe. Depending on the circumference of the track, more than one weldhead may be used on the same weld joint. The mounting tracks used ranged from 42 to 50 inches. The oversized tracks relative to the pipe OD (33.5 to 38 inches) allowed for equipment clearances when mounting on an elbow. Fig. 2 shows a general view of the equipment set up on a pipe.

Fig. 2 Typical Equipment Set Up in the Horizontal Position

The completely equipped carriage weighs 15 pounds not including the service cable. The carriage produces sufficient torque to lift up to 80 pounds of added load while maintaining the specified speed to within 1%. The radial clearance of the weld head varies with the size of track, torch and wire nozzle configuration, and the depth of the weld groove. With special set-ups, it is possible to reduce the radial clearance down to 2 inches. The motorised cross adjustment and width is enclosed in the weldhead carriage and is capable of oscillating the weld torch up to 1¼". The Automatic Voltage Control (AVC) assembly is mounted on the front of the weldhead carriage. This unit carries the water cooled torch block with gas lens, gas cup, tungsten and wire manipulators.

A special 10 pound cold wire feed unit is also attached to the carriage. This design keeps wire spool changes to a minimum while its low profile is advantageous when working near obstructions. Cable extensions allow for weld head operations up to 250 feet from the power supply.

5. ADAPTIONS MADE FOR SIZEWELL 'B' USE

There were several key equipment challenges that had to be addressed before welding could begin at Sizewell 'B'. First, the joint configuration used had a relatively tight included angle. While this benefitted overall production rates, it made the first 1/3 of the weld difficult to complete with a standard equipment configuration. PCI designed and built a special AVC assembly that had 4 inches of usable stroke. Since the loop piping was less than 4 inches in wall thickness, no AVC adjustments were needed throughout the entire weld other than adjusting the torch tilt. Additionally, the wire manipulators on the torch assembly were threaded their entire length to eliminate the need to change this part as the weld progressed.

Purging the inside of the weld joint was another problem that needed to be addressed. Because of parallel activities within the reactor vessel, no argon could be allowed to enter the vessel. Normally purge dams would be installed to cover the entire pipe cross section. However, as it was essential that a ventilation system be installed in the pipe to vent argon safely, another solution was developed. PCI designed and built special lexan purge chambers that only encapsulated the weld area. These devices were in the shape of a ring that allowed air to flow past them yet maintained the argon purge on the weld area. This system maintained a less than 1 percent oxygen content in the weld area but allowed free access to piping, Reactor Vessel and Steam Generator channel heads.

To cope with minor breakdowns, a modular system of spare components was utilised. The four main components to this were the weldhead carriage, wire feed assembly, AVC assembly and the power supply. If any one of these sub-assemblies broke down or needed maintenance, they could be detached within a matter of minutes and another one installed into the system.

6. THE SIZEWELL 'B' APPLICATION

The welding activities on the Sizewell loop were scheduled to take place during an intense period of overall construction activity inside the Reactor Containment Building and a very detailed programme schedule integrating all the installation, setting, welding, inspection and cleaning activities on a day to day basis was drawn up before work started. This reflected a revised target for completing the welding of 24 weeks.

To meet this demanding production schedule four butt welds needed to be welded simultaneously with all work activities carried out on two shifts. With two welding systems on each joint this meant a minimum of eight systems had to be running at all times.

Altogether, twelve systems were staged inside the Containment Building allowing the technicians to install welding equipment not in use on the next available butts as they were released from setting up by GECA-ESL, thus minimising welding down time due to equipment change over. It also allowed for a contingency in case of equipment failure. Because of the lack of space in Containment during construction, special racks were fabricated to stack power supplies two high. The equipment was staged adjacent to the Reactor Vessel in the bottom of the refuel pool. Extension cables were routed through the primary shield wall to the weld areas.

To handle the volume of argon shielding gas that was needed a liquid cryogenic system was used. Containers that held approximately 4000 cubic feet were staged inside Containment near the welding power supplies. A manifold system was connected to the liquid argon container and hoses routed the proper amount of gas to each welding system in use.

Once the fit-up was completed and dimensional surveys recorded, it took an average of one shift to set up the machine welding equipment. After equipment set-up and tacking, welding of the root pass commenced. This activity took about one-half shift. Quality Control would then be called to inspect the root pass visually, then LT the outside of the root. With acceptance verified, welding continued to the 25mm stage. This usually took approximately two shifts. Radiography then took place in the two hour gap between shifts and the film was read as soon as it came from the processor. Again, after the review was completed, welding re-started. The next inspection hold point was at the 50mm stage. Reaching this stage was normally accomplished in four more shifts. The timescale from 50mm to completion varied depending on which part of the piping system was being worked on. Welding to completion of each butt from this stage varied in time to complete from 10 to 18 shifts depending on the actual thickness of the pipes being welded.

Shrinkage was a variable closely monitored. Once the fit up was achieved, data base line marks were vibro-etched onto the pipe on each side of the weld joint. The measurements were recorded in the Quality Control Records for each particular weldment. The actual shrinkage measurements ranged from 7.5mm to 12.5mm after completion of the weld.

Records of the actual welding parameters and bead placements were kept by the welding operators in the work areas. Since several welding operators worked on each joint, their work needed to be tracked for baseline information. The weld log contained information such as Weld Number, Weld Procedure Specification (WPS), Bead Number, Layer Number, Pulse Arc Setting/Sync. Pulse Setting, Amp/Volt Settings, Wire Speed and Travel Speed setting along with the Welders Identification Number. There was also a map of the weld groove that contained the actual placement of the bead. Each bead was numbered on the map to tie this to the Weld Log and there were approximately 190 beads per weld. Such a detailed record needed to be kept so that any defects discovered by the several inspection methods could be compared to the welders log as an aid to determining the cause.

In order to meet daily production goals a great deal of effort went into the close co-ordination of the various groups undertaking component installation, joint set up, equipment movement, welding, inspection, cleaning, grinding, safety and access control to ensure that the overall programme schedule was maintained.

At the peak of the welding activity a total of approximately 72 people were directly involved in the Loop welding operation and services, which included some 28 staff from PCI supporting the welding activities.

In liquidating the work on the 32 large diameter welds associated with Pressure Vessel, Steam Generators and Main Coolant Pump loops, approximately 2,500 pounds of welding wire and 640,000 cubic feet of argon were used.

7. RESULTS ACHIEVED

The first butt to be welded was tacked up on schedule on the 7 October 1991 and the final butt weld was completed ready for final radiography on the 19 February 1992, giving an overall duration of 20 weeks, some 4 weeks inside the original timescale of 24 weeks.

The incorporation within this time frame of the numerous inspection hold points and the application of an average of 40 man hours worth of internal and external scanning of each joint was in itself a major success.

In total over 53,000 linear feet of weld bead was laid into the Primary Circuit and Surge Line welds resulting in 270 linear feet of completed weld. Throughout the programme ultrasonic testing detected defects requiring repair in only some 18 feet of completed weld.

8. CONCLUSIONS

The basis for the successful outcome of this project, as in most others where a mechanised process is to be used, was the careful attention to:

a) The selection of appropriate well proven processes and equipment
b) The careful attention to detail and planning of all supporting requirements for the operation
c) The extensive training of all personnel involved to ensure the achievement of the highest levels of productivity and efficiency

The project has been, we believe, a good example of the application of an applied mechanised welding system to solve what was a complex construction problem in terms of the timescales, locations, inter-relationship of activities and the levels of reliability required to deliver the schedule requirements in these testing circumstances.

I would finally add that the success was also due in no small measure to the will of all involved in this Anglo-American venture to make the project succeed.

Automation and Robotics in Construction X
G.H. Watson, R.L. Tucker and J.K. Walters (Editors)

RLAB: A REMOTELY OPERATED LABORATORY FOR IMPROVED MANAGEMENT OF HAZARDOUS SPILLS

Laura A. Demsetz[a], Elisa C. Bass[b], Eron Ersch[c], James Ash[c]

[a]Department of Civil Engineering, University of California, Berkeley, California, 94720, USA

[b]Department of Mechanical Engineering, University of California, Berkeley, California, 94720, USA

[c]Formerly Department of Civil Engineering, University of California, Berkeley, California, 94720, USA

Abstract

Highway spills of hazardous substances expose emergency response personnel and the public at large to the risk of explosion, fire, and contamination. When the spill is of an unknown substance, response personnel must first determine its identity. Problems of human exposure make automated or remotely controlled sampling and analysis a logical approach for this operation. The State of California's Department of Transportation (Caltrans) has sponsored a pair of projects that demonstrate this approach — a remotely operated vehicle (ROV) to sample an unknown substance and a remotely operated laboratory (RLAB) to identify the substance. This paper describes the conceptual design of the RLAB. Key issues include the selection of a method of identifying unknowns, the tradeoff between complexity and functionality, the division of labor between the RLAB and its human operator, the interaction between the ROV and the RLAB, and the possible use of historical spill data to speed the identification of unknowns.

1. INTRODUCTION

Spills of hazardous materials expose emergency response personnel and the public at large to the risk of explosion, fire, and contamination. In 1988, there were over 6000 hazardous materials transportation incidents in the U.S., resulting in 15 deaths, over 150 injuries, and damages exceeding $20 million. All of the fatalities and over 85% of the damages resulted from hazardous materials spilled on highways. In California alone there were over 300 incidents during 1988, resulting in close to $2 million in damages (1).

The identification of spilled substances is a vital first step in managing a hazardous spill. Identification determines the procedures to be followed in containment and cleanup, and the extent of evacuation required. To effectively manage a hazardous spill, identification, containment, and cleanup must be carried out with minimum human exposure and property damage. Reducing the time required for these activities is critical, particularly for highway spills, where the cost to society due to road closures can be significant.

1.1 Current Practice

Upon arriving at the site of a spill, the first responder (usually Caltrans or the California Highway Patrol) isolates the spill, establishing a safety zone around it. This may require blocking or rerouting traffic. The first responder then attempts to identify the spilled substance. If the substance can be identified by placards, labels, shipping manifests, or interviews with the driver of the transporting vehicle, the first responder can use any of a variety of handbooks, databases, and phone-in services (e.g. 2, 3) to determine appropriate methods for containment, clean up, and disposal.

When an unknown substance is involved – for example, when a substance has fallen or spilled from a vehicle without the driver's knowledge, when a substance is released in an accident that badly damages the transporting vehicle, or when a substance is illegally dumped – the situation must be assumed hazardous until further information is obtained. Additional response personnel are brought in, don protective gear to obtain a sample of the substance, and use a variety of analysis techniques to determine its composition.

In urban areas, although the first responder may be Caltrans or the California Highway Patrol, identification, containment, cleanup, and disposal are carried out by contractors. The first responder will provide the contractor with any information available from a visual inspection of the scene. Depending on the location and time of day, it may take the contractor from 30 minutes to several hours to arrive at the spill site, and another 30 minutes to several hours to identify the spilled substance sufficiently to allow clean up to proceed. During this time, the part or all of the highway is typically closed to traffic. In rural and remote areas, fewer contractors are available, and Caltrans personnel must handle a wider variety of incidents and tasks.

1.2 Identification of Unknowns

At the site of the spill, a qualitative analysis is carried out to identify the unknown. A precise determination of the chemical content of the unknown is not required. Rather, it is necessary to proceed through the analysis only until an appropriate containment and clean up strategy is indicated. In most cases, a more detailed quantitative analysis will be needed prior to disposing of the material. However, this detailed analysis can be carried out in a controlled laboratory environment without the time pressure caused by highway closure.

Three broad categories of equipment and methods may be used in identifying unknowns: field meters, “tailgate” chemistry, and analytical laboratory equipment. Portable hand-held field meters are available to monitor radioactivity and combustibility, to measure pH, and to detect the presence of a variety of organic and inorganic gases. However, these meters do not provide enough information to determine clean-up procedures for most solids and liquids.

The identification of unknowns in the field is most frequently accomplished by "tailgate" chemistry – working through a sequence of relatively simple tests in which small amounts of an unknown are combined with a series of reagents. Several proprietary systems have been developed to lead response personnel through a testing sequence. Typically, the tests can be divided into two types: screen tests and definitive tests. Screen tests are carried out first to detect properties which indicate that a substance is hazardous. Based on the results of screen tests, a specific sequence of definitive tests is carried out to identify common hazardous materials. In some tests, the reaction of an unknown with a reagent is visually observed. In others, the gas given off as the unknown undergoes a reaction is drawn into a prepackaged tube containing a reagent. A change of color in the tube indicates the presence or absence of a particular material.

If "tailgate" chemistry does not provide enough information to determine appropriate clean up procedures, a sample of the spilled material must be sent to a laboratory for analysis. Analytical laboratory equipment such as gas chromatography and mass and infrared spectrometers are used to determine the precise composition of the substance. This process is time consuming (especially if there is no laboratory facility close to the site of the spill) and expensive. Skilled operators are required to run the equipment and interpret the results. The results of this quantitative analysis typically provide far more information than is needed to clean up the site and reopen the highway.

The Berkeley Fire Department provides an example of the state-of-the-art in municipal hazardous spill response. They have a trained emergency response team, and have recently purchased a specially equipped response vehicle which serves as a command station. The vehicle contains communication and computer facilities, handbooks, protective gear for personnel, various systems for chemical analysis and identification, a video monitor, and spotlights. A video camera mounted atop the vehicle provides remote viewing of the spill.

2. IMPROVING RESPONSE THROUGH REMOTE CONTROL

Even with state-of-the-art equipment, current practice puts response personnel at risk. Protective clothing may not be adequate; clothing may protect against one chemical, yet be readily penetrated by another. In addition to the human risk, the dollar cost of current practice is significant. Roads remain closed until a contractor arrives at the site, identifies the spilled substance, and cleans up the spill. Supporting equipment and protective gear are costly, and equipment and gear which cannot be decontaminated must be thrown out and replaced.

Problems of human exposure make a remotely controlled system a logical approach. Remotely controlled vehicles have been developed for use in the cleanup of nuclear and other hazardous waste sites. However, these systems are for the most part too large and too expensive to be used on highway spills. To minimize response time, a system for highway use should be small enough to be stored in a variety of Caltrans vehicles. It should be easily operated by employees whose primary function is not hazardous spill response. In addition, it should be easy to decontaminate, and inexpensive enough to be placed in districts throughout the state and to be discarded when decontamination is not possible. Finally, the system must be intrinsically safe, since many highway spills involve flammable and/or explosive materials.

Identifying an unknown substance can be thought of as a two step process: obtain a sample of the unknown and then analyze the sample. These steps require very different capabilities. The former is primarily physical; whatever is obtaining the sample must get first to the site and then to the exact location of the spill, and must pick up a sample of the spilled substance. Although some initial testing (e.g. for explosivity) may be carried out while obtaining the sample, safety and weight considerations dictate that the bulk of the analysis be carried out at a location slightly removed from the actual spill. The process of analyzing the sample is largely observational. Using existing "tailgate" chemistry procedures, small amounts of the unknown substance and various reagents are combined in a series of test tubes, and the resulting reactions are observed.

Because the requirements for sampling and for analysis are so different, it makes sense to consider the development of separate devices for these tasks. Caltrans has sponsored a pair of projects that demonstrate this approach — a remotely operated vehicle (ROV) to sample an unknown substance and a remotely operated laboratory to identify the substance (RLAB). A prototype ROV has been developed at California State University, Chico, and is described elsewhere (4). A proof-of-concept model of the RLAB is currently under construction. Figure 1 shows how these devices would be used together. Upon arriving at the scene of a spill, Caltrans personnel would deploy the ROV. Using a video camera mounted on the ROV, response personnel would inspect the site, looking for labels or placards, without the need to suit up in protective gear. If no identifying information is found, the ROV would obtain a sample of the spilled substance. At the same time, response personnel would deploy the RLAB in a location slightly removed from the spill, but also removed from people at the site. The ROV would deposit the sample in the RLAB. Testing in the RLAB would be controlled and monitored remotely. Response personnel would not come into contact with the unknown until an appropriate containment and clean up procedure is identified.

3. RLAB SYSTEM DESIGN ISSUES

Key issues in the design of the RLAB included the selection of a method of identifying unknowns, the tradeoff between complexity and functionality, the division of labor between the RLAB and its human operator, the interaction between the ROV and the RLAB, and the possible use of historical spill data to speed the identification of unknowns.

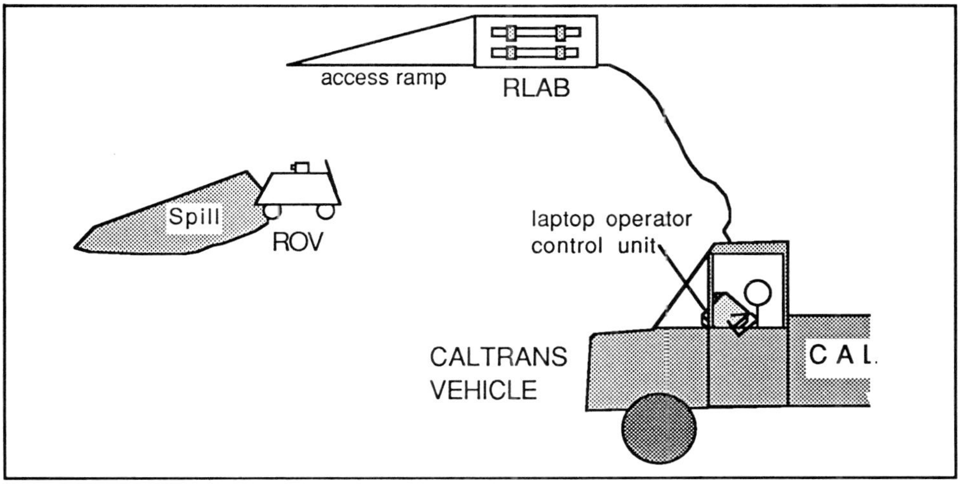

Figure 1. Schematic Illustration of ROV/RLAB Operation.

3.1 Identification Method

The fundamental issue in the design of the RLAB was the method of identification to be used. Current field practice, "tailgate" chemistry carried out manually, could be duplicated in a remotely controlled system. This option was attractive in that, in concept, it would be readily acceptable to field personnel. However, it would have resulted in a very complex and cumbersome system. From a philosophical standpoint, the design of the RLAB provided an opportunity to reevaluate identification methods in the context of increased automation. In the end, two paths were followed. A proof-of-concept model of the RLAB based on a subset of the "tailgate" chemistry tests is being constructed. In addition, specifications are being developed for a new automated testing system that would resemble available laboratory equipment, but would carry out the qualitative analysis required at the spill site. This effort is described in greater detail in Section 4.

3.2 Complexity vs. Functionality

A remotely controlled version of the complete battery of "tailgate" chemistry tests would have allowed the first responder to carry out a complete qualitative analysis of most unknowns. However, the resulting system would have been large, unwieldy, and impractical for placement in many response vehicles. Instead, the RLAB design effort focused on determining a subset of tests that would allow the first responder to provide the follow-up crew with information that would give them a "head start" on identification and clean up.

Discussions with hazardous materials specialists within Caltrans, with local fire departments and contractors, and with the developers of several "tailgate chemistry" systems indicated that the following are important characteristics in assessing the hazards posed by a spilled material: explosivity, volatility, flammability, oxidation potential, water reactivity, corrosivity, and radioactivity. Response time could be reduced if a first responder could provide information on these characteristics to the contractor or organization brought in to contain, clean up, and dispose of the spilled material. Therefore, the screen tests that detect these characteristics provide a minimum set of tests for the RLAB.

Historical spill data indicates that by far the most frequently spilled substances are gasoline, motor oil, and diesel fuel. A test for these substances is therefore included on the RLAB. A char test provides a great deal of information to response personnel, and is included also. Working backward from clean up procedures showed that the presence of ammonia, amines, or cyanide can be important in determining the appropriate clean up strategy. Tests for these substances are included on the RLAB as well. Table 1 summarizes the detection equipment and tests to be included on the ROV and RLAB.

Type	Meter/Test	Characteristic
Meters on ROV	radioactivity meter	radiation
	organic/inorganic gas meter	various gases
	combustibility meter	offgasing, explosivity
RLAB screen tests	evaporation test	volatility
	combustibility test	flammability
	oxidizer test	oxidation potential
	peroxide test	oxidation potential
	water reaction test	water reactivity
	pH test	corrosivity
	char/ignition test	various substances
RLAB definitive tests	iodine crystal test	gasoline, oil, diesel fuel
	ammonia test	ammonia, amines
	cyanide test	cyanide

Table 1. ROV/RLAB Meters and Tests

3.3 Division of Labor

An unknown spill is a potentially dangerous situation. Because each spill is unique, human judgment is important in assessing the situation and taking appropriate response. In considering a more automated approach to managing a spill, it is important to keep the RLAB operator "in the loop" as much a possible without compromising speed or safety. The physical manipulation of the unknown and various reagents is carried out by the RLAB, with the sequences of action required to conduct a particular test carried out automatically. However, higher level control resides with the operator. The operator initiates all tests from a laptop computer, observes the tests through a camera mounted on the RLAB and a monitoring screen in the vehicle. The RLAB controlling software prompts the operator to enter observations and, in the future, may make recommendations based on interim test results. However, all decision making resides with the operator.

3.4 Interaction between RLAB and ROV

The ROV obtains liquid samples with a syringe, and solid (usually powder) samples with a scoop. The syringe and scoop are located at the base of the ROV. To allow gravity to assist in distributing the sample within the RLAB, the ROV should deposit the sample at the top of the RLAB. A lifting mechanism that raises the sample from ground level to the top of the RLAB could be designed, but a simpler solution is to have the ROV drive up a ramp to the top of the RLAB. Once in place over the material delivery system, the ROV releases the sample.

Future modifications of the ROV may allow it to carry out selected containment and clean up operations. To minimize the weight of the mobile ROV, the tools and attachments required for clean up could be mounted to the outside of the RLAB. The ROV would pick up the tools as needed.

3.5 Use of Historical Data

The presence of a computer as the control interface to the RLAB makes it appealing to investigate the use of historical data in the identification process. Different regions of California have different spill histories. For example, in the Imperial Valley, an agricultural area, unknowns frequently turn out to be pesticides. In the Santa Clara Valley, home to computer manufacturers, spills of chemicals used in circuit board manufacturing might be more likely. Additional tests could be added to the RLAB to detect substances that had a historically high spill rate in a particular area.

If extensive spill statistics were available, it might be possible to use them to guide the qualitative analysis. The "optimum" order in which tests should be carried out could be found based on the appearance of the substance, the frequency with which different substances have historically been spilled, and the results of each test as it is carried out. Preliminary consideration of this approach indicated that spill data is too sparse to be useful in this sort of analysis. However, the idea is appealing, and warrants further investigation.

3.6. RLAB Design Summary

The RLAB conceptual design is as follows. The ROV delivers a sample of the unknown to a dispensing system on the top of the RLAB. A carousel houses the tubes in which the tests will be carried out. Depending on the requirements of the test, the tube is either preloaded with a reagent or equipped with a system to dispense a reagent (for some tests, the reagent must be added to the unknown, rather than the unknown being added to the reagent). For several tests, encapsulated reagent tubes are mounted over the test tube. A pump draws a headspace sample into the encapsulated tube. The initiation of each test is remotely controlled by operator; lower level functions (such as the control of the pump) are carried out automatically. A camera mounted on the RLAB allows the operator to view the tests through a remote monitor. Test results are interpreted by the operator with assistance from the operating software.

This design results in a lab that is roughly the size an automobile tire, and requires four actuators (two for the dispenser, one for the char test, one for the encapsulated reagents) and six valves. A proof-of-concept model of the RLAB is current under construction.

4. ANALYTICAL EQUIPMENT FOR QUALITATIVE ANALYSIS

The RLAB conceptual design described above represents one attempt to improve spill response by incrementally improving current practice. A philosophically more appealing approach is to redesign testing technology with automation in mind. The ideal would be a "black box" which would automatically tell the operator the appropriate procedure for containment and clean up of the sample. Standard analytical laboratory equipment can be used to precisely determine the composition of an unknown material. However, such equipment requires a highly trained operator, must be precalibrated based on the anticipated composition of the unknown, and is really designed for quantitative analysis, rather than the qualitative analysis required for containment and clean up of a spill.

Several mobile lab systems have been proposed and are being developed, primarily funded by the Department of Energy. These range in size from a "back pack" to van; anticipated costs range from $150,000 to $1 million. Most of these systems are still under development. For application at hand, these systems are too large, too expensive, and provide more capability than needed.

Conversations with people involved in the development of analytical equipment indicate that the qualitative analysis problem posed by highway spills is one that they have not addressed, but for which there is likely a solution. A survey of Caltrans response personnel is underway to aid in the development of specifications for a portable, field hardened piece of equipment that would automatically carry out qualitative analysis of an unknown.

5. SUMMARY

The ROV/RLAB concept for hazardous spill management has the potential to minimize human exposure and increase response speed while remaining cost effective. Separating response equipment into a mobile ROV and a portable RLAB allows Caltrans to make maximum use of available technology. An RLAB model based on currently used "tailgate" chemistry is being developed. In the future, identification may be automatically carried out by field hardened equipment designed for qualitative analysis.

6. ACKNOWLEDGMENTS

This project has been supported by Caltrans through the AHMCT program.

6. REFERENCES

1. U.S. Department of Transportation, *Annual Report on Hazardous Materials Transportation, Calendar Year 1988.*
2. Research and Special Programs Administration, U.S. Department of Transportation, *1990 Emergency Response Guidebook*, DOT P 5800.5.
3. Chemical Transportation Emergency Center, a chemical industry hotline.
4. P. Hoff and A Critchlow, *Automated Hazardous Spill Pick-Up, Draft Report to Caltrans*, 1992.

Automation and Robotics in Construction X
G.H. Watson, R.L. Tucker and J.K. Walters (Editors)

Study of a New 3-D Measuring Method for Diaphragm Walls

Tetsuya Kikuchi[a] and Satoru Miura[b]

a Research Engineer (Instrumentation)
Mechanical & Process Engineering Dept. Kajima Technical Research Institute, 19-1, Tobitakyuu 2-chome, Chofu-shi, Tokyo, Japan.

b Senior Research Engineer (Instrumentation)
Mechanical & Process Engineering Dept. Kajima Technical Research Institute, 19-1, Tobitakyuu 2-chome, Chofu-shi, Tokyo, Japan.

1. Introduction

Measurement of the shape of an excavated wall, which is one of the control items in diaphragm wall construction, is an absolute necessity during and after excavation in order to detect local surface collapse of the wall and areas of dissipation, and to control the final form.

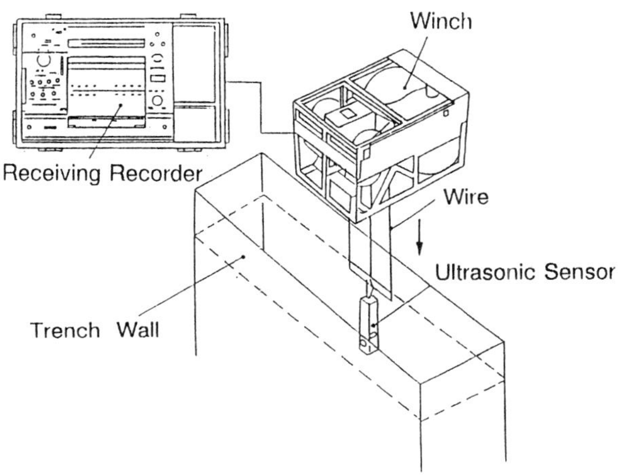

Fig. 1 Previous Equipment

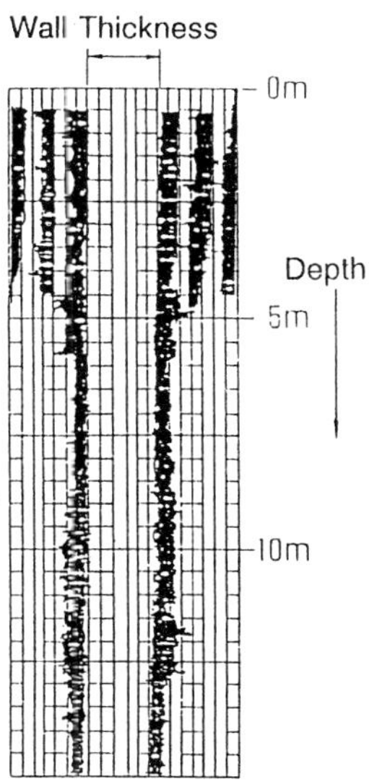

Fig. 2 Previous Output Diagram

As shown in Fig. 1, with previous equipment, a winch was set up on the trench wall and the distance to the wall was measured while lowering an ultrasonic sensor on a wire. The results obtained by this method, output on recording paper, are shown in Fig. 2. The following problems develop when excavation control using this equipment is attempted.

- ○ Since measurements can not be taken during excavation and the excavator must be pulled out of the excavation every 20~30 meters, it takes a lot of time and manpower to take measurements. It is also necessary to rely on the intuition and experience of the excavator operator for control during excavation.
- ○ A single measurement is capable of providing a measurement of the cross section of only the place where the sensor is lowered into the trench. In order to get a grasp of the wall as a surface, it is necessary to take multiple measurements. Records are also made by the distance signals from the ultrasonic sensor being output in analog form on recording paper. In order to determine distance, it must be read by people from the recording paper.
- ○ Ultrasonic signals tend to be attenuated considerably depending on the proportions of sand and clay in the excavated soil and reflected noise from suspended matter results in a poor S/N ratio and makes it difficult to distinguish wall signals.

In this report, we describe the measuring system which we developed for the purposes of eliminating these problems as well as saving energy in excavations and improving the system's reliability. We will also report on an on-site application example.

2. Outline of Measuring System

The measuring system consists of two subsystems. One subsystem is a 3-D measuring device which utilizes ultrasonics to measure the distance to the wall. The other subsystem is an excavator position measuring device which measures the position and posture of the excavator in the ground.

By linking these two sub systems, it is possible to measure the form of a diaphragm wall with respect to design values.

2-1 3-D Measuring Equipment

To enable measuring results to be displayed in real time on a monitor at any time desired during excavation, it is indispensable to mount ultrasonic sensors on the excavator body, make the system capable of recognizing only signals reflected from the wall surface and digitize those signals.

However, during excavation, the trench is in the condition where bentonite slurry and excavated soil are in an agitated state, the specific gravity increases and reflection from muddied suspended matter, etc. makes it difficult to judge signals reflected from the excavated wall. Therefore, we developed a signal processing method, which will be described next, and incorporated it into the system.

(1) Equipment Configuration

Fig. 3 shows the overall configuration of the measuring equipment. The 3-D measuring equipment consists of ultrasonic sensors mounted on the excavator body, a correction sensor (for measuring the velocity of ultrasonic wave propagation), a multiplexer (an ultrasonic wave sensor channel switching device), a relay for transmitting data, a transmitter and receiver on the ground for receiving and sending ultrasonic signals and a signal processing board. (These are in [ ] in the drawing.)

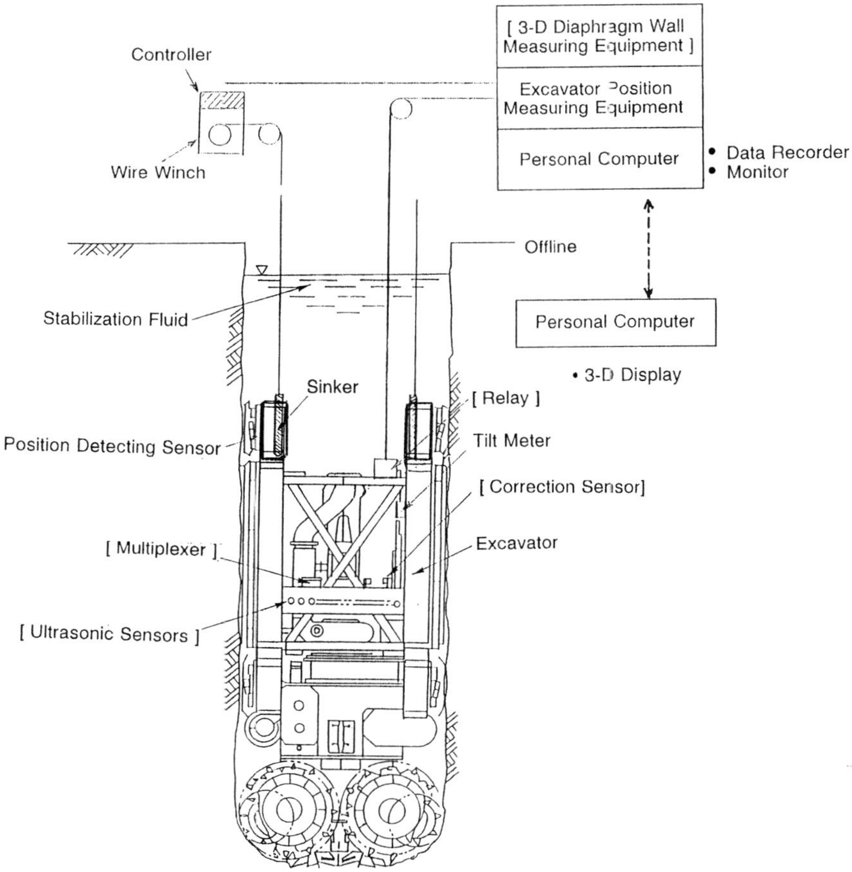

Fig. 3 Configuration of Measuring Equipment

(2) Signal Processing Procedure

For measurements, the multiple ultrasonic sensors mounted on the excavator are switched sequentially by a multiplexer, with each sensor measuring the distance from the sensor to the wall.

Fig. 4 shows the procedure for receiving and processing ultrasonic signals for each ultrasonic sensor in the form of a formula.

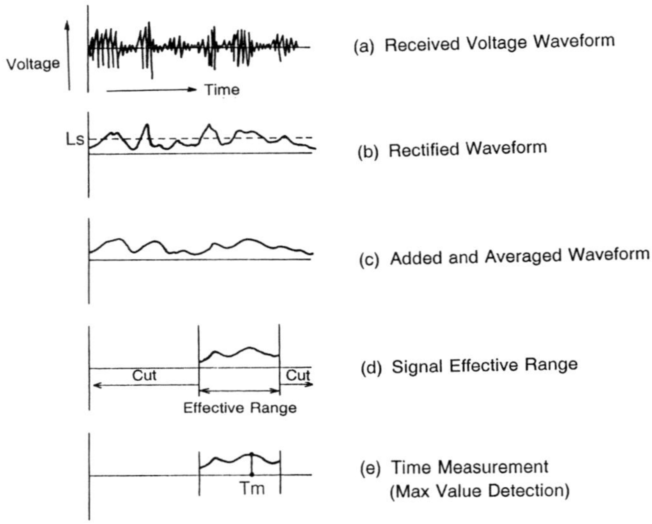

$$\text{Distance} = \frac{\text{Tm} \times (\text{Sound Velocity})}{2} \qquad \text{(f) Distance Calculation}$$

Fig. 4 Signal Processing Procedure

① The voltage waveform (a) which shows the size of the ultrasonic signal received, is rectified in (b). In previous ultrasonic distance sensors, the threshold level (Ls) was set constant and any signals which exceeded this level were regarded as wall signals, but during agitation, judgment of signals in bentonite slurry is difficult.

② Several waveforms are added together and averaged in (c).
Since suspended matter are not all piled up in the same place, the positions of signals reflected from suspended matter show variations. However, wall signals can always be detected from the same position. Therefore, by adding signals together and averaging them, the reflection level from suspended matter, etc. can be made relatively low.

③ The effective range of the reception signal is set according to the mounting locations of ultrasonic sensors on the excavator. (d)

④ Search is made for the maximum amplitude value of the waveform in the effective range (e) and the time Tm of that waveform is detected.
⑤ Using the ultrasonic wave propagation velocity v in slurry determined by the correction sensor and the value Tm, the distance to the wall is calculated.

Through this processing procedure, it becomes easy to judge between signals reflected from the wall and noise signals. That is, even if the level of the signals reflected by the wall is faint, if it is even a little higher than the noise level, then judgment between the wall signals and noise is possible. Digitizing of this processing system has been made possible by making it into a modular board which is mounted in an expansion unit of a personal computer.

2-2 Excavator Position Measuring Equipment

The 3-D wall measuring equipment measures the relative distance from the ultrasonic sensors mounted on the excavator to the wall. Therefore, in order to determine the form of the excavation with respect to the design excavation position for the diaphragm wall (plane and vertical), it is necessary to measure the position of the sensors, that is, the position of the excavator. Next, the equipment developed to solve this problem is described.

(1) Equipment Configuration

The excavator position measuring equipment consists of two sinker position detecting sensors mounted on the excavator body, two tilt meters, two sinkers suspended from ground level, two winch controllers, a data input/output device and display software. (Fig. 3)

(2) Measuring Procedure

Measurement is made by suspending a sinker, which acts as a reference, inside a position detecting sensor tube attached to the top of the excavator body on both sides and determining the relative position using contactless eddy current type distance sensors.
Next, the measuring sequence is described.
① The sinker is suspended in the sensor tube and acts as the reference point. (Fig 6) Up and down control of the sinker is accomplished by eddy current sensors (Z1, Z2). These utilize loop coils on the top and bottom of the sensor tube which sense the vertical position and feed it back to the servomotor driven winch at ground level. Thus this system is constantly following the movement of the excavator, working to keep the sinker in the measuring tube.
② The horizontal position of the excavator is calculated by measuring the position relative to the sinker (reference point) by means of eddy current sensors (X1, X2, Y1, Y2) located on the right and left sides at the front and rear of the measuring tube. (Fig. 5) In this way, the excavator's position with respect to the design value can be calculated.

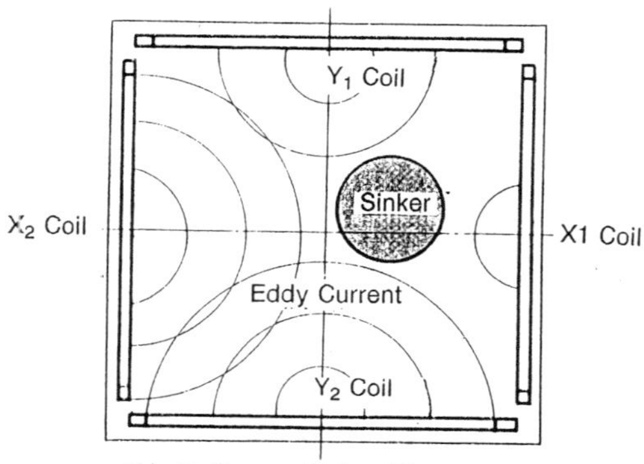

Fig. 5 Sensor Tube Plane View

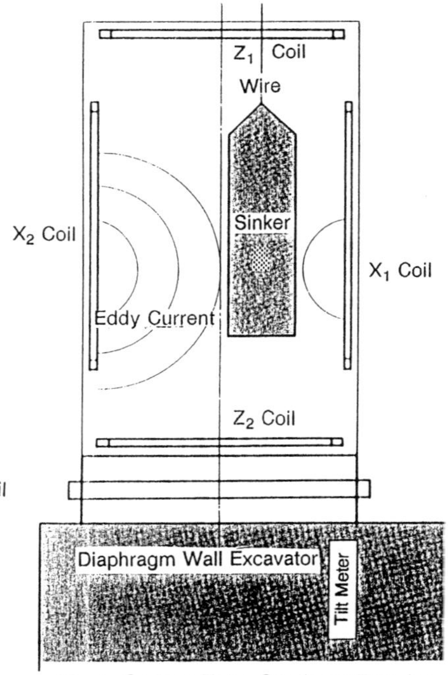

Fig. 6 Sensor Tube Sectional Drawing

2-3 Overall System

By combining 3-D diaphragm wall measuring equipment with excavator position measuring equipment, it is possible to control the form of the diaphragm wall according to the design value.
System specifications are shown in Table 1.

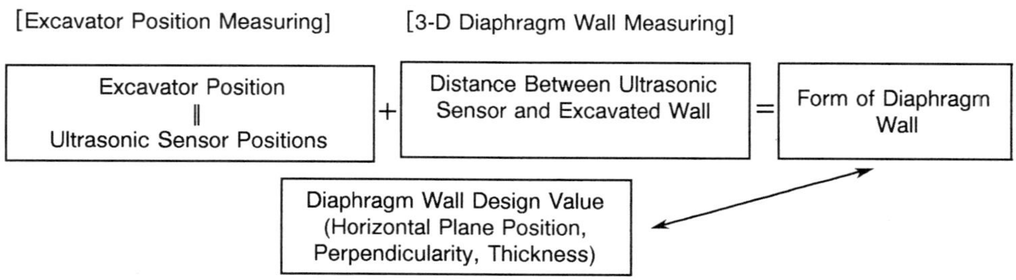

Overall System

Equipment	3-D Diaphragm Wall Measuring	Excavator Position Measuring
Measuring Precision	2cm	2cm
Measuring Range	200 cm (wall ~ sensor)	± 10cm
Possible Measuring Depth	150m	
Wall Thickness	70 cm or greater	
Display Software	Vertical Section, Horizontal Section, Contour Drawings, Bird's-eye Views	
Number of Channels	Max 20	2

Table 1 System Specifications

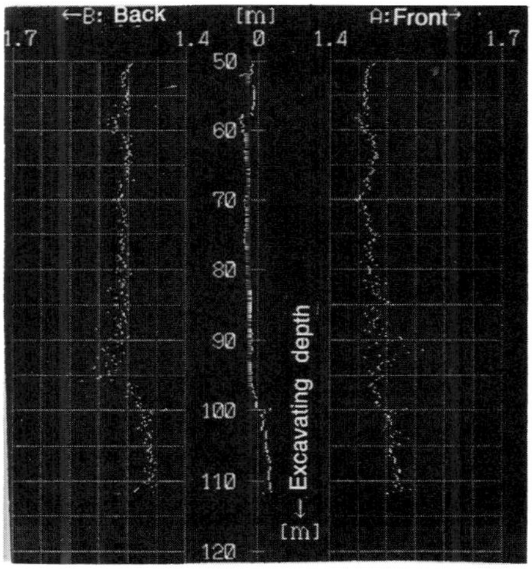

Photo 1 Excavator (Electro mill) Photo 2 Monitor Screen During Excavation

3. Application Example

We undertook studies of site requirements, the number of ultrasonic sensors needed, various display methods and other points and set out to build the optimum system.

This is an example of an application to an actual construction site.

The excavator where the system was installed was an Electro mill excavator (EM-320) and the application was made on a diaphragm wall construction located very deep underground with a design wall thickness of 2800 mm and an excavation depth of 120 m. The system is capable of monitoring the horizontal cross section or vertical cross section and the excavator position from the operation room, and selection can be made according to the needs of the operator. Photo 1 shows the excavator in which the system is installed and Photo 2 shows the monitor screen (displaying the vertical cross section and excavator position) during excavation. The excavation wall thickness and perpendicularity can be determined with good precision. Measuring data can be output offline to diagrams like those shown in Figure 7.

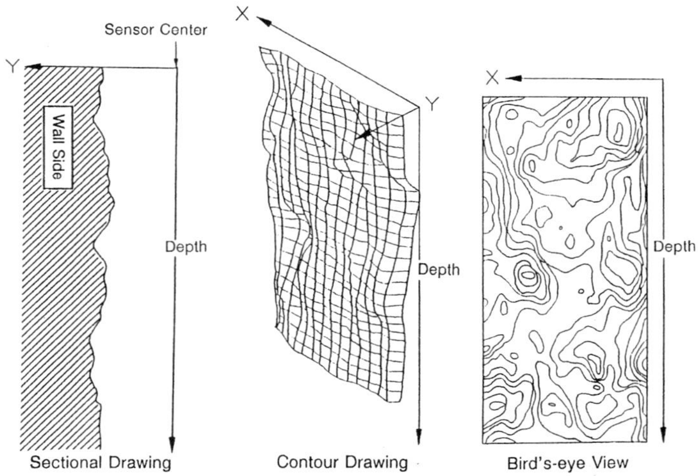

Fig. 7 Drawings

4. Conclusion

From the results of application of the system developed on this occasion, the following points were confirmed.

- The work of detecting work control values for determining the form of a diaphragm wall (indentations, protrusions and perpendicularity of the excavated surface) during excavation was made vastly more efficient.
- Since the measuring data can be digitized and stored on data recording media, sectional drawings, contour drawings, bird's-eye views and other diagrams can be output freely. The volume of earth excavated and other data can now be analyzed.
- Since the system's components were selected for their general purpose applicability, it is possible to build a system that is efficient for the needs of just about any construction job.

This system thus has made it possible to take the measurements needed in work control more efficiently and to automate this task. Thus, we think that the first stage in automating diaphragm wall construction has been completed. In the future, we are planning to study a system for feeding these measuring data back into the excavation operation.

Automation and Robotics in Construction X
G.H. Watson, R.L. Tucker and J.K. Walters (Editors)

Field Applications with CAPSY

Leon DeVos, TNO-BOUW, PO.box 49, 2600 AA Delft, The Netherlands.
Bob Hasara, Spectra Physics Laserplane, 5475Kellenburger Rd. Dayton,OH.

Abstract

In June 1989 CAPSY has been introduced for the first time at the ISARC' 89 in San Francisco. CAPSY (Computer Aided Positioning SYstem) is intended to be a low-cost real-time location tool for construction. Operating in working area's with dimensions like 300 by 500 feet it is capable of locating itself to an accuracy of +/- 1/8 of an inch in X- and Y-direction. Since this introduction CAPSY positioning system has grown from a lab-prototype to a commercially available product now. The initial specifications are met(low cost, accurate, portable) and in some cases even exceeded. Although CAPSY positioning system can be used as just a sophisticated measuring and locating tool there might be an even more important role to fulfil in the future. Because of the onboard computer capabilities CAPSY can be the link between the design stage of the building process (CAD-systems, highly computerised) and the actual building process at the job site(no computerisation at all). Transferring data from the design to CAPSY makes it possible to use CAPSY as a one-man survey instrument. At the same time it can function as a real world digitizer on the job site to feed back the 'as build' situation .

1. INTRODUCTION

In 1989 a project was started at TNO-BOUW to investigate how unhealthy and hazardous working conditions on the construction site could be avoided. The general feeling was that automation and robotization were part of the solution. Not to an extreme situation of eliminating all human labour on the job site (very unrealistic especially in construction) but trying to let the construction worker be the brains and the (automated) equipment to be the muscles. In that case the worker is always in control and the building process can benefit his skills and experience. In order to develop more automated tools and equipment the need for some kind of location device was recognised immediately. Therefor the project was focused in the beginning on getting such a device. After the design of CAPSY and verifying several concepts and techniques with prototypes TNO-BOUW and Spectra Physics Laserplane started a joint venture in order to develop CAPSY from a prototype to a commercially available product. Having this product available now the original goals of the project can continue.
TNO-BOUW is an institute of the Dutch organisation for applied research TNO. There are about 6000 people employed in this non-profit organisation.
Spectra Physics Laserplane is an American division of Spectra Physics, employing 5400 people world wide. SPL is technology and market leader in the field of laser equipment for construction and agriculture applications.

1.1 How CAPSY works.

CAPSY is based on triangulation to calculate its current position. For triangulation it needs to know the exact angles in between three known points as a reference. This is accomplished by rotating a laser beam in a horizontal plane and scanning the environment for the three reference points. These reference points are made of retro-reflective material and each so called 'reflector' is made unique in order to distinguish them from the others. This is done by creating a bar code pattern of the retro-reflective material. (In section 1.2 is explained how these reflectors become known points.) When the laser beam is hitting the reflector the laser light is reflected back into the unit and analysed by the internal computer. This will provide two essential pieces of information. First of all it will measure the exact angle of this reflector in respect to an internal index and secondly it will recognise the specific bar code of this reflector so it knows which target is scanned. The angle is measured with an accuracy of 0.001-0.003 of a degree. This is done by an internal computer controlled angular measuring system which is virtually insensitive to rotational velocity fluctuations. For construction applications the rotational speed will be 300 rpm. Angular errors due to being off level are compensated in software rather then by a mechanical compensator. An in built XY-tilt sensor measures the amount of degrees off level in both axis. The total portable unit is approximately 6 by 6 by 8 inches and weights about 6 lb. Rechargeable battery packs last about 4 hours.

1.2 Getting started with CAPSY.

Before CAPSY can be used as a one-man survey instrument there needs to be done some initial work. The user has to prepare the working area and the unit for operation. This needs to be done only once for every working area and the information will remain in the unit until it is deleted by the user itself. The most important part of initialising a working area is the so called 'set up' procedure. This procedure calculates the still unknown XY-position of the reflectors. As part of preparing the working area the user has placed 3 or more reflectors at the periphery of the area. The triangulation algorithm requires the XY-position of the reference points. It would be very inefficient if it was necessary to survey the positions of the reflectors by a survey crew. Because CAPSY is able to calculate very accurate angles it is capable also to calculate the XY-position of the reflectors. Assuming there are always at least two known reference points marked on the floor by a surveyor, CAPSY uses these points and their co-ordinates as a base for calculating the position of the reflectors. After entering these co-ordinates into the unit the user has to place CAPSY on reference point #1. CAPSY will take about 30 seconds to sample and store the angles of the recognised reflectors. The next step is to place CAPSY on the second reference point. Again CAPSY will sample and store for about 30 seconds the angles of the recognised reflectors. After this CAPSY is able to calculate the XY-positions of the reflectors. The complete set-up procedure is guided to the user by CAPSY itself and takes about 3 to 5 minutes. As long as the reflectors stay in the working area the unit is ready for operation and no new set ups are necessary.

1.3 CAPSY Advisor.

As mentioned before the user has to place at least three reflectors around the working area to make CAPSY work. Reflectors are recognised by CAPSY at 150 feet maximum under a plus or minus 45 degree acceptance angle (horizontal plane). Typically a user will place more than just three reflectors for several reasons. More reflectors means a larger working area. As soon as CAPSY recognise three reflectors it will produce a XY-position. So it is very easy to enlarge the working area by placing more reflectors at certain distances along the periphery of the area. A second reason is to create a full coverage for a irregular shaped working area. Another reason is that sometimes a reflector is blocked. In that case a three reflector system can't calculate a position. CAPSY scans 5 times a second for reflectors. If it recognise more than 3 reflectors it means it has more than one 3 reflector combinations, resulting in more than one solution for this particular position. In that case an evaluation algorithm becomes active. This algorithm will select the best combination of reflectors for this position. Choosing the best reflector combination is very important to get the desired accuracy of +/- 1/8 of an inch. In irregular shaped areas it can be difficult for an inexperienced user to determine where to place the reflectors. A software tool called CAPSY Advisor might be of great help for such a user. After entering the shape of the area CAPSY Advisor will suggest locations for reflectors and present a map of the expected accuracy.

1.4 CAPSY's field book.

CAPSY is used for setting out points and for measuring points. This implies that there is a lot of data involved in the form of XY-coordinates and identifiers. CAD - systems or specific survey software may generate lists of points to be down loaded in CAPSY. Digitised points needs to be up loaded to a computer system. All this data handling is supported through the field book. The field book is a pocket calculator kind of device and communication is performed through a RS232 serial link. It is capable to link to CAPSY and every other computer/device having a RS232 interface. The software in the field book takes care of the data transport between the two devices. On top of that it functions as an extended display and keyboard for CAPSY and as an temporary memory storage for lists of points.

2. GENERAL CONSTRUCTION APPLICATIONS

2.1 XY-positioning.

Calculating a new XY-position is the most basic function of CAPSY. All other functions are derived from this function. The actual position is updated 5 times a second and can be considered there for as a real time position. The simplest way to do XY-positioning is to use the XY-coordinate mode which displays on the LCD screen the current position. By comparing this actual position on the LCD screen with the desired position one should be able to find the right spot on e.g. the floor and mark it. Although this looks easy in practice it is difficult to get any sense for distance and direction based on a comparison of two co-ordinates. More helpful to the user will be when the (list of) co-ordinates are down loaded into the unit. In that case CAPSY is able to guide the user to the desired spot. This is possible because CAPSY not only knows the current position but it calculates it's bearing in respect to the reflectors too. The current position, the desired position and the current bearing is all the information it needs to guide the user to the next position. The way how this is displayed to user can vary. It can be expressed in a distance forward/backward combined with a distance left/right. That is a numerical method. A more intuitive way to present it is some kind of graphical display. CAPSY's LCD screen is a character display (graphic LCD's don't perform well at lower temperatures) but a limited graphic display is made possible. By using a non linear scale and an iconised presentation of CAPSY and the desired position it is possible to position the unit within 1/8 of inch on the desired spot guided by this pseudo graphic display. This can be done without interpreting any numerical information. Basically the position CAPSY calculates is a position of a point inside CAPSY. This internal point is inaccessible and for that reason worthless because it can not be used to mark or point to a

Some field situations with CAPSY

specific location. However, it can be translated to any other external point if the eccentricity is known combined with the bearing in respect to the reflectors.

2.2 Layout on floor and ceiling.

Very often hundreds to thousands of points needs to be marked on the floor and on the ceiling. Points to mark walls, columns, studs, manholes, pipes, supports etc. The dimensions from the drawing are red by the layout carpenter and translated into dimensions in the field. Mis-readings, mis-interpretation of the dimensions and wrong calculations are responsible for errors in the layout. The layout carpenter realises this and will check over and over again his layout. However this is still no guarantee for an error free layout and is time consuming. By down loading a list of points into CAPSY and using the unit to guide him to the points the layout carpenter is freed of a couple of these potential errors. The display will show and tell (literally, by a beep) him when he has positioned the unit within the tolerance on the right spot. A self levelling vertical laser will provide a spot on the floor and on the ceiling to mark the position. This vertical laser is very useful in combination with CAPSY mounted on a mobile tripod. When ever the floor situation allows the use of a mobile tripod it makes it very easy to layout on the ceiling or floor without any distortion. During all this layout work the user can switch to other modes like the '**tape measure**' mode. This provides a distance from a zeroed origin to the current position. Or he can switch to the '**grid line**' mode which helps the user to find redefined positions on a specific line. Layout work on the ceiling can be done by a single person on a mobile scaffold and doesn't require the difficult and sometimes dangerous situations normally necessary for layout work with a measuring tape on the ceiling.

2.3 Layout in the dirt outdoors.

Layout work in the dirt doesn't differ that much from indoor situations on the floor. All the features described in the previous section are valid for the outdoor applications too, with the exception of the use of the mobile tripod of course. Instead of this mobile tripod a bi-pod or similar fixture can be used. There will be practical differences with accessories to mount the targets or to establish reference points. Most of these applications don't require a high level of accuracy. Especially in the situation of a job site just started and hardly any reference points available in the dirt, CAPSY can be useful to layout foundations, sewage's, curbs etc. Also using the '**digitising**' mode (section **2.5)** is an easy way to measure the 'as is' situation with a one man survey crew.

2.4 Curved layouts.

The fact that CAPSY calculates 5 times a second it's actual position is a gateway to a whole range of possibilities. One of them is the so called '**curved**' mode. This mode gives the user the opportunity to layout very irregular shaped lines without the need of the numerous necessary co-ordinates. Simply by entering the shape of the curved line in terms of tangent points and radii into the field book, it will be pre-processed and ready for down loading into CAPSY. When this is done and the '**curved**' mode is active CAPSY will indicate every time it intersects this curve. The user is able to position the unit very accurate on the curve and to make a mark. So the number of marks is not limited and the user can make as many marks as he thinks what is necessary for a proper layout.

2.5 Digitising the real world.

CAPSY can be used as a real world digitiser too. Indoors and outdoors. Digitising an existing situation is very often required to produce up to date drawings as part of the preparations for a construction. Even during the construction process very often already existing situations are digitised in order to verify and monitor the building process. The '**digitise**' mode allows the user to identify and measure more than 2000 points by a one man survey crew. Later on these points can be up loaded into the field book and/or PC for further processing.

3. SPECIAL APPLICATIONS

3.1 Machine Control.

So far CAPSY is discussed as an one man, portable, real time positioning instrument. Because CAPSY is highly computerised and because it provides a real time position it may be used as a positioning sensor for machine control applications. Especially the fact that CAPSY is aware of it's bearing within the co-ordinate system makes it applicable to mount on e.g. earth moving machines, pile foundation machines and AGV's. There are two grades of integrating CAPSY in such a systems. When they are operated by a man , it may provide the information about position and bearing on a screen. It is the operators responsibility how to use this information. The other way is to integrate CAPSY in the control loop of the system and than there will be no human interference.(Overhead cranes,AGV's,Tunnel Drilling Robot). It is very likely that those applications require a special version of CAPSY.(physical and software).

3.2 Industrial.

Basically CAPSY is a real time, accurate, angle measuring device. It is the embedded software which transforms it into a positioning instrument. Although normally the angle measuring part is not visible to the user, it is available by relatively simple software modifications. By mounting reflective tape on specific construction parts or machine parts it can monitor very slight movements over long periods without any human interference. Highly accurate alignments over long distances is another application. It is the users imagination which limits the applications.

4. CONCLUSION

It took 4 years to develop CAPSY from a concept proved by a crude prototype to a real product, ready for mass production as it is now. Still you may consider the product is only just born. There is yet a lot to learn. A difficulty is that this product has no history. There is practically nothing known about how this kind of technology will be accepted on the construction site. Total stations and electronic theodolites are of course no novelty any more, but these are operated by a highly educated category of people. Very important will be the user interface and the easiness of use. Unlike it's technical performance (which can be tested and improved without user input) the user interface needs to be shaped and refined by feedback from the users in the field. It will be of extreme importance to listen to their comments and critics. Fortunately most of the modifications as result of this user feed back will be software modifications. Anticipating to go through several iterations the computer hardware and the embedded sotftware of CAPSY are prepared to handle these software updates very easily. The first field tests (started Jan. 1993) are indicating an excellent technical performance of the instrument. As expected the user interface went through a it's first iteration and is ready for a next sequence of field tests. These uncertainties about the user interface are of course only related to the '**one man positioning crew**' version of CAPSY. A machine control or industrial application will communicate via the standard RS232 interface conform a special protocol. The over all conclusion is that there is definitely a place for CAPSY on the construction site and there may be even more (industrial) applications which can use an accurate, real time, low cost positioning sensor.

5. REFERENCES

1. **[DeVos '89]** The Computer Aided Positioning SYstem CAPSY,
A low cost positioning system for construction.
Leon DeVos, Hans Schouten, TNO-BOUW, The Netherlands.
6 Th. ISARC, June 1989, San Francisco, USA.

2. **[Krom '92]** Automation of Robot Operation Planning Based on Product Modelling.
Ronald Krom, Frits Tolman, TNO-BOUW, The Netherlands.
9 Th. ISARC, June 1992, Tokyo, Japan.

3. **[Krom '93]** The development of a drilling robot for the installation of railway tracks.
Ronald Krom, Ruud Kloek, TNO-BOUW, The Netherlands.
prof. Charles Vos, Delta Marine Consultants, The Netherlands.
10 Th. ISARC, June 1993, Houston, USA.

Automation and Robotics in Construction X
G.H. Watson, R.L. Tucker and J.K. Walters (Editors)

Development of Earthwork Progress Measurement System Using Global Positioning System (GPS)

Mikio Okano /Akimasa Waku /Akira Fujioka/ Katuyuki Kikuta

Technical Research Institute of Fujita Corporation
74 Ohdana-cho kouhokuku Yokohama-City Kanagawa JAPAN 223

ABSTRACT

In this study, a new system for earthwork measurements using the Global Positioning System(GPS) has been developed. This paper describes the development of a device to automatically hold an antenna level and a computer program developed for the transformation of coordinates obtained by GPS surveying to the ordinary plane coordinates.

This paper gives a report on the earthwork progress measurement system, including its effectiveness. This effectiveness was verified a comparison between this new system and the traditional surveying system.

1. INTRODUCTION

In Japan, it is general practice in the carrying out of earthwork, the well balanced cutting and filling of soil is performed at the construction site. This is done in a way that does not generate any surplus or shortage of soil. Therefore, surveying must be executed periodically in correspondence to the progress of the work, in order to control the earth volume.

In many cases, field surveying involves extensive areas that need measurement, therefore surveying methods need to be outstanding in the execution of surveying rapidly, economically, as well as with work-performance ease, including a high-level of accuracy. An Electric Distance Meter (hereafter referred to as EDM surveying) and an aerial photogrammetry are generally used for the field surveying. However, EDM surveying takes time, as well as lack promptness, sometimes failing to follow earthwork progress, while an aerial photogrammetry enables measurement in a very short time, but is inferior to EDM surveying in accuracy. Also, frequent measurement using an aerial photogrammetry is not economical.

On the contrary, in recent years, the Global Positioning System(GPS)[1)~3)], a surveying system that takes advantage of artificial satellites, has attracted attention. The GPS is a system that receives radio wave transmitted from artificial satellites in order to attain three-dimensional coordinates of observation points.

The GPS has led the authors to assume the feasibility of creating such as a system that can perform measurement faster and more economically than the traditional measurement system. On the basis of this concept, earthwork progress control system using the GPS has been developed, for application to large-scale civil engineering projects.

2. OVERVIEW OF GPS SURVEYING

GPS surveying is largely classified into two categories, the point positioning method and the relative positioning method. The former is a surveying method aiming to assist navigation, and has been extensively used, while the latter is a surveying method that can attain the relative coordinates between observation points. The relative positioning outweighs the traditional surveying system in various ways, such as 1) enable highly accurate surveying in relation to long baseline, 2) not requiring through-vision between observation points, and 3) enabling surveying in rainy weather and at night. In this system, kinematic positioning of relative surveying is used.

The kinematic positioning is such that an antenna is fixed at the base point, and another antenna is moved from one observation point to the other successively, in order to attain the relative coordinates between the base point and the observation points. Since observation by this positioning needs only several seconds, it enables the attainment of coordinates for a large number of observation points. The only disadvantage it has, is the requirement for the continued receiving of radio waves from at least four artificial satellites.

3. EARTHWORK PROGRESS MEASUREMENT SYSTEM USING GPS

3.1. System Concept

Fig.1 illustrated the flow of the GPS earthwork progress measurement system developed through this study. This system consists of four constituents, as mentioned below.

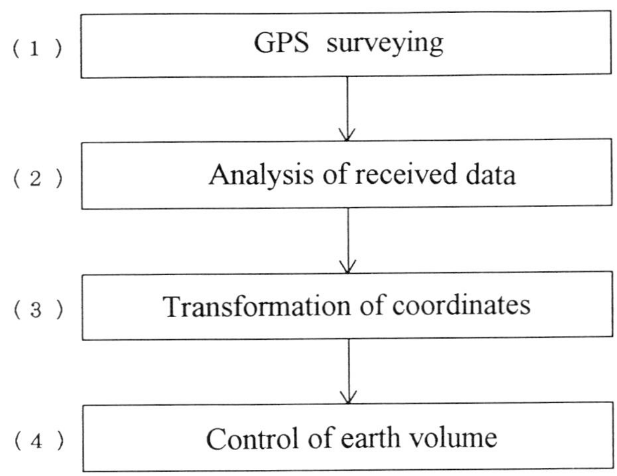

Fig. 1 System Flow

(1) GPS surveying

In earthwork progress measurement, it is necessary to survey observation points quickly in a large site. Therefore, this study used the kinematic positioning that requires only a short time for the measurement at one point. However, the important issues are how quickly the measurement can be taken without causing cycle slips. The meaning of cycle slip is as follows. As some radio waves from artificial satellites are interrupted, they are being received, due to the influence of an obstacle or a noise wave. In this instance, the phase being measured loses continuity. The occurrence of this state is known as the cycle slip.

(2) Analysis of received data

An analysis method for receiving data has been proposed, so this study utilized it[1),4)].

(3) Transformation of coordinates

Since the coordinates of the observation points attained through analysis are derived from the World Geodetic System-1984[1)] (hereafter referred to as WGS-84), using their results directly may give rise to the following problem. The coordinates based on Japanese Geodetic System[1)] are generally used on a construction site, so it is necessary to transform the coordinates attained through GPS surveying, into those in Japanese Geodetic System.

(4) Control of earth volume

Earth volume were calculated using the observation points transformed the coordinates in Japanese Geodetic System.

3.2. Development items

Description of detailed development items in this study are as follows.

1) A device to automatically hold an antenna level

To enable the quick measurement of earthwork progress at an extensive range of sites by using the kinematic positioning, the use of a vehicle that moved from one observation point to the other was contemplated. In addition, if the antenna tilted while was being moved, it might become impossible to receive radio waves from artificial satellites. If cycle slips occur then the job efficiency would drop considerably. All this led to the development of a device that can automatically hold an antenna level in order to enhance the efficiency and reliability of kinematic positioning.

2) Linkage program

The results of GPS surveying are represented in accordance with the WGS-84, so the result, when applied to general earthwork, have to be transformed into coordinates in Japanese Geodetic System. This requires a program capable of transforming of many observation points quickly and accurately. To this end, a program has been developed, that directly links the analytical results of GPS surveying to the earthwork progress program.

4. A DEVICE TO AUTOMATICALLY HOLD AN ANTENNA LEVEL

4.1. Overview of a device

Desired control accuracy of this device was designed as follows.

While vehicle moving	5 degrees
While vehicle at reset	0.5 degrees
Control limit	30 degrees

Fig.2 illustrates this device developed accordance with the target accuracy mentioned above. A clinometer to measure tilt angles of the antenna plane and a gyro to measure angular speed were installed, one each in biaxial direction (at the pitching and the rolling angles) respectively, for a total of four. A DC motor was used in the driving section to hold the antenna plane level.

This automatic device collects the data measured by the clinometers and the gyros, makes correction in an amount corresponding to the tilt angle of the antenna plane, and by driving the DC motor, holds the antenna plane level. In detail, data from the quick-responding gyros is used to control the changing angles of the antenna created by sharp ground undulations felt by the traveling vehicle, while data from the clinometers is used to accurately control the marginally changing angle of the antenna.

4.2. Accuracy Verification tests

For the purpose of verifying the level holding accuracy of this device, this device was mounted on the vehicle(Photo. 1) and tested at the site ((a)smooth gravel road, (b)asphalt-paved surface with a certain gradient, (c)graded site with sharp undulations) at the traveling speed of approximately 5km/h .

Fig. 3 plots some of the results. In the cases of the smooth gravel road and the asphalt-paved surface with a certain gradient, the antenna plane while the vehicle was traveling, was held level with the accuracy of approximately 5 degree. In the case of the roughly graded site with sharp undulations, this device was sometimes unable to follow the motions of the vehicle body. It sometimes exceeded 5 degrees, but was not over 10 degrees in any case. As the above

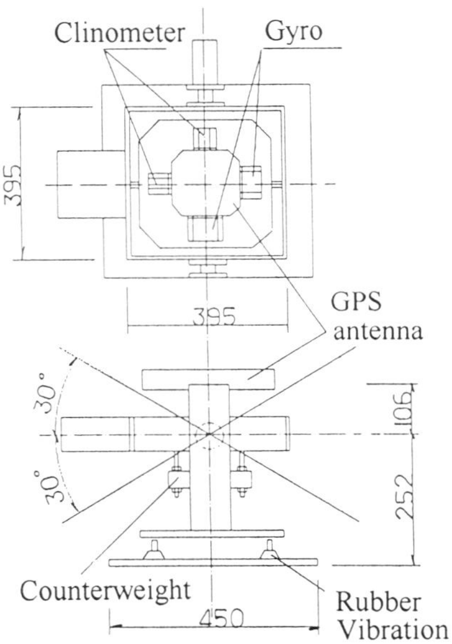

Photo. 1 The Vehicle

Fig. 2 A device to automatically hold an antenna level

mentioned test results indicate, the target accuracy was able to be achieved for the most part within the range of the given limits. However, at the present stage, promptness, and an application to complex topography remain unsolved requiring future modification.

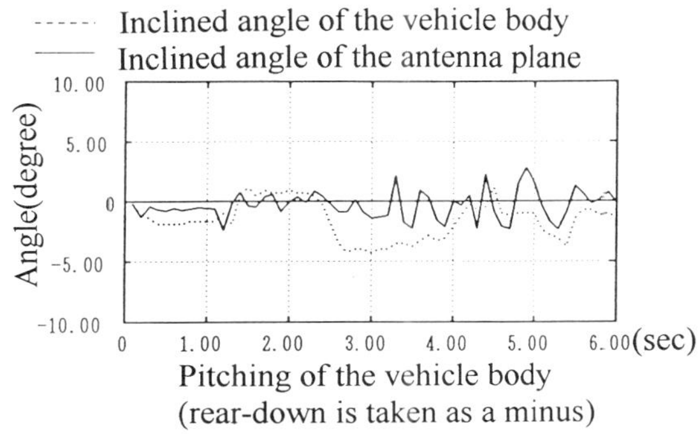

(a)The smooth gravel road

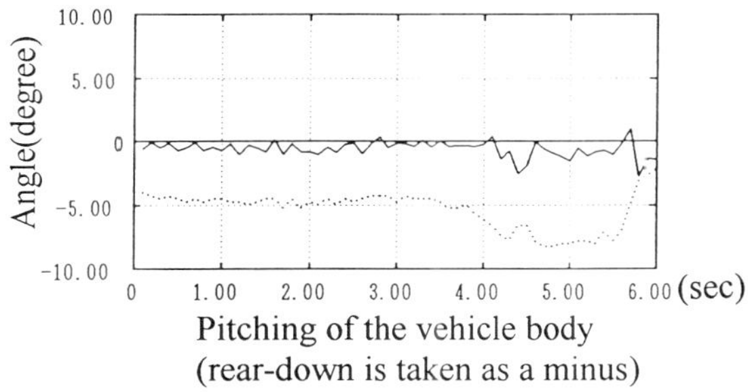

(b)The asphalt-paved surface with a certain gradient

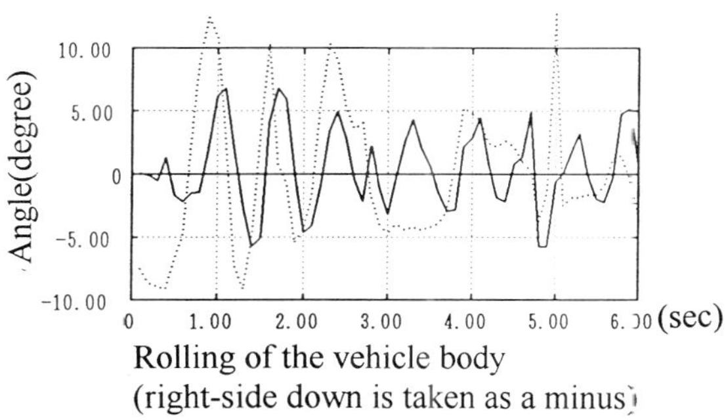

(c)The graded site with sharp undulations

Fig. 3 The results of the accuracy of verification tests

5. DEVELOPMENT OF LINKAGE PROGRAM

A linkage program has been developed to transform coordinates attained through GPS surveying, into plane coordinates in according with the Japanese geodetic system, and to offer the easy attainment of data to be input the earthwork progress program. Its flow is shown in Fig. 4[5]. Using the linkage program, the analysis is accomplished systematically and efficiently.

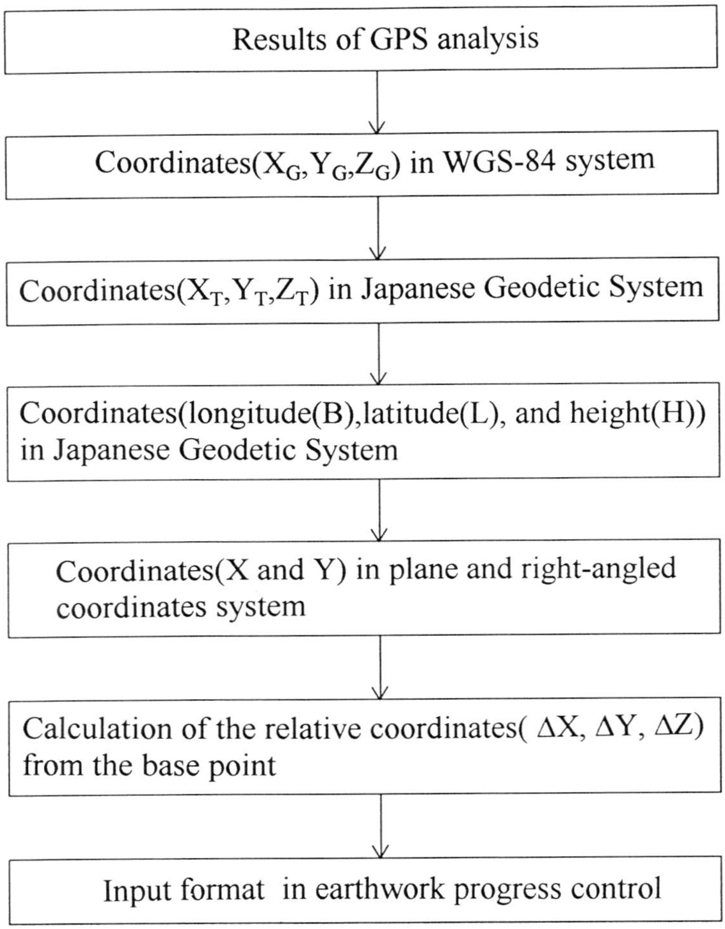

Fig. 4 Flow of the linkage program

6. DEMONSTRATION FIELD TESTS

6.1. Overview

Field tests are done in this study to verify the adequacy of the system developed, after its application to the measurement of earth volume.

The area of test yard was approximately 1 hectare. The earth volume of this site was measured using the system developed through this study and EDM surveying, and both results were compared.

6.2. Measurement result

Fig. 5 is the planar view of the observation points. Table 1 shows the calculated results. The results indicated that the earth volume calculated by this system was merely 0.6% less than that calculated by EDM surveying. This difference would not be a particular problem in the calculation of earth volume.

These tests revealed that the system can offer approximately the same accuracy as the EDM surveying, and that the use of the device to automatically hold an antenna level mounted vehicle was able to facilitate field GPS surveying.

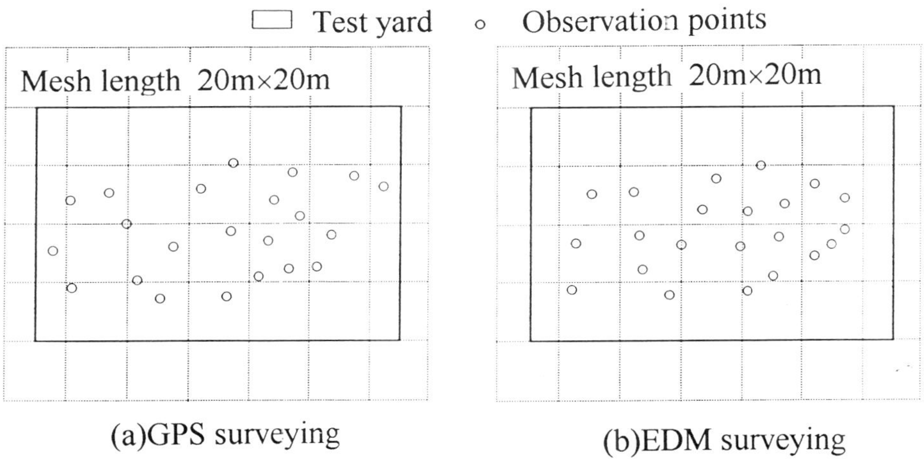

Fig. 5. Observations ground plane

Table 1 Measurement results

	Electric Distance Meter	GPS surveying
Observation points	22	22
Accuracy	31,495m^3	31,672m^3 (0.6%)

7. CONCLUSION

Through this study, the earthwork progress measurement system is developed using the GPS. This system consists of (1)GPS surveying, (2)Analysis of received data,(3)Transformation of coordinates, and (4)Control of earth volume. This basic study specifically focuses on (1) and (3), the development of a device to automatically hold an antenna level mounted vehicle, and a linkage program that links analysis results to the earthwork progress control program.

Field tests conducted at a roughly graded site indicated that the device to automatically hold an antenna level was able to achieve the required accuracy to hold the antenna level. A linkage software is created to link the transformed coordinates with the earthwork progress control program. This was followed by an application of the system to earth volume. The results proved the capability of the system to calculate as the EDM surveying, as well as its excellent features to facilitate the execution of the work.

REFERENCES

1) Geodetic Institute of Japan: Revised edition Precision positioning system by using GPS and artificial satellites, 1989.

2)Wells,D.(ed.): Guide to GPS Positioning, Canadian GPS Associates, 1986.

3)A. Leick: GPS, Satellites Surveying, John Wiley & Sons, 1990.

4)Trimble Navigation: Trimvec-Plus User's Manual(Manual of GPS Surveying Software),1991.

5)Shunsuke Sakurai, Norikazu Shimizu, Shigeya Furutani, Akio Saragai: Surveying from the control point of a large light-emission facility (Spring-8), Japan Society of Civil Engineers Proceedings of Lectures at the 47th Annual Meeting, pp. 648~649,1992.

Automation and Robotics in Construction X
G.H. Watson, R.L. Tucker and J.K. Walters (Editors)

Survey Robot Using Satellite GPS

Tadashi Kanzaki, Shuiti Nishizawa, and Hiroshi Toida

Construction Technology Development Dept., Technology Div., Taisei Corporation, Sanken Bldg., 3-25-1 Hyakunin-cho-, Shinjuku-ku, Tokyo 169, Japan

Abstract

In Japan, many large projects have been undertaken such as the 513-hectare artificial island of Kansai International Airport and the about 1,000-hectare offshore development project of Haneda International Airport. In the case of civil engineering works involving such a wide area, it is important from the viewpoint of construction management that daily topographic survey be made quickly. GPS is an abbreviation for Global Positioning system, which measures the three dimensional coordinates by receiving microwaves from artificial satellites. By loading this receiver on a vehicle, the survey robot three-dimensionally measures the topography at 0.5 second intervals while the vehicle is running. This robot was used in the reclamation work of Kansai International Airport development project. It has accomplished the three-dimensional topographic survey at 4,000 points in an area of 20 ha in only two hours. One further application of this robot is unmanned surveying for golf courses.

1. INTRODUCTION

Highly accurate measurement methods and advanced management techniques are a critical part of the quest to reduce construction period, labor requirements, energy usage, and cost of large projects, such as airports and land reclamation. This is true for all stages of the work, ranging from surveys and planning to actual execution of the work. For instance, at a large-scale land development site, a swift and simple method of surveying can play a significant role in estimating the amount of cut and fill required. Highly systemized management is needed to properly handle information about complex geological structures and various soils measured in surveys. At a large-scale site, such as an offshore airport or reclamation site, the soft ground is often boggy; yet settlement and ground behavior have to be managed through fill geometry measurements to obtain early estimates of any additional fill needed. Over such vast areas, where efficiency and precision are the determining factors of successful work, there is an urgent need for a survey system based on the satellite global positioning system (GPS). A means of integrating such a system into all fields of work management is also being demanded.

This paper outlines part of our response to this need; a robot which can carry out rapid topographical surveys. Its application to the work now in progress at the Kansai International Airport is discussed, along with a brief mention of some other survey robots which may be of interest regarding future unmanned surveying.

2. GLOBAL POSITIONING SYSTEM (GPS)

The global positioning system (GPS) was set up by the U.S. Department of Defense. The ground position is established using a network of 24 satellites about 20,000 km from the earth on six different orbitals, four on each orbital, as shown in Fig. 1.

Two pseudo-microwave signals, called the L1 wave and L2 wave at frequencies of 1.5 GHz and 1.2 GHz, respectively, are continually transmitted from the satellites. These signals carry encoded data giving atomic clock information, ionosphere correction information, and information on the satellites' orbits. This information is coded in two ways: one exclusively for military purposes and the other openly available to the public at no charge.

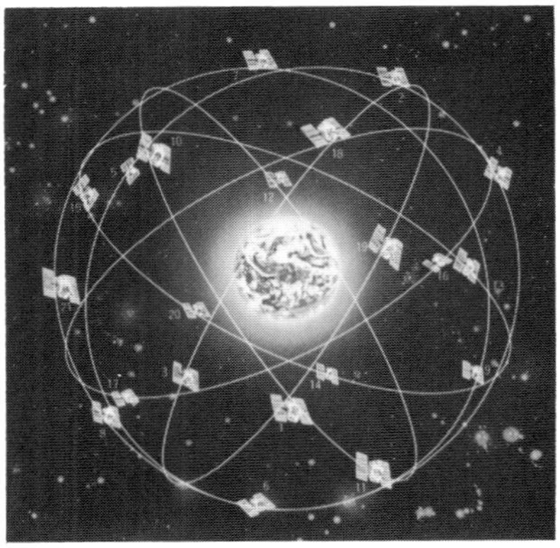

Figure 1. Artist's impression of the GPS satellites and their orbits

3. DYNAMIC GPS SURVEY TECHNIQUE

Conventional survey methods absorb considerable amounts of labor and time because optical surveying equipment requires three to four operators and all analysis has to be done manually. Other survey methods, including the Total Station (Electronic Distance Measurement, EDM) suffers from some constraints:

(1) the need for mutual visual contact between any two survey points;
(2) significant dependence on expertise and experience; and
(3) susceptibility to weather and other factors.

The GPS is currently in use by the Geographical Survey Institute of the Ministry of Construction, which has confirmed the locational accuracy of the system. They are now using it for precise re-mapping of triangulation stations in Japan, however, the static GPS survey methods generally used for reference point surveys take a few hours to measure a single point. In contrast, at a large-scale earthworks site where work is rapidly implemented, topographical surveying of a large area must be completed within a very limited time. For this reason, the normal static GPS method is not considered suitable for earthworks applications.

The dynamic GPS survey technique proposed here, however, enables an instantaneous three-dimensional survey of topography. It has been the subject of various discussions and tests related to survey time and accuracy, and has overcome the conventional restrictions of both software and hardware in this kind of technique. Our new method also reduces the number of survey operators from three or four to only one, and will eventually enable fast, continuous measurements.

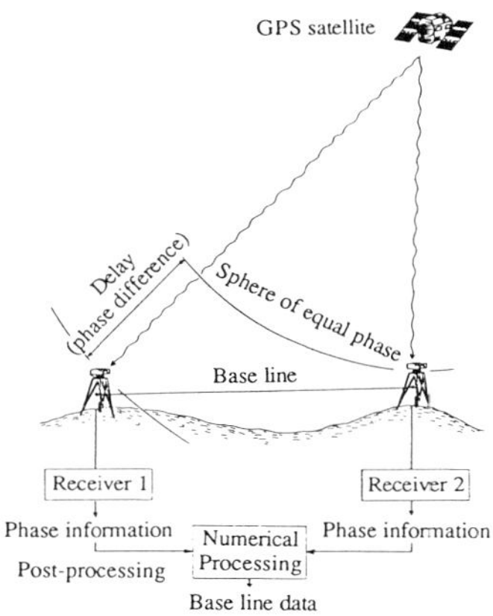

Figure 2. Conceptual outline of relative positioning

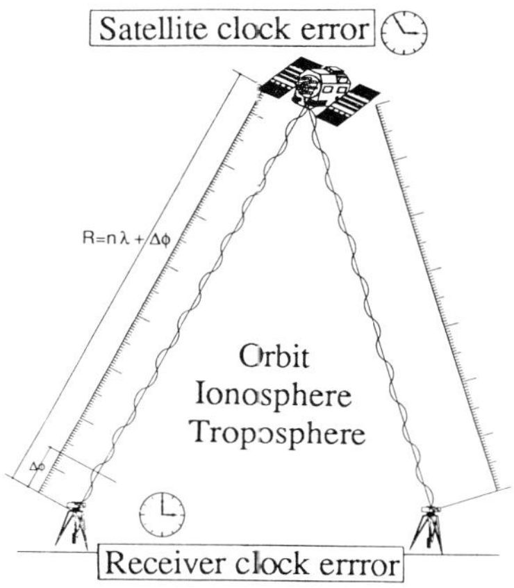

Figure 3. Propagation distance and causes of error

3.1 Principle of measurement

Figure 2 illustrates the idea of relative positioning. This principle is encapsulated in the equation below. ρ_1 and ρ_2 are the distances from the satellite to receivers 1 and 2, respectively.

$$\rho_1 - \rho_2 = \left|\vec{X}_s - \vec{X}_1\right| - \left|\vec{X}_s - \vec{X}_2\right| \tag{1}$$

Where: $\vec{X}_s$: Position vector of the GPS satellite
$\vec{X}_1$: Position vector of the receiver 1
$\vec{X}_2$: Position vector of the receiver 2
$| \ |$: Absolute value of vector

$\vec{X}_1$ and $\vec{X}_2$ on the right hand side of Eq.1 are unknown values with three components, while $\vec{X}_s$ is a known value with three components contained in the orbital information from the GPS satellite. If one of the two receivers is at a known point, the number of unknown values is reduced to three. Since the left hand side is actual data from a satellite, data from three different satellites gives three versions of Eq. 1, each of which can be solved for three unknown values. That is, a receiver in an unknown location can be positioned. However, the collected data contains other information, including the phase difference between the signal received from the GPS satellite at two receivers. This phase difference results from the geometrical relationship of position between the satellite and the two receivers, but it is also influenced by propagation delays in the ionosphere and troposphere, delays within each receiver, and time differences. For this reason, the phase difference measured between receivers 1 and 2 is considered an unknown which depends on the receivers themselves. To obtain its actual value, data from another satellite are required. In practice, many observations are made to analyze and determine this unknown quantity.

3.2 Double-difference processing of carrier phases

The normal accuracy of GPS is around 30 m because of the inability to eradicate errors which are common to all observations,. Any practical application of GPS thus requires the development of a relative positioning system which can measure carrier phase at more than one receiver. Under the double-difference processing system, signals are taken from four or more GPS satellites to determine the absolute coordinates of a survey point.

The measurement errors can be mainly attributed to three effects:

(1) clock errors in the satellite;
(2) clock errors in the receiver; and
(3) wave propagation delays in the ionosphere and atmosphere.

In Fig. 3, the following symbols are used:

R: distance between satellite and receiver, along which signals propagate
n: the number of wavelengths
λ: wavelength
$\Delta\rho_1$: residual phase

To eliminate errors, the following procedure is adopted: First, two receivers receive the carrier phase from the same satellite and the phase difference is calculated. The same measurements are then made using another satellite; the difference in phase measurements is then compared. This double-difference processing allows any errors to be eradicated. Assuming that the absolute phase difference between receiver 1 and 2 (distance from each end of the baseline to the satellite) is $\Delta\rho_1$ and $\Delta\rho_2$, respectively, the difference of positioning, δ, the difference of the phase difference between the two satellites, Σ, is given by the following equation:

$$\Sigma = (n1\lambda + \Delta\rho 1 - \delta) - (n2\lambda + \Delta\rho 2 - \delta) = (n1 - n2)\lambda + \Delta\rho 1 - \Delta\rho 2$$

This double-difference method of processing is essential to high-precision three-dimensional positioning. The positioning data obtained by this method are converted into Japanese geodetic coordinates from WGS-84.

3.3 Characteristics of dynamic GPS surveying

(1) This method allows instantaneous and continuous three-dimensional positioning, significantly improving work efficiency.

(2) This method allows measurements to be made easily by a single person. A quick, simultaneous multiple-point survey of a large area can be implemented by increasing the number of receivers.

(3) This method led to Japan's first case of a continuous survey to 1 cm accuracy.

4. 3-D TOPOGRAPHIC SURVEY ROBOT

The 3-D topographic survey robot, shown in Fig. 4, is a development of GPS technology for use at an artificial island work site. As shown in Photo 1, this robot is a vehicle equipped with a receiving antenna; it carries out continuous topographic surveying while moving around the site. It has performed well at an actual work site, demonstrating its efficiency and precision.

Figure 4. Artist's impression of large-area 3-D topographic surveying using vehicle

Photo 1. Vehicle used for GPS high-speed dynamic positioning

4.1 High-accuracy positioning of dynamic platforms with GPS

In developing a vehicle to carry out highly accurate continuous positioning, a series of tests was implemented to check the basic characteristics of the equipment. Various experiments were then conducted to solve several problems which manifested themselves during this development process.

One characteristic of this system is that only one piece of data is received for each survey point. If signals from the satellite are interrupted for any reason during the survey, errors in the phase data may arise. Such errors, known as cycle slip, must be avoided. The relationship between the velocity of the vehicle and survey accuracy also had to be clarified. The main issues considered in the study were (1) the optimum combination of satellites, (2) the initial setting (selection of a whole number bias), (3) the method of measuring the antenna height, and (4) the optimum vehicle speed.

There are almost no obstacles to reception of GPS satellite transmissions at an airport construction site, but it must be considered that terrestrial refraction could affect high-accuracy positioning.

The refractive index (n) of air for frequencies less than 30 GHz is shown in Eq. 2:

$$n - 1 = \left(0.776 \frac{P}{T} + 3.73 \times 10^3 \frac{e}{T^2} \right) \times 10^{-6} \quad (2)$$

Where P,e: atmospheric pressure and partial pressure of vapor expressed in Pa (1 mbar = 10^2 Pa)

T: temperature in K (0°C = 273K)

Differentiating Eq. 2 by the height, h, gives:

$$\frac{dn}{dh} = \frac{10^{-6}}{T} \left\{ 0.776 \left(\frac{dP}{dh} - \frac{P}{T} \bullet \frac{dT}{dh} \right) + 3.73 \times 10^3 \frac{1}{T} \left(\frac{de}{dh} - \frac{2e}{T} \bullet \frac{dT}{dh} \right) \right\} \quad (3)$$

According to Eq. 3 and the formula for atmospheric density, readings from a satellite at an elevation of 15° or more are required to obtain locational accuracy of better than 1 cm.

Actual use of the new method at an airport construction site has clarified the following:

(1) This method does not require the services of an expert surveyor to obtain the required level of accuracy. Any lay operator can easily and quickly carry out survey work.

(2) This method requires only a single operator for its operation, so it will eventually lead to great savings in labor costs compared with conventional systems that require a group of at least three to four surveyors.

(3) This method allows a single operator to complete a survey of one hectare in five to six minutes, a task which previously took a team of three to four surveyors an hour.

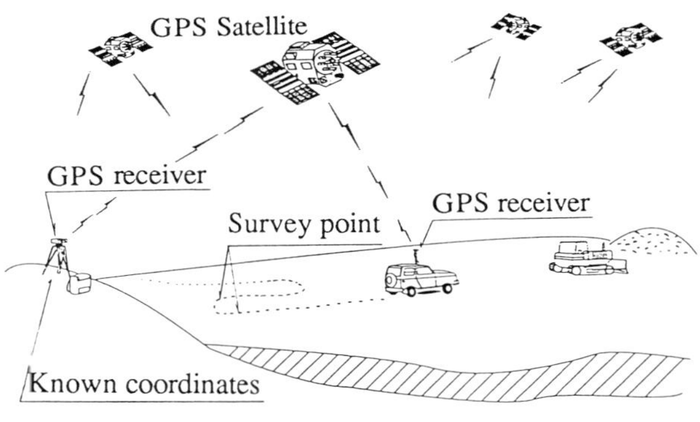

Figure 5. GPS high-speed 3-D topographic survey

4.2 The artificial island project for Kansai International Airport

The high-speed dynamic surveying method described in this paper was put to practical use at the 11th work section of this airport construction site. It was used to survey the height of the fill top over an area of about 21 ha. Before surveying began, the fixed and moving points of the GPS receiver were set at known coordinates. Transmissions were received for a few minutes to determine the number of carrier phases. Then, as shown in Fig. 5, the roving receiver was placed on the roof of a vehicle, which was subsequently driven over unknown points and data taken continuously at intervals of 0.5 to 2.0 seconds.

A total of 5,200 and 1,300 points, respectively, were recorded on the fill top (about 21 ha) and loaded fill (about 9 ha) in a total period of about 3 hours, as shown in Fig. 6. This work at Kansai International Airport quantitatively proved the energy and labor saving effects of this newly developed high-speed dynamic positioning method.

The main features of this new method are summarized below.

(1) A high degree of accuracy, to about 15 mm, is possible, as confirmed in the precision experiments, regardless of the velocity of the vehicle.

(2) The method is especially effective in the planar surveying of a large area.

(3) A single operator is able to complete the topographic surveying of a vast area of land.

(4) The reception interval can be freely chosen according to the purpose of the survey.

(5) Computer processing enables continuous output of analyzed data and coordinates.

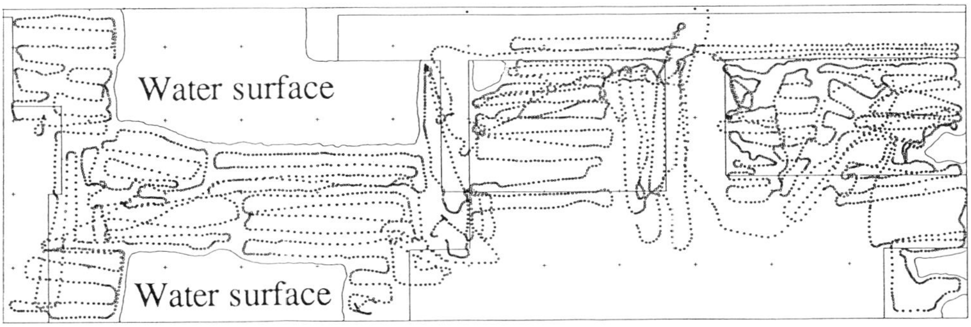

Figure 6. Tracks of survey points (Kansai International Airport: 1,350 x 420 m)

5. SURVEY ROBOT FOR GOLF GREEN CONTOUR MEASUREMENT

A high-accuracy, continuous surveying system using a remote-controlled vehicle, as shown in Photo 2, was used to survey the contours of a golf green. The work entailed equipping a remote control vehicle with a GPS receiving antenna and carrying out a continuous three-dimensional topographical survey of the greens. The system was put into practice at an actual golf course.

This survey enabled contours and bird's-eye views of the golf course to be produced, as shown in Figs. 7 and 8.

This survey robot will enable a data base of well-known golf courses to be created, even allowing them to be duplicated. This function will make it easy to repair existing greens or recover their original form.

Photo 2. Survey robot

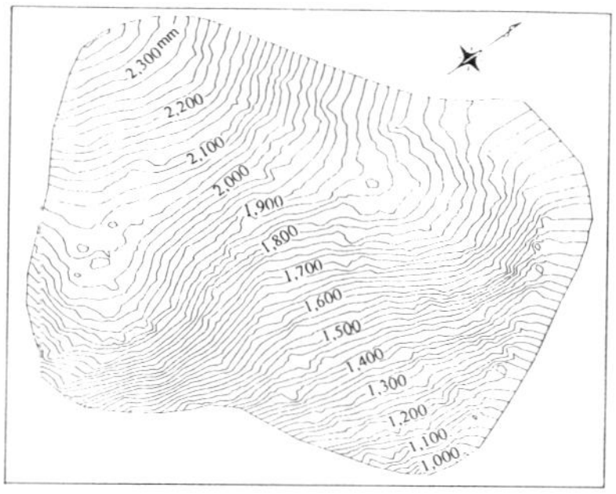

Figure 7. Contours

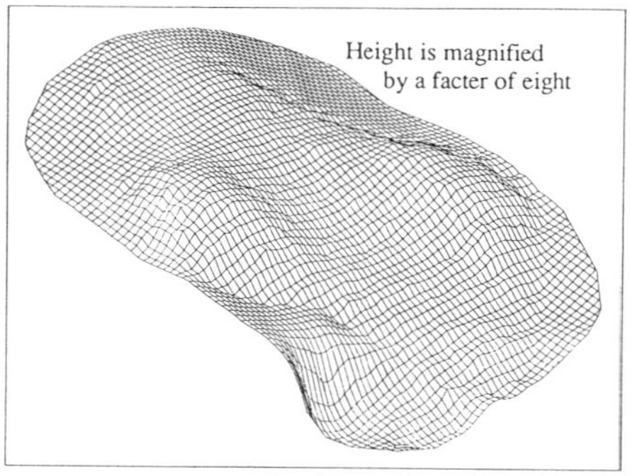

Figure 8. CAD representation in 3-D

6. CONCLUSION

GPS technology is now attracting greater and greater attention from various industries all over the world. Great efforts have been made to widen its range of applications and many of these are now bearing fruit; the system is currently coming into use in a broader range of fields. Practical applications in Japan have also been increasing in number over the past few years. At the same time, Japan's construction industry is moving more and more toward the application of technology in large-scale projects.

The authors' aim was to develop a technique, as proposed in this paper, which would push forward construction technology by using GPS. They hope that this paper will be of use to other researchers and engineers engaged in the same line of work.

Automation and Robotics in Construction X
G.H. Watson, R.L. Tucker and J.K. Walters (Editors)

Application of 3-D CAD and 3-D coordinate meter in frame erection

Pentti Vähä[a], Klaus Känsälä[b], Ilkka Kaisto[c] and Rauno Heikkilä[d]

[a,b]Technical Research Centre of Finland, Electronics Laboratory
P.O.Box 200, SF-90571 Oulu, Finland

[c]Prometrics Ltd., Teknologiantie 6, SF-90570 Oulu, Finland

[d]University of Oulu, Department of Civil Engineering
Kasarmintie 8, SF-90100, Oulu, Finland

Abstract

A method for applying 3-D CAD design tool and 3-D measuring device in frame erection is described. The method links together the design data of the building and the as-built measurement data. The Cartesian measurements for each points on the site can thus easily be defined. The preparatory phase of the element assembly takes the advantage of the check measurements by calculating the clearances between different elements in three dimensions and saves them into the memory. In the assembly this assembly data is further corrected with the element dimension data.

1. INTRODUCTION

Position determination in 3-D coordinates, that is the xyz-position, is needed in many stages of the construction process. Therefore, a measurement system frame for construction sites is needed to provide a possibility to link the design data to the construction process. This kind of unified coordinate system can form a grid over the whole site and thus it can also be used for integration of measurements from different measurement devices. Currently, the design of a building can be performed by using a 3-D CAD design tool running in work stations or even in PCs. These CAD design tools are usually based on rectangular and right-hand coordinate systems. After the design of the building the CAD system is used only to plot the drawings of the building on a paper. These drawings are utilised in different measurements during the building process on the construction site /2/.

At present, coordinate measurement are usually based mainly on traditional devices due to high prices of advanced devices. The measurement results are not usually available immediately and not in xyz-coordinates. In recent years advanced theodolite- as well as laser-based coordinate meters have been developed for determining xyz-coordinate points. The development of these devices has made it possible to apply them in the construction process. Their accuracy is also enough for most construction measurement tasks. Although these devices are used to some extend either in the prefabrication plants or on the construction sites, methods for linking the CAD model of the building to the site measurements are underdeveloped.

In this paper a method on the application of 3-D CAD model and 3-D measuring device in frame erection is described. First, an example building has been designed by using a 3-D CAD software tool which is fully three dimensional and is based on the object oriented design. With the aid of this tool all the element related data both the drawings for production and the assembly schemes for their erection can be produced. A 3-D coordinate measuring device can be linked to the 3-D CAD model to give measuring points of the building in the xyz -coordinates of the 3-D device. By taking the advantage of this connection the operator can prepare a list of measurements. After the foundation work has been completed the columns can be erected and the operator can make quality measurements to check the real position and pose of the erected columns. This as-built measurement data on the real position and pose of columns is fed to the erection software together with the design data obtained from the 3-D CAD model. Based on this information the erection software calculates the instructions for the operator to be utilised in the assembly of wall elements. The operator saves these instructions into a hand computer. These instruction are still corrected with the real element data red from the escort memory inside the element to give more exact erection information on line for the operator.

2. UTILISING 3-D CAD IN 3-D MEASUREMENTS ON THE CONSTRUCTION SITE

2.1 Coordinate frames

Coordinate frames play an essential role in three dimensional design tools as well as in three dimensional measurement systems. A building on the site has to situate on a certain location as well as the rooms, corridors, windows, doors and other components have to situate on defined locations in the building. For the location of the building on the site the municipal authorities will define the position on the site in a three dimensional coordinate frame.

The measurer with his 3-D measuring device usually defines a new coordinate system from the geographical one by using special coordinate transformation. He defines the origin such that on this construction site the values of the coordinates are always positive. The directions of x-coordinate axis is along the line of the building and y is perpendicular to x axis. The direction of the y axis depends on the measuring device, and it can be either a left- or a right-hand coordinate system. Further, he can define more than one coordinate system depending on the shape of the building. .

The building has been designed with the aid of a 3-D CAD design tool applying usually a rectangular and right-hand coordinate system where x-coordinate is along the side of the building, y perpendicular to x axis according to the right-hand system and z grows upwards.

2.2 Tool for 3-D CAD modelling

The 3-D CAD software tool used in this work is based on modular thinking. The tool includes a module for making the frame of the building and an other one for the element design. This program package is called GIDES, and it runs under OS/2 in a standard PC environment /5/. The first step in the model design is to define a suitable module grid for the building. This grid is usually based on the colonnade lines of the building skeleton. The grid is three-dimensional and grows upwards from zero level located on the foundation level. After the grid has been completed the designer can start to design columns. GIDES contains a set of basic construction elements in 3D-primitives. There are for instance different types of columns, beams, slabs and wall elements. The user specifies proper dimensions and coordinates for a certain primitive which then becomes a visible element.

2.3 Instrument for 3-D measuring

The 3-D measurements were carried out with the most advanced optical coordinate meter Acmeter MC, which is especially designed for large scale assembly in shipyards by Prometrics Ltd. Acmeter MC, Fig. 1., measures the coordinates in terms of radial distance with pulsed time-of-light principle and two angles with optical encoders resulting in optimal moving geometry. Being a real one user coordinate meter the measurements are possible to perform without any prism as a target, only the normal diffused surface or cheap diffuse Acman sticker is needed. The accuracy of the coordinate meter is 1 mm (rms) in the measurement range of 3-30 meters and 2 mm (rms) up till 60 meters in xyz-coordinates.

As Acmeter MC is an integrated part of the ACMAN system (Integrated Accuracy Control System) the design data can be used to guide the user during the measurement process. A measurement is an easy task with the automatic focusing of the optical system, which also includes the coaxial red pointing laser for marking and guiding during assembly process. Having efficient methods for calibration to the object coordination, the user gets the difference between the designed and measured coordinate values immediately at the measurement site. The ACMAN system has software with a database and analytical tools for further analysis of the measurement results.

Used Acmeter MC is manually operated system, which limits its use to normal 3-D static measurements. The possibility to use pre-programmed sequences or automatic tracking of the object could extend the feasibility of a 3-D coordinate meter also to dynamic measurements during assembly.

Figure 1. Acmeter MC: An optical coordinate meter.

2.4 Linking 3-D CAD model and 3-D measurement device

CAD-programs are usually designed for right-hand coordinate systems. Some different output formats (DXF, TIFF, IGES) for data transfer between different programs exist, but there is no chance to save specified points in user defined coordinate systems. This saving operation is essential for proper use of different measuring devices. The GIDES 3D-CAD design tool applied in this project contains a function which allows the user to make formatted file of selected points. This function asks for two points in the 3D-model and the corresponding points which have been measured on the construction site. After this have been done the user can digitise all necessary points on the screen with the mouse. The function calculates a conversion matrix from the 3D-model to the user specified coordinate system and uses this matrix to convert digitised points to that system.

3. FRAME ERECTION

3.1 Erection of columns

The measurer defines the exact location and height for the set of bolts of every column in the foundation work phase. After casting the locations for the sets of bolts are checked and corrected if not inside tolerances. Then the columns are erected by lifting them with the aid of a crane to their places. The erectness is set either by using a level, a plumb line or two theodolites. The facade line for the columns is also set as straight as possible. After erecting the whole line of columns the worker uses his eyes for checking the straightness.

3.2 Check measurements

Usually no check measurements are done on the construction site after the worker has erected the columns and they look good for the human eye. However, the columns are never in upright position, but each column is inclined to some direction. Therefore, a check measurement for each column was performed after the erection. Performing the check measurement for high rise columns with ordinary 3-D measurement devices like to prism measuring theodolites is not a simple task especially at the upper end of the column, since someone has to climb to the column to set a measurement point there. In the Acmeter system case, a measuring point, a small piece of paper, can be glued in advance at the upper end of each column. This makes it more easily possible to perform check measurements for the location and erectness of columns. In this way the exact location and pose of columns can be obtained and saved into a file. In checking the location and pose of columns the measurer also measures and saves the height (the z-coordinate) of the foundation for the concrete elements at the points of columns. Hence, all the necessary data except the dimensions of the precast concrete wall elements for preparing the assembly of the elements are at hand.

3.3 Element assembly

Before starting to assemble wall elements into their places, the advantage of the 3-D CAD - and 3-D check measured data was taken. The elements have their designed locations in the building they should also be assembled as accurately as possible to their nominal places. By combining the design data and the as-built data obtained from check measurements, Fig. 2., the gaps between the foundation and the bottom of the element at the point of each column can be calculated giving thus the thickness of wedges to be set on the foundation before assembling the element on the foundation with the aid of the crane. In the same way the gap between the column and the element can be calculated to give the amount of wedges. Correspondingly, before assembling the first element the starting line can be marked to define the third coordinate. Thus, the assembly information for each element can be prepared beforehand by taking the full advantage of the 3-D data. In this way the check measured data is used as a feedback information in the erection process. The prefabrication of concrete elements takes also the advantage of the Acmeter-System. The planning of elements is performed in a CAD system, and this planned data is fed to the Acmeter-System to guide the part fabrication and mould assembly, and after casting for quality control /4/. In this project escort memories were put into the concrete element while pouring the concrete mass into the moulds. After concrete curing a quality control including element check measurements will be performed. The dimension data of the elements obtained in quality control was written into the escort memories inside each element. for late use in the erection process. Thus, the assembly information obtained by combining the 3-D CAD data and the 3-D check measurement data can further be corrected by reading the element dimensions from the escort memory inside the element and correcting the erection data by the element data. In this way the 3D- CAD design data and the real as-built data can be exploited in the frame erection process.

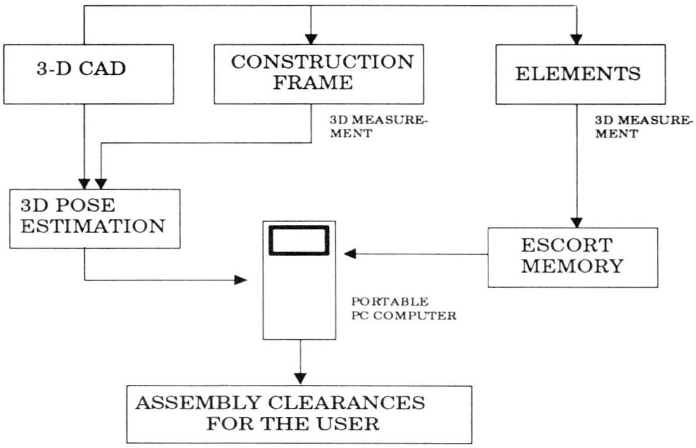

Figure 2. Information flow in assembly.

4. EXPERIMENTATION OF THE ERECTION

4.1 Example building

For testing the method to apply 3-D CAD and 3-D measuring device in the frame erection an example building project was set up. The test environment contains a floor element of concrete, three steel columns and three concrete wall elements. The columns are standing on the floor element and the wall elements are attached to the columns. The 3-D model of the building is seen in Fig. 3. and it was designed by using the GIDES 3-D CAD design tool /5/.

4.2 Frame erection

The columns were erected to their places with the aid of the crane and set to their upright positions by ordinary means with the aid of the level. After erection of the columns their real location and pose were measured together with the height of the element foundation with the 3-D measuring device. Then the design information was combined with the as-built measured data to give the clearances (joints) between the element and the foundation as well as the element and the column for wedge settings. This data was input to the PC for later use in the element assembly. When the elements to be assembled with the aid of the crane come, the element dimension data was read from the escort memory to the PC, and the assembly data was further updated with the element dimension data to give corrected gaps for wedge settings. Then the wedges were set first on the foundation to define the height of the element and after that the element was put down onto the wedges taking at the same time care of the location of the end of the element to the point of the side line. When the element was put down on the

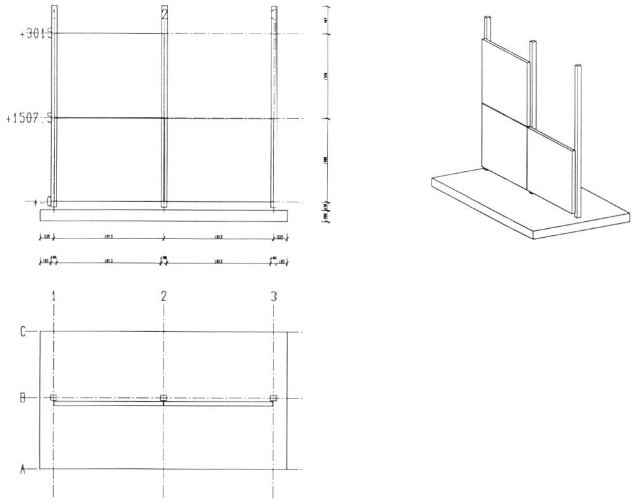

Figure 3. The 3-D CAD design of the example building.

wedges on the foundation the wedges between the columns was set to define final location of the element before fixing it to the columns, Fig. 4. In the assembly of the next element on top of the former assembled one the clearance between the elements is corrected with the element dimension data to give the amount of wedges between elements.

4.3 Evaluation of the assembly

After the element assembly a check measurement was curried out to evaluate the quality of the erection. The location of each corner of the elements was measured with the 3-D measuring device, and compared to the locations designed in the 3-D CAD tool. A tool for checking the Cartesian positions for the corners of the elements was designed and implemented, Fig. 4., to easily curry out the check measurements. The results were then combined to those obtained from the CAD model. In the example erections the assembly accuracy of the elements was within 2 mm.

5. CONCLUSIONS

A method for taking the advantage of the 3-D CAD model and 3-D coordinate meter in the frame erection has been described. The building was modelled by an existing 3-D CAD tool. For on site 3-D measurements a software module was integrated into this tool to make it possible for the measurer to easily prepare a list of measurements and save them into the memory. A method for combining as-built measurement and CAD model data was developed for the preparation of the installation work. This installation data was further corrected by taking into consideration the element dimension data read from escort memories during the installation process.

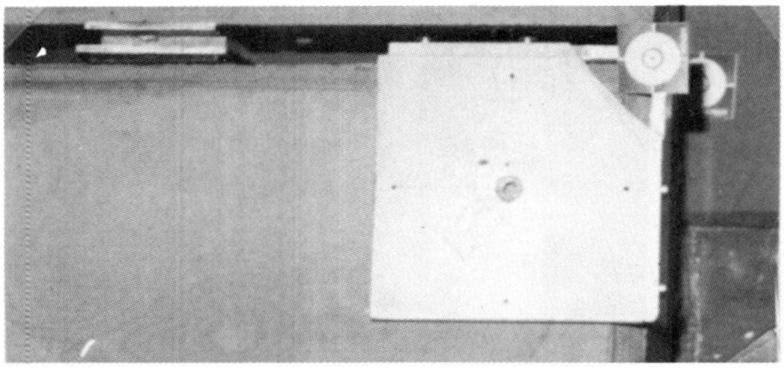

Figure 4. Measurement tool and wedges .

Acknowledgements

The authors want to thank the members of the project group especially Mr. Tapio Jussila from Mitta Ltd. and Mr. Pekka Karjalainen from Cadex Ltd. for their advises and support during the course of this work. The financial support of the Technology Development Center of Finland is gratefully acknowledged

6. REFERENCES

1 S. Pieskä, M. Lindholm, P. Vähä, Ranging measurements in the robotization of construction. ISMCR'90, Houston Texas, USA, June 1990.
2 S. Pieskä, M. Lindholm, L. Koskela, Measurement requirements and advanced techniques for construction robotization. Research Notes 1090, Technical Research Center of Finland, Espoo, Finland 1990.
3 I. Kaisto, J. Kostamovaara, I. Moring, R. Myllylä, Laser range finding techniques in the sensing of 3-D objects. Sensing and Reconstruction of 3-D objects and Scenes, SPIE Vol. 1260, Santa Clara, USA, February 1990, 12 p.
4 P. Vähä, Automation and robotics in construction: State of the art in Finland. Proceedings of the 9th ISARC, 3 - 5 June, 1992, Tokyo, Japan.
5 ConcreteCad Users Manual. Cadex Ltd., 1992, Espoo, Finland.

Automation and Robotics in Construction X
G.H. Watson, R.L. Tucker and J.K. Walters (Editors)

Construction process simulation with rule-based robot path planning

R. Stouffs, R. Krishnamurti, S. Lee, and I. Oppenheim

Department of Architecture, Carnegie Mellon University, Pittsburgh, PA 15213, U.S.A.

Abstract

A rule-based simulation program is developed for application to building construction. From a specified task schedule, the program generates and simulates a motion path for each robot action, avoiding obstacles and incorporating interaction, safety and other considerations.

1. THE RUBICON PROJECT

Simulation refers to the computational modelling of a process described by the activity network of construction tasks, the spatial description of each robot action, and an accompanying representation of the evolving building at each state in time. A simulation can be used to make observations on the feasibility of each task and to maintain a running measure of construction time and cost. This simulation can also provide a graphical display of the building project. In general, a simulation can be used to study the productivity of alternate construction plans and alternate resource mixes.

2. PROJECT EXAMPLE AND SCOPE

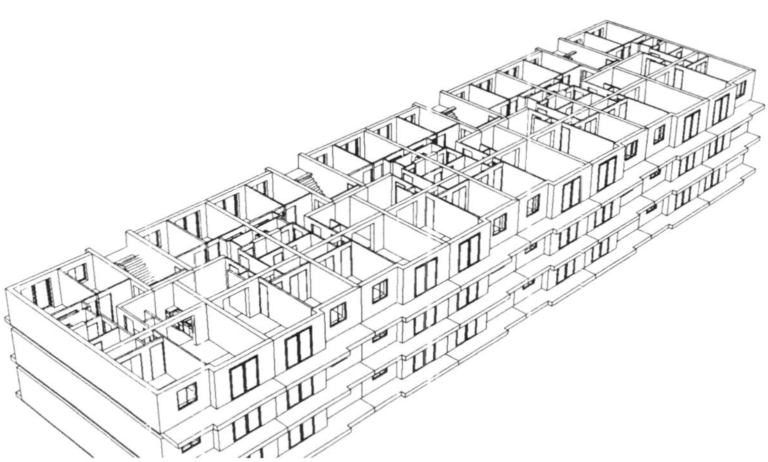

Figure 1. Typical 40 unit building during construction.

For the demonstration of the RUBICON simulator we chose the construction of a typical Japanese precast concrete residential building. Figure 1 illustrates a rendering of three stories of a typical 5 story, 40 unit building; Figure 2 shows the floor plan of a single unit.

The simulation requires a building construction task plan describing the construction elements and the construction process as a task schedule. The simulator translates each task into a robot motion plan using a rule-based description of the robot agents. These motion plans reflect the respective robot's motional capabilities and limits, and they avoid any collision. The simulation also rejects impossible tasks and allows for a variety of different robot types. The result of the simulation consists of a graphical visualization of these motion plans, together with the building under construction.

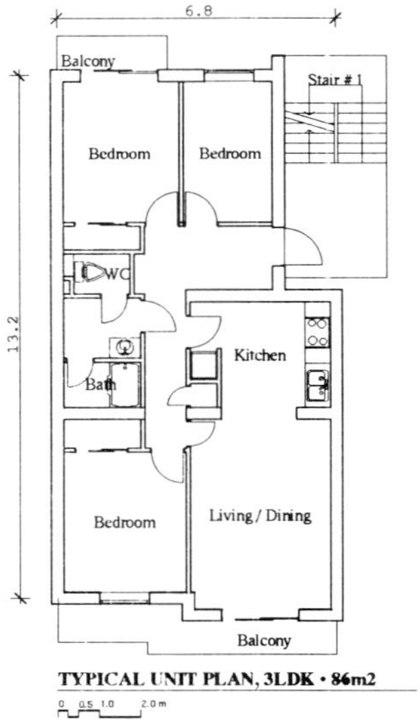

Figure 2. Floor plan of a single unit.

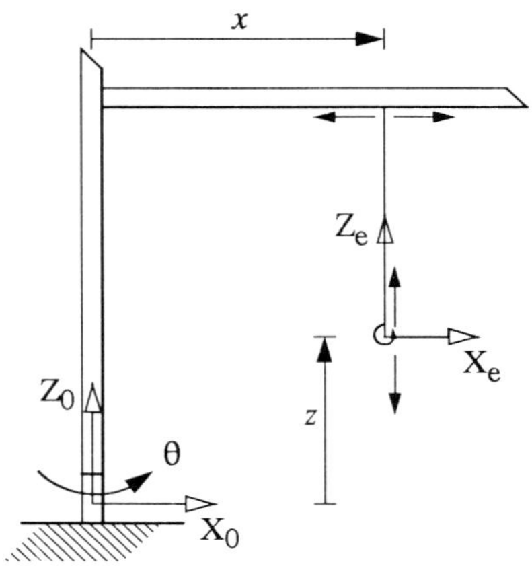

Figure 3. A robot crane.

3. TASK PLANNING

The input to the simulator consists of a task plan. This is a detailed plan describing the construction elements and the construction process as a task schedule. Each task is to be performed by a human or a robot crew. To design a task plan, one must determine the crew resources, both robot and human, and determine the construction elements and the construction sequence, that is, the order in which the different components will be handled. Subsequently, a task file can be built, specifying the construction elements, i.e., their name and geometry, and the sequence of tasks.

Declare an element (type) as

element length width height

An example of a typical construction element may be

WALL-PANEL-1 13.20 0.10 2.40

A task line specifies the estimated duration, the construction component to be handled, its start and final position and orientation and the performing crew or robot type. A component name is specified as

element.floor.part

Declare a robot task as

M *duration component* (*start-position, start-orientation*) *robot-type* (*end-p., end-o.*)

A typical robot task may be: "Move Wall Panel Mark 1, Floor 1, Part #5 (WALL-PANEL-1.1.5) from truck site (0.0, -4.0, 2.0) to position x-y-z (54.4, 6.6, 1.2) with the Robot Overhead Crane. Estimated time is 0.16 days." The corresponding input line reads

M 0.16 WALL-PANEL-1.1.5 (0.0 -4.0 2.0 0 0 0) crane_hoist (54.4 6.6 1.2 0 0 -90)

Declare a human crew task as

P *duration component* (*start-position, start-orientation*) *crew*

A typical human crew task may be: "Attach Wall Panel Mark 1, Floor 1, Part #5 (WALL-PANEL-1.1.5) at truck site (0.0, -4.0, 2.0) to Robot Overhead Crane. Estimated time is 0.03 days." The corresponding input line is

P 0.03 WALL-PANEL-1.1.5 (0.0 -4.0 2.0 0 0 0) connecting_crew

Part of the design work may be automated with the help of a planning software such as PLANEX [1]. PLANEX is a knowledge-based expert system for process planning. It generates a project activity plan that is a general description of the sequence of activities involved in the process. Such an activity plan must be post-processed, either manually or automatically, to deliver a task plan that is detailed enough for the RUBICON program. A purpose-build task planner that would interface between RUBICON and an existing construction management software needs to address the issues of project scheduling, building geometry, construction elements, activity duration, task sequencing and resource allocation.

4. ROBOT MOTIONS

Robots for construction are characterized by their ability to lift, move and place sizable construction components from the delivery location to the placement location. We consider two different robot types: a robot *crane* for handling heavy components; a robot *towmotor* for palettized materials. These two types are illustrated in Figures 3 and 4 respectively.

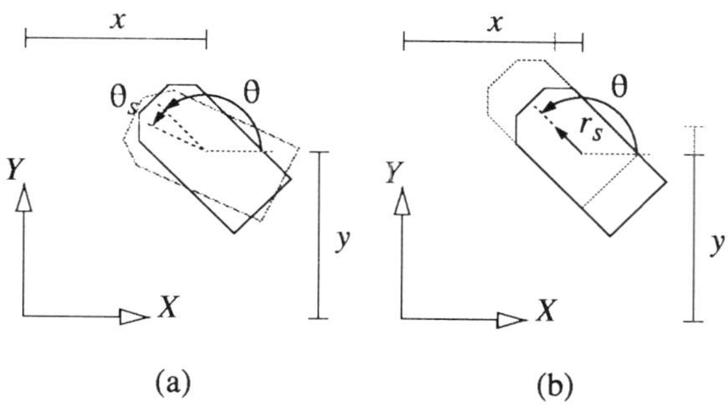

Figure 4. A robot towmotor with (a) stationary rotational motion and (b) translational motion along its axis.

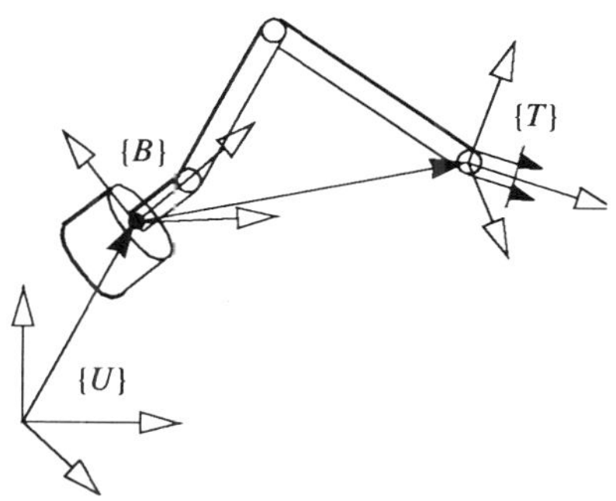

Figure 5. Three frames to describe the configuration of a robot

The description of a robot encompasses a representation of its *configuration*, i.e., position and orientation, over time, its motional capabilities and its configurational limits. We use a fixed number of *frames* and *transforms* to describe a robot's configuration at a particular time, introduce a set of *control parameters* to capture possible configurational changes over time, associate *motion rules* to these control parameters to describe a robot's motional capabilities, and add extremum values to the control parameters to represent the configurational limits.

A robot can be described in terms of the configuration of its components to each of which we rigidly attach a coordinate system or *frame* [2]. See Figure 5. A *transform* describes one frame relative to another. A minimal description of a robot involves the following transforms: ${}^{U}_{B}\mathrm{T}$ describes the base frame relative to the universal frame; ${}^{B}_{T}\mathrm{T}$ describes the tool frame with respect to the base frame; and ${}^{T}_{O}\mathrm{T}$ describes the object frame with respect to the tool frame. The object frame is described relative to the universal frame by the composite transform ${}^{U}_{O}\mathrm{T} = {}^{U}_{B}\mathrm{T}\,{}^{B}_{T}\mathrm{T}\,{}^{T}_{O}\mathrm{T}$.

The ability of these transforms to change over time and, as a result, the relative configuration of the bodies they describe, is captured in a set of *control parameters*. For each transform, these parameters correspond to the respective degrees of freedom (DOF) of the body relative to its reference frame. Physically, the relationship between the end-effector and the base of the robot is defined by a set of links and joints, generally formed into an open kinematic chain. These joints can be either rotational (change of orientation) or prismatic (translational change). The number of DOF of the robot's end-effector, with respect to the base and for an open kinematic chain, corresponds in most cases to the number of joints. The set of valid control parameters consists of x, y, z, ψ, ϕ and θ, where x, y, z define translations parallel to the respective axes and ψ, ϕ, θ define rotations about the X-, Y- and Z-axis, respectively.

The motional capabilities of the base of a robot often are dependent on the current configuration of the base and, thus, defined locally. However, the configuration itself is defined relative to a reference frame that is fixed with respect to the body's motion. *Motion rules* are conceived to describe a robot's motional capabilities possibly independent of its configuration. Foremost, however, they form a tool to embody the discretization of time in the simulation process. A motion rule generally defines the value of a control parameter at time $t+\Delta t$, where Δt denotes the time step, in terms of the control parameter's values at time t and their initial and step values. The initial values are the values of the control parameters at time 0. Under the assumption of constant velocity, for each control parameter we can specify a motion step as a constant value. Each parameter also has a minimum and maximum value specified in accordance with the relative workspace of the body. This workspace is roughly the volume of space which the body can reach in at least one orientation, relative to its reference frame [2].

5. ROBOT BEHAVIOR

In the simulation, we are concerned with the role that a particular robot plays in the construction process and with its place in, and its relation to, the environment, that is, the site as well as the building under construction, the human labor crews and the other robots involved in the construction. This relation and, therefore, the robot's behavior, is constrained by physical obstacles, task-specified and other interactions, and safety and other considerations. A motion language is developed that serves as a means to specify this behavior and, in particular, to specify the order in which the allowed motion steps should be proposed, for each robot or body. The choice of a specific motion step is dependent on the particular situation of the robot at the moment, according to the specified behavior. Then, all motion rules that contain the motion step in question are invoked, and the corresponding control parameters are updated.

The basic building blocks of the motion language are the *motion command elements*. Motion commands may be grouped into series of motion steps using *control structure elements*. The language also allows for the use of variables; it is completed with a set of functions that may be used within assignments or as command arguments. For a more elaborate explanation of the elements of the language, we refer to Stouffs et al. [3].

6. THE RUBICON PROGRAM

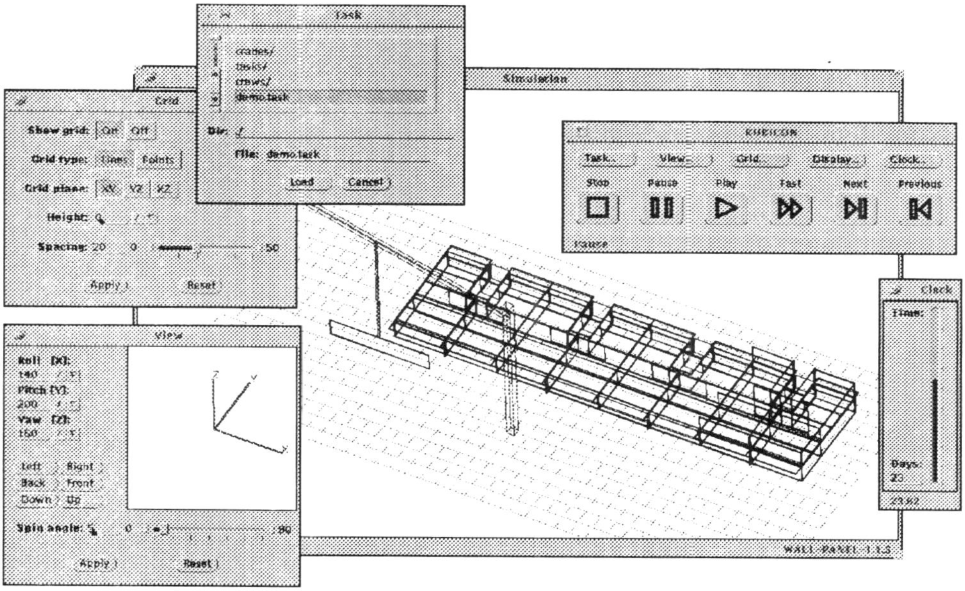

Figure 6. Snapshot of the RUBICON graphical user interface.

The input to the RUBICON program consists of two kinds of files. One file contains the task plan, that is, the description of the components and of the construction process as a sequence of tasks each of which is performed by a human or robot crew. For each robot agent specified in the task plan, a motion file is read in that contains a description of the motional capabilities of the robot and the motion rule set describing its intended behavior, expressed in the motion language specified above.

The output of the RUBICON program is a graphical simulation of the construction process as specified in the task plan, with a visualization of the motional actions of the robot agents and of the transportation of the construction components by these agents, and with a specification of the process time. The output is controlled through an audio/video-like control panel with buttons for *play*, *fast* play, *next*, *previous*, *pause* and *stop*. *Fast* play is achieved by unmaterializing the robot agent; *next* and *previous* instantly jump to the next, respectively previous, construction component.

Other interface panels allow the user to choose and load a new task plan and corresponding motion files, alter the 3D-viewing parameters, alter the displayed grid, and view the current process time (see Figure 6).

7. SIMULATION RESULTS

The RUBICON program serves the project planning engineer in studying alternate task plans, alternate resource mixes and the use of alternate robot types in the construction process. Results of the program include productivity measures, such as total project time and number of human crews and robot agents used. The following are some simple examples illustrating the possibilities of the program to study the results of alternate input.

The simulation demonstrates how alternate task sequences, such as floor-by-floor or staircase-by-staircase construction of residential building examples, result in different robot paths and, therefore, productivity (see Figure 7).

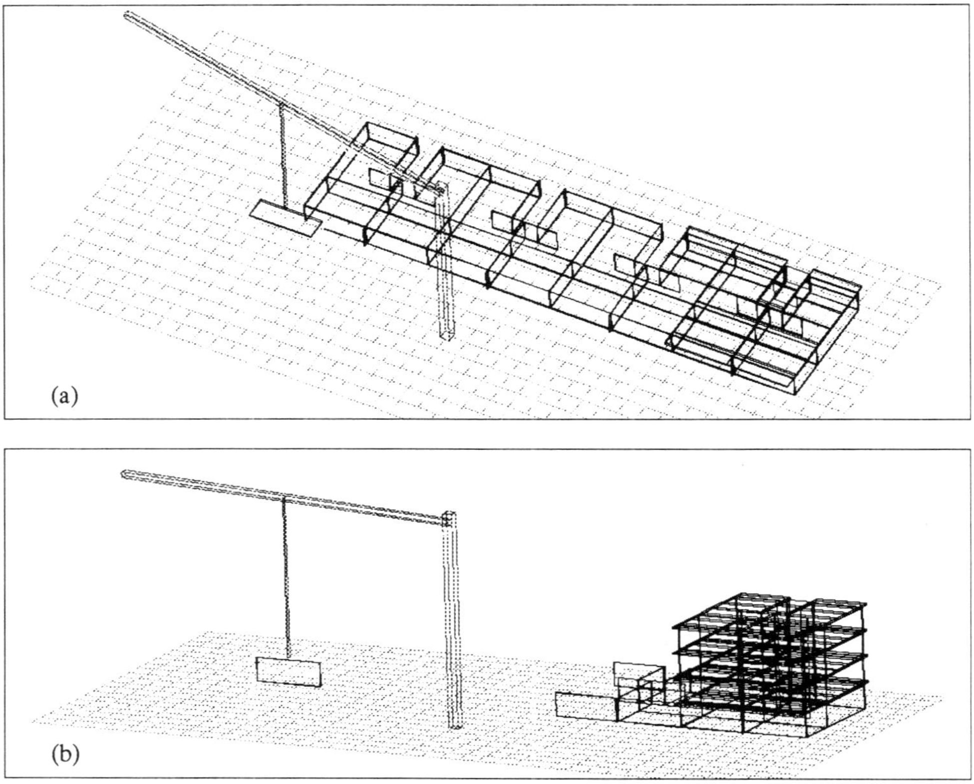

Figure 7. Alternate task sequences (snapshots from the simulation): (a) floor-by-floor and (b) staircase-by-staircase.

Different project durations may result from alternate resource mixes; e.g., given the two tasks of robot panel connection and panel grouting the difference in duration for either one or two human crews - one connecting and one grouting - is obvious.

The productivity of a robot crane depends on the position of the crane on the site, e.g., mid-site or at the edge of the site, and on its ability to be repositioned during construction. Multiple cranes also influence productivity.

The simulation program has the ability to recognize impossible plans and to reveal inefficiencies in task plans; e.g., when using a towmotor, it is imperative that there exists a path from the pick-up location to the delivery location, large enough for the towmotor. Also, whether an efficient path exists for the towmotor may depend on the order of positioning of the construction components, such as wall panels.

Productivity may be further enhanced through the (simultaneous) use of different robot types, e.g., crane robot and towmotor, using each for materials it handles best. Figure 8 illustrates the use of two different robot types in a construction.

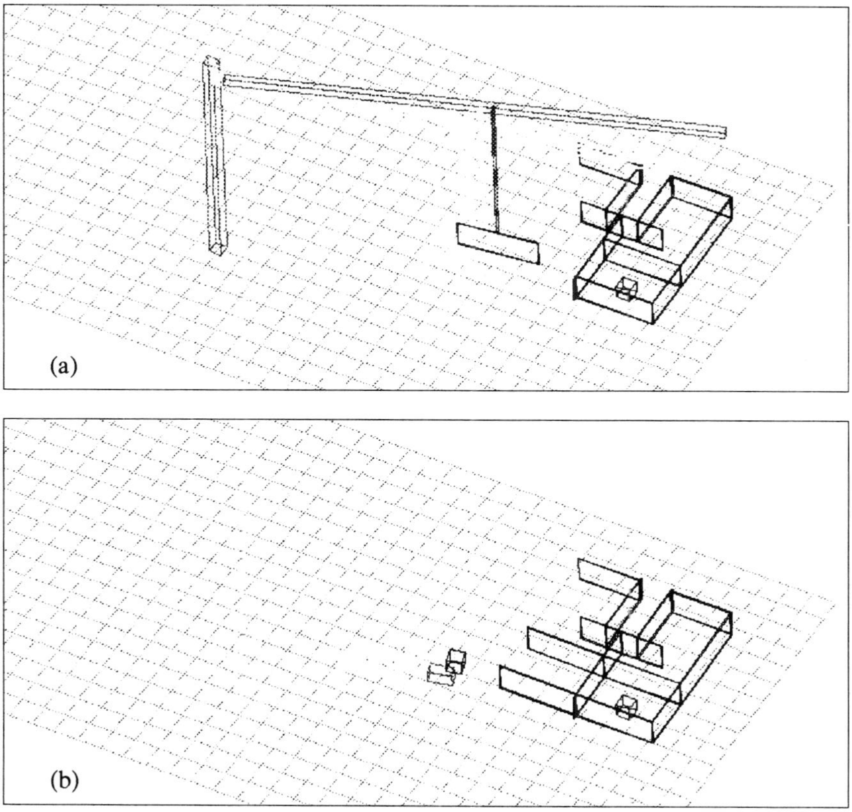

Figure 8. Dual robot task plan (snapshots from the simulation): (a) a crane positions the wall-panels and (b) a towmotor places the palettized materials.

Other considerations that are important are motion constraints such as the reach of a crane, or safety considerations that regulate robot-human crew interaction, e.g., a load may not pass overhead of a grouting crew.

8. CONCLUSION

A distinctive feature of robotic construction is the need for a path to be associated with each robot action. The simulation program described here incorporates, within it, an automatic path planner which returns a path for each action. The path planner was developed as a rule-based planner. The kinematic capability of each robot is described by a motion rule set, and rule chains for finding paths can subsequently be entered. Two simple examples were used within this study: a crane, and a towmotor. The simulation program, RUBICON, is demonstrated for a residential building example constructed with precast concrete panels. Applicable task plans are manually generated; a task plan produced by an knowledge-based expert system, PLANEX, is also examined. Examples are chosen to show that the RUBICON simulation can recognize impossible plans, can reveal inefficiencies in task plans, can illustrate the influence of crane placement, and can simulate influences on productivity of human and robot crew interaction.

RUBICON may serve as a tool for construction company engineers to perform studies on real project task plans; engineers/developers may use RUBICON to develop better task planners.

9. ACKNOWLEDGEMENTS

The development of the RUBICON prototype described in this paper was funded in part by the Japan Research Institute.

10. REFERENCES

1 C. Zozaya-Gorostiza, C. Hendrickson and D.R. Rehak, Knowledge-Based Process Planning for Construction and Manufacturing, Academic Press, San Diego, CA, 1989.
2 J.J. Craig, Introduction to Robotics: Mechanics and Control (2nd ed.), Addison-Wesley Publishing Company, Reading, MA, 1989.
3 R. Stouffs, R. Krishnamurti and I.J. Oppenheim, A behavioral language for motion planning in building construction, manuscript, Dept. of Arch., Carnegie Mellon Univ., Pittsburgh, PA, 1993.

Automation and Robotics in Construction X
G.H. Watson, R.L. Tucker and J.K. Walters (Editors)

Dynamic Interface Simulation for Construction Operations (DISCO)

Rong-Yau Huang[a] and Daniel W. Halpin[b]

[a]Ph.D. Candidate, School of Civil Engineering, Purdue Univ., West Lafayette, Indiana 47907, USA

[b]Professor and Head, Construction Engineering and Management Division, Purdue University, West Lafayette, Indiana 47907, USA

ABSTRACT

Application of the Visual Simulation technique (VS) to assist in the planning and analysis of construction operations has become increasingly important in the research community in recent years. Benefits of applying VS have been reported in several publications. Nevertheless, the effort required to design and construct an "appropriate" representative display for the system to be analyzed in a VS experiment is frequently time consuming and data intensive. The magnitude of this activity is often increased due to certain characteristics of construction operations, such as their complexity and the distribution of work over large areas. An alternative approach, DISCO, which employs a schematic modeling format reflecting the system dynamics on the computer screen during simulation run is presented in this paper.

1. INTRODUCTION

DISCO is an alternative approach for the simulation of construction operations. It employs an abstract model diagram (e.g. CYCLONE model diagrams) of a system as the modeling framework and dynamically shows on the computer screen the associated simulation parameters, such as the simulation clock time, percent of time resources idle, number of resource units idle, percent of time work task busy, etc., during the simulation run-time. System dynamics can be better understood by monitoring the continuously updated information shown on the screen.

The idea of DISCO is conceived from the application of "Visual simulation" and "Schematic Abstraction" for process planning and analysis modified to reflect the characteristics of construction operations. DISCO preserves some of the benefits provided by using VS, while utilizing a schematic model diagram to reduce the amount of data and time required for model building. Both VS and "Schematic abstraction" are briefly reviewed in the paper prior to the further discussion of DISCO.

A prototype program, also named DISCO, is developed to implement the DISCO methodology and is introduced in this paper. A dry batch delivery and concrete placement model is used to demonstrate the application of the DISCO program. Findings and conclusions are presented.

2. VISUAL SIMULATION (VS)

Visual Simulation applies modern computer graphics techniques for facilitating the development and study of simulation experiments. Its utilization of computerized visualization conveys the physical configuration of the system of interest graphically in a format that appears very much like the real thing and allows the user to visualize the dynamic behavior of the system model during the simulation run-time.

Many "expressions" have been used to refer to this type of simulation with each of them featuring its special capabilities. "Animation" (System Modeling Corp. 1990; Johnson and Poorte 1988) is probably the most popular one with its capability of generating cartoon type of visualization. "Visual Interactive Simulation (VIS)" allows the user to "interact" with the model during the simulation (Bell & O'Keefe 1987). Other expressions such as "Graphical Simulation" (CII Source Document 68 1991), or "Visual Simulation" (Rodriguez 1992) are employed to refer to VS systems which feature realistic 3 dimensional visualization.

In general, there are two types of VS applications depending upon the level of display reality - pictorial and iconic. Figure 1 shows an emulated construction job site using a VS system - ICV (Rodriguez 1992). Users are able to see the movement and operation of "cranes", "trucks", "front-end loaders", "fork-lifts" on the screen as the simulation goes on, very much in the way the modeler would see these resources in the real world. This pictorial type of VS frequently involves 3-D CAD modeling and requires the presence of graphical workstations.

Figure 1. An Animated Construction Job Site (Adapted from Rodriguez 1992)

In contrast to pictorial representation, iconic VS models use icons to represent object components in the interested system. The reader is referred to the "CINEMA IV Reference Guide" (Systems Modeling Corp. 1990) for examples.

The benefits of applying VS have been reported in several publications (Rodriguez 1992; CII SD 68 1991; Griffis et. al. 1991; Cleveland & Francisco 1988; Johnson 1988; Bell and O'Keefe 1987). VS modeling provides the following advantages:

- VS enables the evaluation of project design and construction terminology by all members of the team to facilitate the manufacturbility and constructibility review.
- VS enables greater understanding and awareness of the project and helps the site personnel to visualize how and what should be constructed in the coming period.
- VS facilitates the communication among all the parties involved in a project, which helps to alleviate fragmentation during the project realization.
- VS assists in planning the sequence of activities in a process and in organizing people, materials and machines to allow the most productive efforts in a particular area.

- VS helps to prevent safety problems like a collision between two cranes.
- VS serves as a medium for training and education purposes, etc.

Nevertheless, the activity of designing and constructing an appropriate representative display of the interested system is still very data and time intensive. The magnitude of this demand is increased due to certain characteristics of construction operations.

- Construction operations are complex in their nature due to the involvement of dynamic and complicated resource interactions.
- Construction operations are frequently performed on highly diverse one-of-a-kind job sites. The representative display for construction operations can be very different in different job sites.
- Construction operations frequently involve great dimensions, such as in long-distance earth-hauling processes, a long tunnel excavation, a high-rise building construction, and so on. Design and construction of representative visually accurate displays is made more difficult and a schematic diagram may be more informative.
- The dynamic nature of construction sites as the work progresses causes the display of the job site to be continuously changing and generates a large number of viewing perspectives.

3. SCHEMATIC MODEL ABSTRACTION

Model abstraction has become standard practice for engineers in approaching problems in the real world since such problems are often complicated and the underlying interactions involved are better reflected by considering abstract models. An abstract model captures system behavior and enables engineers to perform analysis so as to better understand the targeted problem. In general, the degree of abstraction can range from the reality of a photo image to the abstraction of a mathematical formula (Figure 2).

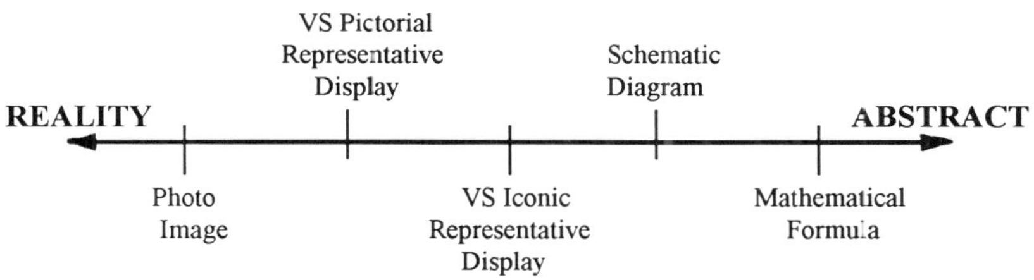

Figure 2. Relationship of "Schematic Diagram" to the Other Model Abstraction

The unique, complex, and dynamic nature of construction operations creates difficulties for the design and construction of representative displays and therefore hinders the application of VS for the purpose of engineering analysis. Schematic abstraction, on the other hand, has long been used for network modeling and analysis in simulation experiments and simplifies model definition. This type of modeling is not data-intensive (no need for geometric information), not dimensionally confined, and logically well defined. In many cases, it is arguable that schematic modeling suits the purposes of engineering planning and analysis of construction operations much better than a pictorial display.

Typical schematic diagrams are like those constructed by using a set of standardized building blocks in general simulation languages, such as GPSS, SLAM II, or construction specific systems, such as MicroCYCLONE (Halpin 1990). A MicroCYCLONE schematic diagram of a dry batch delivery and concrete placement model will be shown later.

4. DISCO

DISCO is a prototype program developed to implement the aforementioned DISCO approach. It is written in Microsoft™ Visual Basic language and runs under the Windows™ operating environment. CYCLONE modeling methodology (Halpin 1973) was selected for the abstract model building due to its emerging popularity in the construction research community. Its set of standard building blocks (namely, NORMAL, COMBI, QUEUE, CONSOLIDATE, COUNTER, and ARC elements) constitute the basic drawing icons in the DISCO graphical menu bar. Readers should refer to "Planning and Analysis of Construction Operations" (Halpin and Riggs 1993) for details regarding the CYCLONE modeling methodology.

Basically, the DISCO program serves as a post-processor of MicroCYCLONE program data because it takes the chronological list file generated by running MicroCYCLONE program as its input and then recreates the entire course of simulation dynamically on the computer screen. The DISCO program performs three major functions as follows:

(1). It builds a schematic model diagram graphically on the screen.

(2). It displays the model dynamics by continuously updating the associated information for each node during the simulation run.

(3). It reports node statistics information graphically and tabularly at any simulation event time point.

4.1 Build the schematic model diagram

DISCO allows the user to create the schematic model diagram, using CYCLONE standard building blocks, for the system being studied. The user can create nodes or connectors between nodes by simply clicking on the appropriate icon from the DISCO graphical menu bar (Figure 3) and dragging it over to the drawing area. A dialogue box will pop up asking for the input of information associated with that particular node, such as label number, node description, initial unit number in the QUEUE node, and so on. Users can always double click on any node to recall the dialogue box for editing the node information. Users can also edit the drawn schematic diagram by moving or deleting nodes and connectors. Whenever a node is deleted or moved, all the connectors associated with that particular node (i.e., which start or end at this node) will be deleted or moved as well.

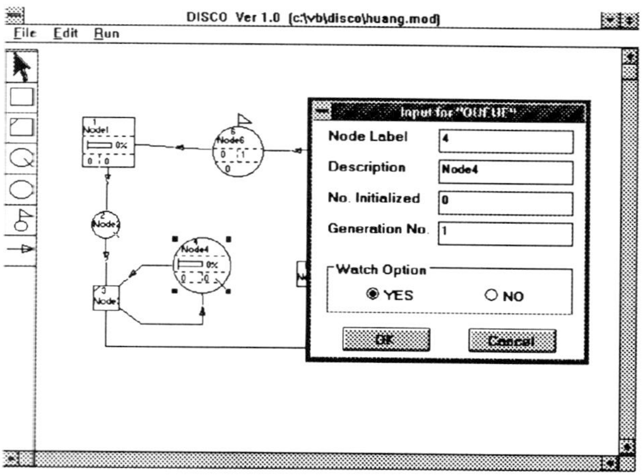

Figure 3. DISCO Menu and Display Screen

It should be noted that the "Watch Option" (Yes or No) information in each node determines whether the simulation run-time information associated with a particular node such as % time idle, will be "watched" (shown) later during the simulation or not. Nodes with "yes" in the "Watch Option" box have a bigger node size in order to display simulation run-time information (e.g., Node 4 in Figure 3). Again, users can always go back and forth to change the option except for the COUNTER node for which the "Watch Option" is automatically invoked and must be used. This mechanism provides the user opportunities to watch nodes alternatively and enables the user to focus on watching certain nodes and not to be distracted by too many "watched" nodes during the simulation run.

The completed schematic model diagram will be used later as the background display during the simulation.

4.2 Display of model dynamics

After the schematic model diagram is drawn on the DISCO drawing area and the chronological list file (.DIS) is generated by running MicroCYCLONE program, the user can trigger the simulation and DISCO will recreate the entire course of simulation dynamically on the screen. A simulation clock will be shown on the screen to indicate the updated simulation time point T-NOW. As the value of simulation time advances, run-time information for those nodes being "watched" are continuously updated throughout the simulation. The display format of the run-time information contained in "watched" nodes is as shown in Figure 4.

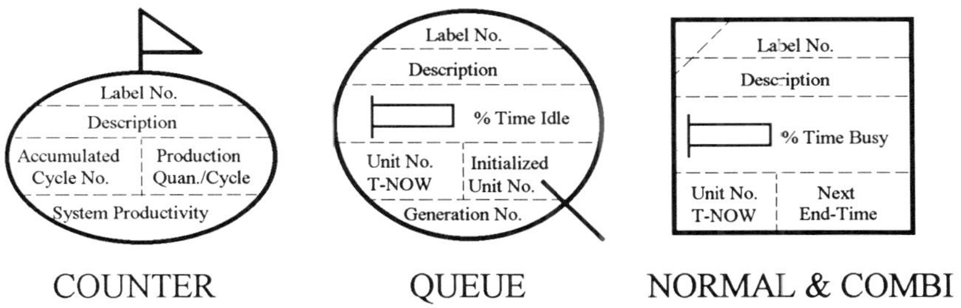

Figure 4. Format of DISCO "watched" Nodes

Most of the legend shown in Figure 4 is self-explanatory except for the two values at the bottom of the NORMAL & COMBI nodes. It is not uncommon for a construction operation to have parallel work tasks. In an earth-hauling process, for instance, two trucks can be loaded at the same time if there are two front-end-loaders available. "Unit No. T-NOW" at the left bottom indicates the number of units presently transiting that particular work task at T-NOW, while "Next End-Time" at the right bottom shows the closest next event time when the realization of that task will be completed.

The DISCO program attempts to create the impression of "animation" during the simulation run by employing colors. Each type of node is assigned a different color (except NORMAL and COMBI which share the same one.) The color goes on if an unit "enters" an empty node. The entering unit can be a truck returning to an empty queue, or the realization of a task which has no parallel ones going on. Similarly, the color goes off when the last unit "exits" a node. Once the simulation starts, the user is able to see colors flashing on and off on the screen, along with continuously updated run-time information. This mechanism allows the user to experience system dynamics during the simulation run.

4.3. Report node statistics

Throughout the simulation, the DISCO program keeps track of all the run-time information associated with each node. The user can request reports on the statistical information of a particular node at any simulation time by double clicking on it. A report box will appear on the screen and the execution of the simulation will be interrupted. The user has to select the type of report to be viewed and the report is then displayed both graphically and in tabular format. Table 1 lists the types of reports provided by DISCO program.

Table 1
DISCO Report Types

	NORMAL	COMBI	QUEUE	CONSOLIDATE	COUNTER
Unit No. Profile	✓	✓	✓	✓	
% Time Idle (Busy)	✓	✓	✓		
Duration Histogram	✓	✓			
System Productivity					✓

The execution of the simulation resumes automatically once the user has completed viewing the reports. The overall statistical information of each node can be obtained in the same manner at the end of simulation.

5. A DRY BATCH AND DELIVERY PLACEMENT MODEL

Figure 5 shows the model diagram of a dry batch delivery and concrete placement model in DISCO format. This model basically involves the interaction of 5 types of resources -- crane, bucket, crew, truck, and concrete mixer. Hauling trucks carry dry concrete batches from batch plant to the paving site and dump them to the skip of the mixer. After the mixer mixes the concrete, wet batches are dumped into a concrete bucket and lifted by a crane to the placement location where it is dumped, spread and finished by a labor crew. Readers are referred to the "MicroCYCLONE System Manual" (Halpin 1990) for more information regarding this model.

In order to monitor the performance of the trucks, mixer, crane and labor crew in this particular simulation run, Node 9, 13, 17, and 23 are set to be "watched." In addition, selective work tasks such as "Load Truck" (Node 2), "Dump to Skip" (Node 6), "Fill Bucket" (Node 12), etc., are "watched" so as to better understand the system dynamics. As shown in Figure 5, prior to the beginning of the simulation, the simulation clock is set to 0.00. Green color (not visible in the figure) in all the QUEUE nodes with initialized resource (e.g., crane, bucket, trucks, etc.) signals the availability of a resource. The % Idle time in these nodes is set to 100 initially. It is immediately apparent that "Load Truck" (Node 2) will be the only task to be scheduled at simulation clock (T-NOW) 0 due to the availability of both required resources -- loading tower and hauling truck.

Once the simulation starts, the node color of "Load Truck" (blue) signals the commencement of this task and the unit No. at the left node bottom is updated to 1. The next-end-time at the right node bottom is updated to 15 to indicate the time loading of the first truck will be completed. Meanwhile, the node color of "Tower" goes off and the unit No. is decreased to 0 since it is busy loading the truck. In addition, the unit No. in the "Truck Queue" (Node 9) is decreased to 3 meaning that 3 trucks are now idle at the truck queue.

T-NOW is advanced to 15 at which time the loading of the first truck is finished. The node color of "Load Truck" goes off. The first truck carries dry concrete batch and travels to the site for mixing. The node color of "Travel" (Node 4) flashes on. The tower is now available (node color on) for loading another truck. Since three trucks are waiting in the truck queue, another "Load Truck" task can be scheduled at this time point. Again, the node color of "Tower" and "Load Truck" flashes off and on respectively. Meanwhile, all the run-time information contained in these nodes is updated accordingly.

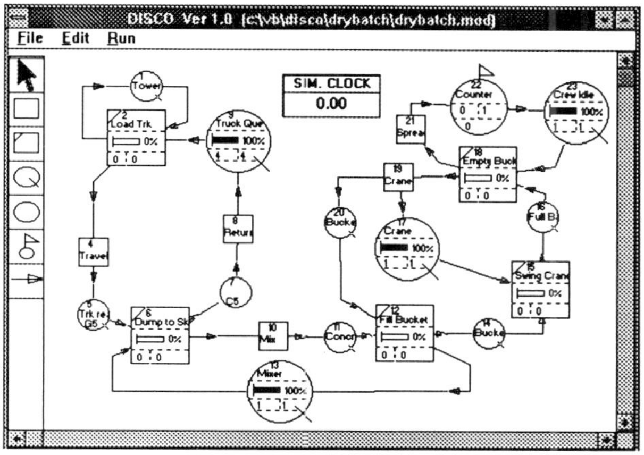

Figure 5. Model Diagram of a Dry Batch Delivery and Concrete Placement Model in DISCO

As the simulation continues, the user can view the loaded truck as it arrives at job site and dumps dry concrete to the mixer skip. The mixer mixes the concrete and fills the crane bucket with wet concrete. The crane then lifts and swings the bucket to the paving site and labor crew empties the bucket, spreads and finishes the concrete while the crane swings back for lifting another concrete bucket.

A crane utilization profile (Figure 6), which records the crane status (busy or idle) through time, is obtained by double-clicking on the "Crane" node (Node 17) at simulation clock time 56.25. It is obvious that, so far, the crane was idle most of the time, which is reflected by its 77% idle time. In fact, it remained idle until time 29.5.

Similar reports on the trucks, the mixer, and the labor crew at different simulation clock times can be obtained in the same manner. By double clicking on the "Counter" node (Node 22), a report on system performance (productivity) is shown in the report box.

6. FINDINGS AND CONCLUSIONS

Instead of pictorial or iconic display, as in Visual Simulation, DISCO employs schematic diagrams as the background display to demonstrate system dynamics during simulation runs and provides an alternative approach for the simulation of construction operations.

Simulation run-time data as well as the end results are more meaningful to analysts when they are displayed on the screen along with the schematic model diagram. System dynamics are better understood through the use of "node color on and off" mechanism and also the continuous updating of the node information. DISCO graphical and tabular reports allow monitoring of chronological information on each node at any simulation time point. This further facilitates the use of the DISCO program for planning and analysis of construction operations.

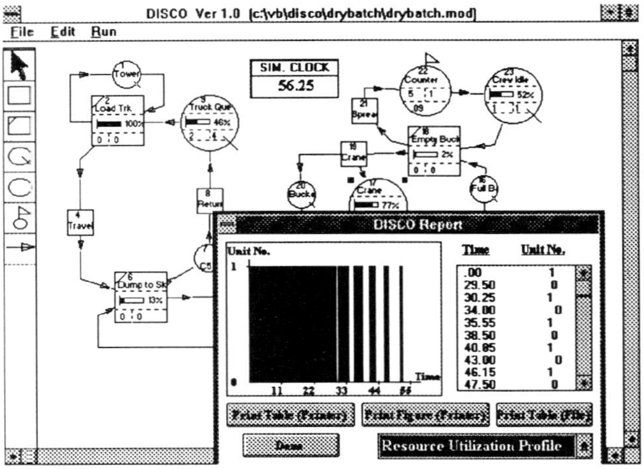

Figure 6. DISCO Report: Crane Utilization Profile

7. REFERENCES

Alciatore, D. G., and O'Connor J. T. and Dharwadkar, P.: A Survey of Graphical Simulation in Construction: Software, Usage and Application, CII Source Document 68, The University of Texas at Austin, Austin, TX, September, 1991.

Bell, P. C. and O'Keefe, R.M.: Visual Interactive Simulation - History, recent developments, and major issues, *Simulation* 49, No. 3, 109 -116, September 1987.

Cleveland, A. B. and Francisco V.: Use of Real-Time Animation in the Construction Process, *Transportation Research Record* 1186, 1988.

Griffis, F. H., Rubinson, D., O'Brien, W. J., Retailleau, S., and Zabilski, R.: Productivity Applications: 3D Models & Simulation, working paper, Columbia University, New York, NY, 1991.

Halpin, D.W.: An investigation of the use of simulation networks for modeling construction operations, PhD thesis, University of Illinois, Urbana, IL, 1973

Halpin, D.W. and Riggs, L.S.: Planning and Analysis of Construction Operations, John Wiley and Sons, Inc., New York, N.Y., 1993

Halpin, D.W.: MicroCYCLONE User's Manual, Division of Construction Engineering and Management, Purdue University, West Lafayette, IN, 1990

Halpin, D.W.: MicroCYCLONE System Manual, Division of Construction Engineering and Management, Purdue University, West Lafayette, IN, 1990

Johnson, M. E. and Poorte, J. P.: A Hierarchical Approach to Computer Animation in Simulation Modeling, *Simulation* 50, No. 1, 30 - 36, January 1988.

Rodriguez, W.: Visual Simulation: Considering the Dynamic Nature of Construction, *Construction Business Review*, Vol. 2, No. 4, July/August 1992.

System Modeling Corp.: CINEMA IV Reference Guide, System Modeling Corp., Sewickley, PA, 1990.

Automation and Robotics in Construction X
G.H. Watson, R.L. Tucker and J.K. Walters (Editors)

The Potential of A "Virtual" Construction Site for Automation Planning and Analysis

Chris Hendrickson and Daniel R. Rehak

Department of Civil Engineering, Carnegie Mellon University, Pittsburgh, PA 15213-3890 USA
Internet: cth@cmu.edu, rehak@cmu.edu, FAX: 412-268-7813

Abstract
Existing simulation models for construction operations typically rely on geometric objects moving through an abstract space or on monte carlo simulation models without specific geometric representations. Physically-based models of such operations coupled with new technologies for user interaction offer tremendous potential for improving the realism and usefulness of simulation models. This paper describes the potential of a "virtual" construction site for automation planning and analysis.

1 Introduction

Concepts for automation and use of robotic equipment on construction sites are evolving to encompass integrated equipment systems and sophisticated, distributed processing architectures. The evolution of construction robotics concepts can best be traced in the ten proceedings from the *International Symposium on Automation and Robotics in Construction.* Initial concepts for construction robotics focused on single-purpose machines based on industrial robot arm configurations. Geometry-based CAD systems were used in early systems for site modeling and planning. Automation schemes were often based upon industrialization of component fabrication [Warszawski 90], or trains of numerical controlled equipment used in repetitive work such as roadway paving [Skibniewski 90]. More recent work in automation design has emphasized changes in process design (to allow more effective automation) and overall integrated systems for tasks such as building assembly. Distributed sensing systems and general site models, such as ConSolve's *SitePlanner*, are also beginning to appear, particularly in response to regulatory requirements on environmental remediation sites.

While more complex automation systems are being developed, there are no comprehensive systematic methods available to assess alternatives for site automation. Prior work has relied on theoretical studies, simple geometric models, or physical experimentation.

A computer-based, "virtual construction site" (VCS) model could be used to investigate alternative designs for sensing, site modeling, distributed processing and automation on construction sites. The virtual construction site model enables a much more realistic assessment of automation and process design alternatives and strategies than is available with existing, geometry-based computer-aided design (CAD) models.

Existing CAD systems, e.g., Bechtel's *Walk-thru* and *AutoCAD* have powerful capabilities for geometric representation, visualization and animation of three-dimensional models, but they do not integrate genuine physical models of motion, gravity, fluid flows or other physical phenomena. For example, a typical CAD model would not recognize that a load swaying from a crane has inertia or that a piece of equipment might overturn and fall under the influence of gravity. Considering realistic issues in automation planning, such as sensor placement and perception range or alternatives for automated system control, requires moving beyond only the considerations of the simple geometric interferences supported by existing CAD systems.

With a realistic model of construction sites, encompassing open user actions and physical phenomena along with geometric representations, we can assess different schemes for site automation more rapidly and cost effectively than would be possible in field trials or physical laboratory simulations. For example, we might wish to evaluate the effectiveness of multiple range sensors placed on site for the purpose of supporting the activity of a robotic excavator. Each range sensor would have a particular viewpoint and capability. The sensor model incorporated into the VCS would have limited access to the description (both geometric and operational state) of the site, based on the sensor capabilities, and data from the VCS model would be filtered to reflect the errors inherent in the sensor operation. As another example, we might evaluate different heavy lift strategies with respect to the possibility of load shifting and allowable cable strains. Incorporating models of physical phenomena in a VCS would allow engineering analysis of gravity and forces on site components.

Development of the VCS model alone may be beneficial for many purposes. In a sense, the VCS would represent an example of the next generation of CAD systems in which engineering information could be accurately simulated. Rather than requiring engineers and designers to explicitly select and model those aspects which are considered, a virtual CAD system for a particular domain would automatically incorporate considerations of applicable phenomena.

The term *virtual* deserves explanation. We use the term in the same form as is traditional in mechanics or computations (e.g., virtual work, virtual memory), an imaginary system which mimics the behavior of the real counterpart which it models. In addition, our vision includes making the user (e.g., site planner, equipment operator) a part of the virtual world, able to directly interact with it. The VCS will be a "virtual reality" world. Connected to typical "virtual reality" interaction devices such as display goggles or instrumented data gloves, the VCS would permit direct interaction with the environment, such as fluid flows or structural strains. For example, the movement of groundwater components in response to different pumping strategies might be represented by different colors, or "felt" through instrumented flow meters. This type of model might be useful in the design and selection of hazardous site remediation strategies. Or the VCS model could be adopted for site logistics planning, in which the planner could "move" equipment and laydown areas, and the feedback would reflect cost or time implications of the selection.

This paper reviews existing work relevant to the development of a virtual construction site, defines system requirements, and considers the overall feasibility and desirability of a VCS system. We begin with an example of a crane pick-and-place operation.

2 An Example of Physically-Based Simulation Modeling: Crane Operations

Suppose that a crane is being used to place beams in a structure. The crane is at a fixed location on the site. A typical pick-and-place operation involves several steps:

1. identify the beam to be placed from those in the laydown area.

2. swing and boom the crane over the beam.
3. drop the hooks to the beam.
4. attach the hooks to the beam.
5. lift, swing and boom the beam into place.
6. position the beam into the final placement position.
7. temporarily attach the beam to the structure.
8. unhook the beam from the crane.
9. move the crane for the next operation.

In a simple model which might be animated in a state-of-the-art computer-aided design software package, the crane hooks and the beam move in a smooth path, from the initial crane location, to the pick, through the lift and to the final position. In the real situation, there are a number of considerations which complicate the operation:

- Beam physics. The beam is a three-dimensional object, which has mass and stiffness. This implies that the inertia of the beam will influence the path of the crane, and the beam as a load will swing from the crane. Likewise, the beam will deflect when lifted, and both the beam's three dimensional shape and stiffness may influence where the hooks can be attached, and the deflection will influence the final placement of the beam in the structure.
- Crane physics. The crane also has mass and stiffness. The crane body, boom, cables and hooks and grappels all are influenced by their relative strength and mass. For example, when unloaded, the hook will drag behind the path of the tip of the boom, and the hooks will swing like pendula. With the load attached, the cables will elongate, the boom will deflect (due to the load and the joint compliance), and the load will again behave with a pendula-like motion.
- Operator Action. The actual crane and beam motion will be driven by the operator actions in swinging and lifting the load. The speed of the engine and velocity of the elements of the crane and timing of the operator actions will drive the system.
- Environment. The crane and the load are not uncoupled from the world in which they are operating. Simple factors, such as wind (varying dynamically during the lift) combined with the mass and aerodynamics of the elements combine to influence the actual motions. Other influence, such as temperature also may be important (picking the beam from a cold shaded area and placing it into a sunny area where the thermal effect on beam length influences fit).

All of these physical constraints will combine to determine the actual path of the crane. A more realistic model of behavior will incorporate these effects:

1. assume the crane is initially at rest in some configuration (influenced by the self-weight and stiffness of the crane).

2. identify the beam to be placed.
3. swing and boom the crane towards the load. In the process, incorporate the mass of the crane, boom, cables, hooks, etc., computing the path of the boom tip and hooks, as driven by the operator's action and the environment.
4. drop the hooks to the beam (this and all other actions are influenced by the physics, etc., describe above).
5. attach the hooks to the beam.
6. lift, swing and boom the beam into place.
7. position the beam into the final placement position.
8. temporarily attach the beam to the structure.
9. unhook the beam from the crane.
10. move the crane for the next operation.

The above sequence just describes the basic mechanics of the pick and place operation. In addition, the processes of overall control and communications between the different people involved in the process may involve significant management or process design issues and thus deserve attention in the model. For example, the crane operator might not be able to see the hooks and load at various times in the operation (e.g., the beam is picked from behind another building or pile of materials, or the placement position is obscured from the operator's view). Thus a realistic model should also incorporate additional factors:

- Project World. The crane operates in a three-dimensional world of the entire construction site. This world is dynamically changing, with some elements changing slowly over days, such as the site and the structure being built, and other moving quickly, such as other cranes.
- Communications. The process of inter-person or electronic communication may involve delays or loss of information.
- Control. Control at a simple level is no more than specifying which element is to be placed and its final location. More detailed operational control consists of planning the path of the crane, or optimizing the power of the crane engine to the hoist and hydraulics to minimize fuel consumption.
- Sensing. Automated control of various aspects of the process or even direct monitoring of the operations will involve sensors that may have limited range, delays or other constraints.

Incorporation of these various physical constraints and control issues may be crucial for developing an accurate portrayal of crane movements and other construction operations. Similar sets of constraints may be developed for other construction operations, including issues such as soil or groundwater movements.

While developing and validating physical models of the type described above may be time-consuming, the basic models should be able to be re-used in numerous construction applications in the same way that standard component representations are re-used in CAD models. Moreover, simulation of automation and control options in a simulation model may be much less expensive and faster than actual field trials. From this example of possibilities, we turn to examine the existing state of the art.

3 Virtual Reality

In most conventional computer applications, the computer and the user are two different participants in an environment with a limited and predefined set of interactions. The actions which the computational agent can perform are planned before the system is implemented, as are the types of output it will produce. The user has a restricted communications channel to the machine, typically just a keyboard and display. Unacceptable and unplanned inputs are ignored, or worse, cause failures.

"Virtual reality" attempts to fully integrate the user with the computer system, bringing the user into the environment of a single modeled world. In the extreme, all of the user's actions are monitored (e.g., he is in a fully instrumented body suit on a sensing platform). All the user's sensory input is provided by the machine, typically a head-mounted display that presents images and blocks the rest of the visual field. Aural feedback is also controlled, along with forces and motions. The actions which the computer performs are determined by the user actions, and all of the information which the user relies on to control his actions is provided by the machine.

While no virtual reality systems exist which completely bring the user into the computer's environment, applications and developments in place are moving towards a tighter coupling of the user and computer. Current applications model a number of phenomena in which the actions dynamically evolve throughout a simulation. One of the most advanced applications is *BattleTech*, a "video game" environment in which each member of a group of players individually controls a robotic "tank" in a battle situation over a large terrain [Grimes 92]. In *BattleTech*, the computer provides the artificial environment (weather, terrain), moves the tanks through the world, and mediates the interactions among the players. Developments are underway in other areas such as entertainment, architecture, and medical and scientific simulations [Grimes 91, Haggerty 91].

One common virtual world is a geometric model in which the user can navigate. Using sensors, the computer interprets user actions in terms of walking and looking in different directions in the world. As the user moves though the space, the display is automatically changed to reflect location and view direction. Such applications require only geometric models of the world, and a simple model of user motion. A head-mounted display or stereo display presents the scene to the user, and an instrumented glove senses the movement and position of the fingers and hand. Different hand gestures indicate movement, speed, and view direction.

Simple virtual worlds only model geometry and visual effects (e.g., lighting), and user interaction is limited to moving in the world. Letting the user modify the world, by moving objects, opening doors, etc., provides the next level of complexity. The behavior and physics of objects and the world are introduced in more complex environments. Displays can range from simple wire-frame models to realistic images, according to the requirements of the domain. Head-mounted displays which block all other vision provide the most realistic interaction, however a single or stereo monitor may be sufficient in some applications. User interaction can be simulated with common devices such as mice or joysticks, while most simulations use three-dimensional input devices such as wands, data gloves, and in the extreme, body suits. Most of the complexity in building a virtual world is in building the models, programming the displays, and controlling user interaction.

4 Virtual Reality Applications in Construction

No virtual reality applications have yet appeared for construction automation and planning, although researchers are beginning to discuss the idea [Kangari 92]. A number of developments over the last two decades lead towards virtual reality applications in the AEC community. While most of these do not qualify as true virtual reality systems, they all share characteristics with current virtual reality work.

One of the first attempts to build a virtual world for visualization dates from the early 1970s, when as part of the plan to locate the I. M. Pei *H. E. Johnson Museum of Art* on the Cornell University campus, a *walk-though* was generated. This consisted of a film sequence of travel throughout the campus, overlaid with a computer-generated display of the facility, permitting different siting options to be explored. While not interactive, it illustrates the same "virtual" concept: the building displays show realistic views of the new structure at the potential sites.

This idea led directly to similar developments, either combining film and computer images, or relying totally on a computer generated environment. A computer-generated display can be combined with video from a video disk. Random access of the video disk lets the user select the viewing sequence, and the display can be generated to properly overlay on the video. More realistic worlds require the computer to display the entire model, and also permit the viewer to chose his path and sight line, rather than following a predefined path.

Such walk-through (or fly-through) animations have been used on many architectural projects. Current systems extend the visualizations to include high-quality renderings of surfaces, lighting, etc. While many applications require artificial means to specify the path (e.g., trace a path on a 2-D plan view), recent systems incorporate head-mounted displays and data gloves.

The state of the art in architectural visualization is represented by the work of Brooks et al., at UNC Chapel Hill, and in AEC, by the *Walk-thru* program developed by Bechtel. These systems have been applied primarily to view alternative designs. In a similar application, Stone & Webster has used a walkthrough system to plan placement of large equipment in a power plant; the equipment follows the path, and interference detections are used to eliminate infeasible moves. Such a system goes beyond simple display and includes interactions of parts of the world. Fully interactive systems of this type can simplify complex planning.

Simple animation systems for viewing construction operations have been in place for many years (e.g., [Kano 84] or [Stouffs 93]). These systems illustrate a construction sequence, displaying an evolving geometric model, following an erection sequence derived from the schedule.

The first commercial AEC virtual reality application is the *Virtual Showroom*, a kitchen planning system developed by Matsushita Electric Works. A three-dimensional model of the kitchen is built from the user's specifications, and by using a headset and data glove, the user can move around the space, "touch" counter-tops, open doors, etc. Physical models here include simple interference detection.

Most of the current AEC work in virtual reality is aimed towards visualization. Virtual worlds which include complex physical phenomena are beginning to appear, as are more realistic displays and more complex modes of user interaction. Virtual reality applications extend the earlier visualization, simulation and animation work by incorporating of all these features.

All of the basic virtual reality technologies needed to build a VCS are in place. Sufficient computing power is available, and significant work is underway in visualization and developing physical models. Details of these evolutions have been presented in the various *SIGGRAPH* proceedings. The *ISARC* proceedings have presented work such as force-cognizant excavation, crane collision avoidance systems, and models of various robotic agents. Together, sufficient information and tools are available to build a VCS.

5 Impact of Physically-Based Virtual Reality for Construction

While statements about the concepts of using virtual worlds in civil engineering have been made [Fenves 91], there has been little substantive effort towards developing a "virtual construction site system" or using a virtual environment for modeling and analysis in construction automation and planning. We believe that this next generation of construction simulation tools should incorporate both enhanced user interaction mechanisms common in "virtual reality" demonstrations and games, and a variety of physically-based models to greatly improve the realism and accuracy of such simulation tools.

At this time, the appropriate representations and software architecture for a "virtual construction site" have not yet been determined. Efforts to develop physically-based models in other domains typically rely on distributed processing and object-oriented representations. With proper attention, particular model types might be capable of being moved in a modular fashion from one application to another. We are already investigating this type of modularity in a crane model. However, the peculiarities of many construction operations and concerns suggest that systems developed for other domains will be either serious deficient or inappropriate for construction applications.

In sum, simulation models are a demonstrated way in which to improve the overall construction process. We anticipate new technologies in computers and communications also playing major roles in construction process improvements. Work on this general problem can help lead the construction industry into the next century.

6 References

[Fenves 1991] Fenves, S. J., "The Potentials of Computer-Based Technologies in Civil Engineering," *Twenty-Ninth Henry M. Shaw Lecture in Civil Engineering*, Department of Civil Engineering, North Carolina State University, Raleigh, NC, December, 1991.

[Kano 1984] Kano, N., and Tamura, Y., "Application of Computer Graphics in Construction Planning and Scheduling," *Proceedings Workshop Conference on Robotics in Construction*, Pittsburgh, PA, Department of Civil Engineering, Carnegie Mellon University, June 17–20 1984.

[Grimes 1991] Grimes, J., "Virtual Reality 91 Anticipates Future Reality," *IEEE Computer Graphics & Applications*, Vol. 11, No. 6, pp. 81–83, November 1991.

[Grimes 1992] Grimes, J., "Virtual Reality Goes Commercial with a Blast," *IEEE Computer Graphics & Applications*, Vol. 12, No. 2, pp. 16–17, March 1992.

[Haggerty 1991] Haggerty, M., "Virtual Reality Dominates SIGGRAPH," *IEEE Computer Graphics & Applications*, Vol. 11, No. 5, pp. 14–15, September 1991.

[Kangari 1992] Kangari, R., and Akimoto, M., "Artificial Reality in Robotics and A/E/C Firms," *Proceedings 9th International Symposium on Automation and Robotics in Construction*, Tokyo, June 1992.

[Skibniewski 1990] Skibniewski, M., and Hendrickson, C., "Automation and Robotics for Road Construction and Maintenance," *ASCE Journal of Transportation Engineering*, American Society of Civil Engineers (ASCE), Vol. 116, No. 3, pp. 261–272, May/June 1990.

[Stouffs 1993] Stouffs, R., Krishnamurti, R., Lee, S., and Oppenheim, I., "Construction Process Simulation with Rule-Based Robot Path Planning," *Proceedings 10th International Symposium on Automation and Robotics in Construction*, Houston, Texas, May 1993.

[Warszawski 1990] Warszawski, A., *Industrialization and Robotics in Building*, Harper and Row, 1990.

Automation and Robotics in Construction X
G.H. Watson, R.L. Tucker and J.K. Walters (Editors)

Simulation Analysis of Process Piping Construction

Min-Yuan Cheng[a] and James T. O'Connor[b]

[a]Department of Construction Engineering & Technology, National Taiwan Institute of Technology, Associate Professor

[b]Department of Civil Engineering, University of Texas at Austin, Associate Professor

Abstract

Previous research has demonstrated that automation can improve industrial pipe lifting productivity. However, there is little information on finding how much productivity can be improved in the other parts of pipe erection processes to coordinate with the productivity improvement in pipe lifting. The processes of piping construction include pipe delivery, pipe lifting, pipe alignment, pipe connection, and pipe inspection & connection. This paper studied the potential productivity improvement by simulating the process of piping operations in order to synchronize with the productivity improvement in piping lifting. This paper is based in part on previous findings from studies performed by Glass and Fisher at the University of Texas at Austin in 1984 and 1989. The semi-automated environment for piping erection assumes a Grove Pipe Manipulator attached to the boom nose of a 22-ton crane as the base piece of equipment for piping erection. A system model is established to simulate and analyze the process of piping construction. A simulation technique is developed in this study to identify the possible critical tasks and the factors of obstruction in the process of piping construction that would eliminate the productivity improvement in piping erection.

1. BACKGROUND

The Manipulator was first brought into construction in 1980. Since then the Pipe Manipulator has been modified several times for the problems of technical utilization and was tested in the construction industry for handling large scale of piping construction. The Pipe Manipulator is an electrohydraulically control, multifunction, large bore pipe handling device which consists of five major components, namely a side mounted gravity leveling operator's basket with control panel, an attaching support frame, a telescoping boom, a pipe grappling and pivot head assembly, and a free standing storage rack for the unit when it is not being used (Glass 1984). Glass developed his study in comparing the productivity of the Pipe Manipulator with conventional piping erection method using a small, 15-ton hydraulic "Cherry Picker"(Glass 1984). In the Glass' study, it was found that the Pipe Manipulator was superior to conventional piping erection when lifting large diameter, vertical and bent pipe configurations. Fisher, in 1989, summarized piping erection activities and the percentage of cycle breakdowns of time, from two of Glass' time lapse films, for horizontal piping erection (film D-3 and D-4). Fisher's findings, from the imitation of the paths in time lapse films D-3 and D-4 to record the process of simulation on plastic and computer models, indicated that the Pipe Manipulator could be improved through four phases of piping erection method. In each phase, constructability issues associated with piping construction was explored and improved to promote a more automated operation.

2. METHODOLOGY

This study was conducted in three steps:

1. establish a cycle model to simulate the piping installation process,
2. identify the critical work tasks (i.e., bottleneck activities), and
3. study sensitivity analysis to determine the optimal combinations of labor crew (i.e., fitter and welder) under different circumstances.

2.1 The Cycle Model

This study used MicroCyclone developed by Professor Daniel Halpin in 1986 to simulate the process of piping construction. Twelve individual flow unit cycles were established in the model: (1) Pipe erection cycle, (2) Crane cycle, (3) Load pipe cycle, (4) Pipe transportation cycle, (5) Manipulator cycle, (6) Manipulator respoting cycle, (7) Fitter cycle, (8) Remove empty cart cycle, (9) Welder cycle, (10) Clear & inspect cycle, (11) Reweld cycle, and (12) Clear and inspect & reweld cycle. Figure 1 shows the integrated model.

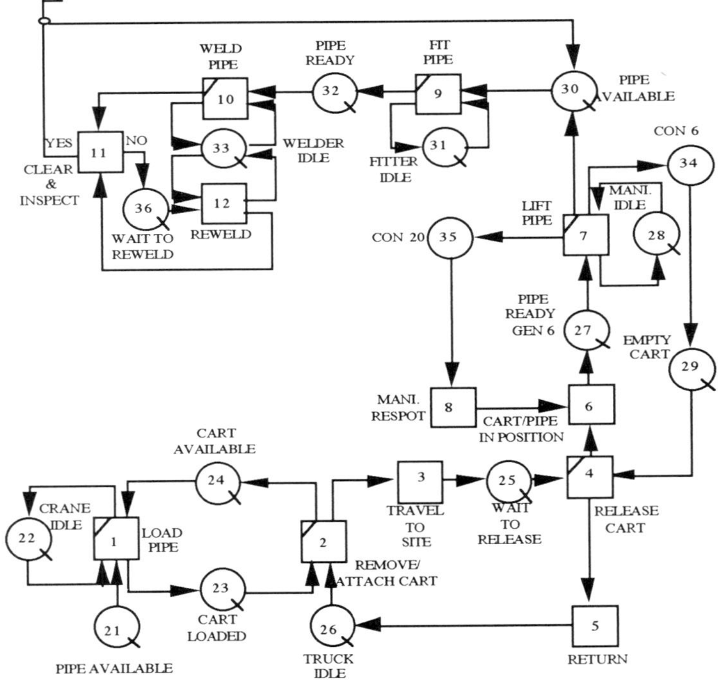

Figure 1 The final integrated model for piping process

2.2 Identification of the Critical Work Tasks

By simulating the developed cycle model, the critical work tasks of process piping construction were identified in this step. From the output reports of simulation, the critical work tasks could be identified by checking the percentage of work task that is busy and the percentage of queue that is occupied. The percentage of work task that is busy is the percentage of time the work task was in operation. The percentage of queue that is occupied is the percentage of time that a unit was waiting (or "idle time") in the queue before moving to the following work task. The last term is a key indicator of the delays that occurred in the process.

2.3 Sensitivity Analysis

This section discusses the objective statements and both the Tree Algorithm and the Time Deviation methods used for analyzing this study. Tree Algorithm method includes a tree combination of resources and comparison of different interest to approach the decision-making process. Time Deviation method analyzes the best range of time control of critical activity that could achieve maximum productivity increase.

2.3.1 Objective Statement

The purposes of this study is to establish a cycle model to simulate the piping installation process, identify the critical work tasks (i.e.., bottleneck activities), analyze the frequency of pipe delivery (# of trucks), and determine the number of labor crew (i.e.., fitter and welder) under different circumstances. This study involves different effects, given different combinations of trucks, fitting crews, welding crews, and productivity rates between conventional and improved piping erection methods.

Three major objectives are concerned in this study:

Objective 1: Use standard and conventional piping erection process in the model to simulate and identify the possible critical work task. Analyze the combination of trucks, fitting crews and welding crews required in coordinating with pipe lifting activity. Different goals such as maximum productivity per hour (Spools/HR), maximum productivity per man-hour (Spools/MH), least idle time of pipe waiting the Manipulator to lift can be achieved with a Tree Algorithm method. Depending on the different interests, the decision-maker would like to trade-off according to the results to make his best decision.

Objective 2: Findings from D.J. Fisher "Piping Erection Constructability Issues In a Semi-Automated Environment" include specific suggestions that would improve the Pipe Manipulator performance by 48% over conventional piping erection methods. Four types of studies to improve Pipe Manipulator installation cycle time were identified in Fisher's study (see Table 1). Based on the improvement of cycle time in each phase, this paper identified the best combination of trucks and labor crews to coordinate with the productivity improvement of the Pipe Manipulator. Also, under each Manipulator cycle time, the sensitivity analysis (Tree Algorithm method) is used to analyze the goals as mentioned in objective 1.

Table 1 Total Cycle Time of Pipe Manipulator

ANALYSIS METHOD	Time Lapse	1. computer 2. Plastic	1. computer 2. Plastic	1. computer 2. Plastic
Condition	Time Lapse	Time Lapse	Nominal	Optimal
Total Cycle Time	**1116 (sec)**	**901.5**	**832.5**	**702**

Objective 3: Under a certain combination derived from objective 2, the model is simulated based on increasing or decreasing the percentage of the operation time of the critical work task. "FIT PIPE" and "WELD PIPE" were analyzed by increasing or decreasing the activity duration twenty percent each time. Calculating and comparing the slope of productivity on each increasing or decreasing phase, the modeler can tell the optimal phase (i.e., the phase with the maximum slope) of improving the duration of activity that may achieve the maximum productivity increment.

2.3.2 Tree Algorithm Method

Tree Algorithm is a method used in sensitivity analysis to determine optimal piping productivity (Spools/HR) by using different combinations of trucks, fitters, and welders (see Figure 2). The method is to test the effect of changing the number of trucks, fitters, and welders based on the behavior of the system. Different approaches such as maximum productivity per hour (Spools/HR), maximum productivity per man-hour (Spools/MH), and least idle time for waiting for the Manipulator to lift could be achieved with this method.

2.3.3 Time Deviation Method

Time deviation is another method used in sensitivity analysis. Some of the work task time in the model are variables. One of the variables can be fixed and the work task time can be increased or decreased by a certain percentage (see Table 2). The results of the Time Deviation method can give us information about time control. The Time Deviation method is presented in a series of diagrams. These diagrams show the impact of productivity (vertical scale) as a function of the percentage of change of a variable (horizontal scale). The steeper the slope of the curve, the more sensitive is the range.

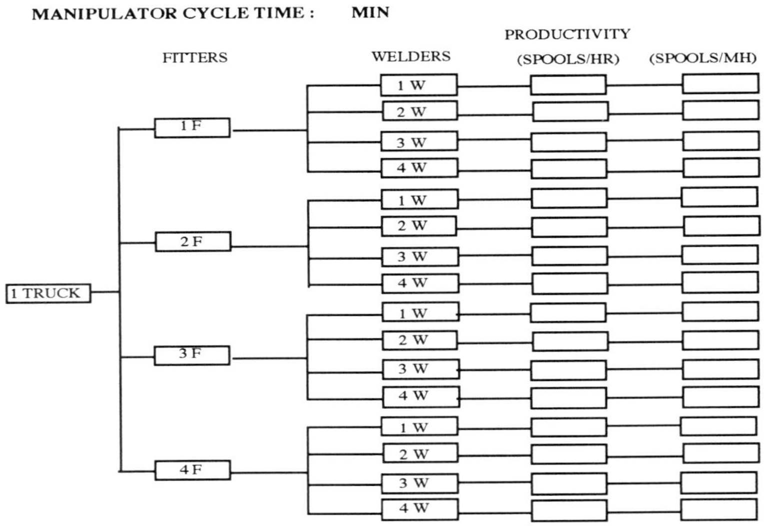

Figure 2 Tree Algorithm Method

Table 2 Time Deviation Method

MANIPULATOR DURATION= min

FITTER 1 WELDER 1

DEVIATION

WORK TASKS	-60%	-50%	-40%	-30%	-20%	-10%	0	20%	40%
FIT PIPE									
WELD PIPE									

FITTER 2 WELDER 2

DEVIATION

WORK TASKS	-60%	-50%	-40%	-30%	-20%	-10%	0	20%	40%
FIT PIPE									
WELD PIPE									

FITTER 2 WELDER 3

DEVIATION

WORK TASKS	-60%	-50%	-40%	-30%	-20%	-10%	0	20%	40%
FIT PIPE									
WELD PIPE									

FITTER 3 WELDER 4

WORK TASKS	-60%	-50%	-40%	-30%	-20%	-10%	0	20%	40%
FIT PIPE									
WELD PIPE									

3. DATA ANALYSIS AND RESULTS

3.1 Description of simulation

The results of simulation were obtained under the assumption that as the limited working space, the maximum number of fitter and welder assigned one Manipulator is no more than four. Under different Manipulator erection cycle time, simulation with 1 truck 1 fitter & 1 welder, 1 truck 1 fitter & 2 welders, 1 truck 1 fitter & 3 welders etc. (also 1 truck 2 fitters &1 welder, 1 truck 2 fitters & 2 welders etc.) is run and the results in terms of productivity per hour, and per man-hour are identified. Referring to Table 1 (different total cycle time of Manipulator), the Tree-Algorithm method were applied to simulate the model. Taking all possible combinations of resources for the work task "LIFT PIPE" and within the limits specified, the program finished simulating 16 different possible combinations under 8 different erection cycle times for the Manipulator. Tables 3 and 4 show the simulation results of five different combinations (1F1W, 1F2W, 2F2W, 2F3W,3F4W).

Table 3 Results of simulation: The productivity per hour

VARIABLE : MANIPULATOR CYCLE TIME

MANI. CYCLE TIME (MIN) (unit: Spools/HR)

CREW COMBI.	11.7	12.55	12.98	13.14	13.88	15.3	18.6
1F1W	1.47	1.47	1.47	1.47	1.47	1.47	1.46
1F2W	2.16	2.16	2.16	2.16	2.15	2.15	2.14
2F2W	2.6	2.6	2.6	2.6	2.59	2.58	2.55
2F3W	3.5	3.48	3.46	3.46	3.44	3.38	3.26
3F4W	4.05	4	3.97	3.96	3.92	3.86	3.74

Table 4 Results of simulation: The productivity per man-hour (Spools/MH)

VARIABLE : MANIPULATOR CYCLE TIME

MANI. CYCLE TIME (MIN) (unit: Spools/MH)

CREW COMBI.	11.7	12.55	12.98	13.14	13.88	15.3	18.6
1F1W	0.735	0.735	0.735	0.735	0.735	0.735	0.73
1F2W	0.72	0.72	0.72	0.72	0.717	0.717	0.713
2F2W	0.65	0.65	0.65	0.65	0.648	0.645	0.638
2F3W	0.7	0.696	0.692	0.692	0.688	0.676	0.652
3F4W	0.579	0.571	0.567	0.566	0.56	0.551	0.534

As stated in objective statements, the purpose of this study is to find the optimal combination to coordinate with the productivity improvement in pipe lifting. Therefore the first thing that needs to be improved is the capability of 1F1W which is not enough to handle the productivity improvement in pipe lifting. Referring to Table 3, the modeler notices that the productivity of 1F1W didn't increase simultaneously as the Manipulator cycle time decreased. Bottleneck activity "FIT PIPE" eliminated the productivity improvement of the Manipulator by increasing the waiting time in QUEUE node 30 "PIPE AVAILABLE" which precedes it. The waiting time of pipe to be welded at QUEUE 32 was also increasing as the Manipulator cycle time decreased. For 1F2W, the increment of productivity stagnated at 2.16 Spools/HR and did not increase as the Manipulator cycle time decreased. The same situation took place in the case of 2F2W (see Table 3). If the modeler wants to increase productivity and bring the Manipulator into full play, he has to add more resources into the process in order to synchronize it with the productivity improvement of the Manipulator.

3.2 Tree-Algorithm Analysis

Case of maximum productivity per hour

Reading through all of the results in the Table 3 which were simulated under different conditions, 3F4W is always the optimal combination to achieve the maximum productivity per hour. The results of simulation conclude that the more the productivity of the Manipulator is improved, the more the productivity of labor (including the number of laborers and motivation factors) needs to be improved.

With regards to cost, another approach was studied to find the best combination. Table 5 shows the productivity slope analysis derived from Table 3. The table was obtained by fixing the cycle time of the Manipulator (the vertical column) and calculating the slope of the productivity between different combinations of working crews (the horizontal row). In the table, the value of productivity increase in the third column is always the largest in comparison with the value in the other columns. Therefore the maximum increase in the rate of productivity can be obtained for additional labor from 2F2W to 2F3W. Considering the factor of profit and cost, 2F3W will become the optimal choice.

Case of maximum productivity per man-hour

In Table 4, the productivity of 1F1W is always the largest of all under different Manipulator cycle times. But considering the synchronization of the Manipulator improvement as discussed above, 1F1W and 2F2W will be ignored because of the flat slope and 2F3W becomes the optimal combination to achieve the goal. For analyzing the slope of productivity per man-hour, Table 6 was created in the same manner in which Table 5 was created. In the Table 6, all of the values are negative except the values in the third column. Therefore considering the synchronization of the Manipulator improvement and the factor of profit and cost just mentioned above, 2F3W is still the optimal choice in this case.

Table 5 Tree-Algorithm : slope analysis

Relation between the increase of productivity and the number of labor

PRODUCTIVITY PER HOUR:
VARIATION OF LABOR

MANIP. CYC.TIME	1F1W-1F2W	1F2W-2F2W	2F2W-2F3W	2F3W-3F4W
18.6 min	0.68	0.41	0.71	0.48
15.3 min	0.68	0.43	0.80	0.48
13.88 min	0.68	0.44	0.85	0.48
11.7 min	0.69	0.44	0.90	0.55

(unit spools/HR)

Table 6 Tree-Algorithm : slope analysis

Relation between the increase of productivity per man-hour and the number of labor

PRODUCTIVITY PER MAN-HOUR:
VARIATION OF LABOR

MANIP. CYC. TIME	1F1W-1F2W	1F2W-2F2W	2F2W-2F3W	2F3W-3F4W
18.6 min	-0.017	-0.075	0.014	-0.118
15.3 min	-0.018	-0.072	0.031	-0.125
13.88 min	-0.018	-0.069	0.040	-0.128
11.7 min	-0.015	-0.070	0.050	-0.121

(unit: spools/MR)

Case of least idle time of pipe waiting manipulator to lift

The least the idle time of pipe waits Manipulator to lift, the more the percentage of the Manipulator is busy. A comparison of the percentage of waiting time in QUEUE 27 "CART/PIPE READY" is identified. When Manipulator cycle time equals 18.6 minutes, the best combination for achieving the goal is 3F4W. As Manipulator cycle time equals 15.3 minutes, the optimal combination switches to 2F3W. In the case of 13.88 and 11.7 minutes, the optimal combination changes back to 3F4W.

3.3 Time-Deviation Analysis

The study of this section was made by fixing the duration of all tasks initially and varying the duration of work task "FIT PIPE" by percentage under different Manipulator cycle times, then calculating the slope of productivity under different increasing or decreasing phases. The same was done for analyzing work task "WELD PIPE". Consequently, there are two objectives to be identified in the study: (1) The maximum productivity increase for the entire working crew, and (2) The maximum productivity increase per man-hour.

(1) The maximum productivity increase for the entire working crew (fitter and welder)

The results of productivity increase owing to the decrease of duration of tasks "FIT PIPE" is illustrated in Table 7. In Table 7, under the condition of different Manipulator cycle times, the modeler studied work task "FIT PIPE" with four different combinations (1F1W, 2F2W, 2F3W, 3F4W) and varied the task duration from increasing the task duration 40% to decreasing the task duration 60%. Afterwards, in each increasing or decreasing phase, the modeler can obtain the different increase or decrease of productivity under these four combinations. In a certain decreasing phase of duration, the modeler would like to choose the combination whose slope is negative and has the maximum absolute value, thus the optimal combination for achieving the maximum increase of productivity can be obtained in the phase. Vice versa, in the increasing phase of duration, the combination whose slope is negative and has the minimum absolute value is preferred to avoid the excessive lost of productivity. The bold numbers shown in Table 7 identify the optimal combination under each increasing or decreasing phase. The study of work task "WELD PIPE" can be made in the same manner.

Table 7 Time Deviation : The productivity per hour slope analysis (WORK TASK FIT PIPE)

(a) MANIPULATOR CYCLE TIME= 18.6 min

	TIME DEVIATION							
COMBI'N	-60~-50%	-50~-40%	-40~-30%	-30~-20%	-20~-10%	-10~0%	0~20%	20~40%
1F1W	0	0	-0.01	0	-0.01	0	**-0.01**	**0**
2F2W	-0.02	-0.01	-0.01	-0.01	-0.02	-0.01	-0.02	-0.02
2F3W	-0.02	-0.02	**-0.02**	**-0.03**	-0.03	-0.04	-0.39	-0.14
3F4W	**-0.05**	**-0.03**	-0.02	-0.03	**-0.06**	**-0.08**	-0.24	-0.15

(b) MANIPULATOR CYCLE TIME= 15.3 min

COMBI	-60~-50%	-50~-40%	-40~-30%	-30~-20%	-20~-10%	-10~0%	0~20%	20~40%
1F1W	0	-0.01	0	-0.01	0	0	**-0.01**	**-0.01**
2F2W	-0.01	-0.01	-0.01	-0.01	-0.01	-0.02	-0.02	-0.02
2F3W	-0.02	-0.02	-0.02	-0.02	-0.03	-0.03	-0.41	-0.16
3F4W	**-0.07**	**-0.06**	**-0.05**	**-0.07**	**-0.05**	**-0.05**	-0.09	-0.32

(c) MANIPULATOR CYCLE TIME= 13.88 min

COMBI	-60~-50%	-50~-40%	-40~-30%	-30~-20%	-20~-10%	-10~0%	0~20%	20~40%
1F1W	0	0	-0.01	0	-0.01	0	**-0.01**	**-0.01**
2F2W	-0.02	-0.01	-0.01	-0.02	-0.01	-0.01	-0.02	-0.03
2F3W	-0.02	-0.02	-0.02	-0.02	-0.02	-0.02	-0.13	-0.3
3F4W	**-0.06**	**-0.07**	**-0.06**	**-0.05**	**-0.04**	**-0.03**	-0.06	-0.44

(d) MANIPULATOR CYCLE TIME= 11.7 min

COMBI	-60~-50%	-50~-40%	-40~-30%	-30~-20%	-20~-10%	-10~0%	0~20%	20~40%
1F1W	-0.01	0	0	-0.01	0	-0.01	**0**	**-0.01**
2F2W	-0.01	-0.02	-0.01	-0.01	**-0.02**	-0.01	-0.02	-0.03
2F3W	-0.03	-0.02	-0.02	**-0.03**	-0.02	-0.02	-0.44	-0.2
3F4W	**-0.05**	**-0.03**	**-0.03**	-0.03	-0.02	**-0.05**	-0.08	-0.44

(2) The maximum productivity increase per man-hour

Under each combination (1F1W, 2F2W, 2F3W, 3F4W), divide the values in Table 7 by the sum of working crew (i.e., fitter & welder). For example, in Table 7, the numbers in the first row were divided by two and the numbers in second row were divided by four. The results were shown in Table 8. The bold numbers are the optimal combination under each increasing or decreasing phase. With the same manner discussed above, in a certain decreasing phase, the combination with the maximum absolute value is the optimal. Vice versa, in the increasing phase of duration, the combination with the minimum absolute is preferred. The optimal results in Table 8 have little change from the results obtained in Table 7. Depending on the different interests, the manager would have to trade-off according to the different results between these two methods in his decision-making process. In the same manner, the study of work task "WELD PIPE" is conducted.

Table 8 Time Deviation : The productivity per man-hour slope analysis (WORK TASK: FIT PIPE)

(a) MANIPULATOR CYCLE TIME= 18.6 min

	TIME DEVIATION							
COMBI'N	-60~-50%	-50~-40%	-40~-30%	-30~-20%	-20~-10%	-10~0%	0~20%	20~40%
1F1W	0	0	**-0.005**	0	-0.005	0	-0.005	**0**
2F2W	-0.005	-0.0025	-0.0025	-0.0025	-0.005	-0.0025	-0.005	-0.005
2F3W	-0.004	-0.004	-0.004	**-0.006**	-0.006	-0.008	-0.078	-0.028
3F4W	**-0.0071**	**-0.0043**	-0.00286	-0.00429	**-0.0086**	**-0.0114**	**-0.0343**	-0.02143

(b) MANIPULATOR CYCLE TIME= 15.3 min

COMBI	-60~-50%	-50~-40%	-40~-30%	-30~-20%	-20~-10%	-10~0%	0~20%	20~40%
1F1W	0	-0.005	0	-0.005	0	0	-0.005	-0.005
2F2W	-0.0025	-0.0025	-0.0025	-0.0025	-0.0025	-0.005	**-0.005**	**-0.005**
2F3W	-0.004	-0.004	-0.004	-0.004	-0.006	-0.006	-0.082	-0.032
3F4W	**-0.01**	**-0.0086**	**-0.0071**	**-0.01**	**-0.0071**	**-0.0071**	-0.01286	-0.04571

(c) MANIPULATOR CYCLE TIME= 13.88 min

COMBI	-60~-50%	-50~-40%	-40~-30%	-30~-20%	-20~-10%	-10~0%	0~20%	20~40%
1F1W	0	0	-0.005	0	-0.005	0	-0.005	**-0.005**
2F2W	-0.005	-0.0025	-0.0025	-0.005	-0.0025	-0.0025	**-0.005**	-0.0075
2F3W	-0.004	-0.004	-0.004	-0.004	-0.004	-0.004	-0.026	-0.06
3F4W	**-0.0086**	**-0.01**	**-0.0086**	**-0.0071**	**-0.0057**	**-0.0043**	-0.00857	-0.06286

(d) MANIPULATOR CYCLE TIME= 11.7 min

COMBI	-60~-50%	-50~-40%	-40~-30%	-30~-20%	-20~-10%	-10~0%	0~20%	20~40%
1F1W	-0.005	0	0	-0.005	0	-0.005	**0**	**-0.005**
2F2W	-0.0025	**-0.005**	-0.0025	-0.0025	**-0.005**	-0.0025	-0.005	-0.0075
2F3W	-0.006	-0.004	-0.004	**-0.006**	-0.004	-0.004	-0.088	-0.04
3F4W	**-0.0071**	-0.00429	**-0.0043**	-0.00429	-0.00286	**-0.0071**	-0.01143	-0.06286

4. CONCLUSIONS

This paper proves that Computer Simulation Method (CSM) is an appropriate method for the evaluation of construction productivity improvement by automation. It is known that productivity improvement of each work task by automation might cost money. The higher the degree of automation applied, the more cost increases. This study gives the decision-maker the information necessary to determine whether the effort is worthwhile by comaring investment of capital to improvement in productivity. That is, to analyze by how much productivity can be improved when the activity is improved by automation.

5. REFERENCES

1. Cheng, M.Y., "Simulation Analysis of Process Piping Construction", MS thesis, University of Texas, December 1989.
2. Fisher, Deborah J. "Piping Erection Constructability Issues in a Semi-Automated Environment." Ph.D. dissertation, University of Texas, May 1989.
3. Glass, Clinton C. "The Pipe Manipulator: A Complete Assessment of a New Idea in Construction Equipment Technology." MS thesis, University of Texas, Dec. 1984.
4. Halpin, Daniel W. "MicroCyclone user's Manual."; 1986, 1.0 version and 1988, V 2.0.

Automation and Robotics in Construction X
G.H. Watson, R.L. Tucker and J.K. Walters (Editors)

Development of Segment Bolt Tightening Robot

Keizo Kazama, Hisateru Uyeda, Akihiro Higashide
OBAYASHI CORPORATION

Susumu Nunomura, Toshiaki Morii, Hirotoshi Shimoda
HITACHI ZOSEN CORPORATION

Abstract

(1) We have developed a manipulator for handling large, heavy objects for construction work, and applied it to bolt tightening work for shielded tunnel segments.
(2) The manipulator is vertical multi-articulated type, with 6 axis of freedom, and a handling capacity of 500kgf by a 2.5m long arm.
(3) The robot equipped two set of manipulator with bolt tightener. 120 bolts (ø30mm) of one segment ring which is 11.3m in diameter will be tighten by this robot in one hour.
(4) The manipulator performs positioning automatically, by recognizing the position of each bolts with an image sensor mounted at the top of its arm.

1. Introduction

In Japan, there is very active utilization of underground space, so the shielded tunneling method is indispensable to the development of social infrastructure. In recent years, construction projects have become increasingly large scale, and demand for automation and greater sophistication has increased due to the need to ensure safety, improve quality and achieve labor-savings etc.

For these reasons, automation has become a fixture of tunneling work, and active attempts are currently being made to automate tunneling direction control, and provide automation for tunnel lining work including the segment assembly. Other than this, however, no attempts are being made to automate incidental work.

We have developed a large-scale, heavy object manipulator to automatic these incidental tasks, and have applied it to automation of segment bolt tightening work. This paper reports mainly on the development of a tightening robot for segments which outer diameter 11.3m.

2. Current state of automation for the shielded tunneling method

In the shielded tunneling method, the primary tasks are excavation and lining work like segment assembly. On the other hand, there are various types of incidental work, like material transport, rail laying and segment bolt tightening after assembly. Much of this work involves repetition of simple tasks, and progress is being made with automation in many areas. Fig. 1 gives a flowchart of the main types of work in the shielded tunneling method, and Table 1 shows the results of a survey concerning the current state of automation for each type of work.

As you can see from Table 1 , automation is being actively employed in excavation, segment assembly and material transport, but automation is still lacking for incidental work like tie and rail laying and segment bolt tightening.

The main reason for this is that incident work has low importance and is difficult to automate due to its complexity. However, the final goal is a work system involving only a minimal number of maintenance personnel (2-3 persons), so automating these incidental works is an important topic for study and development.

As a common element for automating these incident works, we have developed a large and heavy object handling manipulator, and have applied it to segment bolt tightening work.

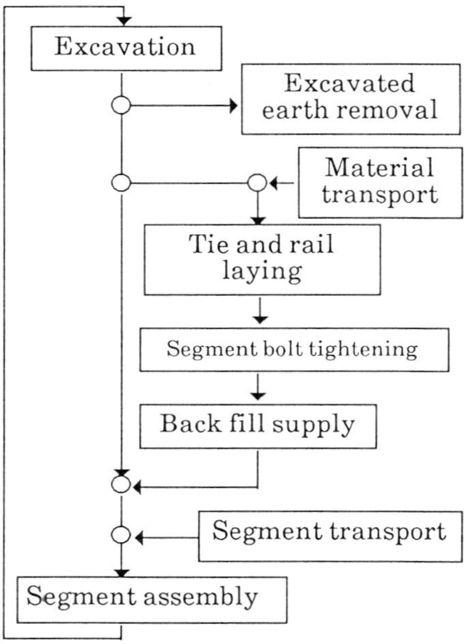

Fig. 1. Shielded tunneling method work flowchart

Table 1. Current state of automation

	Implemented	Under development	Not planned
Excavation	○		
Tunneling direction control	○		
Excavating ground control	○		
Segment assembly	○		
Segment bolt tightening		○	
Material transport	○		
Tie and rail laying			○
pipe extension	○		
wire extension			○

3. Structure of segment bolt tightening robot

Fig. 2 shows the overall mechanism of the segment bolt tightening robot. The robot support cart can travel on the rails and two manipulators installed on the lifting lowering masts which fixed on the both side of the support cart. The bolt tightener is set at the top of the manipulator arm. The sensors at the tip of the manipulators recognize bolts to enable tightening. The robot automates the entire process, from bolt detection to positioning and bolt tightening.

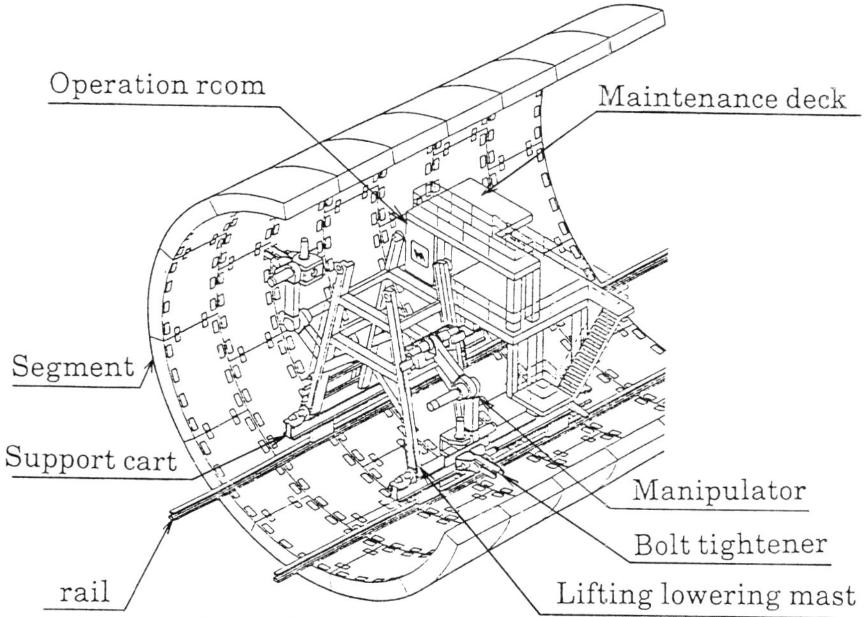

Fig. 2. Segment bolt Tightening Robot

3.1 Support cart

The support cart is the robot's base frame, and is equipped with manipulators.

The cart propels itself through the tunnel, up to the segment tightening position which is the current work point. The support cart has lifting-lowering masts on both sides to raise and lower the manipulator, and an operation room and maintenance deck.

3.2 Manipulators

Fig. 3 shows the manipulators which form the core of this robot. Each manipulator has 6 joints and 6 degrees of freedom, so the basic structure is the same as an industrial robot. Each joint has a high torque servo motor designed for construction with large, heavy objects. In particular, large special-purpose motors are not used to drive the 1st arm (which requires the most torque). Instead, a synchronous operation using 2 servo-motors is employed in order to lower costs and increase the handling load capacity.

A bolt tightener (composed of a gripper which holds the bolt, and a nut runner which turns the nut) and a bolt sensor which detects the bolt position are installed at the tip of the manipulator.

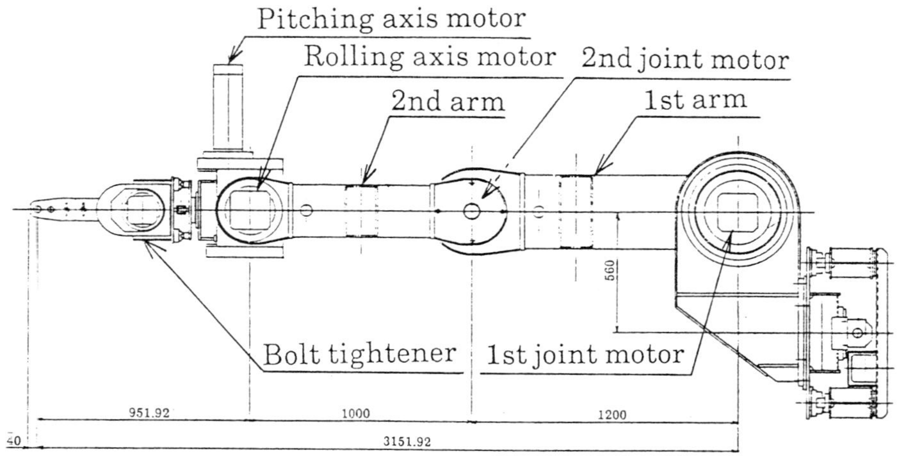

Fig. 3. Manipulator

3.3 Control system

The control system is composed of sensors arranged at various parts of the robots, controllers which control each of the actuators, and a monitoring system which manages and displays data relating to sensors and controllers. Fig. 4 shows the control system block diagram. The top-level preside controller, servo controllers for actuator sensors and other devices are connected by communication lines. Fig. 5 shows the bolt tightening operation flowchart.
The flow of bolt tightening operation is as follows :

(1) The support cart advances by 1 ring, and moves to the segment position to be tightened next. → Support cart advance
(2) The manipulator is moved to near the bolt coordinates through pre-programmed sequential numerical control. → Rough positioning
(3) The bolt-box is detected by a television camera, and the center of the box is found through image processing. Then the bolt tightener is inserted. → Bolt box sensing
(4) A bolt is detected in the bolt box by an ultrasonic sensor, and the bolt is grasped by the gripper. → Bolt sensing
(5) Tightener is performed by a nut runner.
(6) The manipulator returns to rough positioning.

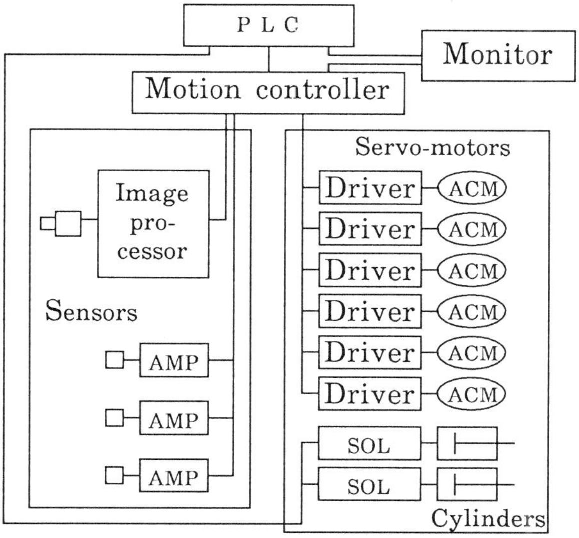

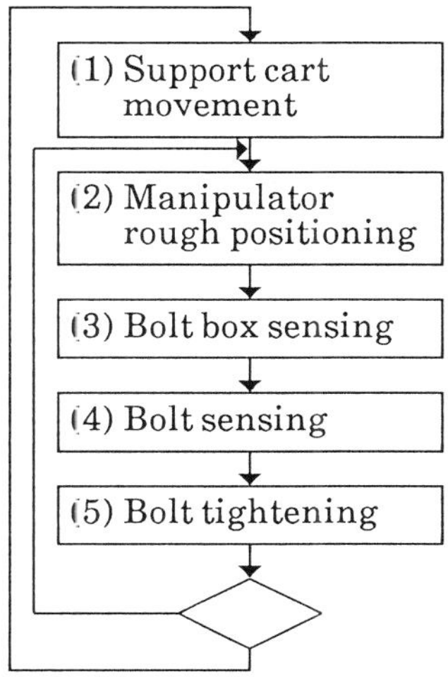

Fig. 4. Control block diagram   Fig. 5. Fastening operation flowchart

4. Bolt tightening robot performance confirmation

Critical points in the development of this robot were a sensing system for detecting bolts, and a manipulator capable of supporting the weight of the bolt tightener and the tightening reactive force, and performing high precision positioning.

Therefore in 1991 we tested the sensing elements, and in 1992 we confirmed performance of the manipulator through experiments using the actual equipment. At present we are conducting validation tests of function as the bolt tightening robot at our factory.

4.1 Bolt sensing experiments

The bolt sensing described above is composed of two stages. First rough positioning is performed through sequence control so that the box is captured on the television screen. The center of the box is extracted from the television image in primary sensing for bolt tightener insertion. Then a bolt in the box is detected using an ultrasonic sensor in secondary sensing so the gripper can grasp the bolt.

Fig. 6 shows the sensing experiment equipment. An orthogonal table with 3 degrees of freedom (X, Y and Z directions) is set-up over a model of the bolt box, and a bolt gripper, television camera and ultrasonic sensor are mounted at the end of the Z-axis (vertical axis). Sensing data is fed back to the computer, and after calculating the relative distance between the gripper and work, a position command is issued to the orthogonal table drive servo motor and the bolt is grasped.

Fig. 7 summarizes the image sensing results of the sensing experiment. The eccentricity in the diagram expresses the distance between the measured object in the image and the center of the image. As you can see, the measurement precision improves to the extent that the eccentricity is low.

Fig. 6. Sensing experiment equipment

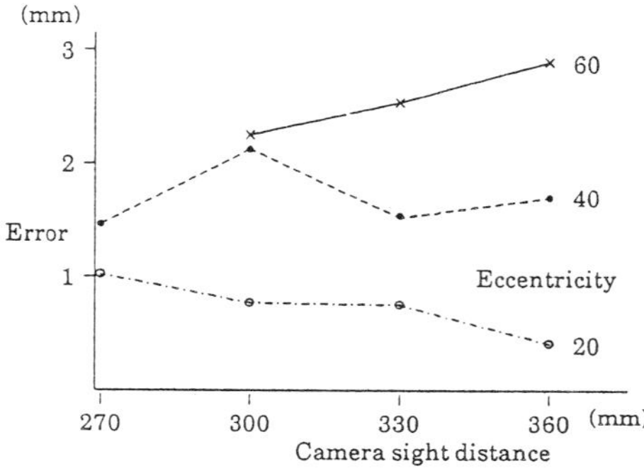

Fig. 7. Image sensing result

4.2 Actual manipulator experiments

Experiments concerning static and dynamic characteristics were conducted using 3-joint manipulator with the same specifications as the actual manipulator. The scene of the experiment is shown in Fig. 8. This picture shows the manipulator and measuring equipment used in the experiment.

The main parts of the experiment and their results are shown in Table 5. Even though the manipulator is designed for heavy objects, experiments confirmed that the system has high precision (with a positioning repeatability of ± 0.1mm) and performance greatly exceeding design specifications regrading the maximum permissible loading capacity.

In tests of dynamic characteristics it was found that residual vibration of the arm tip can be suppressed by appropriately controlling the arm speed pattern, and the system has satisfactory characteristics for use in heavy load handling.

In addition to experimental confirmation, the system was analyzed (including the robot mechanism and servo control system) using CAE(computer aided engineering) technology to ensure optimal design. As examples of robot arm motion simulation by computer, Fig. 9 shows an animation of step response, and Fig. 10 shows response of the arm tip at this time.

Table 2. Experiment Result

	Item	Main Result
1	Arm vibration characteristic	Natural frequency 4.3 Hz Residual vibration 0.5mm
2	Positional repeatability	Within +/– 0.1mm
3	Maximum permissible load capacity	Max. 970kgf, 18.8 sec.
4	Interpolation precision	Deviation 1.1mm (Arc interpolation)

Fig. 8. Manipulator experimental set-up

4.3 Bolt tightening validation test

The validation test of robot shall be conducted until the middle of May,and after data analysis and evaluation,we plan to start on-site operation in October of this year.

Prior to operation of the robot at the site, bolt tightening validation testing is conducted at the factory using actual segments in order to determine adjustment of the final model and applicability to the site, and to certify software etc.

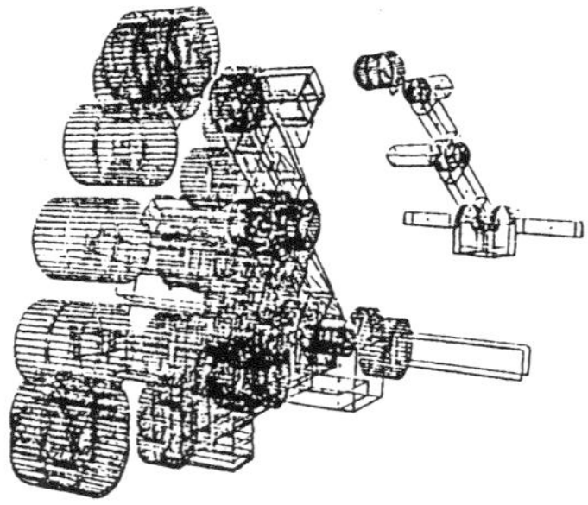

Fig. 9. Animation of step response

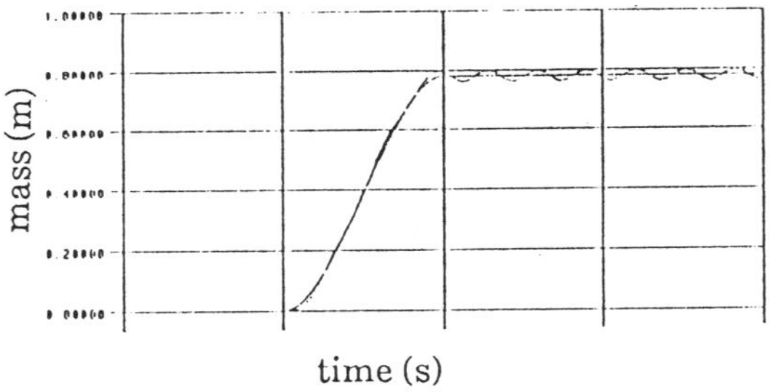

Fig. 10. Response at step input

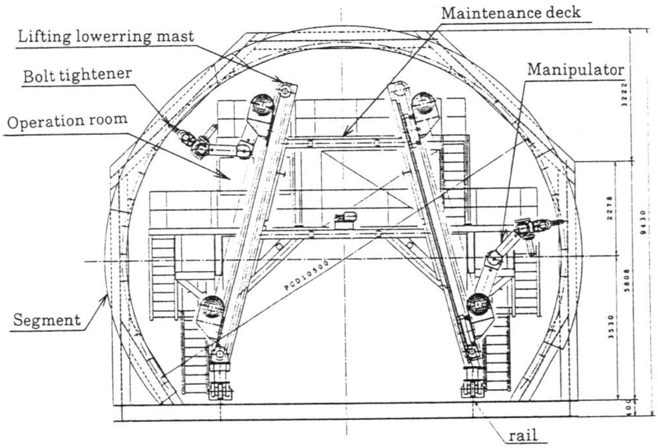

Fig. 11. In-factory validation test

5. Conclusion

This paper has described development of a general-purpose, large, high-precision manipulator, and an example of its application in a segment tightening robot.

Automation and labor-saving will continue to be critical topics at construction sites, and this general-purpose robot can play an effective role in the field. The following are the conditions which distinguish construction robots from industrial robots :

(1) Applicability to highly diverse types of workpieces
(2) Handing of large, heavy objects
(3) Self-propulsion

In the future we aim to contribute to development of automation for the construction site by striving to develop practical construction robots meeting the above needs.

Automation and Robotics in Construction X
G.H. Watson, R.L. Tucker and J.K. Walters (Editors)

Self-Supported Segment Assembly Robot for the Shield Tunneling Method

Keizo Kazama[a], Hisateru Ueda[a], Akihiro Higashide[a] and Yoshimitsu Kumao[b]

[a]OBAYASHI CORPORATION, 2-3, Kandatsukasa-cho, Chiyoda-ku, Tokyo 101, Japan

[b]MITSUBISHI HEAVY INDUSTRIES, LTD., Kobe Shipyard & Engine Works, 1-1-1, Wadasaki-Cho, Hyogo-ku, Kobe 652, Japan

Abstract

The self-supported segment assembly robot, [O-SERO]*, has been developed through several fundamental experiments and a series of assembly tests in the factory, then applied to a real project. 70 no.s of segment ring (inside diameter: 6.0 meters) were assembled in the tunnel. The assembly process is full automatically done by this robot instead of skilled workers. One of distinctive features is that the self-supported robot is independent from a shield machine while usual assembly equipments are supported from shield machines. The robot is supported from the inside of a primary lining (assembled segment rings) and its attitude follows the attitude of a primary lining. The paper is intended to present 1) the features of the self-supported segment assembly robot and its control system 2) the structure and mechanism of the robot 3) the results of the assembly tests and application to the real project.
(*O-SERO ; Obayashi - Segment Erection RObot)

1.CIRCUMSTANCES AND TARGETS OF DEVELOPMENT

In the conventional method, as a primary lining, several pieces of segments are manually assembled and bolted up by skilled workers with a erector equipment inside tail of a shield machine. However, the automation system for this assembly work has been expected to be developed to resolve the following problems; 1) the workability and the quality have been poor because of the shortage of skilled workers. 2) it is dangerous to assemble segments on a high deck within a limited space. 3) the required torque for bolt-up is quite big in case of a large diameter tunnel and it's impossible to manually bolt up segments. Several automation systems have been reported to be developed, however, the assembly time by a robot was more than twice as it by workers. And also, the development cost and production cost were suspected to be very expensive. Therefore, it was considered that these systems could not be practically applied to a project.

Before the development started, we decided the following three points as the targets considering the defects of other pre-developed assembly systems:

① The assembly time should be shortened so far as it by skilled workers.
② The robot should have such a structure that enables itself to be reusable for more than one project.
③ Some parts of the assembly process by skilled workers should be introduced to it by the robot.

With regard to the first target, the robot is difficult to be applied to a project if the assembly time is comparatively longer than it by the conventional method because the assembling segments is a critical work for a shield project.

About the second target, it is poor economy to full depreciate an assembly robot for just one project because it costs quite expensive. From this point of view, in order to encourage the possibility of re-use, the segments for the system are determined to be the most widely used type in Japan even though this type may not be suitable for the automation.

In point of the third target, it is considered that there are two alternatives to design an automatic system for such an assembly job. One way is that the automatic process is designed to simulate the assembly process by skilled workers. The other is that it is designed without employing any know-how of skilled workers. Most of the pre-developed systems was designed on the basis of the second way. For example, a segment piece was positioned in order to align all of its bolt holes to their counterparts at the same time. The positioning mechanism of the robot has six degrees of freedom, so the process for positioning should become quite complex and require long time. Therefore, we have attempted to simulate the assembly process by skilled workers to make the positioning control simple.

2.OUTLINE OF ASSEMBLY SYSTEM

The Figure 1. shows the sketch of the robot in a shield tunnel. One of the reasons why the self-supported type is selected is that this type of structure enables itself to be reused for another project with minimum modification. Another reason is that during the assembly process, the relative positions and attitudes between the robot and the primary lining can be kept constant unlike an ordinary erector equipment. So, once the rotation plane of the robot is adjusted and locked before the assembly starts, two of six freedoms, pitching and yawing, need not to be controlled again until the assembly of one ring is finished.

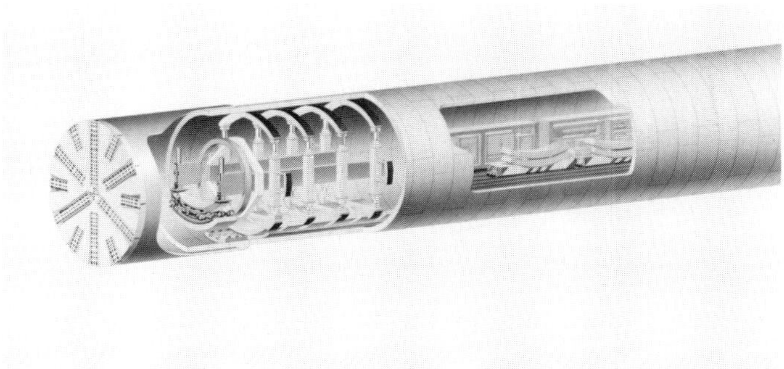

Figure 1. Sketch of Self-Supported Segment Assembly Robot in Tunnel

2.1.Elector Equipment

The elector equipment includes the mechanisms to grip a segment piece, to position it, to bolt it up, and to adjust the rotation plane.

The mechanism for griping a segment piece is shown in Figure 2. The gripping device is buried in the segment piece, and the robot inserts its gripping hook into the gripping device and rotate it. Then, the gripping hook is pulled up until the segment piece is gripped tightly. The relative position is adjusted by inserting guide pins into guide holes on the surface of the segment.

The positioning mechanism has six degrees of freedom, and controls the position and attitude of the segment piece by means of hydraulic actuators.

The bolt-up device, totally 13 no.s equipped, are arranged for respective bolt boxes of three types of segment pieces. This device positions a set of hexagon-headed bolt and nut, fed in advance, to the center of a bolt hole, inserts the bolt, and then tightens them.

The mechanism for adjusting the rotation plane consists of a pitching and a yawing frame. The pitching frame is supported from main beams of the self - support equipment, the yawing frame is supported from the pitching frame through pin connections and the bearing for rotation is installed in the yawing frame. The rotation plane is adjusted to be parallel with the last assembled segment ring by controlling the attitudes of both frames.

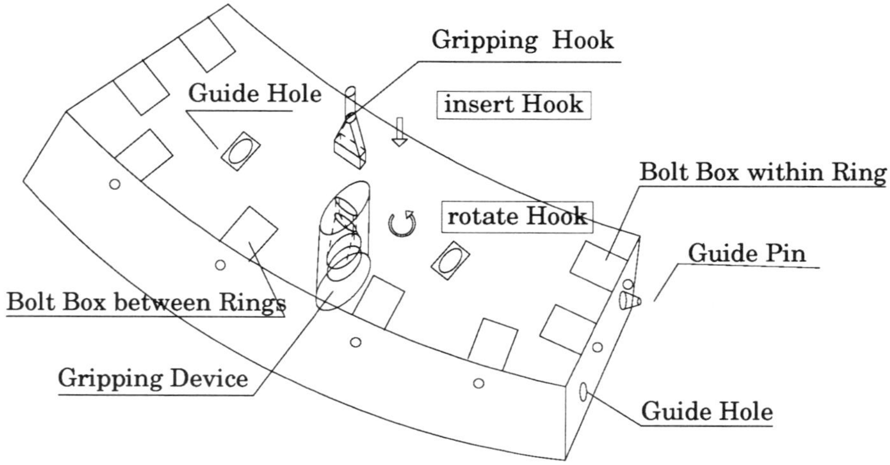

Figure 2. Sketch of Gripping Mechanism

2.2. Segment Feeder

The segment feeder is a conveyor suspended from the main beams. It stocks six pieces of the segments, feeds the foremost segment piece to the gripping position of the erector equipment and transport segments forward by one piece length at each time. And also, the position and attitude of the foremost segment is adjusted to be constant before it is fed to the erector equipment.

2.3. Self-Support Equipment

The self-support equipment consists of main beams, four sets of supporting frames including upper supporting arches and lower supporting shoes and a transportation mechanism. During the assembly process, the main beams are supported by the foremost and aftermost supporting frames. When the robot is transported forward, it is supported by middle two supporting frames through four sets of rollers and is slid by a transportation mechanism.

2.4. Specifications

The specifications of the assembly robot and the segment is described in the Table 1. The front view of [O-SERO] in the factory is shown in Figure 3. When it is employed in a tunnel, it is located behind a shield machine.

Table 1.
Specifications

[O-SERO]		Segment	
Dimensions	H=6m * L=7m	Type	RC
Weight	70 ton	Dimensions	O.D.=6.5m, W=0.9m
Lift-Up Force	23 tonf	No. of Divisions	7 (A*4+B*2+K)
Bolt-Up Force	30 kgf*m	Max. Weight	2 ton
No. of Bolt-Up Devices	13	No. of Bolts	39 pieces/ring

Figure 3. Front View of [O-SERO]

3. FLOW CHART OF ASSEMBLY PROCESS

The flow chart of the assembly process is shown in Figure 4.

"Pre-Sensing" is carried out only once before the assembly of the segments starts. It detects the data to analyze and to calculate 1) the relative attitude and position between the rotation plane of the robot and the the last assembled segment ring 2) the cross-section of the last assembled segment ring.

"Primary Positioning" is to bring the gripped segment piece in the adjacent position with pre-assembled segment pieces by the numerical control until the sensing devices can detect their surface.

"Fine Positioning" is to precisely align the segment by detecting its relative attitude and position with the adjoining segments. Then, the segment is rotated against the adjoining pre-assembled piece of the same ring so as to force the gripped segment to copy the attitude and position of the adjoining segment because a guide mechanism is installed between them (see Figure 2.).

Next, only the joint of two pieces of segments within a ring are bolted up first, then the relative position between the rings is detected again and adjusted by controlling only the movement in the radial direction. Finally, the joint between the rings are bolted up.

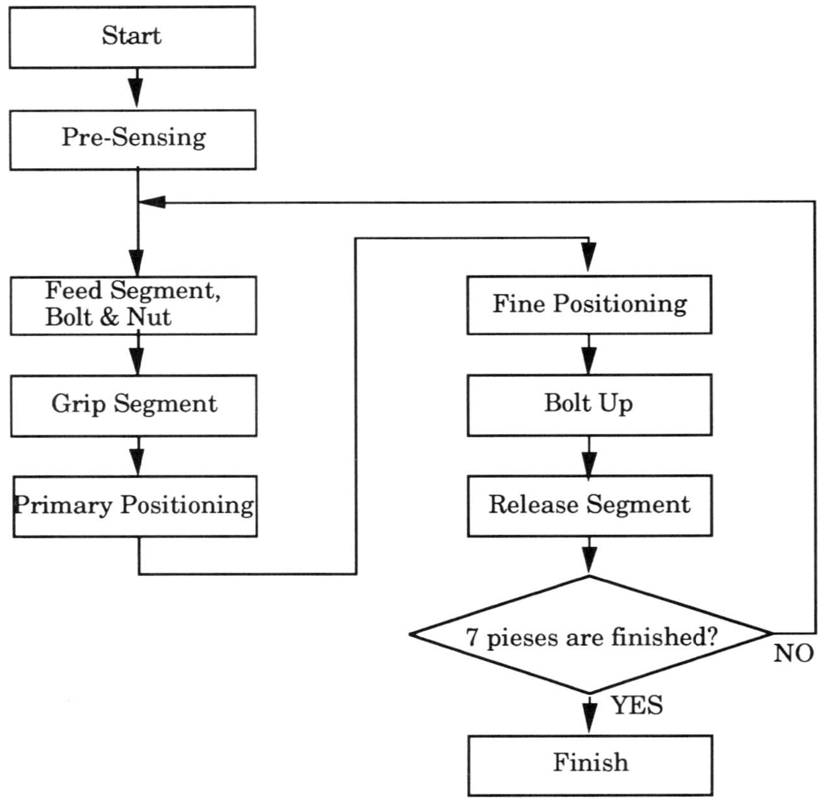

Figure 4. Flow Chart of Assembly Process

4. CIRCULAR CONTROL SYSTEM

The system, integrated with the artificial intelligence, is to analyse the cross-sectional shape of the segment ring and to control it to be a circle. The cross-section of the segment ring is designed to be a circle, however, the shape of the primary lining is transformed to some extent similar in shape to the transverse cross section of an egg because the unsymmetrical external force, such as earth pressure and back-filling grouting pressure, works on the tunnel. The cross-sectional shape of the last assembled segment ring greatly affects the automatic assembly process because a segment piece is assembled on the basis of it. If the shape is transformed so far as a certain degree, it should be impossible to assemble segments automatically.

The host computer, connected with the master controller of the robot through the Ether Network, calculates the cross-sectional shape of the previous ring by analysing the data collected during pre-sensing process, then it determines the compensatory values for each segment and transmits the values to the master controller. The robot assembles the segments by using these compensatory values, and the shape of the assembled ring becomes slightly different from it of the previous ring. After the next excavation, the host computer examines the shape of the segment ring again, which was assembled considering the compensatory values, then it studies how the compensatory values were effective to cause the ring to become close to a circle. The results of the study were accumulated and utilized when it determines new compensatory values.

5.RESULTS OF ASSEMBLY TEST AND FIELD TRIAL

5.1. Assembly Test in Factory

The assembly test was conducted in the factory for six months. The first three months was spent to confirm that the functions worked properly and performances were achieved as designed. During the next three months, the automatic processes were revised to shorten the assembly time and the circular control system was developed by way of trial.

Finally, the assembly time per piece became six to seven minutes on average, and the accumulated assembly time per ring took 40 to 50 minutes. This assembly time is almost equal to it by skilled workers.

5.2. Field Trial

From the beginning of October, 1992, the field trial of the assembly system was conducted for about one month, and 70 No.s of segment rings were automatically assembled in the tunnel.

One of the main purposes of the field trial was to examine the effectiveness of the circular control system. The max. deviation of the inside diameter of the first segment ring was more than 20 mm obtained from the recursive solution calculated by the circular control system (the inside diameter of the segment ring could not be directly measured because of the obstruction of the screw conveyor of the shield machine). In this case, it was quite difficult to automatically align all of the bolt holes of a segment piece to their counterparts. The deviation were reduced to less than 10 mm after about ten rings were assembled while modifying their cross-sectional shapes by the circular control system. Then, the automatic process was smoothly proceeded and the assembly time became almost the same as it of the assembly test in the factory. The shape of the segment rings had been properly controlled to the end of the trial with the deviation of around 10 mm.

Another main purpose was to examine the durability and the reliability of the system under the sever conditions in the tunnel. In the first ten days, the automatic assembly was sometimes interrupted by the failures of electric devices. It was considered that the humidity in the tunnel was higher than expected. This kind of troubles was resolved by improving the waterproof ability of electric devices, and any critical troubles except this did not occur during the field trial.

6.CONCLUSION

Compared with other automatic assembly systems, [O-SERO] has various kinds of characteristics due to not only its self-supported mechanism but also its unique assembly processes to which the know-how of skilled workers are introduced. Through the field trial, it was confirmed that this automation system satisfied the targets of the development and indicated the high reliability.

Automation and Robotics in Construction X
G.H. Watson, R.L. Tucker and J.K. Walters (Editors)

Hyper-Shield System Achieves Automated Excavation

Kiyosi Sato, Katutosi Sekiya, Kazuhiro Higaki, Research Engineers,

Presidential Research Laboratory "ICHIKEN", Fujita Corporation

2-17 Kanda Suda-cho, Chiyoda-ku, Tokyo 101, Japan c/o SAGAMI Building

Abstract

The hyper-shield system developed by Fujita Corporation is used to perform automatic excavation of tunnels. The system, applicable to either the slurry or the earth-pressure method, employs the pivotal technologies of artificial intelligence and fuzzy logic to automatically control excavation of tunnels. Also the system allows for its travel along a planned path, in ways which can prevent the ground surface immediately above the tunnel being excavated, from subsidence and upheaval. This paper outlines the hyper-shield system, together with the report of its first, field application as a slurry system. The major purpose of this study, which was to successfully control the position and attitude of the shield machine and also to support the natural ground in order to prevent it from subsidence and uplifting, was achieved with satisfaction through the execution results.

1. INTRODUCTION

In recent years, in Japan, shield work involves such severe conditions as sharper curves, longer distances and greater underground depth, therefore more advanced tunneling technologies are in demand. At the same time, the shortage of skilled operators and the increasing number of aging field workers have given rise to social issues. Through the recognition of these outstanding issues, Fujita Corporation has, since several years ago, taken a progressive attitude toward the development of information-based and automated tunneling methods using a shield machine. The results have provided for the industrialization of this hyper-shield system.

The primary objectives of this hyper-shield system are to enable a shield machine to automatically start and stop, as well as, through the application of artificial intelligence and fuzzy logic, to automatically control the position and attitude of the shield machine and also to support the natural ground. Although this hyper-shield system is compatible with both the slurry method and the earth-pressure method, which currently maintains the lead among all the shield tunneling methods, this paper focuses on description of the slurry hyper-shield system.

2. SYSTEM OVERVIEW

This section outlines the configuration of the slurry hyper-shield system and discusses the technologies of controlling the position and attitude of the shield machine, and of supporting the natural ground and controlling the excavation velocity.

2.1. System configuration

Figure 1 provides the configuration of the slurry hyper-shield system. In the figure, the system consists of three computers linked with associated control panels. These computers have the following functions:

(1) Excavation control system

The excavation control system collects various data, calculates data for position and attitude control, and executes such basic tasks as the preparation of daily reports.

(2) Position and attitude control system

The position and attitude control system determines a jack pattern to hold the shield machine with an optimal attitude through the collation of the machine's attitude data with the expert system rule.

(3) Automatic control system

The automatic control system automatically controls the start and stop of the treatment plant, the equipment of the shield machine, and the transportation of fluid. It also employs fuzzy logic to control the supporting of the facing, and the excavation speed.

These systems are provided in the central control room, which are collectively controlled by integrated management. It is the normal practice of the operator that from the central control room, he confirms the status of in-tunnel operating conditions through the monitor television and speaker, and that by pressing the single button on the control panel front, allows the shield machine to automatically travel on excavating.

2.2. Position and attitude control

2.2.1. Automatic surveying system

To control the attitude of the shield machine it is necessary to make a real-time measurement of both the position and attitude of the shield machine. With this system, the position and attitude of the shield machine are calculated based on the data derived from the surveying conducted by the system manager using four sensors (a gyro sensor, a level sensor, a pitching sensor and a stroke sensor) mounted on the shield machine. The data is used to calculate the position of the shield machine by the excavation control computer through the local micro computer provided in the tunnel, and transmitted to the position and attitude control computer.

2.2.2. Position and attitude control system

This system uses artificial intelligence to automatically select jack

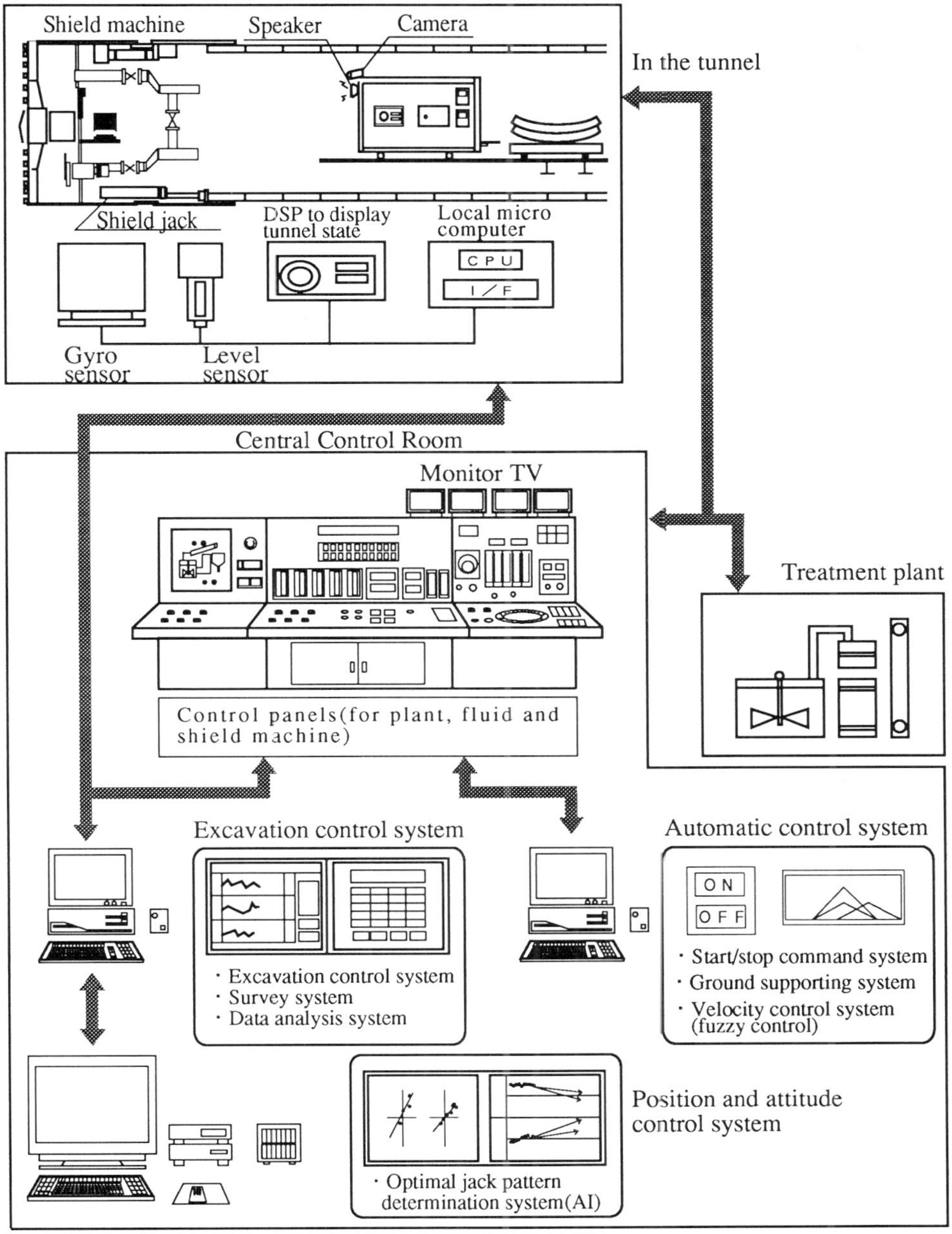

Figure 1. Slurry hyper-shield system outline.

patterns in order to move the shield machine in an optimal direction. Previously this was done based on the survey results by the machine operator. This optimal jack pattern is calculated in the following sequence:

(1) Confirming the position and attitude of the shield machine and calculating adjusted declination

Firstly, the error between the position of the shield machine and the planned path is calculated based on the position (at coordinates X, Y and Z) of the shield machine. From that position, the desired path with which the shield machine should be aligned, is established in accordance with the specified rule. Subsequently, from the route on which the shield machine has traveled, and from the orientation of the shield machine, the adjusted attitude along which the shield machine will travel from the path as it has advanced and from its orientation, is calculated. The error between both, known as an adjusted declination, is calculated in a horizontal direction and a vertical direction.

(2) Calculating the amount of torque vector

Next, the amount of torque vector required to make the adjusted declination zero is calculated. The calculation is done, based on one hundred pieces of up-to-date data on the actual jack pattern and the amount of changes in the direction of the shield machine. The relationship between a turn and torque is equated using the method of least squares. By inserting the adjusted declination into the equation, the required amount of each torque vector is determined.

(3) Correcting the amount of torque vector

In this stage, the behavior of the shield machine and other various data are permitted to conform with the expert system which was preliminarily created based on empirical knowledge, and calculation is made to obtain the corrected value on the amount of torque vector.

(4) Determining an optimal jack pattern

Finally, the jack pattern which corresponds to the corrected amount of torque vector is chosen from among the jack patterns (7,478 kinds for 16 jacks). This is shown on the table which has been prepared in advance, and this optimal jack pattern is transmitted to the excavation control computer.

This system has the functions of automatically determining the required number of jacks by taking into account the total thrust, and by diagnosing the systems and the sensors.

2.3. Natural ground supporting control

2.3.1. Method of supporting the natural ground

The slurry shield method, while excavating the tunnel, prevents the facing from collapsing by pushing the slurry with the properties corresponding to the natural ground conditions, thereby making the amounts of slurry being supplied and discharged well balanced. With this system, the following methods are used to support the natural ground against the conditions mentioned above.

With the slurry shield method, to conduct real-time assessment of the balanced volume of excavated material, there is the "rate of deviation flow" which acts as an index. When this value is zero, it means that well balanced control is underway. In this regard, we assumed that, with this system, the facing's hydraulic pressure is controlled so that this index be zero.

As well, with consideration given to the stabilization of the facing, the facing's hydraulic pressure was assumed to be controlled within the range of natural hydraulic pressure < the facing's hydraulic pressure < natural hydraulic pressure + 0.5kgf/cm^2. Another assumption was that when deviating from this range, the rate of slurry discharged would be automatically modified to provide for other settings.

2.3.2. Method of controlling the facing's hydraulic pressure

The conventional method used to control the facing's hydraulic pressure, is to make the rate of deviation flow zero while the operator is watching it. In other words, when the rate of deviation flow is negative, the operator judges that the slurry is slightly deviating from the normal route. He accordingly reduces the facing's hydraulic pressure settings to a value smaller than the current settings. In lieu of such a conventional method, we have established a system to automatically control the facing's hydraulic pressure by using fuzzy logic, as mentioned below:

(1) Fuzzy-based control mode

The fuzzy-based control mode used for this system is the if-then mode as mentioned below:

if X is the fuzzy value X, and Y is the fuzzy value Y, then Z is the fuzzy value Z

in which the input variable X is the rate of deviation flow during previous control, and Y is the current rate of deviation flow. And the output variable Z is the changing range of the facing's hydraulic pressure.

(2) Membership functions

Figure 2 shows the membership functions representing the fuzzy values X, Y and Z referred to in (1) above. In the figure, there are five fuzzy functions, each being triangular.

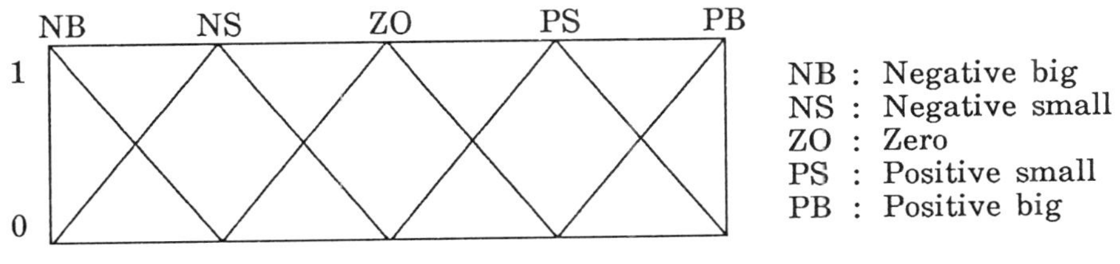

Figure 2. Membership functions.

(3) Fuzzy rule

Table 1 shows the fuzzy rule governing the control of the facing's hydraulic pressure.

Table 1
Fuzzy rule governing the control of the facing's hydraulic pressure

		Current rate of deviation flow				
		NB	NS	ZO	PS	PB
Previous rate of deviation flow	NB	NB	NS	ZO		
	NS	NB	NS	ZO		
	ZO	NS	NS	ZO	PS	PS
	PS			ZO	PS	PB
	PB			ZO	PS	PB

(4) Inference

For the control mentioned above, the inference method employs the method of algebraic product - addition - center of gravity. The control of the facing's hydraulic pressure is performed about at 1-minute intervals, which was established by taking into account the reaction time until the facing's hydraulic pressure reaches the value set against the control performed by the associated computer.

2.4. Excavation velocity control

In view of the collapse of the natural ground, advancing the shield machine faster would be more advantageous in shield excavation. Based on this concept, we decided that the velocity control of the traveling shield machine would be performed in a range which imposes less of a load on the various pieces of the machine's equipment. Velocity control includes two ways; velocity-up control and velocity-down control. The former control uses fuzzy logic while the latter uses a simple rule. This velocity control deals with the shield jack pressure, the cutter current, the density of discharged slurry and the revolutions of a P2 pump, as the control factors.

3. EXECUTION EXPERIENCE AND RESULTS

To date the hyper-shield system has experienced its application to the slurry shield method at five jobsites, and to the earth-pressure method at three jobsites. The following gives the information of the Toyotomi tunneling project, for which the slurry hyper-shield system was first employed.

3.1. Project rundown

(1) Shield outer diameter : ø4,480mm

(2) Total excavation distance : 2,270 meters
(3) Geological properties : sandy diluvium (sand: 88 ~ 94%)
(4) Construction period : May 1991 ~ June 1992
(5) Location : Onoda-cho, Funabashi City, Chiba Prefecture, Japan

3.2. Alignment and inner geometry of the tunnel route

This section classifies the route into sharp curves, linear or slow curves which were found through geometrical survey using segments. For the purpose of this project, sharp curve means a section with the radius of curvature of 50 meters while the linear or slow curve means the other sections. Table 2 summarizes the results.

Table 2
Measurement results of errors (mm)

	Number of data	Horizontal direction			Vertical direction		
		Average	Typical deviation	Max. value	Average	Typical deviation	Max. value
Sharp curve	180	4.1	10.5	28.0	3.6	10.0	-22.0
Linear or slow curve	1730	4.6	8.1	25.0	0.1	9.8	28.0

From the table above, it is known that the errors of curves and linear or slow curves in both the horizontal and vertical directions are within ±30mm.

3.3. Volumetric measurement results of excavated material

This section compares excavation data between the sections (561 ~ 625R: 75 pieces of data) where the parameters shown in the table below received automatic control, and those (626 ~ 650R: 25 pieces of data) where likewise the parameters received manual control. Table 3 shows the results.

Table 3
Comparison between automatic control and manual control

	Automatic control		Manual control	
	Average	Typical deviation	Average	Typical deviation
Facing's hydraulic pressure (kg/cm^2)	1.91	0.09	1.82	0.01
Integrated rate of deviation flow (m^3)	-0.12	0.10	0.72	0.59
Volume of excavated material (%)	95.5	0.80	96.2	1.5

In the table above, the integrated rate of deviation flow means the value attained by integrating the rate of deviation flow attained at real time, by the time interval at which the deviation flow occurs. The volume of excavated material (%) represents the ratio of the volume of excavated material to the cubage of the excavated material.

The table above provides an insight that the typical deviation of the facing's hydraulic pressure under automatic control is greater than that under manual control. Likewise, the average and the typical deviation of the integrated rates of deviation flow and the volumes of excavated material, under automatic control, are greater than those under manual control. This can prove that automatic control enabled greater sophistification of the control of the facing's hydraulic pressure than manual control did, and judging from the volume of excavated material, permitted more stable excavation by far.

3.4. Measurement results of geometrical changes on road surface

Throughout the shield execution period, survey points remained located at 20-meter intervals directly above the shield path in order to measure any subsidence of the natural ground. The results indicated that at sections with deep overburden, no geometrical change occurred, while at those with shallow overburden (G.L.-9 meters), changes in the range of 0 to -3mm were observed. This verifies the excellent effectiveness of the natural ground supporting system.

4. CONCLUSIONS

This article has provided an overview of the slurry hyper-shield system, which is an automatic excavation system using a shied machine developed by Fujita. The execution results mentioned indicated that the system enabled a tunnel to be excavated with the planned path geometrically achieved within the deviation of plus or minus 30mm. The results also showed that automatic control obviously allowed for greater stability in excavation than manual control did as a comparison between both excavation data proved, and that subsidence of the road surface did not exceed 3mm. This well confirms that the developed hyper-shield system has thoroughly satisfied the target accuracy in controlling the position and attitude of a shield machine and the supporting of the natural ground. The incorporation of the system with the control panels clustered in the central control room led to success in the reduction of operators from the conventional two down to one, and also made it possible for a less skilled operator to operate the system very easily. We believe this also achieved labor savings, which was one of the established goals.

Currently, research and development of a control system is underway at Fujita, which will give consideration to clearance, and a position and attitude control system which can automatically control a collapsible mechanism and act as a copy cutter. In the future, we will also take up the challenge to realize a natural ground supporting control system which can volumetrically feed back excavated material.

Automation and Robotics in Construction X
G.H. Watson, R.L. Tucker and J.K. Walters (Editors)

The development of "Gyro Runner", a totally unmanned shield guiding system

T. Nozawa, F. Sato and N. Nishino

Mechatronics Research Office, Technical Research Institute, Aoki Corp.
36-1 Oaza Kaname, Tsukuba City, Ibaragi Pref., JAPAN

Abstract

This paper reports on the development of 'Gyro Runner', a totally unmanned shield guiding system, consisting of the automatic survey system and the automatic shield guiding system.

More than 90% of current shield tunnels are small-bore and of curved geometry, where a gyrocompass mounted in the machine have been frequently used. However, the gyrocompass cannot detect the potential side slip of the shield machine, and the accumulated positional error can only be adjusted by a conventional tunnel survey. A system that minimizes such survey work has been developed.

The new survey method comprises an autotracking total station, a laser gyro and an image processor, and performs accurate shield position/posture detection by referring to the survey points at every measurement cycle.

The shield guiding system, on the other hand, selects the optimum jack pattern and directional control procedure that conducts the shield machine on the planned line.

On site, good accuracy and labor saving effect have been verified.

1. Introduction

Many developments aiming at total shield work automation have been carried out by contractor firms. Among these, cases of position/posture survey technology development have been often reported. Most of these, however, are for large cross section shields. Small-bore, and especially curved geometry tunnels have little internal work space , reason why automatic survey in such cases have been considered difficult to perform.

Aoki Corp. has developed the 'Gyro Runner', a guiding system for small-bore, curved geometry shield tunnels, capable of accurate positional detection and guiding. The system has already been implemented on site, and shown satisfactory results.

2. Conventional automatic survey methods

Automatic survey methods in use today can mainly be classified in three types: (1)the optical method(fig.1), which uses light wave survey instruments ; (2)the gyrocompass method(fig.2); and (3)the optical/gyro method(fig. 3),which is the combination of both. To each method, some problems can be pointed out:

(1)Optical
- -light path necessary
- -influence of segment shift
- -constant re-establishment of survey points required

(2)Gyrocompass
- -machine side slip error
- -cumulative error
- -check survey (once or twice a day) required

Yet, (3)gyro/optical method, which combines the advantages of both, meets the same problems in a small tunnel environment.

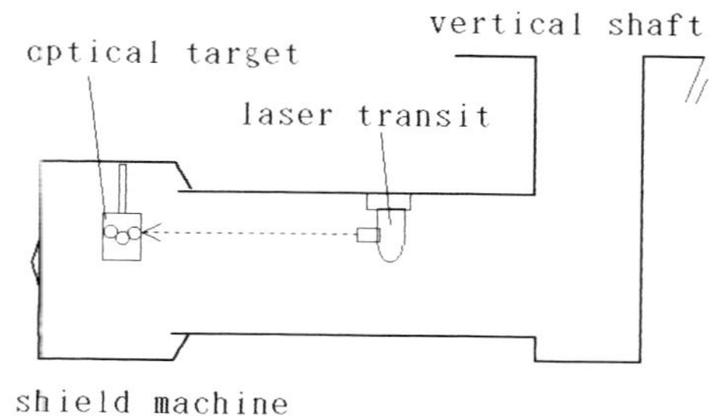

Figure 1. The optical method

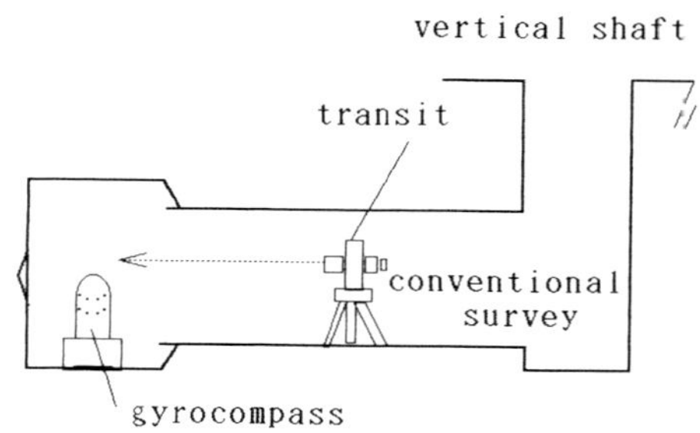

Figure 2. The gyro compass method

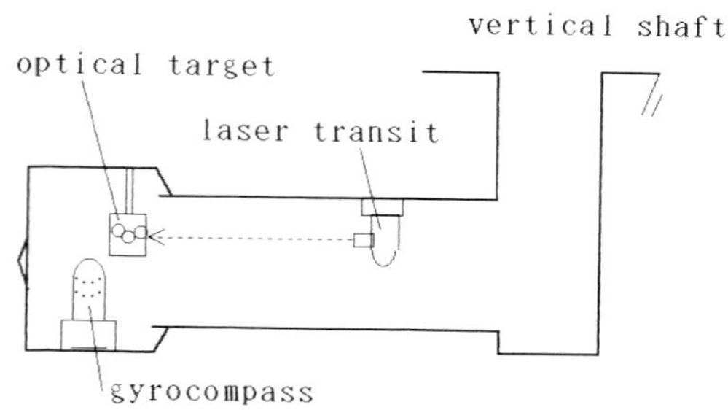

Figure 3. The gyro/optical method

3. Development directions on the new survey method

An accurate shield survey is possible if the machine can be collimated by a survey instrument placed on a steady reference point, usually located far back in the tunnel. In small tunnels, however, backup trains obstruct the necessary light path.

To overcome this, a gyro sensor, instead of being placed on the shield machine, was put to run to and fro between the both ends of the backup train, in a way that survey can be performed even when light path's not available. An autotracking total station, placed on the backmost carriage of the backup train, was used to detect the posture and the position of the back end of the route followed by the running gyro. At the backup train's front end, an image processor was placed in order to follow the movement of the shield.

By adopting this method, the labor required for re-establishing reference survey points can be minimized, and also survey errors due to segment shift can be avoided.

4. The Automatic Survey System

4.1 OVERALL DESCRIPTION

The above described directions were materialized in the form of three subsystems, distributed in the tunnel as shown in figure 4.

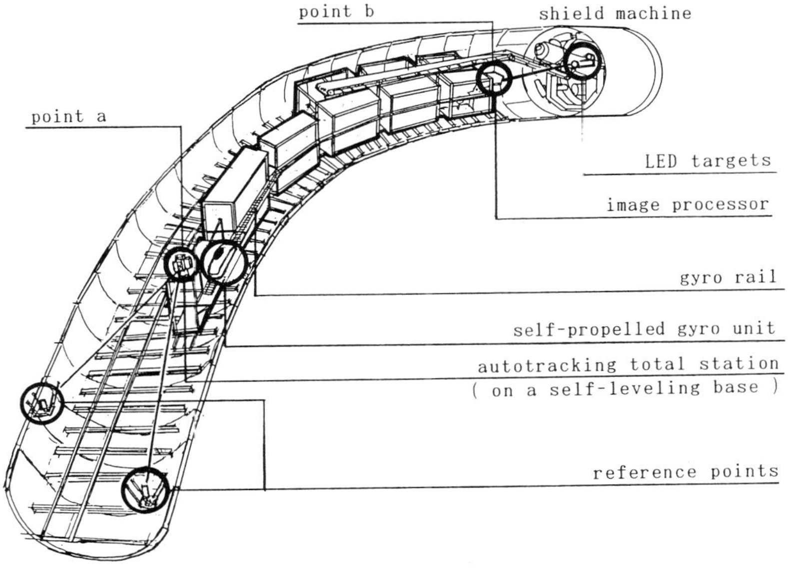

Figure 4. Overall configuration of the automatic survey system

4.2 MEASUREMENT METHODS AND SYSTEM CONFIGURATION

(1) Rear Measurement Subsystem

Figure 5 shows the configuration of this subsystem.

The light wave total station at the backmost carriage of the backup train can automatically seek for the target prisms. This instrument was installed on a self leveling table, which keeps the base at level posture with an accuracy of 10 (ten) seconds.

The measuring point 'a''s position and posture are calculated from the direction and the azimuth to each of the two prisms. The adopted mathematical methods were the Reduction and the Weisbach Triangle methods.

(2) The Gyro Car Subsystem

The Gyro Car Subsystem's configuration can be seen in figure 6.

A self-propelled car bearing a Laser Gyro travels between the two ends of the backup train, on a rail placed throughout its extension. While running, the subsystem continuously measures distance and azimuth, from which the route line, and thus the relative position/posture of the end points 'a' and 'b', are calculated.

(3) The Shield Detection Subsystem

The configuration of the Shield Detection Subsystem is shown in figure 7.

Two circle-arranged LED targets are fixed to the previously known points in the shield machine. They are detected by an image processor with its two cameras placed at the front end of the backup train. The centroid coordinates of the LED targets' image provide data for shield position/posture calculation, based on the previously surveyed initial relative position of the cameras and the targets.

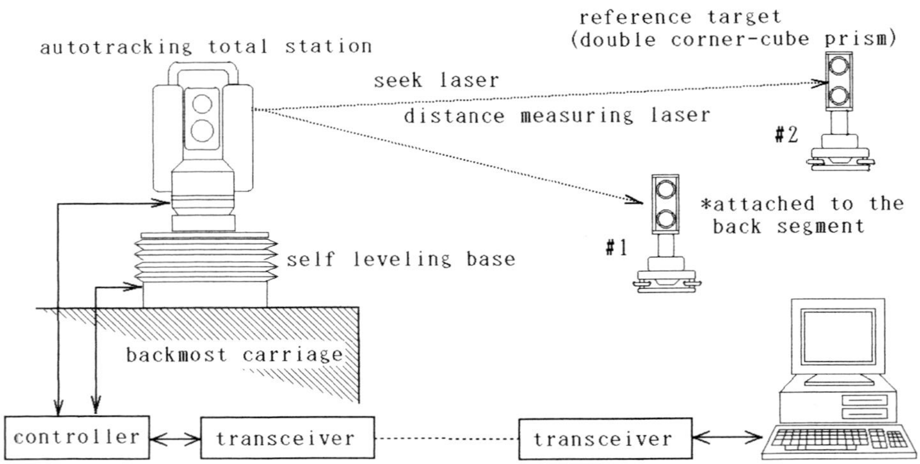

Figure 5. The Rear Measurement Subsystem

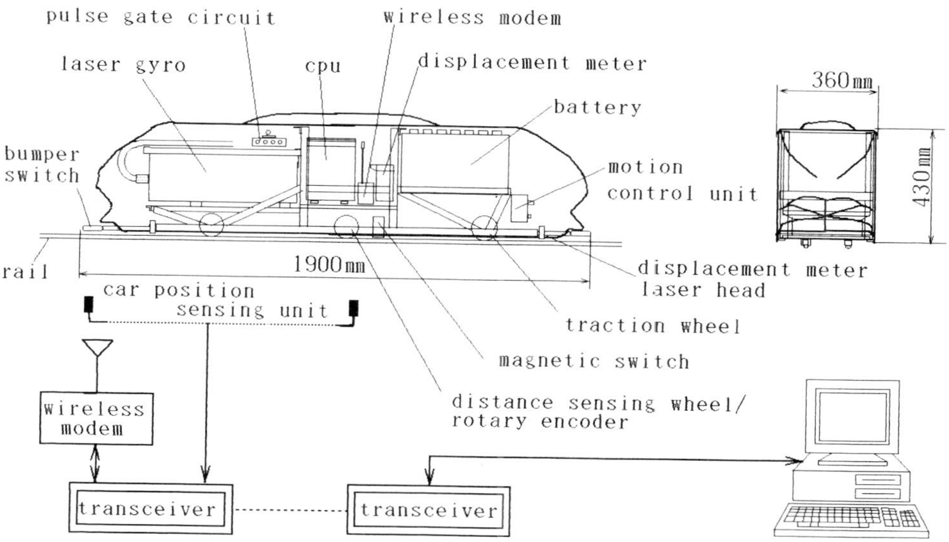

Figure 6. The Gyro Car Subsystem

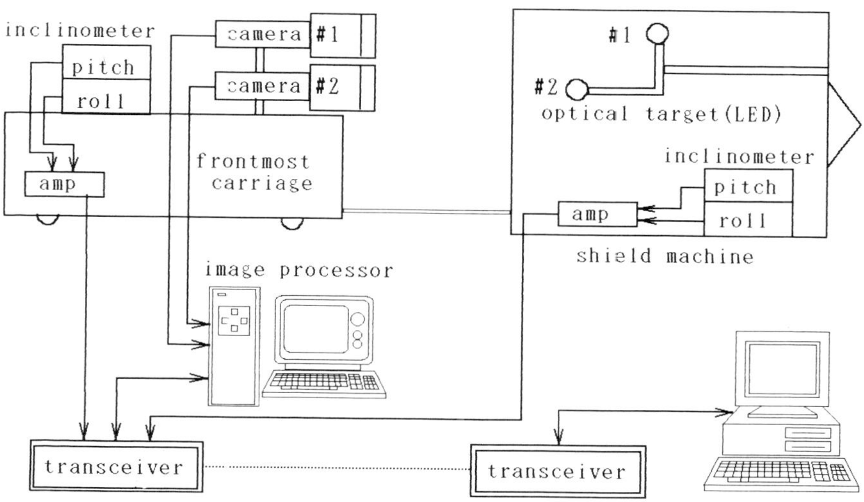

Figure 7. The Shield Detection Subsystem

The three subsystems' measurement results are combined to obtain the shield head coordinates (northing x, easting y) and azimuth deviation (yawing angle), having the back located survey points as reference. The height information (z) is provided by a water level meter.

4.3 SURVEY TIMING

The automatic survey is performed while the machine stops, as during segment assembly or any break in excavation. A data acquisition system allows automatic detection of the end of an excavation cycle, when the main computer generates the "start" command to each subsystem. The whole survey is concluded within 5 minutes.

5. Automatic shield guiding system

Figure 8 depicts the outline of the shield guiding system.

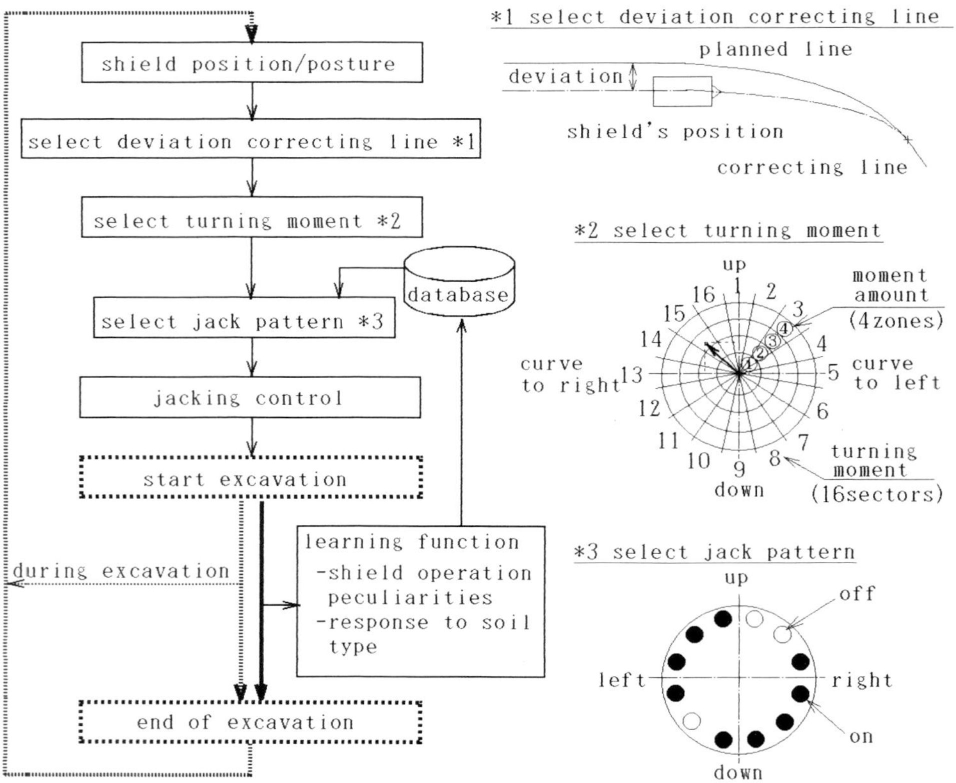

Figure 8. Jack pattern selection and jacking cotrol flow

In conducting the shield machine on the planned course, adequate jack pattern and jack control procedure must be selected, taking into consideration factors like terrain condition and machine operating peculiarities. Currently, operators make such decisions. This system, by using its learning function, automatically selects the optimum jack pattern and the jack control procedure, with a decision level comparable to a human operator.

By the learning function, data as soil pressure(which influences machine behavior), the utilized jack pattern and the actual course taken by the shield machine, are accumulated in form of a database. With the required turning moment (calculated from the current shield machine deviation from the planned course), and environment condition data as soil pressure, a similar situation is sought in the database, and by this means a suitable jack operation procudure is chosen. From the alternation of the jacking control and the evaluation of the resulting route, this system learns how to operate the shield machine in an optimum way.

6. Features of the system

(1) Suitability to small diameters (2 m) and sharp curves(30R)
(2) Survey without side slip effects, commonly seen in shield machines carrying a gyrocompass
(3) Survey free from cumulative error, due to constant reference to back points (in straight lines, unmanned operation is possible for more than a month)
(4) Automatic decision making similar to skilled operators, allow excavation according to the soil type and the machine peculiarities.
(5) The combination of the automatic survey and guiding systems makes unmanned shield work possible

7. Field results

The automatic survey system has been implemented on a real shield excavation site, and has shown precision suitable for excavation management (machine head position error within +/-20 mm). It was confirmed that unmanned survey for a period within a week and a month is possible.

The shield guiding system has also been introduced to site, where evaluation works have been carried out.

Figure 9 shows Gyro Runner in the shield site.

Figure 9. The Gyro Runner introduced to a shield site

8. Conclusion

To achieve a fully automatic shield work, element tasks as excavation supervision, machine position supervision, directional control, segment assembly and material conveyance, require automation technology development. Among these, Gyro Runner represents accomplishments in directional control and automatic survey. Also, Aoki Corp. has already had practical results on excavation supervision, by developing the "Soil Volume-Controlled Shield Method", and the "Soil Pressure Shield Expert System". Our future efforts shall be focused on the remaining segment assembly and material conveyance automation.

We believe that the technology developed with Gyro Runner is applicable not only to shields but also to other kinds of construction works.